Ten Years of TerraSAR-X—Scientific Results

Ten Years of TerraSAR-X—Scientific Results

Special Issue Editors

Michael Eineder
Achim Roth
Alberto Moreira

MDPI • Basel • Beijing • Wuhan • Barcelona • Belgrade

MDPI

Special Issue Editors

Michael Eineder
Remote Sensing Technology Institute
German Aerospace Center (DLR)
Germany

Achim Roth
Remote Sensing Data Center DFD
German Aerospace Center (DLR)
Germany

Alberto Moreira
Microwaves and Radar Institute
German Aerospace Center (DLR)
Germany

Editorial Office
MDPI
St. Alban-Anlage 66
4052 Basel, Switzerland

This is a reprint of articles from the Special Issue published online in the open access journal *Remote Sensing* (ISSN 2072-4292) from 2018 to 2019 (available at: https://www.mdpi.com/journal/remotesensing/special_issues/TerraSAR-X)

For citation purposes, cite each article independently as indicated on the article page online and as indicated below:

LastName, A.A.; LastName, B.B.; LastName, C.C. Article Title. *Journal Name* **Year**, *Article Number*, Page Range.

ISBN 978-3-03897-724-7 (Pbk)
ISBN 978-3-03897-725-4 (PDF)

Cover image courtesy of TerraSAR-X satellite over Markakol, Kazakhstan. Courtesy of the German Aerospace Center (DLR).

Contents

About the Special Issue Editors

Michael Eineder, Prof. Dr., is with the Remote Sensing Technology Institute at DLR and an Honorary Professor with the Technical University of Munich (TUM). He contributed to the TerraSAR-X mission with SAR signal processing algorithms and product design. His interests encompass SAR interferometry and precision positioning using SAR imaging geodesy.

Achim Roth, is with the German Remote Sensing Data Center at DLR. He is the Scientific Coordinator of the TerraSAR-X Mission. He contributed to the TerraSAR-X mission with SAR processor and application developments.

Alberto Moreira, Prof. Dr., is the Director of the Microwaves and Radar Institute at DLR and Professor with the Karlsruhe Institute of Technology (KIT). He has contributed to the approval process of the TerraSAR-X mission and later on to its mission performance and engineering, signal processing and the demonstration of new techniques and technologies. He is also the principal investigator for the TanDEM-X mission.

remote sensing

MDPI

Editorial

Ten Years of TerraSAR-X—Scientific Results

Michael Eineder [1,*], Alberto Moreira [2] and Achim Roth [3]

[1]   Remote Sensing Technology Institute, German Aerospace Center (DLR), Münchenerstr. 20, 82234 Wessling, Germany

[2]   Microwaves and Radar Institute, German Aerospace Center (DLR), Münchenerstr. 20, 82234 Wessling, Germany; Alberto.Moreira@dlr.de

[3]   German Remote Sensing Data Center, German Aerospace Center (DLR), Münchenerstr. 20, 82234 Wessling, Germany; Achim.Roth@dlr.de

*   Correspondence: Michael.Eineder@dlr.de

Received: 31 January 2019; Accepted: 1 February 2019; Published: 11 February 2019

Abstract: This special issue is a collection of papers addressing the scientific utilization of data acquired in the course of the TerraSAR-X mission. The articles deal with the mission itself, the accuracy of the products, with differential interferometry, and with applications in the domains cryosphere, oceans, wetlands, and urban areas. This editorial summarizes the content.

Keywords: synthetic aperture radar; TerraSAR-X; SAR interferometry; land subsidence; precise orbit determination; geometric and radiometric calibration; PSI

1. Introduction

We commemorated the 10th anniversary of the TerraSAR-X mission on June 2017. TerraSAR-X is a German SAR satellite operating in X-Band. Its twin satellite TanDEM-X was launched three years later in June 2010. Both satellites serve the TerraSAR-X and the TanDEM-X missions and have provided images of superb quality since then. For the TerraSAR-X mission just one of the satellites is employed while the TanDEM-X mission always supplies a simultaneously acquired image pair, serving the goal of a global digital elevation model. The satellites were developed in the scope of a public–private partnership between the German Aerospace Center (DLR) and Airbus Defense and Space.

This special issue addresses the scientific utilization of data acquired in the course of the TerraSAR-X mission.

2. Peculiarities of the TerraSAR-X Mission

TerraSAR-X is characterized by numerous operational and experimental modes of its SAR instrument, unfolding a wide parameter space, e.g., multiple polarizations, along-track interferometry and a variety of SAR imaging modes with resolutions down to 0.5 m in range and 0.25 m in azimuth. Despite being well beyond their design lifetime of 5.5 years, both satellites are still fully functional, deliver high-quality images without any performance deterioration, and have enough consumables for operation into the 2020s.

From the very beginning, the high resolution data of the TerraSAR-X mission complemented the European medium resolution missions in C-Band and the Japanese L-Band missions. Not quite unexpectedly, it was especially the high resolution that revealed many new insights into SAR scattering, which led to a large number of scientific publications. The expectation that the interferometric coherence over distributed scatterers and longer time spans would be low in X-band was confirmed. But it was more than compensated by a real surprise: the high resolution and the small wavelength revealed a very high number of point-like reflectors in urban and even in natural environments, which gave rise to some ground breaking new methods such as urban SAR tomography and new applications in 3D point localization and deformation mapping of buildings and of infrastructure.

3. Contents of this Special Issue

The 20 papers of this special issue cover six major disciplines. Five papers [1–5] deal with interferometric applications for land surface deformation mapping. A key success factor here is the high resolution, which enables the mapping of smaller landslides, of single buildings and of infrastructure.

Another five papers [6–10] demonstrate the success of TerraSAR-X in the field of oceans and wetlands exploiting the high resolution and the multi-polarization capabilities of TerraSAR-X. As it turned out, not only the space segment but also the ground segment performance such as reliability and fast near real-time services contributed to the success of TerraSAR-X in maritime applications. Three papers [11–13] cover such operational aspects and the scientific and operational use of TerraSAR-X data.

Four papers [14–17] deal with the radiometric and geometric performance validation and with the new methods which made TerraSAR-X the first SAR sensor with geodetic accuracy. Two papers about mapping urban environments [18,19] and one about the cryosphere [20] finalize the different topics covered in this special issue.

From the 25 submissions for this special issue, 20 could be accepted and 5 had to be rejected or were withdrawn.

Of course, this collection is only a snapshot in time and many more papers have been published in the years before, demonstrating the success of this mission. A quick search for TerraSAR-X on Google scholar delivers more than 20,000 results!

4. Summary and Outlook

At the time of writing more than 235,000 individual data takes were acquired. Of those, 100,000 were initiated for scientific purposes and 1636 scientific proposals have been submitted to DLR to work with these data. Numerous scientists benefitted from TerraSAR-X which is a benchmark in geometric and radiometric accuracy—getting even better after ten years of operation.

Funding: This research received no external funding.

Acknowledgments: We thank all reviewers for their careful work in this review process. Finally, we thank all the authors for their submissions and the editor Nelson Peng for guiding us guest editors efficiently through the process and making this special issue possible.

Conflicts of Interest: The authors declare no conflict of interest.

References

1. Hosseini, F.; Pichierri, M.; Eppler, J.; Rabus, B. Staring Spotlight TerraSAR-X SAR Interferometry for Identification and Monitoring of Small-Scale Landslide Deformation. *Remote Sens.* **2018**, *10*, 844. [CrossRef]
2. Rabus, B.; Pichierri, M. A New InSAR Phase Demodulation Technique Developed for a Typical Example of a Complex, Multi-Lobed Landslide Displacement Field, Fels Glacier Slide, Alaska. *Remote Sens.* **2018**, *10*, 995. [CrossRef]
3. Tosi, L.; Lio, C.; Teatini, P.; Strozzi, T. Land Subsidence in Coastal Environments: Knowledge Advance in the Venice Coastland by TerraSAR-X PSI. *Remote Sens.* **2018**, *10*, 1191. [CrossRef]
4. Zhao, F.; Mallorqui, J.; Iglesias, R.; Gili, J.; Corominas, J. Landslide Monitoring Using Multi-Temporal SAR Interferometry with Advanced Persistent Scatterers Identification Methods and Super High-Spatial Resolution TerraSAR-X Images. *Remote Sens.* **2018**, *10*, 921. [CrossRef]
5. Zhu, X.; Wang, Y.; Montazeri, S.; Ge, N. A Review of Ten-Year Advances of Multi-Baseline SAR Interferometry Using TerraSAR-X Data. *Remote Sens.* **2018**, *10*, 1374. [CrossRef]
6. Adolph, W.; Farke, H.; Lehner, S.; Ehlers, M. Remote Sensing Intertidal Flats with TerraSAR-X. A SAR Perspective of the Structural Elements of a Tidal Basin for Monitoring the Wadden Sea. *Remote Sens.* **2018**, *10*, 1085. [CrossRef]
7. Ermakov, S.; Sergievskaya, I.; da Silva, J.; Kapustin, I.; Shomina, O.; Kupaev, A.; Molkov, A. Remote Sensing of Organic Films on the Water Surface Using Dual Co-Polarized Ship-Based X-/C-/S-Band Radar and TerraSAR-X. *Remote Sens.* **2018**, *10*, 1097. [CrossRef]

8. Irwin, K.; Braun, A.; Fotopoulos, G.; Roth, A.; Wessel, B. Assessing Single-Polarization and Dual-Polarization TerraSAR-X Data for Surface Water Monitoring. *Remote Sens.* **2018**, *10*, 949. [CrossRef]
9. Magalhaes, J.; da Silva, J. Internal Solitary Waves in the Andaman Sea: New Insights from SAR Imagery. *Remote Sens.* **2018**, *10*, 861. [CrossRef]
10. Wohlfart, C.; Winkler, K.; Wendleder, A.; Roth, A. TerraSAR-X and Wetlands: A Review. *Remote Sens.* **2018**, *10*, 916. [CrossRef]
11. Buckreuss, S.; Schättler, B.; Fritz, T.; Mittermayer, J.; Kahle, R.; Maurer, E.; Böer, J.; Bachmann, M.; Mrowka, F.; Schwarz, E.; et al. Ten Years of TerraSAR-X Operations. *Remote Sens.* **2018**, *10*, 873. [CrossRef]
12. Lang, O.; Lumsdon, P.; Walter, D.; Anderssohn, J.; Koppe, W.; Janoth, J.; Koban, T.; Stahl, C. Development of Operational Applications for TerraSAR-X. *Remote Sens.* **2018**, *10*, 1535. [CrossRef]
13. Roth, A.; Marschalk, U.; Winkler, K.; Schättler, B.; Huber, M.; Georg, I.; Künzer, C.; Dech, S. Ten Years of Experience with Scientific TerraSAR-X Data Utilization. *Remote Sens.* **2018**, *10*, 1170. [CrossRef]
14. Balss, U.; Gisinger, C.; Eineder, M. Measurements on the Absolute 2-D and 3-D Localization Accuracy of TerraSAR-X. *Remote Sens.* **2018**, *10*, 656. [CrossRef]
15. Cong, X.; Balss, U.; Rodriguez Gonzalez, F.; Eineder, M. Mitigation of Tropospheric Delay in SAR and InSAR Using NWP Data: Its Validation and Application Examples. *Remote Sens.* **2018**, *10*, 1515. [CrossRef]
16. Hackel, S.; Gisinger, C.; Balss, U.; Wermuth, M.; Montenbruck, O. Long-Term Validation of TerraSAR-X and TanDEM-X Orbit Solutions with Laser and Radar Measurements. *Remote Sens.* **2018**, *10*, 762. [CrossRef]
17. Schwerdt, M.; Schmidt, K.; Klenk, P.; Tous Ramon, N.; Rudolf, D.; Raab, S.; Weidenhaupt, K.; Reimann, J.; Zink, M. Radiometric Performance of the TerraSAR-X Mission over More Than Ten Years of Operation. *Remote Sens.* **2018**, *10*, 754. [CrossRef]
18. Esch, T.; Bachofer, F.; Heldens, W.; Hirner, A.; Marconcini, M.; Palacios-Lopez, D.; Roth, A.; Üreyen, S.; Zeidler, J.; Dech, S.; et al. Where We Live—A Summary of the Achievements and Planned Evolution of the Global Urban Footprint. *Remote Sens.* **2018**, *10*, 895. [CrossRef]
19. Havivi, S.; Schvartzman, I.; Maman, S.; Rotman, S.; Blumberg, D. Combining TerraSAR-X and Landsat Images for Emergency Response in Urban Environments. *Remote Sens.* **2018**, *10*, 802. [CrossRef]
20. Stettner, S.; Lantuit, H.; Heim, B.; Eppler, J.; Roth, A.; Bartsch, A.; Rabus, B. TerraSAR-X Time Series Fill a Gap in Spaceborne Snowmelt Monitoring of Small Arctic Catchments—A Case Study on Qikiqtaruk (Herschel Island), Canada. *Remote Sens.* **2018**, *10*, 1155. [CrossRef]

remote sensing

MDPI

Review

Ten Years of Experience with Scientific TerraSAR-X Data Utilization

Achim Roth [1,*], Ursula Marschalk [1], Karina Winkler [2], Birgit Schättler [3], Martin Huber [1], Isabel Georg [2], Claudia Künzer [1] and Stefan Dech [1]

[1] German Remote Sensing Data Center (DFD), German Aerospace Center (DLR), Münchener Strasse 20, D-82234 Weßling, Germany; Ursula.Marschalk@dlr.de (U.M.); Martin.Huber@dlr.de (M.H.); Claudia.Kuenzer@dlr.de (C.K.); Stefan.Dech@dlr.de (S.D.)
[2] SLU-Sachverständigenbüro für Luftbildauswertung und Umweltfragen, Kohlsteiner Str. 5, D-81243 Munich, Germany; Karina.Winkler@slu-web.de (K.W.); Isabel.Georg@slu-web.de (I.G.)
[3] Remote Sensing Technology Institute (IMF), German Aerospace Center (DLR), Münchener Strasse 20, D-82234 Weßling, Germany; Birgit.Schaettler@dlr.de
* Correspondence: Achim.Roth@dlr.de; Tel.: +49-8153-28-2706

Received: 22 June 2018; Accepted: 21 July 2018; Published: 24 July 2018

Abstract: This paper presents the first comprehensive review on the scientific utilization of earth observation data provided by the German TerraSAR-X mission. It considers the different application fields and technical capabilities to identify the key applications and the preferred technical capabilities of this high-resolution SAR satellite system from a scientific point of view. The TerraSAR-X mission is conducted in a close cooperation with industry. Over the past decade, scientists have gained access to data through a proposal submission and evaluation process. For this review, we have considered 1636 data utilization proposals and analyzed 2850 publications. In general, TerraSAR-X data is used in a wide range of geoscientific research areas comprising anthroposphere, biosphere, cryosphere, geosphere, and hydrosphere. Methodological and technical research is a cross-cutting issue that supports all geoscientific fields. Most of the proposals address research questions concerning the geosphere, whereas the majority of the publications focused on research regarding "methods and techniques". All geoscientific fields involve systematic observations for the establishment of time series in support of monitoring activities. High-resolution SAR data are mainly used for the determination and investigation of surface movements, where SAR interferometry in its different variants is the predominant technology. However, feature tracking techniques also benefit from the high spatial resolution. Researchers make use of polarimetric SAR capabilities, although they are not a key feature of the TerraSAR-X system. The StripMap mode with three meter spatial resolution is the preferred SAR imaging mode, accounting for 60 percent of all scientific data acquisitions. The Spotlight modes with the highest spatial resolution of less than one meter are requested by only approximately 30 percent of the newly acquired TerraSAR-X data.

Keywords: TerraSAR-X; synthetic aperture radar (SAR); radar mission; remote sensing

---

1. Introduction

On 15 June 2007, Germany's first operational radar satellite TerraSAR-X was launched into space. The TerraSAR-X mission is organized in a public–private partnership (PPP) between the German Aerospace Center (DLR) and the European Aeronautic Defence and Space Company (EADS) Astrium GmbH (since 2014 Airbus Defence and Space (DS)) [1]. Representing the German government, DLR solely owns and operates the satellite and coordinates the scientific utilization of the TerraSAR-X data and products. The commercial exploitation rights are exclusively granted to Airbus DS. The X-band

SAR can be operated in Spotlight, StripMap and ScanSAR mode providing different spatial resolutions and areal coverages with two polarizations in various combinations.

The TerraSAR-X mission pursues two main goals. The first is the provision of high-quality, multimode X-band SAR-data for scientific research and applications. The second goal is to support the establishment of a commercial earth observation (EO) market and to develop a sustainable EO-service business, based on derived information products. This paper reviews the TerraSAR-X mission solely from the perspective of scientific data utilization. It considers technical and operational issues and provides a resume of the mission's utilization and a background for future EO mission designs.

In general, earth observation data are an invaluable source of information for geoscientific research which, in turn, has resulted in a growing body of associated publications. NASA's Seasat was the first successful civilian SAR project in space, operating for 100 days in 1978 [2]. The SIR-C/X-SAR mission provided the unique capability for investigating the suitability of different frequencies and polarizations for applications in ecology, geology, hydrology, and oceanography [3]. Europe, in particular the European Space Agency, Canada and Japan have run SAR missions since the early 1990s. In this regard, these SAR missions [4–7], their objectives [8,9] and design, the sensors and other technical aspects were of interest [1,2,9–14]. Also the summary of a research call [15] and the potential for specific geoscientific applications are presented [16–20]. To date however, a comprehensive review of a high-resolution SAR system such as TerraSAR-X considering the technical capabilities, individual user needs in the relevant application fields and the system utilization associated with it has not been published in the peer-reviewed literature yet. Hence, this paper aims to answer the following questions: What are the key applications and modes of TerraSAR-X and how does the scientific user community generally utilize the mission? Therefore, we briefly summarize the TerraSAR-X mission, its technical capabilities and the product provision mechanism for scientific purposes. Based on the analysis of 1636 data utilization proposals, we determine the thematic scope as well as user motivations and expectations for using high resolution X-band SAR data. We examine all data takes acquired on requests by scientific users to assess the share of the different modes and polarizations and to identify the preferred settings and geographic areas of interest. We analyze the published results based on TerraSAR-X data in peer-reviewed scientific literature. Finally, we compare and discuss the revealed findings.

1.1. The TerraSAR-X Mission

Since the late 1970s, the German Aerospace Center (DLR) has supported a long and successful technology development line for high frequency X-band SAR systems. On 15 June 2007, a key milestone was achieved when Germany's first operational radar satellite TerraSAR-X was launched into space. Operational product ordering and delivery was opened to scientific and commercial users on 1 January 2008 after a six month commissioning phase. TanDEM-X, a twin-satellite to TerraSAR-X, was launched in June 2010. Both satellites fly in a close formation enabling single pass interferometry [21] and serving two purposes. The TerraSAR-X mission utilizes one of the satellites when just a single acquisition is needed, while the TanDEM-X mission simultaneously acquires two images from the same area on the ground using both satellites. The main goal of the latter has been the generation of global, consistent and high-quality digital elevation models (DEMs) [21].

The SAR sensors of TerraSAR-X and TanDEM-X operate at X-band with 9.65 GHz center frequency. Depending on the SAR imaging mode, the transmitted bandwidth is 55 to 100, 150, or 300 MHz [22]. The satellite was positioned in a polar sun-synchronous dawn-dusk orbit with an inclination of 97.44 degrees and an altitude of 515 km. The repeat cycle is 11 days, with ground track repeatability within ±250 m for the TerraSAR-X satellite. One repeat period consists of 167 consecutive orbits. TerraSAR-X features different modes and polarizations. The SpotLight, StripMap, and ScanSAR-modes provide high-resolution SAR images for detailed analysis as well as wide swath capability whenever a larger coverage is required. Operational imaging is possible in single and dual polarization.

Quad-polarized data are available on an experimental basis only. Beam steering enables observations at different incidence angles and double side access can be realized by satellite roll maneuvers.

Within the StripMap and ScanSAR modes, a continuous data take of up to 1500 km until the flight direction can be realized. This is not possible in the Spotlight modes. Here, the sensor rather focusses on a predefined position on the ground during the overflight. This enhances the synthetic aperture and, thus, provides a high spatial resolution in azimuth. The range bandwidth resolution is 150 MHz for Spotlight mode. The High Resolution Spotlight and the Staring Spotlight are acquired with the 300 MHz variant. The mission's potential was upgraded in 2013 when two new modes were implemented. The Staring Spotlight capability increases the spatial resolution in azimuth to 24 cm. The new WideScanSAR mode (6 ScanSAR beams) extends the spatial coverage to 200 km. The characteristics of the operational ScanSAR and StripMap modes are provided in Table 1, the specifications of the Spotlight mode are displayed in Table 2.

Table 1. TerraSAR-X's StripMap and ScanSAR mode parameters (modified from [23]).

Parameter	StripMap	4 Beams ScanSAR	6 Beams ScanSAR
Number of subswaths	1	4	6
Swath width (ground range)	30 km [1], 15 km [2]	100 km	194 to 266 km
Nominal L1b product length	50 km	150 km	200 km
Full performance incid. angle range	20°–45°	20°–45°	15.6°–49°
Data access incidence angle range	15°–60°	15°–60°	15.6°–49°
Azimuth resolution	3.3 m [1], 6.6. m [2]	18.5 m	40 m
Ground range resolution	1.7 m–3.49 m	1.7 m–3.49 m	<7 m
Polarizations	HH or VV [1] HH/VV, HH/HV, VV/VH [2]	HH or VV	HH or VV

[1] single pol, [2] dual pol.

Table 2. TerraSAR-X Spotlight mode parameters (modified from [23]).

Parameter	Spotlight	High Resolution Spotlight	Staring Spotlight
Number of subswaths	1	4	6
Scene extent (azimuth x ground range)	5 km × 10 km	5 km × 10 km	2.5 km × 6 km
Full performance incid. angle range	20°–55°	20°–55°	20°–45°
Data access incidence angle range	15°–60°	15°–60°	15°–60°
Azimuth steering angle	Up to ± 0.75°	Up to ± 0.75°	±2.2°
Azimuth resolution	1.7 m [1], 3.4. m [2]	1.1 m [1], 2.2. m [2]	0.24 m [1]
Ground range resolution	1.48 m–3.49 m	1.48 m–3.49 m (150 MHz) 0.74 m–1.77 m (300 MHz)	0.85 m–1.77 m
Polarizations	HH or VV [1] HH/VV [2]	HH or VV [1] HH/VV [2]	HH or VV [1]

[1] single pol, [2] dual pol.

The TerraSAR-X Ground Segment [24] controls and operates the satellite platform as well as the SAR sensor [25], performs instrument calibration [26], acquisition and archiving of the SAR data and the generation of user-defined products on request [22].

The Mission Control Center is located at DLR in Oberpfaffenhofen, Germany. The central receiving station for the TerraSAR-X mission is at DLR in Neustrelitz, Germany. Additionally, commercial data requests are downloaded at direct-access stations. The demand for TerraSAR-X products increased over the years and data downlink became a limiting factor. Since 2015, extra downlink capacity has been purchased at other receiving stations. Near-real-time access was realized at the DLR stations in Neustrelitz and O'Higgins in Antarctica.

All data acquired from the TerraSAR-X and TanDEM-X missions are stored in the multimission archive of DLR. By the end of June 2017, more than 235,000 individual data takes were ingested corresponding to almost 290 TBytes of SAR raw data. The raw data are processed to SAR Level 1b products following user requests.

1.2. Scientific Access to TerraSAR-X Data

Access to TerraSAR-X data for scientific purposes needs to be gained via a proposal submission and evaluation process. Here, interested users describe the intended research, their project partners and team organization, the expected schedule, an estimate of the number of TerraSAR-X products needed and the location of the test site. A principle investigator (PI) submits the proposal and acts as project lead and single point of contact to DLR. The evaluation is usually completed within two months. The evaluation includes an assessment whether the data is used for scientific purposes and would, therefore, conform to the mandate of DLR.

DLR issued several calls for proposals, so-called announcement of opportunities (AO). The first was the pre-launch AO, released already in 2004. Seven additional calls followed, four of which addressing specific issues such as promoting the new modes or enabling collaborations with the Canadian Space Agency (CSA) as an international partner were of limited duration. Three calls remain open for submission at any time during the mission. The archived data AO provides simplified access to archived TerraSAR-X data older than 18 months at the time of ordering. Archived experimental data, e.g., in full-polarimetric mode, can be requested through the Special Products AO. New acquisitions or more recently acquired data can be accessed through the general AO.

Users with accepted proposals get an account with access to DLR's EOWEB GeoPortal EGP [27]. The account's functionality is configured according to the corresponding AO, enabling for example access to future acquisitions or ordering products from the archive. When new observations are needed, the PI gets full flexibility to select and order the best suited mode, polarization and acquisition date. In case of acquisition conflicts, a priority scheme is applied by DLR's mission planning in order to automatically generate the acquisition timeline [25].

The TerraSAR-X mission is governed by national security regulations because of the high spatial resolution capability. In particular, these take effect when users request the Spotlight modes or StripMap Single Look Slant range Complex products (SSC). The provision of such products may become sensitive depending on the geographical region or the urgency of the data delivery. The latter can usually be simplified if the default latency of five days is acceptable for the user. Otherwise, DLR as the provider has to apply for the data release from the relevant state authority.

The vast majority of the scientific orders could generally be delivered without any extra actions. Altogether, a total of 154,000 scientific orders requested TerraSAR-X products, either from the archive or new data acquisitions. Only 306 of them needed further clarification. From these, 279 requests were accepted for delivery and, in only 27 cases, the corresponding products could not be provided.

2. TerraSAR-X Proposals, Scientific Orders and Literature Review

2.1. TerraSAR-X Proposals

The access to TerraSAR-X data for scientific purposes is granted via a proposal mechanism. TerraSAR-X data are usually exploited by a team of scientists headed by a PI, with partners acting as Co-Investigators (Co-Is). In total, 955 Principle Investigators from 617 institutions in 55 different countries submitted requests for TerraSAR-X products until the end of June 2017. Figure 1 shows countries of the institution with which the PIs were affiliated.

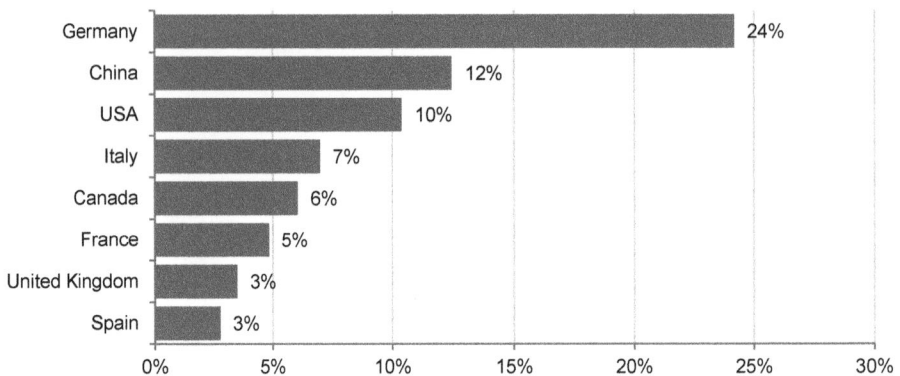

Figure 1. Countries of the principle investigator's (PI) institutions: 955 PIs submitted proposals. All countries with more than 20 PIs were considered.

Please note that the location of the investigator's affiliated institution but not the PI's nationality is considered. A further 1396 Co-Is from 1062 institutions in 79 countries collaborate with the PIs in the related projects.

Some of the PIs handed in several proposals so that in total 1636 proposals were submitted until the end of June 2017. A keyword scan was applied to all of the 1636 proposals in order to investigate which user community uses TerraSAR-X data for what purposes. Here, 55 thematic and 25 technique-related keywords were considered. On that basis, we assigned every proposal to one of the following six areas of research (=sphere):

- Anthroposphere (e.g., urban areas, infrastructure, mining, vulnerability studies).
- Biosphere (e.g., agriculture, forestry, grassland, wetland).
- Cryosphere (e.g., glacier, snow, permafrost, sea ice).
- Hydrosphere (e.g., sea conditions, ocean current, tidal flats and coastal areas, soil moisture, hydrological cycle).
- Geosphere (e.g., earthquakes, volcanos, landslides, soil conditions).
- Methods and techniques (e.g., new image analysis and processing techniques, preparation of new imaging modes).

Note that a proposal can just belong to one sphere so that the sum of proposals over all spheres is 100 percent. Each sphere is subdivided into characteristic thematic fields. Thus, one keyword may occur multiple times within title and keyword list. These keywords are used to characterize the corresponding research and application more specifically. Here, the number of keywords used within the respective sphere is given as a percentage, though the sum is not necessarily 100 percent. A more detailed analysis of the proposals is provided in Sections 3.1–3.7, which are supplemented by references to published results.

2.2. Scientific TerraSAR-X Acquisitions

By the end of June 2017, more than 235,000 individual data takes were acquired and stored in the TerraSAR-X mission archive. A data take is a continuously acquired strip of SAR raw data which can consist of several individual scenes. 100,711 of them were initiated for scientific and mission control purposes. Figure 2 shows their yearly distribution between 19 June 2007 and 30 June 2017. General ordering was opened up on 1 January 2008. Hence, all acquisitions in 2007 were exclusively related to commissioning activities. Therefore, they are not considered in the following analyses.

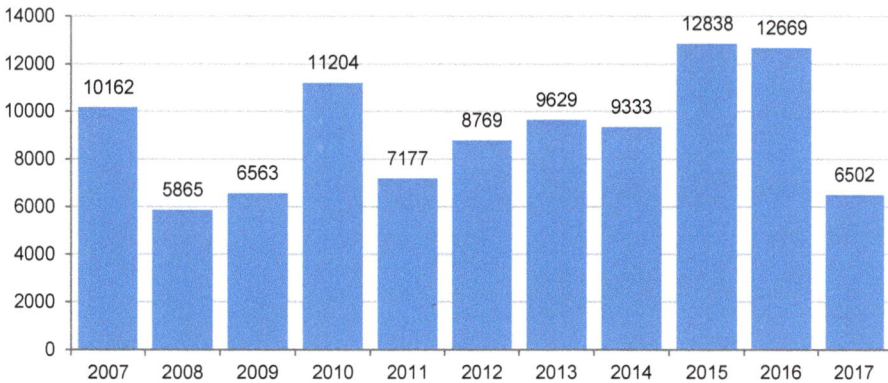

Figure 2. Number of scientific data takes of the TerraSAR-X mission between 19 June 2007 and 30 June 2017.

For the purpose of this review, all data takes acquired on request of scientific users after 1 January 2008 are examined to examine the role of different modes and polarizations, to identify the preferred settings and geographic areas of interest.

2.3. Scientific Publications

The publication of research results represents an important indicator for the assessment of key applications of a high-resolution SAR system and the scientific utilization of the TerraSAR-X mission. For this purpose, an analysis of renowned journals was performed based on Elsevier's Scopus data base, a comprehensive abstract and citation data base of peer-reviewed literature. For this study, we only considered papers written in English and published in scientific journals and books. Conference proceedings were not taken into account. In order to avoid confusion with the TanDEM-X mission, papers addressing a bistatic use of SAR data were excluded from this analysis. In total, 2850 publications refer to the TerraSAR-X mission.

Similar to the analysis of the TerraSAR-X proposals, a keyword scan was applied to all titles, authors and index keywords within the Scopus data base. Here, the 80 keywords of the proposal analysis were supplemented by another 36. Those 116 keywords were grouped in a specific way to represent the same fields of research as applied for the proposals. The publications were then assigned to the research sphere with the majority of hits revealed from the keywords search. Like the proposals, a publication can only belong to one sphere so that the sum of all proposals is 100 percent.

3. Results

3.1. Proposal Analysis Overview

All 1636 proposals were classified into one of the six spheres as a result of the keyword scan. Generally, a largely homogeneous distribution over the different fields of research can be noticed, as shown in Figure 3. The share ranges from 13 percent for methodological and technical developments to 20 percent for research in the geosphere domain.

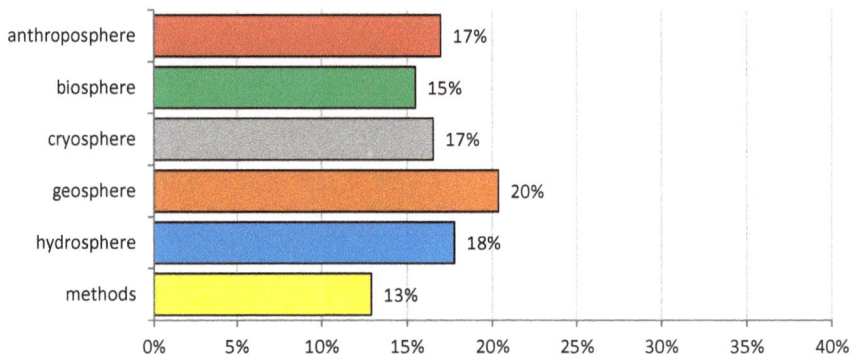

Figure 3. Histogram of TerraSAR-X related fields of research, expressed in percent based on the analysis of 1636 proposals.

3.1.1. Anthroposphere

The focus of 17 percent of all TerraSAR-X proposals is the environment created by humans (see also Figure 4). Today, more than half of the world's population live in cities and the trend of migration into urban areas is still increasing. Urban sprawl, expanding, and increasing exposure to hazards are only some of the associated challenges. High-resolution TerraSAR-X data are utilized for mapping the urban extent [28–30] and its temporal evolution [31]. Especially the StripMap mode's ground resolution of approximately three meters enables an automated derivation of built-up areas on a global scale [32]. Furthermore, TerraSAR-X StripMap time series are employed for investigating the impact of urban growth and associated anthropogenic activities like tunnel excavation, or ground water pumping [33].

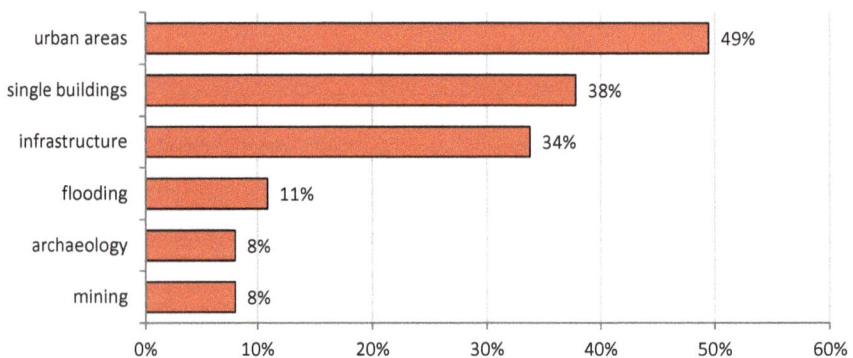

Figure 4. Thematic fields of anthroposphere-related TerraSAR-X proposals sorted by frequency, addressing the built-up area in general, individual buildings and infrastructure (e.g., dams, bridges, roads, railway tracks), the impact of natural disasters (in particular flooding), archaeology and manmade activities like mining.

The very high-resolution Spotlight modes are utilized on a local scale. Here, the reconstruction of individual buildings and the observation of settlements are of interest [34,35]. Not only buildings but also infrastructure like expressways [36], railway tracks [37], bridges [38], and dams [39] are investigated. Very high-resolution data are even employed to measure thermal expansion of buildings and infrastructure [40,41]. Mostly, interferometric techniques are applied.

All TerraSAR-X modes are employed for investigating flooding events [42–44] and identifying areas with a higher risk of being flooded [45]. Damage assessment [46] as well as the monitoring of reconstruction activities are research fields utilizing high-resolution SAR data [47]. A new application field arises with the use of remote sensing data for archaeological purposes. This includes the investigation of historical sites and their surrounding areas [48–50] as well as the monitoring of historical sites endangered by natural or human-made hazards or under pressure from settlement expansion [51,52]. Moreover, looting activities are observed in areas which are not accessible due to armed conflicts [53]. Most of the archaeological applications benefit from the high spatial resolution of TerraSAR-X.

SAR interferometric techniques are applied to monitor mining activities [54,55]. This includes active mining but also the effects of already sealed underground mines on above-ground settlements. Furthermore, the impact of geothermal drilling is investigated [56].

More than 50 percent of all proposals in the anthroposphere domain address SAR interferometry in general or a particular interferometric measurement technique. Due to the high spatial resolution, more scatterers with a stable backscattering behavior over time can be identified, enabling for example the persistent scatterer technique. Further, SAR tomography employs SAR data acquired from different incidence angles. It allows not only the reconstruction of the three-dimensional position of stable scatterers along a building facade but also their spatial dynamics over time when time series are available. Notably, SAR polarimetry is not a driver when SAR data are used for mapping and monitoring of urban areas.

3.1.2. Biosphere

Geoscientific research in the biosphere domain is addressed by 15 percent of the proposals (see Figure 5), the majority of which is concerned with agricultural issues, specifically crop type mapping and monitoring. Most commonly, spatio-temporal backscatter variations are employed for crop type discrimination [57], crop production forecast [58], and mapping of phenological conditions [59]. In addition, surface characteristics like soil moisture [60] and their change due to tillage and harvesting activities are of interest [61]. Further studies map [62] and monitor rice growth [63–65].

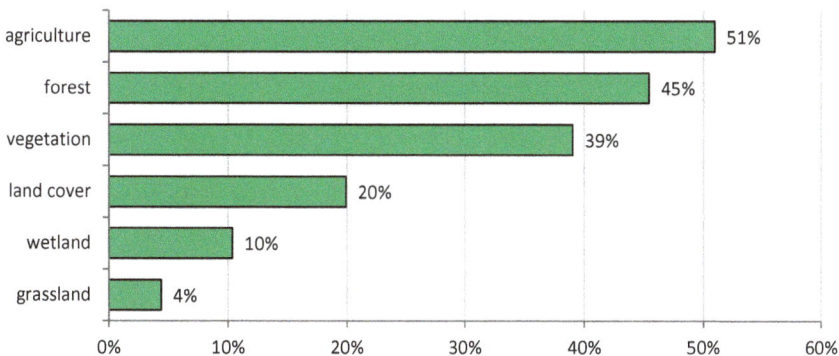

Figure 5. Thematic fields of biosphere-related TerraSAR-X proposals, sorted by frequency, specifically addressing agriculture, forest, wetland and grassland issues or, more generally, vegetation and land cover.

Mapping and monitoring of forests is another essential issue in biosphere-related applications. Here, high-resolution SAR data are used for the derivation of forest stand attributes, forest structure and biomass estimations. Generally, three basic sources of information from SAR data are utilized: Backscatter, coherence and phase-based approaches [66]. Besides, radargrammetry is used for biomass estimation by determining tree heights through surface and terrain models [67,68]. Also, changes of the forest cover over time are investigated, such as the impact of forest fires [69], degradation, deforestation and forest health [70].

Several proposals target rather general issues of vegetation and land cover mapping such as the assessment of soil erosion processes [71], shrub growth monitoring [72], the detection and characterization of hedgerows [73], or biodiversity monitoring [73,74].

Wetlands are threatened by population growth, extension of agricultural land, and general human interventions in the water budget. Goals of the research community are, for example, to assess and better understand the wetland functions to improve land and water management [75]. Proposals and research with focus on reed belts [76] and other wetland vegetation [77] belong to this group. Additionally, the possibility of applying differential interferometry for measuring water level and water level changes is investigated [78].

A relatively small group of proposals addresses mapping and monitoring of grassland. The identification of grassland and especially its conversion into cropland is a relevant topic for studies on sustainable land use [79]. El Hajj et al. [80] identify the economic use of grassland for hay production. SAR data are used for biomass estimation and monitoring of cutting activities [81].

Most of the applications in the biosphere group use TerraSAR-X data for monitoring (65 percent of the proposals) and mapping purposes. Compared to 30 percent of the users requesting polarimetric data, SAR interferometry plays a minor role, but still 11 percent utilize InSAR products. Furthermore, change detection is addressed in 9 percent of the proposals. TerraSAR-X data are often synergistically used with other EO data in order to benefit from both the higher spatial resolution and higher frequency of the X-band in combination with other SAR and optical systems. This is evident through the 24 percent of the biosphere proposals describing multisensoral approaches.

3.1.3. Cryosphere

The influence of climate and the climate variability on ice- and snow-covered areas of the Earth (see Figure 6) is investigated by 17 percent of the proposals. Almost 50 percent concentrate on the observation of ice caps and glaciers including monitoring of glacier outlines [82] and their spatiotemporal variability [83,84] and, changing of calving rates at the shore line [85]. The major application field, however, is the determination of the glacier flow velocity [86,87], an essential factor for the assessment and understanding of mountain [88–90], rock [91] and polar glacier dynamics. The latter is often combined with the determination of the glacier's grounding line [92,93]. TerraSAR-X data are used to monitor the collapse of ice shelves [94–96] and to investigate their internal and basal structure [97].

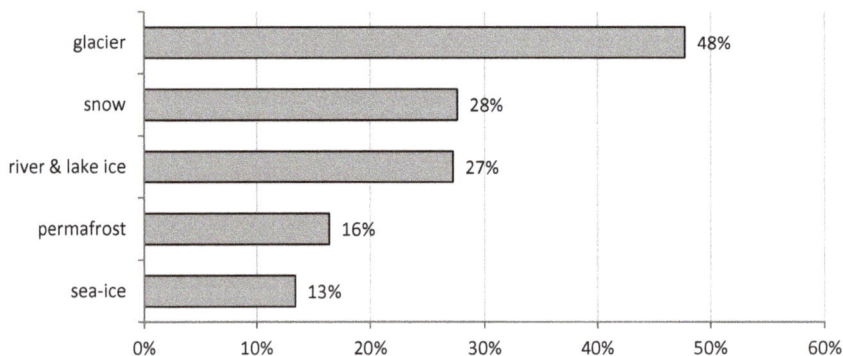

Figure 6. Thematic fields of cryosphere-related TerraSAR-X proposals, sorted by frequency, addressing polar and mountain glaciers, snow, permafrost, sea-, lake- and river-ice.

Mapping of snow-covered areas is a key aspect of 28 percent of the cryosphere proposals. A detailed knowledge of snow-covered areas, snow height [98] and the structure and physical properties of the snow pack [99] are important for many applications such as water supply, flood and avalanche risk prediction, weather forecast as well as the understanding of climate change in the long term. Not only snow cover over land is observed but also its impact on sea ice cover and breakup dynamics [100].

About 27 percent of the proposals monitor the river and lake ice phenology [101–105]. Water resource distribution and accessibility, the use of frozen rivers and lakes for travel and transportation purposes as well as an improved hydropower management are some of the motivations for research. Additionally, thermokarst lakes are key elements of the permafrost landscape. With permafrost thawing and associated greenhouse gas production limited to the period of unfrozen lakes, the ice phenology can be regarded as an indicator for climate change.

Permafrost research is focus of 16 percent of the proposals. This ecosystem faces dramatic changes caused by climate warming and increasing human activities in the Arctic. TerraSAR-X data are used to determine the thickness [103] of the so-called active layer, the soil above the permafrost where most of the ecological, hydrological and biological activities take place. The seasonal thawing and refreezing of the active layer causes surface deformations, posing a challenge for infrastructure and engineering [104]. Also surface changes of the permafrost landscape [105,106] and arctic wetlands [107] as well as the erosion of permafrost coastlines and riverbanks [108] are monitored.

The retreat of sea ice in the Arctic is a clear indicator and a visible sign of global warming. The determination of the sea ice thickness [109], composition and extent [110,111] as well as iceberg monitoring [112] provides essential information for operational ice services as well as the modelling of the interaction between the polar oceans and the atmosphere. Information derived from TerraSAR-X data about the sea ice state and in particular the dynamic processes forcing the sea ice breakup are used for ice breaker support [113,114]. Also, the formation and dynamics of polynyas, which are persistent areas of open water even during the polar winter, are monitored [115]. Sea-ice related issues are investigated by 13 percent of the cryosphere proposals.

TerraSAR-X data are mainly used for mapping purposes. Apart from that, the derivation of glacier flow velocity and velocity change is significant. About a third of the cryosphere proposals address this issue. Here, both interferometric and feature tracking techniques are applied. The latter especially benefits from the high spatial resolution of TerraSAR-X. For 18 percent of the proposals, polarimetric data are requested.

3.1.4. Geosphere

Most of the TerraSAR-X proposals are related to the geosphere—one-fifth intends to use the data to improve the understanding of the dynamics of the Earth's crust as well as soil properties and erosion processes (see Figure 7). 37 percent of the geosphere proposals address tectonic issues. Co- and post-seismic surface deformations after an earthquake event are investigated by providing displacement fields and trends [116,117]. The findings help, for example, to improve the understanding and identification of the local tectonic settings [118] and even the interaction between two subsequent earthquake events [119] by combining the TerraSAR-X derived displacement field with GPS measurements and elastic dislocation models. Earthquakes can also be triggered by ground water extraction [120].

Observation of active volcanos is addressed by 27 percent of the geosphere-related proposals. Since their surface is constantly and rapidly changing, TerraSAR-X data are used to determine surface deformation caused by emptying and refilling of magma reservoirs [121]. Also, the growth and collapse of lava domes is monitored [122]. High-resolution X-band data are particularly utilized for the identification and analysis of small scale surface deformations at the caldera's rim and walls [123]. Moreover, the rapidly changing geomorphology and deposits of eruptive volcanos is investigated by evaluating the backscatter changes [124]. TerraSAR-X data are further used for the monitoring of geothermal fields [125–127]

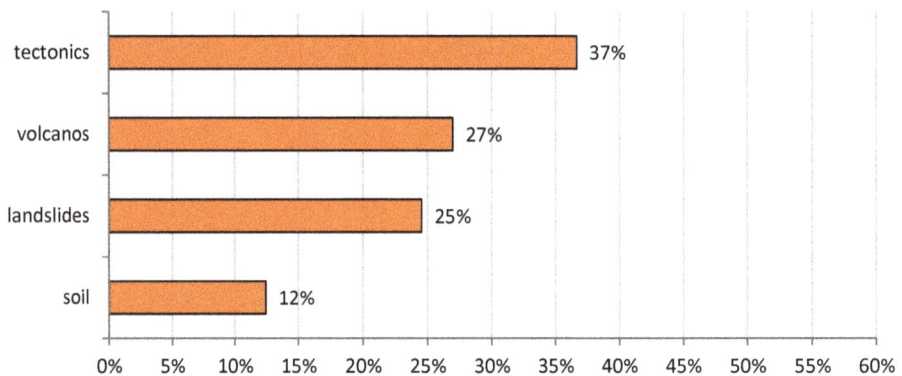

Figure 7. Thematic fields of geosphere related TerraSAR-X proposals, sorted by frequency addressing small- and large-scale surface deformations caused by the Earth crust's dynamic (tectonics), volcanic activities, landslides and soil mapping including erosion.

The identification of landslides, their spatial extent and dynamics is focus of 25 percent of the proposals. Landslides mostly appear on hill slopes dominated by weakly consolidated sediments [128]. They can be triggered by earthquakes, hydrological factors [129], melting of permafrost [130] and other erosion processes [131], but also human activities like cutting steep slopes to accommodate roads [132]. Inventories help to assess landslide hazards to protect people, properties and infrastructure [133,134].

A smaller share of proposals (12 percent) concentrate on soil mapping. Here, high-resolution X-band data are utilized for top-soil mapping [135], soil surface texture [136] and the investigation of erosion and sedimentation processes [137]. Other proposals assess the impact of ground water extraction [138]. A special case is the formation of sinkholes when subsurface cavities are formed which suddenly collapse [139].

All these applications benefit from the high spatial resolution and the short wavelength of TerraSAR-X. These features, together with the high observation frequency, provide a very sensitive observation and mapping tool. Accordingly, the higher resolution StripMap and Spotlight modes are requested. SAR interferometry is by far the dominant measurement technique of this research field and is mentioned in 84 percent of the proposals. SAR polarimetry is rarely applied; only 4 percent request polarimetric data.

3.1.5. Hydrosphere

The hydrosphere is the second largest group, accounting for 18 percent of all proposals. Here, research questions referring to oceans, coastal areas, limnology, soil moisture and overarching the hydrological cycle in general are addressed (see Figure 8).

Wind field and sea state retrieval is the major application, with 53 percent of the hydrosphere-related proposals. The latter mainly benefits from the higher spatial resolution which enables the derivation of shorter ocean waves with the shorter wavelength X-band system [140,141]. High radiometric performance is essential for wind field retrieval [142,143]. The high precision is utilized for offshore wind farming support [144]. The relatively low orbit of TerraSAR-X supports an improved observation of features on the sea surface compared to other L- and C-band sensors. Extreme weather conditions like storm observations and rogue waves generation [145] are investigated as well as the wind and wave interactions with ice [146].

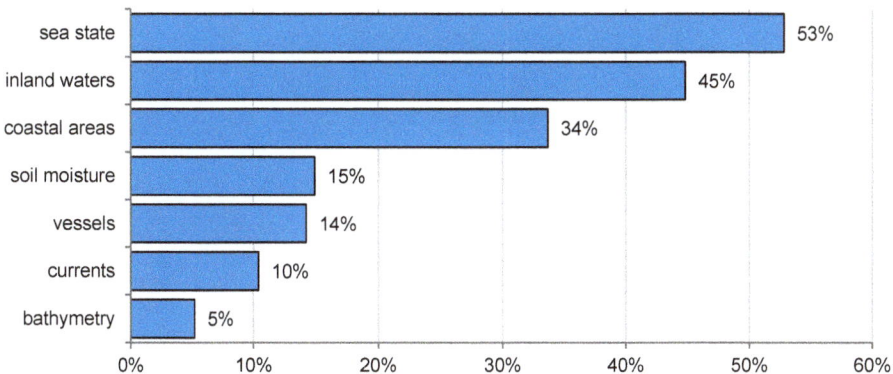

Figure 8. Thematic fields of hydrosphere-related TerraSAR-X proposals, sorted by frequency addressing oceanographic applications, coastal zones, inland waters, soil moisture and the identifications and observation of ships and ship traffic.

The mapping of inland waters is focus of 45 percent of the proposals. Seasonal changes [147,148] are monitored and the available amount of water is estimated [149]. Special attention is paid to water management [150].

A third of the proposals deal with coastal applications. Tidal flats [151,152], salt marshes [153] and submarine ground water discharge [154] are monitored. Intertidal bedform shifts are investigated and the impact on sediments and benthic macrofauna [155] is assessed. For coastal protection purposes, oil spill detection and monitoring is performed [156–158].

Mapping and monitoring of soil moisture in arid and semi-arid environment is another key application (15 percent of the proposals) [159,160]. Here, soil moisture is important for water resources and irrigation management, but also for runoff estimations and land surface degradation processes.

Another 14 percent are engaged in the identification and observation of vessels. This includes not only ships, but oil platforms and other sea based infrastructure as well. Ship surveillance is performed to increase security [161] and safety [162]. Apart from the ship identification and classification, ship heading and velocity estimation [163] is dealt with. Also, near real time services have been developed [164].

Furthermore, ocean currents are of interest for 10 percent of the proposals. Here, the capability of Along Track Interferometry (ATI) is evaluated [165]. Variations in the backscatter intensity are likewise utilized, e.g., for the investigation of bottom topography-induced current front in a tidal channel [166].

The bathymetry, especially of coastal areas, is of interest for 5 percent of the proposals. The bottom topography is derived by observing the wave behavior in shallow waters [167] and by evaluating the temporal correlation between the backscatter intensity and the water level [168].

Only 20 percent of the proposals mention interferometric measurement techniques. 6 percent of these intend to apply along-track interferometry, while 19 percent request polarimetric data.

3.1.6. Methods and Techniques

A small share of the proposals—13 percent—addresses methodological and technical developments rather than specific application fields. Figure 9 shows the major techniques where further developments were proposed.

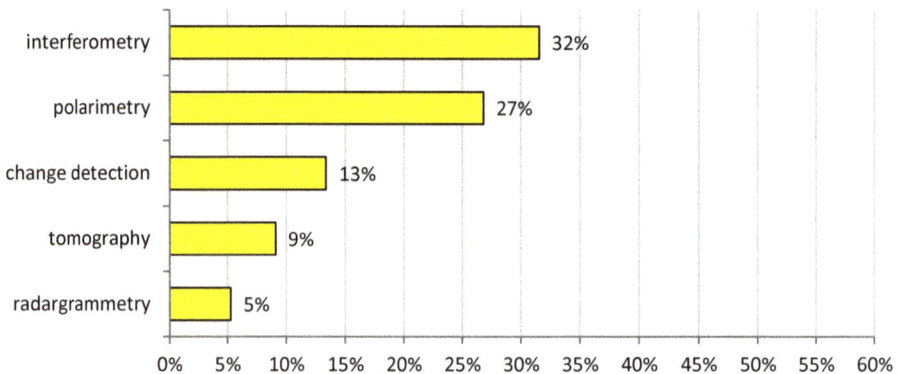

Figure 9. Thematic fields of methodological and technical research related TerraSAR-X proposals, sorted by frequency addressing differential SAR interferometry including its further developments such as PSInSAR, polarimetry, automated detection of changes, SAR tomography, and radargrammetry for 3d mapping.

Close to a third (32 percent) of the proposals target improvements of techniques for measuring surface movements like persistent scatterer interferometry [169], distributed scatterer interferometry [170] and subpixel offset techniques [171]. The common goal is to overcome the limitations of the differential interferometry like geometric and temporal decorrelation, atmospheric artifacts and the one-dimensional line of sight sensitivity. Along track interferometry combines data sets acquired during a single overflight and enable the observation of internal waves [172] and tidal energy resource assessment [173].

The goal of 27 percent of the proposals is to assess the polarimetric information in TerraSAR-X data. Some examples are the derivation of polarimetric information for mapping purposes [174], the improvement of the image segmentation of high-resolution polarimetric data [175,176], novel classification approaches [177] or texture tracking in polarimetric SAR data [178].

The development of techniques for the identification of changes is focus of 13 percent of the methods related TerraSAR-X proposals. The high spatial resolution and the SAR image speckle cause significant backscatter intensity variations even within a single object on ground. In order to simplify and optimize change detection, techniques such as image segmentation [179,180] and similarity cross tests based on polarimetric features [181] are applied.

SAR tomography is another development of the persistent scatterer interferometry that utilizes multi-incidence angle observations of the same targets on ground. It thereby enables the 3d-localization and motion of scattering objects [182,183]. The 3D coordinates can also be derived by radargrammetry [184,185]. Here, an image set acquired from at least two different positions in space is processed similar to the elevation model generation from optical data with a stereo-like view. The so called SAR imaging geodesy enables solid Earth tide measurements [186] by utilizing the very high spatial resolution and range measurement accuracy.

In summary, SAR interferometry is the main application for TerraSAR-X data. Fifty-two percent of all proposals investigate various kinds of surface movements. Different differential interferometric measurement and feature tracking techniques are employed. Twenty-seven percent request single polarized TerraSAR-X data for mapping purposes. Nineteenpercent ask for dual- and quad-polarized data. The remaining 2 percent are interested in SAR experiments and the development of new modes.

3.2. Analysis of the Scientific TerraSAR-X Acquisitions

The predominant share of the acquisitions (59 percent) throughout the entire operational phase of the TerraSAR-X mission was ordered in the three to five meter resolution StripMap mode. Only 14 percent were acquired in ScanSAR mode, benefitting from the coverage of larger areas. The highest spatial resolution is offered by the different Spotlight modes and 28 percent are allotted to these modes. This report does not distinguish between the different polarizations. Single, dual and quad-polarized data are jointly considered. The trends and the share of the different modes over time are presented in Figure 10.

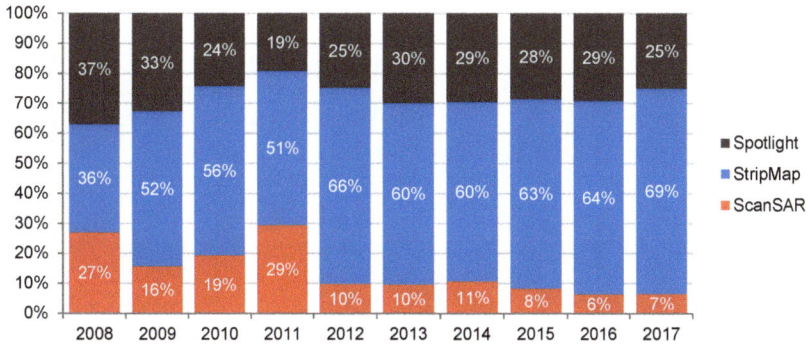

Figure 10. Proportion (in percent) of TerraSAR-X modes on an annual basis from 1 January 2008 to 30 June 2017: High-resolution Spotlight mode displayed on top, the medium resolution StripMap mode in the middle and the coarser resolution ScanSAR mode at the bottom.

The interest in the different modes was quite uniform at the beginning of the mission. In the following years the demand varied but has been stable since 2012. Approximately two-thirds of all acquisitions are performed in StripMap mode, 30 percent require the high spatial resolution of the Spotlight modes and the remaining 5 percent are interested in a larger coverage and order ScanSAR products. Figure 11 provides more detail about the utilization of the Spotlight modes.

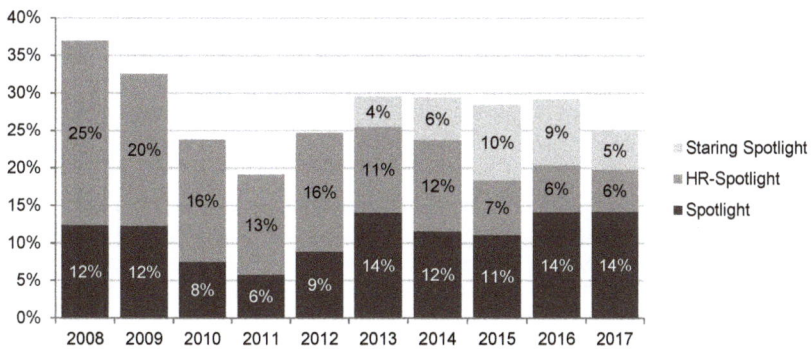

Figure 11. Proportion (in percent) of TerraSAR-X Spotlight modes on an annual basis from 1st January 2008 to 30th June 2017: The highest resolution Staring Spotlight mode displayed on top, the high-resolution HR-Spotlight mode in the middle and the standard resolution Spotlight mode at the bottom.

The interest in the highest resolution modes was the highest at the beginning of the mission. Almost 40 percent of all acquisitions were performed in the Spotlight modes. This demand dropped until 2011, but then increased again to a relatively constant level of 30 percent. In 2013, the Staring Spotlight mode was introduced. Remarkably, the very high-resolution potential is well accepted but did not lead to a systematic increase of the percentage of Spotlight orders in relation to the other modes.

The proportion of single and dual polarized acquisitions has been relatively constant throughout the mission's operating phase. However, an increase in the request of dual polarized data since 2016 is noticeable. On average, 79 percent of all scientific data takes were acquired in either HH or VV polarization. 21 percent requested dual polarization (HH/VV, HH/HV, VV/VH) and a few experimental data sets acquired in full polarization (0.4 percent). In 2017, the proportion changed to 60 percent of the acquisitions in single and 40 percent in dual polarization.

Figure 12 shows the geographic and quantitative distribution of the 100,711 scientific acquisitions of the TerraSAR-X mission. Note that the number of acquired scenes and not the number of individual PIs was taken into account for this map. Hence, areas where more scenes were acquired for scientific studies are represented to a higher extent. The number of scenes available in the archive per $1° \times 1°$ tile decreases from red (representing intensively observed areas) via orange, yellow and green to blue.

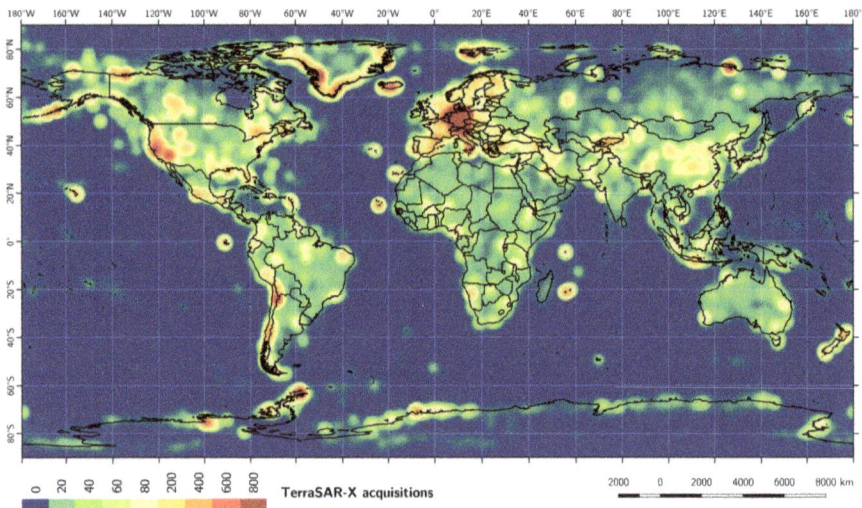

Figure 12. Geographic and quantitative distribution of scientific areas of interest observed by TerraSAR-X between 2007 and 2017, the color indicates the number of acquisitions per $1° \times 1°$ tile.

The research sites are globally distributed, with a large concentration on central Europe. However, other sites are regularly monitored whenever the system is not fully utilized. TerraSAR-X data are provided in the frame of several internationally coordinated activities as well, including the GEOHazard initiative [187], which monitors volcanos and earthquake-threatened areas, the Global Forest Observations Initiative (GFOI) [188] and the Polar Space Task Group [189]. In these cases, TerraSAR-X provides high-resolution SAR data of hot spots for local analyses.

3.3. Analysis of the Scientific Publications based on TerraSAR-X Data

Analogous to the analysis of the 1636 TerraSAR-X proposals, a keyword scan was performed in Scopus to all titles, author and index keywords of the 2850 publications referring to TerraSAR-X. In total, 116 keywords were selected and grouped in a way so that they represent the same six fields of research as for the proposals. The publications were then assigned to the research area (sphere) which best represents the keyword search (see also Section 2.1). Figure 13 shows the corresponding results.

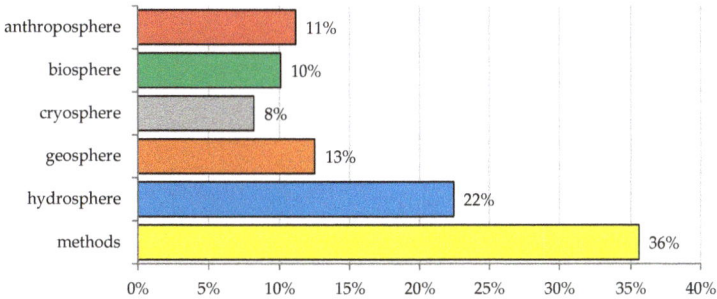

Figure 13. Histogram of TerraSAR-X related fields of research, expressed in percent based on the analysis of 2850 publications.

A noticeably high percentage of papers deal with methodological and technical issues (36 percent), followed by a large number of papers in the hydrosphere group (22 percent). The geosphere (13 percent) represents the second largest geoscientific group. Whereas the fractions of anthroposphere (11 percent) and biosphere (10 percent) are very similar, the cryosphere (8 percent) shows a relative small number of publications. It should be noted that there are overlaps between the different research fields, particularly between hydrosphere and cryosphere as well as anthroposphere and geosphere. Snow and glaciers are at the center of attention in cryosphere research but at the same time they also play an important role in the hydrological cycle, e.g., as water source and reservoir. The observations of volcanos and earthquakes improve the understanding of the tectonically active areas of the Earth in terms of location and areal extent, dynamics and intensity. Many of those areas are densely populated and, thus, relevant for anthropogenic research as well.

Figure 14 presents the number of TerraSAR-X related publications from 2003 to 2016. Prior to the launch of the TerraSAR-X satellite in 2007, only a small number of papers about preparatory work were published. Three years after the launch, a significant increase of publication activities can be noticed, with an increase from 153 publications in 2010 to 510 in 2016. The number of articles continuously increased in all research fields until 2015. This trend continues for the anthroposphere, hydrosphere and geosphere. The number of publications related to cryosphere issues remains fairly constant, while those related to biosphere and methods decreased between 2015 and 2016.

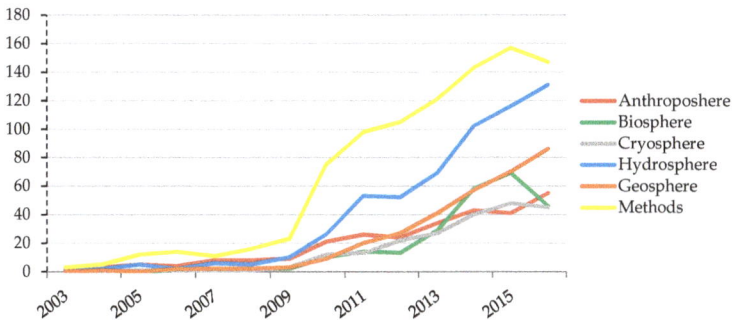

Figure 14. Accumulation of TerraSAR-X related research papers for thematic areas between 2003 and 2016.

Analysis of the scientific literature suggests that the dominant use of high-resolution SAR data is the derivation of surface movements. 1044 publications address this topic, 834 (29 percent of all TerraSAR-X related publications) of them deal with deformation and displacement of the solid land surface including glacier flow and 210 with movements of the ocean surface like currents, wave and wind fields. 720 articles (25 percent) used SAR interferometry including its different processing variants like PSInSAR or SAR-tomography. Feature tracking techniques were applied in 214 publications (8 percent), in 72 cases in combination with SAR interferometry. SAR polarimetry is the focus of 256 papers (9 percent). Change detection is carried out in 8 percent of the published articles and proposals.

The analysis further reveals monitoring as one of the main applications, referred to in 425 articles from the keyword search. Mostly monitoring comes along with the detection and understanding of changes of the land cover composition and extent. A regular observation of the area of interest is required, and longer time series are employed to analyze seasonal and multi-year dynamics. A significant number of publications (237 papers) deal with changes and change detection. Not only longer-term but also short-term events like natural disasters are part of scientific investigations. 122 papers used TerraSAR-X data for hazard mapping and vulnerability analyses.

4. Discussion

This paper assesses the utilization of the TerraSAR-X mission by the international scientific user community with respect to research themes. Furthermore it identifies key applications and preferred imaging modes and settings. The study is mainly based upon 1636 scientific TerraSAR-X data proposals. This number corresponds to the experience from other missions [15]. Here, more than 1000 projects were initiated based on European remote sensing satellite (ERS) data. In response to the Envisat prelaunch AO, 577 scientific proposals requested access to ASAR data. While researchers from 40 countries participated in the Envisat-AO in 2001, the TerraSAR-X user community increased to 57 countries in 2017. This indicates the growing interest in SAR data in general and is a sign of the positive development of the EO market that all SAR missions jointly serve. During its 17 years of operation, RADARSAT-1 provided images to governmental, commercial, and scientific users in 60 countries worldwide [5].

The analysis of the proposals and publications reveals that high-resolution X-band data are requested from all fields of geo-research. It even shows a largely homogenous distribution over the different spheres. Science projects supported by ALOS-2 are grouped into biosphere, geosphere, cryosphere, marine and disaster [9]. Similar categories were defined for RADARSAT-1 and RADARSAT-2 [18]. The potential of Sentinel-1 for scientific ocean, cryosphere, and land applications is summarized in [7]. All three publications describe the required capabilities, established by intensive studies and expert consultations before the launch. The approach of this TerraSAR-X study is to analyze the entirety of individual proposals and related publications and thus directly the utilization by the scientific user community when the mission is in the operational phase. The thematic fields of this study (see Figures 4–8) correspond very well to those of other SAR missions, although the fields of research (Figure 3) slightly differ from the ones in [7,9,18]. This confirms the general applicability and appreciation of SAR data for these applications.

Every SAR mission has its assets and advantages. They depend on the technical layout like wave length and spatial resolution, but also on the mission operation design. The features of TerraSAR-X are the high spatial resolution, the short wave-length, a high orbit repetition rate of 11 days with a short revisit time of less than three days and the multimode capability. The high spatial resolution is the main asset of TerraSAR-X as it enables investigations in all fields of research at a high level of detail. The evaluation of surface displacements is the prime application of a high-resolution SAR system like TerraSAR-X and is the principal topic of 63 percent of all proposals. The preferred method is SAR interferometry and its different variants, with feature tracking techniques being employed as well. SAR interferometry benefits from the higher number and density of long-term stable scatterers, but also from the high repeat cycle of 11 days. Feature tracking can be applied to data sets with a longer temporal baseline, but strongly depends on the identification of terrain features and structures to be tracked.

The majority of proposals request single polarized data. The investigation and utilization of multipolarized SAR data is of interest for nine percent of the proposals. TerraSAR-X can only be ordered in dual-polarization mode which is sufficient for applications like snow and ice mapping. However, the development and application of polarimetric decomposition techniques is limited. Nevertheless, the demand for dual-polarized TerraSAR-X data has increased since 2014.

TerraSAR-X is designed to provide users with the highest level of flexibility for selecting suitable resolution, coverage, and polarization for their research. As a consequence, the data are mostly acquired for a specific need and not necessarily suitable for multi-use purposes. Other missions like Sentinel-1 and ALOS-2 strictly follow a systematic and predefined observation plan [4,10], aiming at global mapping and optimized multi-use at the expense of flexibility. The most requested mode in the TerraSAR-X mission for scientific applications is StripMap (59 percent) in single polarization, followed by the high-resolution Spotlight (28 percent) and ScanSAR modes (13 percent). StripMap data are easy to process and provide the best ratio of resolution versus coverage. The Spotlight modes are appreciated by the user community for very detailed spatial analyses. The introduction of the even higher resolution Staring Spotlight mode in 2013 initially led to a slight increase of the percentage of Spotlight orders, but then stabilized at 30 percent. ScanSAR offers larger area observation capability. A rough estimate based on the coverages per mode of RADARSAT-2 [6] indicate a share of 20 percent for the highest spatial resolution modes, 30 percent for medium and 50 percent for the coarsest ScanSAR modes. However, these numbers are dominated by commercial use and the needs of operational services like the Canadian Ice Service. Often, the X-band is synergistically evaluated with other sensors either by adding high frequency information or by shortening the revisit time for a specific point on ground by for example placing a TerraSAR-X acquisition between two consecutive RADARSAT-2 observations.

5. Conclusions

The main aim of this paper was the identification of key applications and technical modes and settings for a high-resolution X-band system like TerraSAR-X. A further goal was to determine how the mission is utilized by the scientific user community. Generally, data are requested from all fields of geo-research with a largely homogeneous distribution of the corresponding number of proposals over the different spheres. The general applicability and appreciation of SAR data was confirmed. However, the evaluation of surface displacements is clearly revealed as the key application of a high-resolution SAR system. Mostly, SAR interferometry is applied, but also feature tracking techniques are of increasing importance. Future high-resolution SAR systems should support and strengthen this application, e.g., by maintaining and improving the orbit precision and interferometric baseline stability.

The three-meter spatial resolution of the StripMap mode is well-suited for systematic monitoring on a regional scale while meter and submeter resolution of the Spotlight mode enables detailed but spatially limited analyses. A future mission would benefit from an extended across track coverage of the three meter StripMap mode. Spatially higher resolution of less than one meter would improve local studies and the evaluation of surface displacements.

Scientific users make use of polarimetric SAR capabilities, although these are not a main feature of TerraSAR-X. The demand has been growing over the recent years and a future system would benefit from the availability of fully polarized SAR data.

All geoscientific fields request systematic monitoring and the creation of time series for the investigation of seasonal, interannual, and even decadal variations. The use of single scenes for analyzing local and singular events is a minor issue.

Author Contributions: A.R. as the TerraSAR-X Science Coordinator reviewed all TerraSAR-X proposals and wrote the paper. U.M. performed the keyword analysis of the TerraSAR-X scientific proposals. K.W. carried out the literature review and the corresponding keyword analysis. B.S. evaluated the TerraSAR-X archive and prepared the scientific acquisitions statistics. M.H. prepared the geographic distribution of scientific acquisitions. I.G. performed the English editing. C.K. and S.D. strengthened the paper through critical discussion and review. All authors contributed to the paper by discussions and revisions of the manuscript.

Funding: This research received no external funding.

Acknowledgments: In this section you can acknowledge any support given which is not covered by the author contribution or funding sections. This may include administrative and technical support, or donations in kind (e.g., materials used for experiments).

Conflicts of Interest: The authors declare no conflict of interest.

References

1. Werninghaus, R.; Buckreuss, S. The TerraSAR-X mission and system design. *IEEE Trans. Geosci. Remote Sens.* **2010**, *48*, 606–614. [CrossRef]
2. Evans, D.; Alpers, W.; Cazenave, A.; Elachi, C.; Farr, T.; Glackin, D.; Holt, B.; Jones, L.; Liu, W.; McCandless, W.; et al. Seasat—A legacy of success. *Remote Sens. Environ.* **2005**, *94*, 384–404. [CrossRef]
3. Schmullius, C.; Evans, D. Review article Synthetic aperture radar (SAR) frequency and polarization requirements for applications in ecology, geology, hydrology, and oceanography: A tabular status quo after SIR-C/X-SAR. *Int. J. Remote Sens.* **1997**, *18*, 2713–2722. [CrossRef]
4. eoPortal Directory. Copernicus: Sentinel-1. Available online: https://directory.eoportal.org/web/eoportal/satellite-missions/c-missions/copernicus-sentinel-1 (accessed on 4 July 2018).
5. eoPortal Directory. RADARSAT-1. Available online: https://directory.eoportal.org/web/eoportal/satellite-missions/r/radarsat-1 (accessed on 4 July 2018).
6. eoPortal Directory. RADARSAT-2. Available online: https://directory.eoportal.org/web/eoportal/satellite-missions/r/radarsat-2 (accessed on 4 July 2018).
7. eoPortal Directory. ALOS-2. Available online: https://directory.eoportal.org/web/eoportal/satellite-missions/a/alos-2 (accessed on 4 July 2018).
8. Malenovský, Z.; Rott, H.; Cihlar, J.; Schaepman, M.; García-Santos, G.; Fernandes, R.; Berger, M. Sentinels for Science: Potential of Sentinel-1, -2 and -3 missions for scientific observations of ocean, cryosphere and land. *Remote Sens. Environ.* **2012**, *120*, 91–101. [CrossRef]
9. Shimada, M. ALOS-2 science program. In Proceedings of the International Geoscience and Remote Sensing Symposium (IGARSS), Melbourne, Australia, 21–26 July 2013; pp. 2400–2403. [CrossRef]
10. Shinichi, S.; Kankaku, Y.; Shimada, M. ALOS-2 acquisition strategy. In Proceedings of the International Geoscience and Remote Sensing Symposium (IGARSS), Melbourne, Australia, 21–26 July 2013; pp. 2412–2415. [CrossRef]
11. Torres, R.; Snoeij, P.; Geudtner, D.; Bibby, D.; Davidson, M.; Attema, E.; Potin, P.; Rommen, B.; Floury, N.; Brown, M.; et al. GMES Sentinel-1 mission. *Remote Sens. Environ.* **2012**, *120*, 9–20. [CrossRef]
12. Hillman, A.; Rolland, P.; Chabot, M.; Périard, R.; Ledantec, P.; Martens, N. RADARSAT-2 mission operations status. In Proceedings of the International Geoscience and Remote Sensing Symposium (IGARSS), Vancouver, BC, Canada, 24–29 July 2011; pp. 3480–3483. [CrossRef]
13. Covello, F.; Battazza, F.; Coletta, A.; Manoni, G.; Valentini, G. COSMO-SkyMed mission status: Three out of four satellites in orbit. In Proceedings of the International Geoscience and Remote Sensing Symposium (IGARSS), Cape Town, South Africa, 12–17 July 2009; pp. II773–II776. [CrossRef]
14. Rosenqvist, A.; Shimada, M.; Ito, N.; Watanabe, M. ALOS PALSAR: A pathfinder mission for global-scale monitoring of the environment. *IEEE Trans. Geosci. Remote Sens.* **2007**, *45*, 3307–3316. [CrossRef]
15. Desnos, Y.-L.; Palazzo, F.; Regner, P.; Zehner, C.; Benveniste, J.; Doherty, M. Research activities in response to the Envisat Announcement of Opportunity. *ESA Bull.* **2001**, *106*, 118–127. Available online: http://www.esa.int/esapub/bulletin/bullet106/bul106_11.pdf (accessed on 19 June 2018).
16. Nagler, T.; Rott, H.; Hetzenecker, M.; Wuite, J.; Potin, P. The Sentinel-1 mission: New opportunities for ice sheet observations. *Remote Sens.* **2015**, *7*, 9371–9389. [CrossRef]

17. Salvi, S.; Stramondo, S.; Funning, G.J.; Ferretti, A.; Sarti, F.; Mouratidis, A. The Sentinel-1 mission for the improvement of the scientific understanding and the operational monitoring of the seismic cycle. *Remote Sens. Environ.* **2012**, *120*, 164–174. [CrossRef]

18. Van der Sanden, J. Anticipated applications potential of RADARSAT-2 data. *Can. J. Remote Sens.* **2004**, *30*, 369–379. [CrossRef]

19. Attema, A.; Alpers, W.; Askne, J.; Gray, L.; Herland, E.; Hounam, D.; Keyte, G.; Le Toan, T.; Rocca, F.; Rott, H.; et al. *Envisat ASAR Science and Applications*; European Space Agency (ESA): Noordwijk, The Netherlands, 1998; Volume SP-1225, pp. 1–59. Available online: https://earth.esa.int/c/document_library/get_file?folderId=13019&name=DLFE-615.pdf (accessed on 19 June 2018).

20. Fletcher, K. *ERS Missions 20 Years of Observing Earth*; European Space Agency (ESA): Noordwijk, The Netherlands, 2013; Volume SP-1326, Available online: https://ftp.space.dtu.dk/pub/Altimetry/CLS/SP-1326_ERS_lores.pdf (accessed on 19 June 2018).

21. Krieger, G.; Moreira, A.; Fiedler, H.; Hajnsek, I.; Werner, M.; Younis, M.; Zink, M. TanDEM-X: A satellite formation for high-resolution SAR interferometry. *IEEE Trans. Geosci. Remote Sens.* **2007**, *45*, 3317–3341. [CrossRef]

22. Breit, H.; Fritz, T.; Balss, U.; Lachaise, M.; Niedermeier, A.; Vonavka, M. TerraSAR-X Processing and products. *IEEE Trans. Geosci. Remote Sens.* **2010**, *48*, 727–740. [CrossRef]

23. Eineder, M.; Fritz, T. *TerraSAR-X Ground Segment Basic Product Specification Document*; Doc: TX-GS-DD-3302; German Aerospace Center (DLR): Wessling, Germany, 2009; Issue 1.6; p. 108. Available online: https://tandemx-science.dlr.de/pdfs/TX-GS-DD-3302_Basic-Products-Specification-Document_V1.9.pdf (accessed on 22 September 2017).

24. Buckreuss, S.; Schättler, B. The TerraSAR-X Ground Segment. *IEEE Trans. Geosci. Remote Sens.* **2010**, *48*, 623–632. [CrossRef]

25. Maurer, E.; Mrowka, F.; Braun, A.; Geyer, M.P.; Wasser, Y.; Wickler, M. The Mission Planning System: Automated command generation for spacecraft operations. *IEEE Trans. Geosci. Remote Sens.* **2010**, *48*, 642–648. [CrossRef]

26. Steinbrecher, U.; Schulze, D.; Böer, J.; Mittermayer, J. TerraSAR-X instrument operations rooted in the system engineering and calibration project. *IEEE Trans. Geosci. Remote Sens.* **2010**, *48*, 633–640. [CrossRef]

27. Rotzoll, H.; Dietrich, D.; Dengler, K.; Buckl, B.; Kiemle, S.; Heinen, T. From discovery to download. In Proceedings of the PV Conference, Darmstadt, Germany, 3–5 November 2015.

28. Esch, T.; Marconcini, M.; Felbier, A.; Roth, A.; Heldens, W.; Huber, M.; Schwinger, M.; Taubenböck, H.; Müller, A.; Dech, S. Urban footprint processor—Fully automated processing chain generating settlement masks from global data of the TanDEM-X mission. *IEEE Geosci. Remote Sens. Lett.* **2013**, *10*, 1617–1621. [CrossRef]

29. Gamba, P.; Aldrighi, M.; Stasolla, M. Robust Extraction of urban area extents in HR and VHR SAR images. *IEEE J. Sel. Top. Appl. Earth Obs. Remote Sens.* **2011**, *4*, 27–34. [CrossRef]

30. Leinenkugel, P.; Esch, T.; Künzer, C. Settlement detection and impervious surface estimation in the Mekong Delta using optical and SAR remote sensing data. *Remote Sens. Environ.* **2011**, *115*, 3007–3019. [CrossRef]

31. Taubenböck, H.; Esch, T.; Felbier, A.; Wiesner, M.; Roth, A.; Dech, S. Monitoring urbanization in mega cities from space. *Remote Sens. Environ.* **2012**, *117*, 162–176. [CrossRef]

32. Esch, T.; Heldens, W.; Hirner, A.; Keil, M.; Marconcini, M.; Roth, A.; Zeidler, J.; Dech, S.; Strano, E. Breaking new ground in mapping human settlements from space—The Global Urban Footprint. *ISPRS J. Photogramm. Remote Sens.* **2017**, *134*, 30–42. [CrossRef]

33. Bai, L.; Jiang, L.; Wang, H.; Sun, Q. Spatiotemporal characterization of land subsidence and uplift (2009-2010) over Wuhan in central China revealed by TerraSAR-X InSAR analysis. *Remote Sens.* **2016**, *8*, 350. [CrossRef]

34. Gernhardt, S.; Bamler, R. Deformation monitoring of single buildings using meter-resolution SAR data in psi. *ISPRS J. Photogramm. Remote Sens.* **2012**, *73*, 68–79. [CrossRef]

35. Lan, H.; Li, L.; Liu, H.; Yang, Z. Complex urban infrastructure deformation monitoring using high resolution psi. *IEEE J. Sel. Top. Appl. Earth Obs. Remote Sens.* **2012**, *5*, 643–651. [CrossRef]

36. Shi, X.; Liao, M.; Wang, T.; Zhang, L.; Shan, W.; Wang, C. Expressway deformation mapping using high-resolution TerraSAR-X images. *IEEE Geosci. Remote Sens. Lett.* **2014**, *5*, 194–203. [CrossRef]

37. Liu, G.; Jia, H.; Zhang, R.; Zhang, H.; Jia, H.; Yu, B.; Sang, M. Exploration of subsidence estimation by persistent scatterer InSAR on time series of high resolution TerraSAR-X images. *IEEE J. Sel. Top. Appl. Earth Obs. Remote Sens.* **2011**, *4*, 159–170. [CrossRef]

38. Lazecky, M.; Hlavacova, I.; Bakon, M.; Sousa, J.J.; Perissin, D.; Patricio, G. Bridge displacements monitoring using space-borne X-band SAR interferometry. *IEEE J. Sel. Top. Appl. Earth Obs. Remote Sens.* **2017**, *10*, 205–210. [CrossRef]

39. Emadali, L.; Motagh, M.; Haghshenas Haghighi, M. Characterizing post-construction settlement of the Masjed-Soleyman embankment dam, southwest Iran, using TerraSAR-X spotlight radar imagery. *Eng. Struct.* **2017**, *143*, 261–273. [CrossRef]

40. Crosetto, M.; Monserrat, O.; Cuevas-González, M.; Devanthéry, N.; Luzi, G.; Crippa, B. Measuring thermal expansion using x-band persistent scatterer interferometry. *ISPRS J. Photogramm. Remote Sens.* **2015**, *100*, 84–91. [CrossRef]

41. Anghel, A.; Vasile, G.; Boudon, R.; d'Urso, G.; Girard, A.; Boldo, D.; Bost, V. Combining spaceborne SAR images with 3D point clouds for infrastructure monitoring applications. *ISPRS J. Photogramm. Remote Sens.* **2016**, *111*, 45–61. [CrossRef]

42. Martinis, S.; Kersten, J.; Twele, A. A fully automated TerraSAR-X based flood service. *ISPRS J. Photogramm. Remote Sens.* **2015**, *104*, 203–212. [CrossRef]

43. Liu, W.; Yamazaki, F.; Gokon, H.; Koshimura, S.-I. Extraction of tsunami-flooded areas and damaged buildings in the 2011 Tohoku-Oki earthquake from TerraSAR-X intensity images. *Earthq. Spectra* **2013**, *29* (Suppl. 1), S183–S200. [CrossRef]

44. Mason, D.C.; Giustarini, L.; Garcia-Pintado, J.; Cloke, H.L. Detection of flooded urban areas in high resolution synthetic aperture radar images using double scattering. *Int. J. Appl. Earth Obs. Geoinf.* **2014**, *28*, 150–159. [CrossRef]

45. Künzer, C.; Guo, H.; Schlegel, I.; Tuan, V.Q.; Li, X.; Dech, S. Varying scale and capability of Envisat ASAR-WSM, TerraSAR-X ScanSAR and TerraSAR-X StripMap data to assess urban flood situations: A case study of the Mekong delta in Can Tho province. *Remote Sens.* **2013**, *5*, 5122–5142. [CrossRef]

46. Dekker, R.J. High-resolution radar damage assessment after the earthquake in Haiti on 12 January 2010. *IEEE J. Sel. Top. Appl. Earth Obs. Remote Sens.* **2011**, *4*, 960–970. [CrossRef]

47. Liu, W.; Yamazaki, F.; Sasagawa, T. Monitoring of the recovery process of the Fukushima Daiichi nuclear power plant from VHR SAR images. *J. Disaster Res.* **2016**, *11*, 236–245. [CrossRef]

48. Bachofer, F.; Quénéhervé, G.; Märker, M. The delineation of paleo-shorelines in the Lake Manyara Basin using TerraSAR-X data. *Remote Sens.* **2014**, *6*, 2195–2212. [CrossRef]

49. Linck, R.; Busche, T.; Buckreuss, S. Visual analysis of TerraSAR-X backscatter imagery for archaeological prospection. *Photogramm. Fernerkund. Geoinf.* **2014**, *1*, 55–65. [CrossRef]

50. Balz, T.; Caspari, G.; Fu, B.; Liao, M. Discernibility of burial mounds in high-resolution X-band SAR images for archaeological prospections in the Altai Mountains. *Remote Sens.* **2016**, *8*, 817. [CrossRef]

51. Tomás, R.; García-Barba, J.; Cano, M.; Sanabria, M.P.; Ivorra, S.; Duro, J.; Herrera, G. Subsidence damage assessment of a gothic church using differential interferometry and field data. *Struct. Health Monit.* **2012**, *11*, 751–762. [CrossRef]

52. Confuorto, P.; Plank, S.; Novellino, A.; Tessitore, S.; Ramondini, M. Implementation of DInSAR methods for the monitoring of the archaeological site of Hera Lacinia in Crotone (southern Italy). *Rend. Online Soc. Geol. Ital.* **2016**, *41*, 231–234. [CrossRef]

53. Tapete, D.; Cigna, F.; Donoghue, D.N.M. 'Looting marks' in space-borne SAR imagery: Measuring rates of archaeological looting in Apamea (Syria) with TerraSAR-X Staring Spotlight. *Remote Sens. Environ.* **2016**, *178*, 42–58. [CrossRef]

54. Chen, B.; Deng, K.; Fan, H.; Yu, Y. Combining SAR interferometric phase and intensity information for monitoring of large gradient deformation in coal mining area. *Eur. J. Remote Sens.* **2015**, *48*, 701–717. [CrossRef]

55. Fan, H.; Xu, Q.; Hu, Z.; Du, S. Using temporarily coherent point interferometric synthetic aperture radar for land subsidence monitoring in a mining region of western China. *J. Appl. Remote Sens.* **2017**, *11*, 1–17. [CrossRef]

56. Lubitz, C.; Motagh, M.; Kaufmann, C. Ground surface response to geothermal drilling and the following counteractions in Staufen im Breisgau (Germany) investigated by TerraSAR-X time series analysis and geophysical modeling. *Remote Sens.* **2014**, *6*, 10571–10592. [CrossRef]

57. Kenduiywo, B.K.; Bargiel, D.; Soergel, U. Higher order dynamic conditional random fields ensemble for crop type classification in radar images. *IEEE Trans. Geosci. Remote Sen.* **2017**, *55*, 4638–4654. [CrossRef]

58. McNairn, H.; Kross, A.; Lapen, D.; Caves, R.; Shang, J. Early season monitoring of corn and soybeans with TerraSAR-X and RADARSAT-2. *Int. J. Appl. Earth Obs. Geoinf.* **2014**, *28*, 252–259. [CrossRef]

59. Sonobe, R.; Tani, H.; Wang, X.; Kobayashi, N.; Shimamura, H. Discrimination of crop types with terrasar-x-derived information. *Phys. Chem. Earth Parts A/B/C* **2015**, *83*, 2–13. [CrossRef]

60. Baghdadi, N.; Boyer, N.; Todoroff, P.; El Hajj, M.; Bégué, A. Potential of SAR sensors TerraSAR-X, ASAR/Envisat and Palsar/ALOS for monitoring sugarcane crops on Reunion Island. *Remote Sens. Environ.* **2009**, *113*, 1724–1738. [CrossRef]

61. Zhao, L.; Yang, J.; Li, P.; Zhang, L. Characteristics analysis and classification of crop harvest patterns by exploiting high-frequency multipolarization SAR data. *IEEE J. Sel. Top. Appl. Earth Obs. Remote Sens.* **2014**, *7*, 3773–3783. [CrossRef]

62. Gebhardt, S.; Huth, J.; Nguyen, L.D.; Roth, A.; Künzer, C. A comparison of TerraSAR-X quadpol backscattering with RapidEye multispectral vegetation indices over rices fields in the Mekong Delta, Vietnam. *Int. J. Remote Sens.* **2012**, *33*, 7644–7661. [CrossRef]

63. Yuzugullu, O.; Erten, E.; Hajnsek, I. Rice growth monitoring by means of X-band co-polar SAR: Feature clustering and bbch scale. *IEEE Geosci. Remote Sens. Lett.* **2015**, *12*, 1218–1222. [CrossRef]

64. Lopez-Sanchez, J.M.; Cloude, S.R.; Ballester-Berman, J.D. Rice phenology monitoring by means of SAR polarimetry at X-band. *IEEE Trans. Geosci. Remote Sen.* **2012**, *50*, 2695–2709. [CrossRef]

65. Koppe, W.; Gnyp, M.L.; Hütt, C.; Yao, Y.; Miao, Y.; Chen, X.; Bareth, G. Rice monitoring with multi-temporal and dual-polarimetric TerraSAR-X data. *Int. J. Appl. Earth Obs. Geoinf.* **2012**, *21*, 568–576. [CrossRef]

66. Koch, B. Status and future of laser scanning, synthetic aperture radar and hyperspectral remote sensing data for forest biomass assessment. *ISPRS J. Photogramm. Remote Sens.* **2010**, *65*, 581–590. [CrossRef]

67. Solberg, S.; Riegler, G.; Nonin, P. Estimating forest biomass from TerraSAR-X StripMap radargrammetry. *IEEE Trans. Geosci. Remote Sens.* **2015**, *53*, 154–161. [CrossRef]

68. Perko, R.; Raggam, H.; Deutscher, J.; Gutjahr, K.; Schardt, M. Forest assessment using high resolution SAR data in X-band. *Remote Sens.* **2011**, *3*, 792–815. [CrossRef]

69. Tanase, M.A.; Santoro, M.; Wegmüller, U.; de la Riva, J.; Pérez-Cabello, F. Properties of X-, C- and L-band repeat-pass interferometric SAR coherence in Mediterranean pine forests affected by fires. *Remote Sens. Environ.* **2010**, *114*, 2182–2194. [CrossRef]

70. Lausch, A.; Erasmi, S.; King, D.J.; Magdon, P.; Heurich, M. Understanding forest health with remote sensing-part I—A review of spectral traits, processes and remote-sensing characteristics. *Remote Sens.* **2016**, *8*, 1029. [CrossRef]

71. Bargiel, D.; Herrmann, S.; Jadczyszyn, J. Using high-resolution radar images to determine vegetation cover for soil erosion assessments. *J. Environ. Manag.* **2013**, *124*, 82–90. [CrossRef] [PubMed]

72. Duguay, Y.; Bernier, M.; Lévesque, E.; Tremblay, B. Potential of C and X band SAR for shrub growth monitoring in sub-arctic environments. *Remote Sens.* **2015**, *7*, 9410–9430. [CrossRef]

73. Betbeder, J.; Nabucet, J.; Pottier, E.; Baudry, J.; Corgne, S.; Hubert-Moy, L. Detection and characterization of hedgerows using TerraSAR-X imagery. *Remote Sens.* **2014**, *6*, 3752–3769. [CrossRef]

74. Künzer, C.; Ottinger, M.; Wegmann, M.; Guo, H.; Wang, C.; Zhang, J.; Dech, S.; Wikelski, M. Earth observation satellite sensors for biodiversity monitoring: Potentials and bottlenecks. *Int. J. Remote Sens.* **2014**, *35*, 6599–6647. [CrossRef]

75. Betbeder, J.; Rapinel, S.; Corgne, S.; Pottier, E.; Hubert-Moy, L. TerraSAR-X dual-pol time-series for mapping of wetland vegetation. *ISPRS J. Photogramm. Remote Sens.* **2015**, *107*, 90–98. [CrossRef]

76. Heine, I.; Jagdhuber, T.; Itzerott, I. Classification and monitoring of reed belts using dual-polarimetric TerraSAR-X time series. *Remote Sens.* **2016**, *8*, 552. [CrossRef]

77. Schmitt, A.; Brisco, B. Wetland monitoring using the curvelet-based change detection method on polarimetric SAR imagery. *Water* **2013**, *5*, 1036–1052. [CrossRef]

78. Hong, S.-H.; Wdowinski, S.; Kim, S.-W. Evalution of TerraSAR-X observations for wetland InSAR applications. *IEEE Trans. Geosci. Remote Sens.* **2010**, *48*, 864–873. [CrossRef]

79. Dusseux, P.; Corpetti, T.; Hubert-Moy, L.; Corgne, S. Combined use of multi-temporal optical and radar satellite images for grassland monitoring. *Remote Sens.* **2014**, *6*, 6163–6182. [CrossRef]
80. El Hajj, M.; Baghdadi, N.; Belaud, G.; Zribi, M.; Cheviron, B.; Courault, D.; Hagolle, O.; Charron, F. Irrigated grassland monitoring using a time series of TerraSAR-X and Cosmo-Skymed X-band SAR data. *Remote Sens.* **2014**, *6*, 10002–10032. [CrossRef]
81. Voormansik, K.; Jagdhuber, T.; Zalite, K.; Noorma, M.; Hajnsek, I. Observations of cutting practices in agricultural grasslands using polarimetric SAR. *IEEE J. Sel. Top. Appl. Earth Obs. Remote Sens.* **2017**, *9*, 1382–1396. [CrossRef]
82. Yang, Y.; Li, Z.; Huang, L.; Tian, B.; Chen, Q. Extraction of glacier outlines and water-eroded stripes using high-resolution SAR imagery. *Int. J. Remote Sens.* **2016**, *37*, 1016–1034. [CrossRef]
83. Neelmeijer, J.; Motagh, M.; Wetzel, H.-U. Estimating spatial and temporal variability in surface kinematics of the Inylchek Glacier, Central Asia, using TerraSAR-X data. *Remote Sens.* **2014**, *6*, 9239–9259. [CrossRef]
84. Ponton, F.; Trouvé, E.; Gay, M.; Walpersdorf, A.; Fallourd, R.; Nicolas, J.-M.; Vernier, F.; Mugnier, J.-L. Observation of the Argentière glacier flow variability from 2009 to 2011 by TerraSAR-X and GPS displacement measurements. *IEEE J. Sel. Top. Appl. Earth Obs. Remote Sens.* **2014**, *7*, 3274–3284. [CrossRef]
85. Luckman, A.; Benn, D.I.; Cottier, F.; Bevan, S.; Nilsen, F.; Inall, M. Calving rates at tidewater glaciers vary strongly with ocean temperature. *Nat. Commun.* **2015**, *6*, 1–7. [CrossRef] [PubMed]
86. Schubert, A.; Faes, A.; Kääb, A.; Meier, E. Glacier surface velocity estimation using repeat TerraSAR-X images: Wavelet- vs. correlation-based image matching. *ISPRS J. Photogramm. Remote Sens.* **2013**, *82*, 49–62. [CrossRef]
87. Fang, L.; Xu, Y.; Yao, W.; Stilla, U. Estimation of glacier surface motion by robust phase correlation and point like features of SAR intensity images. *ISPRS J. Photogramm. Remote Sens.* **2016**, *121*, 92–112. [CrossRef]
88. Fallourd, R.; Trouvé, E.; Roşu, D.; Vernier, F.; Bolon, P.; Harant, O.; Gay, M.; Bombrun, L.; Vasile, G.; Nicolas, J.-M.; et al. Monitoring temperate glaciers by multi-temporal TerraSAR-X images and continuous GPS measurements. *IEEE J. Sel. Top. Appl. Earth Obs. Remote Sens.* **2011**, *4*, 372–386. [CrossRef]
89. Godon, C.; Mugnier, J.L.; Fallourd, R.; Paquette, J.L.; Pohl, A.; Buoncristiani, J.F. The Bossons Glacier protects Europe's summit from erosion. *Earth Planet. Sci. Lett.* **2013**, *375*, 135–147. [CrossRef]
90. Benoit, L.; Dehecq, A.; Pham, H.-T.; Vernier, F.; Trouvé, E.; Moreau, L.; Martin, O.; Thom, C.; Pierrot-Deseilligny, M.; Briole, P. Multi-method monitoring of glacier d'Argentière dynamics. *Ann. Glaciol.* **2015**, *56*, 118–128. [CrossRef]
91. Neckel, N.; Loibl, D.; Rankl, M. Recent slowdown and thinning of debris-covered glaciers in south-eastern Tibet. *Earth Planet. Sci. Lett.* **2017**, *464*, 95–102. [CrossRef]
92. Joughin, I.; Shean, D.E.; Smith, B.E.; Dutrieux, P. Grounding line variability and subglacial lake drainage on Pine Island Glacier, Antarctica. *Geophys. Res. Lett.* **2016**, *43*, 9093–9102. [CrossRef]
93. Marsh, O.J.; Rack, W.; Floricioiu, D.; Golledge, N.R.; Lawson, W. Tidally induced velocity variations of the Beardmore Glacier, Antarctica, and their representation in satellite measurements of ice velocity. *Cryosphere* **2013**, *7*, 1375–1384. [CrossRef]
94. Willis, M.J.; Melkonian, A.K.; Pritchard, M.E. Outlet glacier response to the 2012 collapse of the Matusevich Ice Shelf, Severnaya Zemlya, Russian Arctic. *J. Geophys. Res. Earth Surf.* **2015**, *120*, 2040–2055. [CrossRef]
95. Rott, H.; Müller, F.; Nagler, T.; Floricioiu, D. The imbalance of glaciers after disintegration of Larsen-b ice shelf, Antarctic Peninsula. *Cryosphere* **2011**, *5*, 125–134. [CrossRef]
96. Braun, M.; Humbert, A. Recent retreat of Wilkins Ice Shelf reveals new insights in ice shelf breakup mechanisms. *IEEE Geosci. Remote Sens. Lett.* **2009**, *6*, 263–267. [CrossRef]
97. Humbert, A.; Steinhage, D.; Helm, V.; Hoerz, S.; Berendt, J.; Leipprand, E.; Christmann, J.; Plate, C.; Müller, R. On the link between surface and basal structures of the Jelbart Ice Shelf, Antarctica. *J. Glaciol.* **2015**, *61*, 975–986. [CrossRef]
98. Leinss, S.; Parrella, G.; Hajnsek, I. Snow height determination by polarimetric phase differences in X-band SAR data. *IEEE J. Sel. Top. Appl. Earth Obs. Remote Sens.* **2014**, *7*, 3794–3810. [CrossRef]
99. Phan, X.V.; Ferro-Famil, L.; Gay, M.; Durand, Y.; Dumont, M.; Morin, S.; Allain, S.; D'Urso, G.; Girard, A. 1D-Var multilayer assimilation of X-band SAR data into a detailed snowpack model. *Cryosphere* **2014**, *8*, 1975–1987. [CrossRef]

100. Paul, S.; Willmes, S.; Hoppmann, M.; Hunkeler, P.A.; Wesche, C.; Nicolaus, M.; Heinemann, G.; Timmermann, R. The impact of early-summer snow properties on Antarctic landfast sea-ice X-band backscatter. *Ann. Glaciol.* **2015**, *56*, 263–273. [CrossRef]

101. Antonova, S.; Duguay, C.R.; Kääb, A.; Heim, B.; Langer, M.; Westermann, S.; Boike, J. Monitoring bedfast ice and ice phenology in lakes of the Lena river delta using TerraSAR-X backscatter and coherence time series. *Remote Sens.* **2016**, *8*, 903. [CrossRef]

102. Sobiech, J.; Dierking, W. Observing lake- and river-ice decay with SAR: Advantages and limitations of the unsupervised k-means classification approach. *Ann. Glaciol.* **2013**, *54*, 65–72. [CrossRef]

103. Wildham, B.; Bartsch, A.; Leibman, M.; Khomutov, A. Active-layer thickness from X-band SAR backscatter intensity. *Cryosphere* **2017**, *11*, 483–496. [CrossRef]

104. Short, N.; Brisco, B.; Couture, N.; Pollard, W.; Murnaghan, K.; Budkewitsch, P. A comparison of TerraSAR-X, Radarsat-2 and ALOS-Palsar interferometry for monitoring permafrost environments, case study from Herschel Island. *Can. J. Remote Sens.* **2011**, *115*, 3491–3506. [CrossRef]

105. Antonova, S.; Kääb, A.; Heim, B.; Langer, M.; Boike, J. Spatio-temporal variability of X-band radar backscatter and coherence over the Lena river delta, Siberia. *Remote Sens. Environ.* **2016**, *182*, 169–191. [CrossRef]

106. Ullmann, T.; Schmitt, A.; Roth, A.; Duffe, J.; Dech, S.; Hubberten, H.-W.; Baumhauer, R. Land cover characterization and classification of arctic tundra environments by means of polarized synthetic aperture X- and C-band radar (polsar) and Landsat 8 multispectral imagery—Richards Island, Canada. *Remote Sens.* **2014**, *6*, 8565–8593. [CrossRef]

107. Muster, S.; Heim, B.; Abnizova, A.; Boike, J. Water body distributions across scales: A remote sensing based comparison of three arctic tundra wetlands. *Remote Sens.* **2013**, *5*, 1498–1523. [CrossRef]

108. Stettner, S.; Beamish, A.L.; Bartsch, A.; Heim, B.; Grosse, G.; Roth, A.; Lantuit, H. Monitoring Inter- and Intra-Seasonal Dynamics of Rapidly Degrading Ice-Rich Permafrost Riverbanks in the Lena Delta with TerraSAR-X Time Series. *Remote Sens.* **2018**, *10*, 51. [CrossRef]

109. Kim, J.-W.; Kim, D.-J.; Hwang, B.J. Characterization of arctic sea ice thickness using high-resolution spaceborne polarimetric SAR data. *IEEE Trans. Geosci. Remote Sens.* **2012**, *50*, 13–22. [CrossRef]

110. Ressel, R.; Singha, S.; Lehner, S.; Rosel, A.; Spreen, G. Investigation into different polarimetric features for sea ice classification using X-band synthetic aperture radar. *IEEE J. Sel. Top. Appl. Earth Obs. Remote Sens.* **2016**, *9*, 3131–3143. [CrossRef]

111. Eriksson, L.E.B.; Borenäs, K.; Dierking, W.; Berg, A.; Santoro, M.; Pemberton, P.; Lindh, H.; Karlson, B. Evaluation of new spaceborne SAR sensors for sea-ice monitoring in the Baltic Sea. *Can. J. Remote Sens.* **2010**, *35* (Suppl. 1), S56–S73. [CrossRef]

112. Frost, A.; Ressel, R.; Lehner, S. Automated iceberg detection using high-resolution X-band SAR images. *Can. J. Remote Sens.* **2016**, *42*, 354–366. [CrossRef]

113. Liu, H.; Li, X.-M.; Guo, H. The dynamic processes of sea ice on the east coast of Antarctica-a case study based on spaceborne synthetic aperture radar data from TerraSAR-X. *IEEE J. Sel. Top. Appl. Earth Obs. Remote Sens.* **2016**, *9*, 1187–1198. [CrossRef]

114. Park, J.-W.; Kim, H.-C.; Hong, S.-H.; Kang, S.-H.; Graber, H.C.; Hwang, B.; Lee, C.M. Radar backscattering changes in arctic sea ice from late summer to early autumn observed by space-borne X-band HH-polarization SAR. *IEEE Geosci. Remote Sens. Lett.* **2016**, *7*, 551–560. [CrossRef]

115. Dmitrenko, I.A.; Wegner, C.; Kassens, H.; Kirillov, S.A.; Krumpen, T.; Heinemann, G.; Helbig, A.; Schröder, D.; Hölemann, J.A.; Klagge, T.; et al. Observations of supercooling and frazil ice formation in the Laptev sea coastal polynya. *J. Geophys. Res. Oceans* **2010**, *115*. [CrossRef]

116. Yague-Martinez, N.; Eineder, M.; Cong, X.Y.; Minet, C. Ground displacement measurement by TerraSAR-X image correlation: The 2011 Tohoku-Oki earthquake. *IEEE Geosci. Remote Sens. Lett.* **2012**, *9*, 539–543. [CrossRef]

117. Liu, W.; Yamazaki, F.; Matsuoka, M.; Nonaka, T.; Sasagawa, T. Estimation of three-dimensional crustal movements in the 2011 Tohoku-Oki, Japan, earthquake from TerraSAR-X intensity images. *Nat. Hazards Earth Syst. Sci.* **2015**, *15*, 637–645. [CrossRef]

118. Benekos, G.; Derdelakos, K.; Bountzouklis, C.; Kourkouli, P.; Parcharidis, I. Surface displacements of the 2014 Cephalonia (Greece) earthquake using high resolution SAR interferometry. *Earth Sci. Inf.* **2015**, *8*, 309–315. [CrossRef]

119. Hamling, I.J.; D'Anastasio, E.; Wallace, L.M.; Ellis, S.; Motagh, M.; Samsonov, S.; Palmer, N.; Hreinsdōttir, S. Crustal deformation and stress transfer during a propagating earthquake sequence: The 2013 Cook Strait sequence, central New Zealand. *J. Geophys. Res. B: Solid Earth.* **2014**, *119*, 6080–6092. [CrossRef]
120. Frontera, T.; Concha, A.; Blanco, P.; Echeverria, A.; Goula, X.; Arbiol, R.; Khazaradze, G.; Pérez, F.; Suriñach, E. Dinsar coseismic deformation of the May 2011 mw 5.1 Lorca earthquake (southeastern Spain). *Solid Earth* **2012**, *3*, 111–119. [CrossRef]
121. Delgado, F.; Pritchard, M.E.; Basualto, D.; Lazo, J.; Córdova, L.; Lara, L.E. Rapid reinflation following the 2011–2012 rhyodacite eruption at Cordón Caulle volcano (Southern Andes) imaged by Insar: Evidence for magma reservoir refill. *Geophys. Res. Lett.* **2016**, *43*, 9552–9562. [CrossRef]
122. Walter, T.R.; Subandriyo, J.; Kirbani, S.; Bathke, H.; Suryanto, W.; Aisyah, N.; Darmawan, H.; Jousset, P.; Luehr, B.-G.; Dahm, T. Volcano-tectonic control of Merapi's lava dome splitting: The November 2013 fracture observed from high resolution TerraSAR-X data. *Tectonophysics* **2015**, *639*, 23–33. [CrossRef]
123. Richter, N.; Poland, M.P.; Lundgren, P.R. TerraSAR-X interferometry reveals small-scale deformation associated with the summit eruption of Kilauea Volcano, Hawai'i. *Geophys. Res. Lett.* **2013**, *40*, 1279–1283. [CrossRef]
124. Wadge, G.; Cole, P.; Stinton, A.; Komorowski, J.-C.; Stewart, R.; Toombs, A.C.; Legendre, Y. Rapid topographic change measured by high-resolution satellite radar at Soufriere Hills volcano, Montserrat, 2008–2010. *J. Volcanol. Geotherm. Res.* **2011**, *199*, 142–152. [CrossRef]
125. Feigl, K.L.; Le Mével, H.; Ali, S.T.; Córdova, L.; Andersen, N.L.; DeMets, C.; Singer, B.S. Rapid uplift in Laguna del Maule volcanic field of the Andean southern volcanic zone (Chile) 2007–2012. *Geophys. J. Int.* **2013**, *196*, 885–901. [CrossRef]
126. Vasco, D.W.; Rutqvist, J.; Ferretti, A.; Rucci, A.; Bellotti, F.; Dobson, P.; Oldenburg, C.; Garcia, J.; Walters, M.; Hartline, C. Monitoring deformation at the Geysers geothermal field, California using C-band and X-band interferometric synthetic aperture radar. *Geophys. Res. Lett.* **2013**, *40*, 2567–2572. [CrossRef]
127. Ali, S.T.; Akerley, J.; Baluyut, E.C.; Cardiff, M.; Davatzes, N.C.; Feigl, K.L.; Foxall, W.; Fratta, D.; Mellors, R.J.; Spielman, P.; et al. Time-series analysis of surface deformation at Brady Hot Springs geothermal field (Nevada) using interferometric synthetic aperture radar. *Geothermics* **2016**, *61*, 114–120. [CrossRef]
128. Motagh, M.; Wetzel, H.-U.; Roessner, S.; Kaufmann, H. A TerraSAR-X Insar study of landslides in southern Kyrgyzstan, Central Asia. *IEEE Geosci. Remote Sens. Lett.* **2013**, *4*, 657–666. [CrossRef]
129. Jiang, Y.; Liao, M.; Zhou, Z.; Shi, X.; Zhang, L.; Balz, T. Landslide deformation analysis by coupling deformation time series from SAR data with hydrological factors through data assimilation. *Remote Sens.* **2016**, *8*, 179. [CrossRef]
130. Meng, Y.; Lan, H.; Li, L.; Wu, Y.; Li, Q. Characteristics of surface deformation detected by X-band SAR interferometry over Sichuan-Tibet grid connection project area, China. *Remote Sens.* **2015**, *7*, 12265–12281. [CrossRef]
131. Barboux, C.; Strozzi, T.; Delaloye, R.; Wegmüller, U.; Collet, C. Mapping slope movements in alpine environments using TerraSAR-X interferometric methods. *ISPRS J. Photogramm. Remote Sens.* **2015**, *109*, 178–192. [CrossRef]
132. Rauste, Y.; Lateh, H.B.; Jefriza, J.; Wan Mohd, M.W.I.; Lönnqvist, A.; Häme, T. TerraSAR-X data in cut slope soil stability monitoring in Malaysia. *IEEE Trans. Geosci. Remote Sens.* **2012**, *50*, 3354–3363. [CrossRef]
133. Oliveira, S.C.; Zêzere, J.L.; Catalão, J.; Nico, G. The contribution of PSInsar interferometry to landslide hazard in weak rock-dominated areas. *Landslides* **2015**, *12*, 703–719. [CrossRef]
134. Liu, G.; Guo, H.; Perski, Z.; Fan, J.; Bai, S.; Yan, S.; Song, R. Monitoring the slope movement of the Shuping landslide in the Three Gorges reservoir of China, using X-band time series SAR interferometry. *Adv. Space Res.* **2016**, *57*, 2487–2495. [CrossRef]
135. Bachofer, F.; Quénéhervé, G.; Hochschild, V.; Maerker, M. Multisensoral topsoil mapping in the semiarid Lake Manyara region, northern Tanzania. *Remote Sens.* **2015**, *7*, 9563–9586. [CrossRef]
136. Zribi, M.; Kotti, F.; Lili-Chabaane, Z.; Baghdadi, N.; Ben Issa, N.; Amri, R.; Duchemin, B.; Chehbouni, A. Soil texture estimation over a semiarid area using TerraSAR-X radar data. *IEEE Geosci. Remote Sens. Lett.* **2012**, *9*, 353–357. [CrossRef]
137. Baade, J.; Schmullius, C. Interferometric microrelief sensing with TerraSAR-X-first results. *IEEE Trans. Geosci. Remote Sens.* **2010**, *48*, 965–970. [CrossRef]

138. Chen, M.; Tomás, R.; Li, Z.; Motagh, M.; Li, T.; Hu, L.; Gong, H.; Li, X.; Yu, J.; Gong, X. Imaging land subsidence induced by groundwater extraction in Beijing (China) using satellite radar interferometry. *Remote Sens.* **2016**, *8*, 468. [CrossRef]

139. Theron, A.; Engelbrecht, J.; Kemp, J.; Kleynhans, W.; Turnbull, T. Detection of sinkhole precursors through SAR interferometry: Radar and geological considerations. *IEEE Geosci. Remote Sens. Lett.* **2017**, *14*, 871–875. [CrossRef]

140. Bruck, M.; Lehner, S. Coastal wave field extraction using TerraSAR-X data. *J. Appl. Remote Sens.* **2013**, *7*, 1–19. [CrossRef]

141. Thompson, D.R.; Horstmann, J.; Mouche, A.; Winstead, N.S.; Sterner, R.; Monaldo, F.M. Comparison of high-resolution wind fields extracted from TerraSAR-X SAR imagery with predictions from the WRF mesoscale model. *J. Geophys. Res. Oceans* **2012**, *117*, 1–17. [CrossRef]

142. Kuzmić, M.; Grisogono, B.; Li, X.; Lehner, S. Examining deep and shallow Adriatic bora events. *Q. J. R. Meteorol. Soc.* **2015**, *141*, 3434–3438. [CrossRef]

143. Díaz Méndez, G.M.; Lehner, S.; Ocampo-Torres, F.J.; Li, X.M.; Brusch, S. Wind and wave observations off the south pacific coast of Mexico using TerraSAR-X imagery. *Int. J. Remote Sens.* **2010**, *31*, 4933–4955. [CrossRef]

144. Li, X.M.; Lehner, S. Observation of TerraSAR-X for studies on offshore wind turbine wake in near and far fields. *IEEE J. Sel. Top. Appl. Earth Obs. Remote Sens.* **2013**, *6*, 1757–1768. [CrossRef]

145. Pleskachevsky, A.L.; Lehner, S.; Rosenthal, W. Storm observations by remote sensing and influences of gustiness on ocean waves and on generation of rogue waves. *Ocean Dyn.* **2012**, *62*, 1335–1351. [CrossRef]

146. Gebhardt, C.; Bidlot, J.; Jacobsen, S.; Lehner, S.; Persson, P.O.G.; Pleskachevsky, A.L. The potential of TerraSAR-X to observe wind wave interaction at the ice edge. *IEEE J. Sel. Top. Appl. Earth Obs. Remote Sens.* **2017**, *10*, 2799–2809. [CrossRef]

147. Moser, L.; Schmitt, A.; Wendleder, A.; Roth, A. Monitoring of the Lac Bam Wetland extent using dual-polarized X-band SAR data. *Remote Sens.* **2016**, *8*, 302. [CrossRef]

148. Heine, I.; Francke, T.; Rogass, C.; Medeiros, P.H.A.; Bronstert, A.; Foerster, S. Monitoring seasonal changes in the water surface areas of reservoirs using TerraSAR-X time series data in semiarid northeastern Brazil. *IEEE J. Sel. Top. Appl. Earth Obs. Remote Sens.* **2014**, *7*, 3190–3199. [CrossRef]

149. Baup, F.; Frappart, F.; Maubant, J. Combining high-resolution satellite images and altimetry to estimate the volume of small lakes. *Hydrol. Earth Syst. Sci.* **2014**, *18*, 2007–2020. [CrossRef]

150. Klemenjak, S.; Waske, B.; Valero, S.; Chanussot, J. Automatic detection of rivers in high-resolution SAR data. *IEEE J. Sel. Top. Appl. Earth Obs. Remote Sens.* **2012**, *5*, 1364–1372. [CrossRef]

151. Gade, M.; Melchionna, S. Joint use of multiple synthetic aperture radar imagery for the detection of bivalve beds and morphological changes on intertidal flats. *Estuar. Coast. Shelf Sci.* **2016**, *171*, 1–10. [CrossRef]

152. Choe, B.-H.; Kim, D.-J. Retrieval of surface parameters in tidal flats using radar backscattering model and multi-frequency SAR data. *Korean J. Remote Sens.* **2011**, *27*, 225–234. [CrossRef]

153. Lee, Y.-K.; Park, J.-W.; Choi, J.-K.; Oh, Y.; Won, J.-S. Potential uses of TerraSAR-X for mapping herbaceous halophytes over salt marsh and tidal flats. *Estuar. Coast. Shelf Sci.* **2012**, *115*, 366–376. [CrossRef]

154. Kim, D.-J.; Moon, W.M.; Kim, G.; Park, S.-E.; Lee, H. Submarine groundwater discharge in tidal flats revealed by space-borne synthetic aperture radar. *Remote Sens. Environ.* **2011**, *115*, 793–800. [CrossRef]

155. Adolph, W.; Jung, R.; Schmidt, A.; Ehlers, M.; Heipke, C.; Bartholomä, A.; Farke, H. Integration of TerraSAR-X, Rapideye and airborne lidar for remote sensing of intertidal bedforms on the upper flats of Norderney (German Wadden Sea). *Geo-Mar. Lett.* **2017**, *37*, 193–205. [CrossRef]

156. Li, X.-M.; Jia, T.; Velotto, D. Spatial and temporal variations of oil spills in the North Sea observed by the satellite constellation of TerraSAR-X and TanDEM-X. *IEEE J. Sel. Top. Appl. Earth Obs. Remote Sens.* **2016**, *9*, 4941–4947. [CrossRef]

157. Velotto, D.; Migliaccio, M.; Nunziata, F.; Lehner, S. Dual-polarized TerraSAR-X data for oil-spill observation. *IEEE Trans. Geosci. Remote Sens.* **2011**, *49*, 4751–4762. [CrossRef]

158. Ivanov, A.Y. The oil spill from a shipwreck in Kerch Strait: Radar monitoring and numerical modeling. *Int. J. Remote Sens.* **2010**, *31*, 4853–4868. [CrossRef]

159. Gorrab, A.; Zribi, M.; Baghdadi, N.; Mougenot, B.; Fanise, P.; Chabaane, Z.L. Retrieval of both soil moisture and texture using TerraSAR-X images. *Remote Sens.* **2015**, *7*, 10098–10116. [CrossRef]

160. Baghdadi, N.; Camus, P.; Beaugendre, N.; Issa, O.M.; Zribi, M.; Desprats, J.F.; Rajot, J.L.; Abdallah, C.; Sannier, C. Estimating surface soil moisture from TerraSAR-X data over two small catchments in the Sahelian part of western Niger. *Remote Sens.* **2011**, *3*, 66–1283. [CrossRef]

161. Brusch, S.; Lehner, S.; Fritz, T.; Soccorsi, M.; Soloviev, A.; Van Schie, B. Ship surveillance with TerraSAR-X. *IEEE Trans. Geosci. Remote Sens.* **2011**, *49*, 1092–1103. [CrossRef]

162. Vespe, M.; Greidanus, H. SAR image quality assessment and indicators for vessel and oil spill detection. *IEEE Trans. Geosci. Remote Sens.* **2012**, *50*, 4726–4734. [CrossRef]

163. Graziano, M.D.; D'Errico, M.; Rufino, G. Ship heading and velocity analysis by wake detection in SAR images. *Acta Astronaut.* **2016**, *128*, 72–82. [CrossRef]

164. Chaturvedi, S.K.; Yang, C.-S.; Ouchi, K.; Shanmugam, P. Ship recognition by integration of SAR and AIS. *J. Navig.* **2012**, *65*, 323–337. [CrossRef]

165. Romeiser, R.; Suchandt, S.; Runge, H.; Steinbrecher, U.; Grünler, S. First analysis of TerraSAR-X along-track Insar-derived current fields. *IEEE Trans. Geosci. Remote Sens.* **2010**, *48*, 820–829. [CrossRef]

166. Wang, X.; Zhang, H.; Li, X.; Fu, B.; Guan, W. SAR imaging of a topography-induced current front in a tidal channel. *Int. J. Remote Sens.* **2015**, *36*, 3563–3574. [CrossRef]

167. Brusch, S.; Held, P.; Lehner, S.; Rosenthal, W.; Pleskachevsky, A. Underwater bottom topography in coastal areas from TerraSAR-X data. *Int. J. Remote Sens.* **2011**, *32*, 4527–4543. [CrossRef]

168. Catalao, J.; Nico, G. Multitemporal backscattering logistic analysis for intertidal bathymetry. *IEEE Trans. Geosci. Remote Sens.* **2017**, *55*, 1066–1073. [CrossRef]

169. Devanthéry, N.; Crosetto, M.; Monserrat, O.; Cuevas-González, M.; Crippa, B. An approach to persistent scatterer interferometry. *Remote Sens.* **2014**, *6*, 6662–6679. [CrossRef]

170. Goel, K.; Adam, N. A distributed scatterer interferometry approach for precision monitoring of known surface deformation phenomena. *IEEE Trans. Geosci. Remote Sens.* **2014**, *52*, 5454–5468. [CrossRef]

171. Singleton, A.; Li, Z.; Hoey, T.; Muller, J.-P. Evaluating sub-pixel offset techniques as an alternative to D-Insar for monitoring episodic landslide movements in vegetated terrain. *Remote Sens. Environ.* **2014**, *147*, 133–144. [CrossRef]

172. Romeiser, R.; Graber, H.C. Advanced remote sensing of internal waves by spaceborne along-track Insar-a demonstration with TerraSAR-X. *IEEE Trans. Geosci. Remote Sens.* **2015**, *53*, 6735–6751. [CrossRef]

173. Ferreira, R.M.; Estefen, S.F.; Romeiser, R. Under what conditions sar along-track interferometry is suitable for assessment of tidal energy resource. *IEEE J. Sel. Top. Appl. Earth Obs. Remote Sens.* **2016**, *9*, 5011–5022. [CrossRef]

174. Schmitt, A.; Wendleder, A.; Hinz, S. The Kennaugh element framework for multi-scale, multi-polarized, multi-temporal and multi-frequency SAR image preparation. *ISPRS J. Photogramm. Remote Sens.* **2015**, *102*, 122–139. [CrossRef]

175. Chen, Q.; Li, L.; Xu, Q.; Yang, S.; Shi, X.; Liu, X. Multi-feature segmentation for high-resolution polarimetric SAR data based on fractal net evolution approach. *Remote Sens.* **2017**, *9*, 570. [CrossRef]

176. Dabboor, M.; Karathanassi, V.; Braun, A. A multi-level segmentation methodology for dual-polarized SAR data. *Int. J. Appl. Earth Obs. Geoinf.* **2011**, *13*, 376–385. [CrossRef]

177. Krylov, V.A.; Moser, G.; Serpico, S.B.; Zerubia, J. Supervised high-resolution dual-polarization SAR image classification by finite mixtures and copulas. *IEEE J. Sel. Top. Signal Process.* **2011**, *5*, 554–566. [CrossRef]

178. Harant, O.; Bombrun, L.; Vasile, G.; Ferro-Famil, L.; Gay, M. Displacement estimation by maximum-likelihood texture tracking. *IEEE J. Sel. Top. Signal Process.* **2011**, *5*, 398–407. [CrossRef]

179. Yousif, O.; Ban, Y. A novel approach for object-based change image generation using multitemporal high-resolution SAR images. *Int. J. Remote Sens.* **2016**, *38*, 1765–1787. [CrossRef]

180. Schmitt, A.; Wessel, B.; Roth, A. An innovative curvelet-only-based approach for automated change detection in multi-temporal SAR imagery. *Remote Sens.* **2014**, *6*, 2435–2462. [CrossRef]

181. Lê, T.T.; Atto, A.M.; Trouvé, E.; Solikhin, A.; Pinel, V. Change detection matrix for multitemporal filtering and change analysis of SAR and polsar image time series. *ISPRS J. Photogramm. Remote Sens.* **2015**, *107*, 64–76. [CrossRef]

182. Zhu, X.X.; Montazeri, S.; Gisinger, C.; Hanssen, R.F.; Bamler, R. Geodetic SAR tomography. *IEEE Trans. Geosci. Remote Sens.* **2016**, *54*, 18–35. [CrossRef]

183. Siddique, M.A.; Wegmüller, U.; Hajnsek, I.; Frey, O. Single-look SAR tomography as an add-on to PSI for improved deformation analysis in urban areas. *IEEE Trans. Geosci. Remote Sens.* **2016**, *54*, 6119–6137. [CrossRef]

184. Gutjahr, K.; Perko, R.; Raggam, H.; Schardt, M. The epipolarity constraint in stereo-radargrammetric DEM generation. *IEEE Trans. Geosci. Remote Sens.* **2014**, *52*, 5014–5022. [CrossRef]

185. Goel, K.; Adam, N. Three-dimensional positioning of point scatterers based on radargrammetry. *IEEE Trans. Geosci. Remote Sens.* **2012**, *50*, 2355–2363. [CrossRef]

186. Eineder, M.; Minet, C.; Steigenberger, P.; Cong, X.; Fritz, T. Imaging geodesy—Toward centimeter-level ranging accuracy with TerraSAR-X. *IEEE Trans. Geosci. Remote Sens.* **2011**, *49*, 661–671. [CrossRef]

187. GEO. Group on Earth Observations: Geohazard Supersites & Natural Laboratories. Available online: http://supersites.earthobservations.org/ (accessed on 9 April 2018).

188. SDCG Element-3 Strategy: Satellite Data in Support of Research & Development (R&D) Activities for the Global Forest Observations Initiative for CEOS SIT-31. April 2016. Available online: http://www.gfoi.org/documents (accessed on 9 April 2018).

189. Polar Space Task Group—SAR Coordination Working Group. Data Compendium—A Summary Documentation of SAR Satellite Data Collections, Plans and Activities. April 2016. Available online: http://www.wmo.int/pages/prog/sat/pstg-sarcwg_en.php (accessed on 9 April 2018).

remote sensing

MDPI

Review

Ten Years of TerraSAR-X Operations

Stefan Buckreuss [1,*], **Birgit Schättler** [2], **Thomas Fritz** [2], **Josef Mittermayer** [1], **Ralph Kahle** [3], **Edith Maurer** [3], **Johannes Böer** [1], **Markus Bachmann** [1], **Falk Mrowka** [3], **Egbert Schwarz** [4], **Helko Breit** [2] and **Ulrich Steinbrecher** [1]

[1] DLR, Microwaves and Radar Institute, 82234 Wessling, Germany; Josef.Mittermayer@dlr.de (J.M.); Johannes.Boeer@dlr.de (J.B.); Markus.Bachmann@dlr.de (M.B.); Ulrich.Steinbrecher@dlr.de (U.S.)
[2] DLR, Remote Sensing Technology Institute, 82234 Wessling, Germany; Birgit.Schaettler@dlr.de (B.S.); Thomas.Fritz@dlr.de (T.F.); Helko.Breit@dlr.de (H.B.)
[3] DLR, German Space Operations Center, 82234 Wessling, Germany; Ralph.Kahle@dlr.de (R.K.); Edith.Maurer@dlr.de (E.M.); Falk.Mrowka@dlr.de (F.M.)
[4] DLR, German Remote Sensing Data Center, 82234 Wessling, Germany; Egbert.Schwarz@dlr.de
* Correspondence: Stefan.Buckreuss@dlr.de; Tel.: +49-81-5328-2344

Received: 4 May 2018; Accepted: 28 May 2018; Published: 5 June 2018

Abstract: The satellite of the TerraSAR-X mission, called TSX, was launched on 15 June 2007 and its identically constructed twin satellite TDX, which is required by the mission TanDEM-X, launched on 21 June 2010. Together they supply high-quality radar data in order to serve two mission goals: Scientific observation of Earth and the provisioning of remote sensing data for the commercial market (TerraSAR-X mission) and the generation of a global digital elevation model (DEM) of Earth's surface (TanDEM-X mission). On the occasion of the 10th anniversary of the mission, the focus will be on the development of the TerraSAR-X system during this period, including the extension of the ground segment, the evolution of the product portfolio, dedicated mission campaigns, radar experiments, refinement of the satellite operations and orbit control, and the results of the performance monitoring. Despite numerous interventions in the overall system, we managed to incorporate new scientific and commercial requirements and to improve and enhance the overall system in order to fulfill the increasing demand for Earth observation data without noticeable interruptions to ongoing operations.

Keywords: radar; satellite; remote sensing; SAR; TerraSAR-X; operations; ground segment; orbit; mission

1. Introduction

On 15 June 2007, the first national German radar satellite TSX was launched from the Russian Baikonur Cosmodrome in Kazakhstan, which marked the starting point for the highly successful TerraSAR-X remote sensing mission [1]. This mission has now lasted for more than 10 years and achieved remarkable results. Despite a satellite age well beyond its nominal life time of 5.5 years, no significant technical flaws were encountered. The objective of providing high-quality SAR products in Spotlight, Stripmap, and ScanSAR mode variants with various polarizations [2] has since been met without any restrictions and the image quality is just as good as it was at the beginning of the mission. The reasons for the outstanding long-term performance are both the excellent manufacturing of the satellite systems [3] and the design of the ground segment [4] tuned to it, i.e., which is able to exploit all of the capabilities of the satellite bus and the radar instrument. This could already be seen during the launch and early orbit phase. Only four days after launch, the first SAR images were successfully acquired, downlinked, and processed on ground using nominal ground segment workflows. At the end of 2007, the complete SAR system was characterized and calibrated, the SAR products were verified and the ground segment was operationally qualified [5].

Over ten years of operations, the TerraSAR-X system underwent to a certain extent an evolutionary process. Although the TerraSAR-X ground segment, as described in [4], has maintained its basic structure throughout the mission, it has been steadily improved and expanded upon to meet new requirements. The biggest impact had its upgrade to serve a second mission on top, the TerraSAR-X Add-On for Digital Elevation Measurements TanDEM-X, starting in 2010. The primary TanDEM-X mission goal was the generation of a global digital elevation model (DEM) of Earth's total land mass within two years. A second satellite TDX, which is almost identical to TSX, was launched on 21 June 2010. Both satellites since then fly in a close configuration with a minimum distance of about 250 m and perform together bistatic radar acquisitions of the Earth's surface, as required for the DEM generation. Both satellites are also used, either TSX or TDX, to acquire the monostatic acquisitions for the on-going TerraSAR-X mission. The necessary modifications, extensions and tests to implement the TanDEM-X functionalities into the ground segment meant a massive intervention comparable with open-heart surgery, since the ongoing operation of the TerraSAR-X mission was not allowed to be interrupted or disturbed. Commissioning of the TDX system was accomplished in October 2010 and the joint TerraSAR-X/TanDEM-X ground segment was successfully released for TerraSAR-X operations that are based on two satellites flying in a close formation.

Driven by science and commercial business cases, new products, services and capabilities were required, leading to a number of operational extensions and improvements.

- The SAR imaging mode and thus product portfolio has been extended in 2013 by Staring Spotlight and Wide ScanSAR. The first of these improves the obtainable azimuth resolution to about 24 cm, while the second provides an extended swath width of about 200 km.
- The experimental dual-receive antenna configuration provides full polarimetric imagery and along-track interferometry acquisitions.
- The overall receiving station capacity and flexibility has been increased.
- Near-real time support has been improved by shortening the turn-around times and by introducing new data sets for maritime applications.

Radar science is one of the driving forces that aim to extend existing or to test new radar modes and to put theoretical considerations into practice. The flexible design of the radar instrument provides the conditions for numerous experiments. An important proof of technology was the acquisition and processing of data in TOPSAR mode, which—like the ScanSAR mode—acquires data of several sub-swathes in parallel, but also performs an along-track azimuth steering, and thus overcomes severe ScanSAR limitations, like the scalloping effect. The new bi-directional SAR mode BiDiSAR was developed. It shapes an antenna pattern with a forward and a backward oriented lobe, and thus records two time-shifted images of the same scene on ground into only one receiving channel. In summary, we describe in this paper the TerraSAR-X ground segment extensions and service improvements, as well as the evolution of the SAR product portfolio since the TerraSAR-X mission operations started in 2008. We report on dedicated mission campaigns necessitating specific ground segments adaptations on the one hand, on the other hand opening new research opportunities by providing additional products. Details of selected radar experiments performed within the last decade are given. Specific challenges to be met by the satellite operations are discussed. The orbit control philosophy and the operational experience is described. System performance and mission utilization aspects are addressed.

2. Extension of the TerraSAR-X Ground Segment

2.1. Common TerraSAR-X/TanDEM-X Ground Segment

The TerraSAR-X ground segment underwent major extensions when compared to its starting point, as described in [4]. One essential upgrade driver has been the TanDEM-X mission, the TerraSAR-X Add-On for Digital Elevation Measurements, which added a second, mostly identical satellite TDX in 2010. Flying TDX in a close helix formation around the TSX satellite (the one being kept in a reference orbital tube) enables a bistatic acquisition of SAR data, as needed for global DEM generation. Both TSX and TDX are used together for such acquisitions where only one satellite is actively transmitting SAR pulses and both satellites are receiving the echoes. To counterbalance this interferometric usage of the TSX satellite for TanDEM-X, the TerraSAR-X monostatic data acquisitions have to be performed by either TSX or TDX since then. Consequently, essential ground segment subsystems had to be upgraded from their "one mission one satellite" behavior into a combined "two missions two satellites" one even to fulfill their TerraSAR-X mission purpose only, foremost the flight dynamics and the mission planning subsystems. Their operational integration and commissioning had been achieved without disturbance of the on-going operational TerraSAR-X mission. In October 2010, the joint TerraSAR-X/TanDEM-X ground segment had been released for the TerraSAR-X operations based on two satellites that are flying in a close formation [6,7].

Maintaining not only a reference orbit for TSX, but also the challenging helix formation of TDX relative to TSX in an extremely short distance posed new needs for spacecraft navigation and control, e.g., a common maneuver planning and execution. The reference orbit control, as described in Section 7, details on the formation flight are given in [8,9]. Verification of the upgraded flight dynamic system components and specifically validation of their interaction with the operational TSX system was performed for about one year before launch using the in-orbit TSX satellite and on-ground simulated TDX GPS navigation data [10].

Conflict-free timelines for both satellites are obtained only if the acquisitions for both missions are jointly planned inside the mission planning system. TerraSAR-X data takes are assigned by mission planning to either one of the satellites, for TanDEM-X data takes the transmitting satellite has to be chosen. The resulting load distribution has to balance the usage between the two missions. Irreparable damages that are caused by a mutual illumination of the instruments have to be prevented. The operational qualification of the new "two missions, two satellites" version of the mission planning system under realistic conditions within the TDX commissioning phase had to be performed, while the former "one mission one satellite" version was still supporting the on-going TerraSAR-X operations. Operational roll-out, i.e., final replacement of the former version, was mastered within a short maintenance period of a few days only, services were continued with the same robustness and reliability as before [11].

On the payload data handling side, a new production chain for the reception and processing of TanDEM-X data has been established. DLR's existing German Antarctica Receiving Station (GARS) O'Higgins was complemented in 2010 by a new satellite receiving station, the Inuvik Satellite Station Facility (ISSF) in Canada. Both of the stations form the backbone of the TanDEM-X Ground Station Network, which was further supported in the years 2010–2015 by the partner ground station SSC Esrange in Kiruna, Sweden, to handle the TanDEM-X global DEM data amount. Like the main TerraSAR-X ground station at DLR Neustrelitz (NSG), also these stations not only support X-band data reception, but also S-band up- and downlink for telemetry, tracking, and commanding (TT&C) for both the TSX and TDX satellites and thus can be used for both missions, TerraSAR-X and TanDEM-X. Their association to a given mission is configurable and is mostly a question of downlink capacity needs and requested data transfer times between a given station and the TerraSAR-X processing hosted at DLR. GARS and ISSF are supporting TanDEM-X, GARS had been actually used for the downlink of TerraSAR-X background mission data before TDX launch.

Since the beginning of TSX mission operations the density of the S-Band motoring network was significantly increased. At the beginning, a 12-h rhythm was applied. In later stages the monitoring of the satellites was intensified. The S-Band network benefits from the expansion of the X-Band network by GARS and ISSF. The KSAT ground station SvalSat, Svalbard, supports two passes for TSX and TDX monitoring at midday.

The current ground segment layout is depicted in Figure 1.

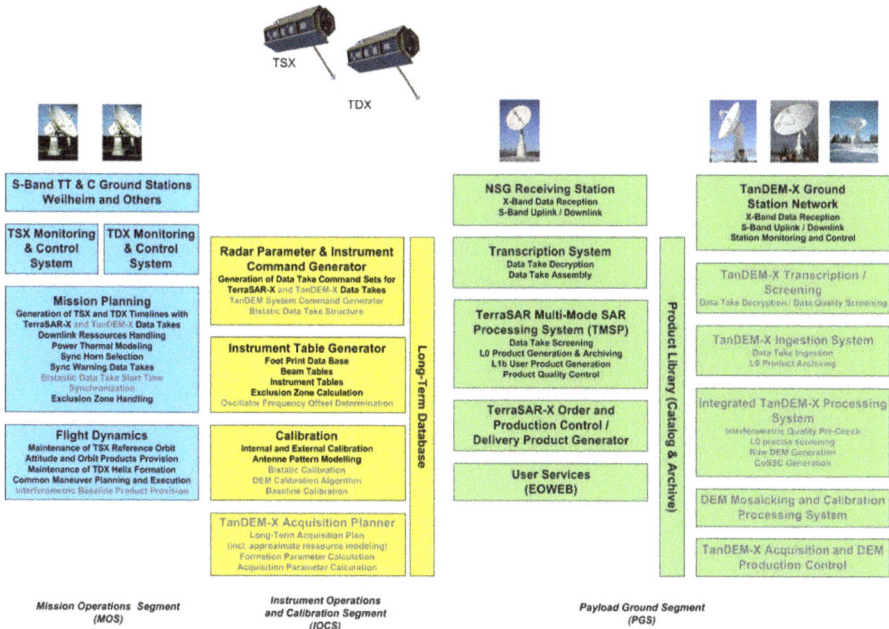

Figure 1. Sketch of main subsystems and components of the common TerraSAR-X/TanDEM-X ground segment. TanDEM-X mission specific parts are in grey.

2.2. Receiving Capacities and Flexibilities

Already in 2009, a TerraSAR-X data reception system was integrated at SvalSat, primarily to shorten near-real time (NRT) product latencies by using selected contact opportunities outside of the NSG visibility range. The data that were received at SvalSat are transferred in an encrypted raw data format to NSG, where they are processed in the same way as those that were received at the station itself. In answer to the increasing demand in TerraSAR-X products, the contacts that are used for the S-band support at SvalSat are also used for X-band data reception, thus enlarging the overall downlink capacity.

In 2015, the receiving station pool concept was introduced. Before then, acquisitions had to be explicitly ordered for downlink at a given station. This allowed for using the GARS for background acquisitions, to place test orders for ISSF during station validation and to explicitly order NRT data takes either at SvalSat or NSG, depending on the nearest downlink opportunity. Grouping stations into one receiving pool and leaving the decision regarding which station to use for the downlink of a given data take up to mission planning, results in a better exploitation of limited resources like on-board memory and downlink capacity [12]. Mission planning uses the next free slot for a downlink. If necessary, pool reconfigurations can be done up to three days before the first downlink to a new station in the pool, in urgent cases, only six hours are needed for the event.

Both the pool concept and the flexibility to use a given station either for TerraSAR-X or TanDEM-X or for both play an important role in exploiting downlink resources and to handle specific constraints that are given by specific mission campaigns.

2.3. Near-Real Time Capabilities

In the beginning, Near-Real Time (NRT) support had not been a strong TerraSAR-X stake holder requirement. Nevertheless, all of the basic products have been offered in a NRT flavor since the beginning of the mission. Sampling and resolution in the NRT products are identical to those of the standard product. The product latency after downlink is typically about 10 min. Processing has to rely upon auxiliary information, which is available at the same time as the raw data. The usage of a predicted orbit results in a reduced geolocation accuracy, the non-availability of instrument house-keeping parameters in a reduced radiometric accuracy. Data takes foreseen for NRT processing are marked as such both in space and on ground, and thus allow for a completely data-driven privileged handling from reception into processing and dissemination.

Within the seven years, the NRT capabilities have been further improved. Receiving stations with online data connection to NSG are grouped into the NRT station pool (currently consisting of NSG and SvalSat). Mission planning uses the next possible contact from the station pool for the downlink of a NRT data take and schedules it as soon as possible within this contact, i.e., at the beginning for already completed acquisitions and—if the remaining contact time allows—within for acquisitions that were taken during the contact.

NRT support for maritime applications has been introduced in 2012 starting with a ship detection service. Wind and wave products that are based on Stripmap multi-look ground range detected products followed in 2016 [13]. In case of ship detection, Automatic Identification System (AIS) data are processed in real time and the obtained information is merged into the SAR based detection results. Ship detection information is provided in a number of different formats, e.g., as ESRI shape file layer, GML, or Google KMZ files assembled with a geocoded quicklook image. Dissemination options include secure FTP variants and electronic mail. Whereas, the ship detection service is already available for external users, wind and wave products are currently provided for selected science projects only. The wind product provides derived wind field information for the estimated wind speed and direction calculated relative to a standard reference height of 10 m (U10) on a raster of two by two kilometers. The wave product contains sea state information about the significant wave height on a raster of three by three kilometers. The core processor to derive those value added products is the SAR AIS Integrated NRT Toolbox [14]. The SAINT processor calls the processes for the various maritime products in parallel. Thus, obtained information is made available instantaneously for the other information extraction processes.

For selected NRT applications, e.g., vessel navigation support in polar-regions, quicklook images of the high-resolution TerraSAR-X products, or further derived information of limited data amount are sufficient. Consequently, TerraSAR-X processing systems were installed both at GARS and ISSF in 2015 enabling the identical NRT product generation, as done at the central processing system. Following a nominal NRT production run, a dedicated station NRT quicklook process generates geocoded quicklook products of a size smaller than 5 Mbyte, which are e.g., suitable for e-mail transfer to NRT users on-board vessels or at research stations with Internet access. Handling of station NRT orders is fully integrated into the payload ground segment. Orders may be placed by authorized users, the derived production requests are routed accordingly into the appropriate station [12].

3. Evolution of SAR Product Portfolio

In its ten years of operation, the TerraSAR-X ground segment extended and improved the SAR product portfolio continuously. The flexibility in commanding the versatile TerraSAR-X and TanDEM-X SAR instruments allowed for numerous radar experiments to test and demonstrate new SAR acquisition modes or mode variants, which are detailed in the following section. The most prominent extension was the operational release of Staring Spotlight (ST) and a six-beam wide ScanSAR mode in April 2013. But, not all evolutionary steps in the product development are related to innovative SAR mode design and instrument commanding, and not all of such experiments lead to new product releases. Since the TerraSAR-X mission is user driven and is not restricted with regards to the future product ordering to specific acquisition conditions, only those modes and product variants that fulfill the high product quality requirements on a global scale were selected, calibrated, and verified as operational products. Some mode variants like the experimental 300 MHz range bandwidth option for the high resolution spotlight were already added to the product portfolio right after the end of the commissioning phase because of their solid performance. Modes, like the experimental aperture switching ATI (ATIS) mode and TOPSAR, however, are limited due to the complexity in their commanding sequences to restricted acquisition scenarios, and hence are not included in the SAR product portfolio. All basic products as well as the operationally provided experimental products are generated by the TerraSAR-X Multi-mode SAR Processor (TMSP). Details are given in [2,14]. As of 2018, the portfolio of basic [2] and experimental products [15] is characterized in Table 1.

Table 1. TerraSAR-X basic and experimental (grey) products.

Mode	Config.	Pol. Mode	Pol. Channel	Resolution (Az. × Ground-Rg.) (m²)	Extent (Az. × Ground-Rg.) (km²)	Product Variants
Stripmap		Single	HH, VV	3.3 × 1.7 … 3.5	50 × 30	complex & detected in SE/RE
		Dual	HH/VV HH/HV VV/VH	6.6 × 1.7 … 3.5	50 × 15	complex & detected in SE/RE
		Twin [1]	HH/VV	9.5 … 13.5 × 9.5 … 13.5	50 × 30	detected in RE only
	DRA [2]	Quad	HH/HV/VH/VV	6.6 × 1.7 … 3.5	50 × 15	complex & detected in SE/RE
	DRA ATI [2]	Single	HH, VV	<3.3 × 1.7 … 3.5	50 × 15 or 30	complex only [3]
ScanSAR	4-Beam	Single	HH, VV	18.5 × 1.7 … 3.5	150 × 100	complex & detected in RE only
	6-Beam	Single	HH, VV, HV, VH	40 × 30 … 42 (detected)	200 × 194 … 266	complex & detected in RE only
Spotlight	Sliding (SL)	Single	HH, VV	1.7 × 1.5 … 3.5	10 × 10	complex & detected in SE/RE
		Dual	HH/VV	3.4 × 1.5 … 3.5	10 × 10	complex & detected in SE/RE
	High Res. (HS)	Single	HH, VV	1.1 × 1.5 … 3.5 (1.1 × 0.7 … 1.8; 300 MHz)	5 × 10 (5 × 6 … 10; 300 MHz)	complex & detected in SE/RE
		Dual	HH/VV	2.2 × 1.5 … 3.5	5 × 10	complex & detected in SE/RE
	Staring (ST)	Single	HH, VV	0.24 × 0.85 … 1.7 (300 MHz only)	2.5 … 2.8 × 4.6 … 7.5	complex & detected in SE/RE

[1]: The Twin mode is a one-beam burst mode with alternating polarizations not suited for interferometry. [2]: Dual Receive Antenna (DRA) modes are only available in dedicated campaigns. [3]: No azimuth spectral weighting applied on ATI DRA channels. The level 1 product variants are SSC = Single-look Slant-range Complex and the detected variants MGD = Multi-looked Ground-range Detected, GEC = Geocoded Ellipsoid Corrected, EEC = Enhanced Ellipsoid Corrected. The detected products are available as spatially enhanced (SE) or radiometrically enhanced (RE) multi-look flavors. RE variants are noise corrected.

3.1. Extension of Basic Products

The ST mode specifically pushes the limits of instrument azimuth beam steering and spectral signal processing, as pointed out in Section 5, but it is also most demanding in terms of reconstructing the "true" geometric range history for focusing each target. With synthetic aperture lengths that correspond to more than 60 km orbit path, the inclusion of higher order terms in the focusing algorithm is required, as well as the intrinsic bistatic tropospheric path delay corrections implemented in the TMSP to keep the reconstructed range history within mm of the geometrical one [16]. The ST azimuth resolution is specified as 24 cm which includes some cm margin since the actual height on ground of each focused scatterer has to be considered for focusing parameter calculation. The TMSP uses a coarse built-in DEM for this, but deviations from the local focal plane of merely several tens of meters may cause resolution loss in the range of centimeters. In turn, we could demonstrate that measuring this very small defocusing effect with methods that are similar to auto-focusing can be used to actually determine the absolute elevation of all sufficiently bright individual scatterers in a single ST image [17].

The six-beam wide-ScanSAR mode seems like a straight forward extension of the established four-beam ScanSAR product. It did however introduce completely new approaches in TerraSAR-X mode and product design. Not only the use of newly created wide-swath sub-beams that differ in pattern tapering from the nominal Stripmap beams allowed for achieving a total swath width of up to 266 km, also variable range bandwidths and processing look factors are used to keep an optimal performance over the whole swath. Its chirp bandwidth is selected for each individual sub-swath, such that the projected ground range resolution of the detected product is kept mostly below 35 m—in the same order as the azimuth resolution—with a nearly constant number of radiometric looks. This allows for a consistent performance in applications like ship or oil spill detection over the full multi-beam swath. Since this mode is mainly designed for maritime applications, a special cross-polar (HV or VH) order option is also provided that increases the contrast between man-made targets and ocean surface at the price of lower sensitivity.

Continuous research activities and TMSP upgrades with respect to processing performance, timing, and phase accuracy, as well as geometric precision also improved the SAR product performance and allowed for turning experimental products into operational ones. Additionally, the product characteristics and the level 1 format have been extended to facilitate higher level processing and geo-information retrieval. Some major achievements were:

- The introduction of operational image noise correction for all radiometrically enhanced (RE) products in December 2008. It is based on a sophisticated statistical evaluation of the thermal noise distribution in SAR images when considering the processing filter noise gains [18].
- Optimized ScanSAR SSC oversampling approaches resulting from InSAR experiments activated in April 2009, which enabled the re-classification of the ScanSAR SSC product from an experimental one into a basic product.
- Also, in April 2009, additional quicklook and image files were added to the level 1 product to support quick evaluation and image interpretation based on the statistical properties of the image (see Figure 2 for an example).
- In preparation of the TanDEM-X mission with a second satellite to provide TerraSAR-X mission products, an antenna phase pattern correction scheme was introduced in March 2010. Similar to the radiometric beam pattern, the individual phase patterns of the numerous beams of the two instruments differ. The individual patterns are projected onto a digital elevation model of the scene and corrected for in order to generate interferometrically exchangeable SAR products from both. The patterns are also annotated in all of the SSC products.
- With the extreme Doppler spectral behavior encountered specifically in the raw data of the newly introduced ST modes and its complex mapping into the focused data, additional annotation files describing the Doppler centroid in the focused image have been added to all complex products in September 2013. These files also facilitate the interferometric processing of burst modes (i.e., ScanSAR).
- ST mode images are basically illuminated by the beam footprint that is fixed to one location on the ground. Hence, they show a strong radiometric variation and strong increase of the image noise towards their azimuth borders. To ease this effect, the operational noise correction is also applied by default to the highest resolution spatially enhanced (SE) products of the ST mode since May 2014.
- The upgrade of May 2014 also added a new azimuth time tag correction method to the processor, which analyses the absolute Radar frequency and drifts of the internal clock w.r.t. GPS time tags and corrects them in the processed data [19].

Figure 2. Example for the additional quicklook files with a color representation of the Synthetic Aperture Radar (SAR) image pixel statistics that supplement each TerraSAR-X level 1 product since April 2009: Single polarization Stripmap GEC product of Auckland, New Zeeland acquired in June 2008. These quicklooks are adaptively "tuned" by the TerraSAR-X Multi-mode SAR Processor (TMSP) to support visual interpretation of the image content and not meant for tasks like scientific land cover classification.

Specifically processing enhancements, like the last point, helped to improve the geometric location accuracy and to turn TerraSAR-X into a geodetic measurement instrument [20]. These developments were supported by the tremendous effort put into phase and signal delay calibration for the TanDEM-X mission, the intrinsic bistatic focusing approach of the TMSP (even applied for mono-static TerraSAR-X data), and the elimination of algorithmic approximations in its focusing kernel. Once atmospheric and ionospheric signal propagation effects are properly considered, pixel location accuracies in the 1–2 cm range are achievable with the precisely determined orbit products [20].

3.2. Experimental Products

The SAR instrument on both satellites TSX and TDX allows for the simultaneous reception of the transmitted SAR signals with the fore and aft segments of the antenna when using the redundant receiver chain in parallel with the primary one. The signals received by either antenna part are recorded in a dedicated channel and are then exploited accordingly on ground. Thus, data acquisition either in full polarization mode to obtain quad pol images or with separated phase centers to allow for along-track interferometry (ATI) is enabled in this dual-receive antenna (DRA) configuration mode. The way in which DRA data are recorded on-board necessitates complex signal demultiplexing operations on ground before any raw data processing can take place. The needed compensation of the relative instrument electronics phase drifts results in a challenging calibration approach. Due to this complexity, the DRA configuration is offered for Stripmap mode acquisitions only.

In full polarization mode, the transmit polarization is toggled between V and H from pulse to pulse. The fore and aft antenna segments use different polarizations for the parallel receiving of the returned signals. The pulse repetition frequency (PRF) is doubled as for the dual polarization mode in the single-receive antenna configuration (SRA), leading to the same half range swath extent. Full (quad) polarization products are available in the same flavors as the SM basic products, i.e., as SSC and MGD, GEC, EEC SE/RE variants [14].

In case of ATI acquisitions, the same polarization is used for transmitting and receiving. The separated phase centers of the fore and aft antenna segments generate a baseline in flight direction, which can be exploited e.g., for ground moving target indication (GMTI). Acquisitions may either use a high PRF setting, and thus a reduced swath width as for fully polarized acquisitions or a low PRF setting resulting in a range extent similar to the single polarization acquisitions. The high-PRF variant is more suitable for GMTI applications, such as traffic monitoring due to its wider unambiguous velocity measurement range. Thereby, the signal-to-clutter ratio is increased allowing for an improved tracking of moving objects in stationary clutter. The ATI products are available as SSC only. Besides the two separate images for the fore and aft antenna segments, the data are also processed to a nominal SRA SSC, and the delivered products thus contain three SSC channels, a SRA image, and the two DRA images.

The effective along-track baseline between the two DRA phase centers of this mode is in the order of 1.2 m only. Thus, its unambiguous ATI phase cycle corresponds to up to 200 m/s Doppler velocity, and it is hence most suited for velocity measurements of fast moving objects. Besides being designed for traffic and vessel detection, it is still sensitive enough to also derive large scale ocean surface current velocity fields (see e.g., [21]). Experimental products from the DRA configuration are offered in DLR's Earth Observation Data Service EOWEB under the heading *TSX-1 Experimental Products* in the collections *TSX-1.SAR.L1b-Stripmap-Quadpol* (830 acquisitions) and *TSX-1.SAR.L1b-Stripmap-ATI* (128 acquisitions), respectively.

3.3. TerraSAR-X-Like Products

In the pursuit monostatic campaign of the TanDEM-X mission in 2014/2015 (see Section 4.2), scientific and operational TanDEM-X mission acquisitions were commanded on both satellites to acquire the same swath on ground with a time lag of about 10 s corresponding to their along-track separation in space. Unlike the bistatic acquisitions that were performed in the nominal TanDEM-X close formation that require specific commanding with synchronization pulse exchange and additional calibration sequences, this raw data is very similar to the one of nominal TerraSAR-X mission acquisitions. The TSX/TDX pairs were thus screened and processed by an upgraded TMSP to generate the so-called TerraSAR-X-like products and make them accessible to the TerraSAR-X user community.

The TerraSAR-X-like products are not subject to the basic product specification, since they do slightly differ in terms of product performance. TerraSAR-X-like products are available in all nominal TerraSAR-X mode variants, and in most cases, the deviations in performance parameters are negligible. For Stripmap data, the TanDEM-X elevation beams have been used in some conditions that are optimized for coverage and not for radiometric performance. In TanDEM-X commanding, also the near range beams use a 100 MHz range bandwidth instead of the nominal 150 MHz. In combination with different approaches for radar timing (PRF selection), these products may thus miss the specified radiometric looks and ground range resolution of 3 m for small incidence angles. Despite these small discrepancies, the available data opens new fields of application exploiting the time lag with nominal TerraSAR-X product analysis tools e.g., for ship motion measurements and other change detection methods.

4. Dedicated Mission Campaigns

4.1. DRA Campaigns

The parallel activation of the primary and redundant receiver chain to obtain experimental quad polarization products and complex ATI products is performed within dedicated campaigns. Since the DRA operation imposes specific further constraints w.r.t. the on-board data acquisition, recording, and downlink when compared to the SRA operation, ordering and planning of DRA acquisitions is done centrally inside the ground segment for pre-selected test sites. The resulting data are then made available for user product ordering from the catalog.

The first DRA campaign lasted for three cycles (33 days) from 11 April until 13 May 2010. Since no SRA data can be acquired while downlinking data in DRA mode, the by then newly built-up station ISSF with the DLR antenna and equipment was used for data reception. Thus, the impact on data acquisition (both SRA and DRA) over Europe and the downlink of SRA data to NSG was kept at a minimum. In total, 356 SM quad pol and 163 SM single pol (ATI) acquisitions are available for user catalog ordering from this time period.

The second DRA campaign was conducted as part of the TanDEM-X Science Phase, starting at 18 November 2015 and ending at 11 February 2016.

4.2. Modifications Induced by TanDEM-X Science Phase

After completion of the TanDEM-X global DEM data acquisition phase, the time period from 9 October 2014 until 2 February 2016 was dedicated to the achievement of the secondary TanDEM-X mission goal, the generation of radar products for a number of new science and technology related applications. Specific flight formation configurations, which realized various along-track and across-track baseline conditions, were chosen in response to the needs of given applications. A second DRA campaign was conducted. This had a number of direct impacts onto the TerraSAR-X mission operation and led to further modifications inside the ground segment [12,22].

4.2.1. Pursuit Monostatic Phase

From 19 September 2014 until 17 March 2015, TSX and TDX were flying in pursuit formation with an along-track separation of about 76 km (10 s). In such a configuration, the two SAR instruments can transmit radar signals in parallel without mutual interferences. For images that were taken by TDX, appropriate time delays and roll angles are applied during commanding to correct for the Earth rotation effects within the 10 s time lag. The parallel operation of both the instruments opens a number of new opportunities, e.g., imaging the same scene within 10 s or enlarging the imaged area based on neighboring or overlapping (either in along-track or across-track direction) data takes.

Also, the downlink of both satellites may be operated in parallel. By using two receiving antennas, the downlinked data amount could be nearly doubled at NSG. However, for stations that were equipped with one antenna only, data downlink has to be restricted to one satellite, since TSX and TDX are too far apart to be tracked together, but too close to be tracked one after the other. Either the satellites were used in an alternating manner or a given station was fixed to a given satellite.

4.2.2. Large Horizontal Baseline Formation

Following the pursuit monostatic phase, TSX and TDX returned to a close formation for bistatic data taking under large cross-track baseline conditions. A maximum horizontal separation of about 3.6 km was reached in the extreme at the equator. Thus, TerraSAR-X acquisitions that were performed by TDX were possibly taken outside the reference orbit tube leading to slightly different product characteristics. Therefore, the so-called preferred satellite concept was introduced. In case that the nominal perpendicular baseline between TSX and TDX exceeds a given (configurable) threshold and if the available resources like memory and downlink capacity allow for it, a scene is taken by TSX (presuming that the user order allows for both, TSX or TDX for the acquisition). In summary, four different satellite selection options are offered for ordering. Orders can specify the TSX satellite for data taking when the observation of the reference orbit tube is needed. TDX is chosen if one wants to explicitly exploit large baseline situations. No satellite is pre-selected, but preference should be given to TSX. Both TSX and TDX are accepted, if no TSX preference is needed. Hence, the user needs were observed as best as possible, while still allowing for mission planning a good load balancing between TSX and TDX.

5. Radar Experiments with TerraSAR-X

The high flexibility provided by TerraSAR-X allowed for research in and demonstration of new SAR modes. The Staring Spotlight mode and the wide-ScanSAR mode were demonstrated and were further developed to fully operational TerraSAR-X modes. Other examples are TOPSAR, BiDiSAR, and Wrapped Staring Spotlight, which were demonstrated by TerraSAR-X.

5.1. Prerequisite: Flexible Instrument Commanding

The flexible instrument commanding in combination with a homogenous order chain for standard SAR data takes, as well as for experimental data takes, also called system data takes, enables the mixed operation for both types of data take within the same timeline, see Figure 3. The configuration of the SAR instrument does not have to be changed, since both, nominal and experimental data takes contain similar building blocks. This approach enables the development, test, and optimization of a new SAR mode with the final operational settings.

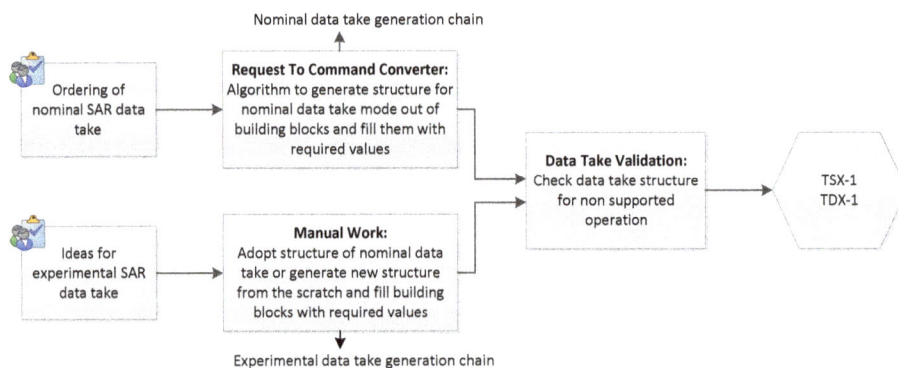

Figure 3. Data take generation chain for nominal and experimental data takes.

The flexible data take commanding is provided by the combination of four main building blocks. One building block describes the pulse to be transmitted (see Figure 4 for a set of parameters describing the pulse). Up to eight different pulses can be used within one data take. Another building block controls the echo window w.r.t. position, length, receive gain. Transmit and receive parameters can be changed from pulse to pulse. The third building block controls the antenna beam direction, pattern, and polarization. In general, the shortest change interval between two antenna settings is some milliseconds, but one selected set of antenna parameters can be changed from pulse to pulse. Up to 127 different antenna settings can be applied during one data take. A switching between H and V antenna polarization from pulse to pulse (toggling) is controlled by a parameter within the antenna properties. The controlling flag for polarization toggling can be set for transmit and receive independently. The fourth building block contains the control parameters for the structure of the data take, e.g., the number of pulses for which the radar is operating with the same radar settings. Sequences of radar settings can be repeated several times. Therefore, this repetition feature enables a convenient commanding of sub-swathes in ScanSAR data takes. The following chapters illustrate the wide range of useful applications utilizing the flexible instrument commanding.

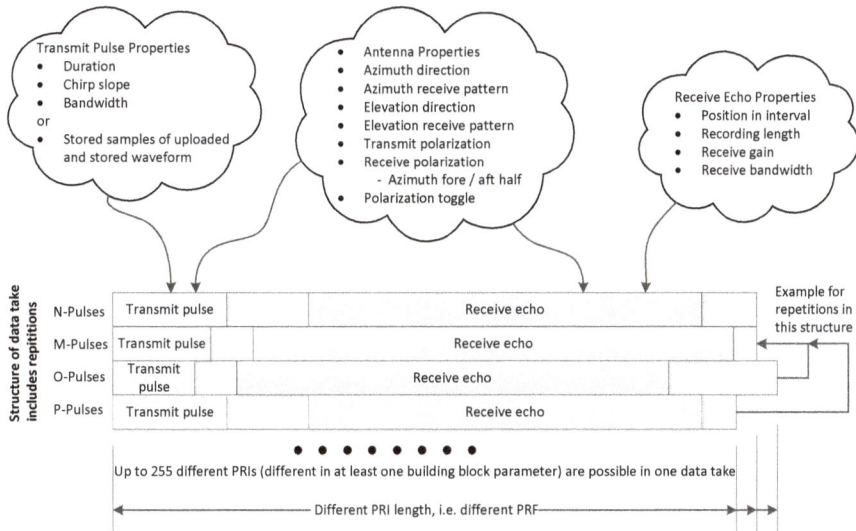

Figure 4. Building blocks and structure of a data take.

To build an along track interferometer with just one receiving chain, the ability of pulse to pulse switching between two antenna settings with different receive azimuth apertures is used. One pulse is received with a dedicated part of the antenna in flight direction (fore part); the next pulse is received with the part that is opposite to the flight direction (aft part). The echoes that were received by both parts are SAR processed separately and result in two SAR images with a baseline in flight direction. This mode is called Along Track Interferometer Switching (ATIS).

The six-beam wide-ScanSAR mode (see Section 3.1) uses the ability to transmit dedicated pulses for each sub-swath, where the transmit bandwidth is optimized for constant radiometric performance over the full incidence range. With this, only the receiving noise of the transmitted bandwidth contributes to the image noise and the signal to noise ratio is increased.

TOPSAR (see Section 5.3) data takes demonstrate the flexibility in commanding both, azimuth and elevation antenna steering, and the repetition of this structure. This mode reaches the limit of the current TerraSAR-X instrument design in using the maximum number of antenna settings, since each azimuth and elevation combination counts as one setting for the data take.

5.2. Staring Spotlight

The azimuth steering angles of TerraSAR-X were initially kept below ±0.75°, especially in the sliding spotlight modes. Up to this limit, the impact of the grating lobes is tolerable as their antenna gain keeps 20 dB below the gain into the desired steering direction. The azimuth resolution for TerraSAR-X has been enhanced by commanding staring spotlight with steering angles up to ±2.2°. In this post-launch introduced operational TerraSAR-X mode, the azimuth resolution is about 20 cm and it improved by a factor of 5 when compared to the sliding mode HS of TerraSAR-X. The problem that was connected to the enlargement of the steering angles is the grating lobes increase. As Figure 5 shows, for a steering angle of 2.2°, the largest grating lobe in red shows equal gain as the steered lobe in green. In order to maintain the SAR performance in the presence of strong grating lobes, two main incidence angle dependent adaptions were introduced, i.e., a PRF optimization in commanding, and a scene size optimization in the SAR processing [23–25].

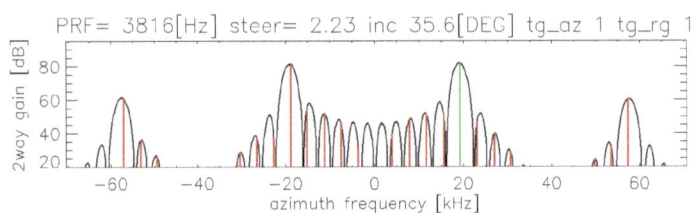

Figure 5. Azimuth pattern steered to 2.2°. Azimuth frequency corresponding to the direction towards the scene center (green) and ambiguous frequencies (red).

Figure 6 shows an example staring spotlight image that was acquired at the test site at DLR in Oberpfaffenhofen, Germany. The corner reflectors in the scene were used to verify the expected geometric resolution, which was found to be within 2% of the expected values. From the comparison of a zoom into the staring spotlight image with the corresponding zoom into a 300 MHz HS sliding spotlight image, the improvement that was achieved by multilooking five azimuth looks is obvious, and the higher level of detail is apparent.

5.3. TOPSAR

TOPSAR is a burst mode like ScanSAR that allows for the quasi-parallel acquisition of several sub-swathes and by this the acquisition of a much wider total ground swath at the expense of a reduced azimuth resolution. The TOPSAR mode overcomes limitations that are imposed by the ScanSAR—the most prominent is the scalloping effect—by steering the antenna along-track during the acquisition of a burst. In this way, all for the targets are illuminated with the complete azimuth antenna pattern, and thus, scalloping is circumvented and an azimuth dependency of signal to noise ratio (SNR) and distributed target ambiguity ratio (DTAR) is avoided, or at least considerably reduced. TOPSAR was demonstrated with TerraSAR-X in the year 2007 [26].

Different TerraSAR-X images have been acquired over Toulouse, France, in order to perform a scalloping analysis in the TOPSAR and ScanSAR modes. Figure 7 presents one sub-swath from the TOPSAR and ScanSAR acquisitions that were processed *without* the nominally applied scalloping correction in the processing. The azimuth resolution in the images is 16 m. About five range looks were processed, resulting in 16 m ground range resolution. The sub-swath size is approximately 90 km in azimuth and 25 km in slant range. No weighting has been applied in the burst image overlap area.

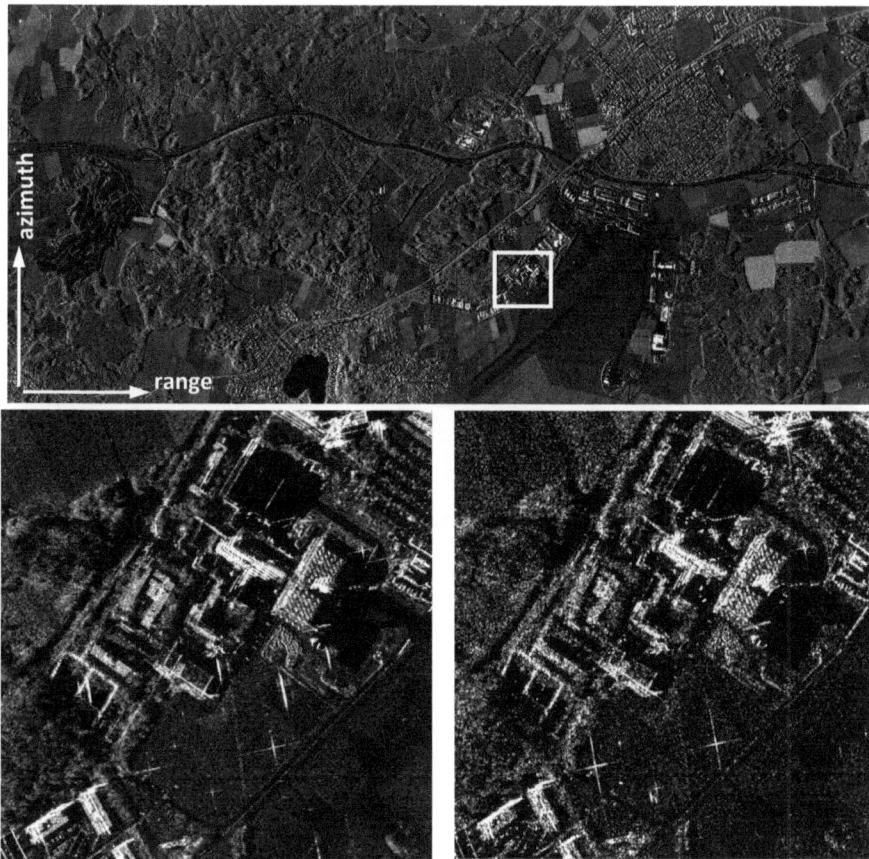

Figure 6. (**Top**) TerraSAR-X staring spotlight (ST) experimental acquisition example over Oberpfaffenhofen, Germany with five azimuth looks, incidence angle 35.6°, PRF 3786 Hz, scene extent 8.8 km × 3.6 km in ground range and azimuth, respectively. The single look resolution is 1.00 m in ground range and 0.21 m in azimuth. (**Bottom**) Zoom into staring image (**left**) and HS image (**right**). The geometric resolution in both zooms is about 1 m in azimuth and ground range, i.e., the staring image is composed of five looks, while the HS sliding image comprises only one.

The TOPSAR image sub-swath that is shown in Figure 7a was acquired with nine bursts. The commanded steering angle is in-between ±0.52°. Due to the low variation of the steering angle, the scalloping effect is hardly visible and it was quantified to be approximately 0.3 dB. It is not fully zero as electronic azimuth steering is implemented in TerraSAR-X, i.e., the element pattern is not steered and it causes a residual scalloping. The ScanSAR image was acquired with 33 bursts. The measured scalloping in the ScanSAR intensity profile is around 1.2 dB.

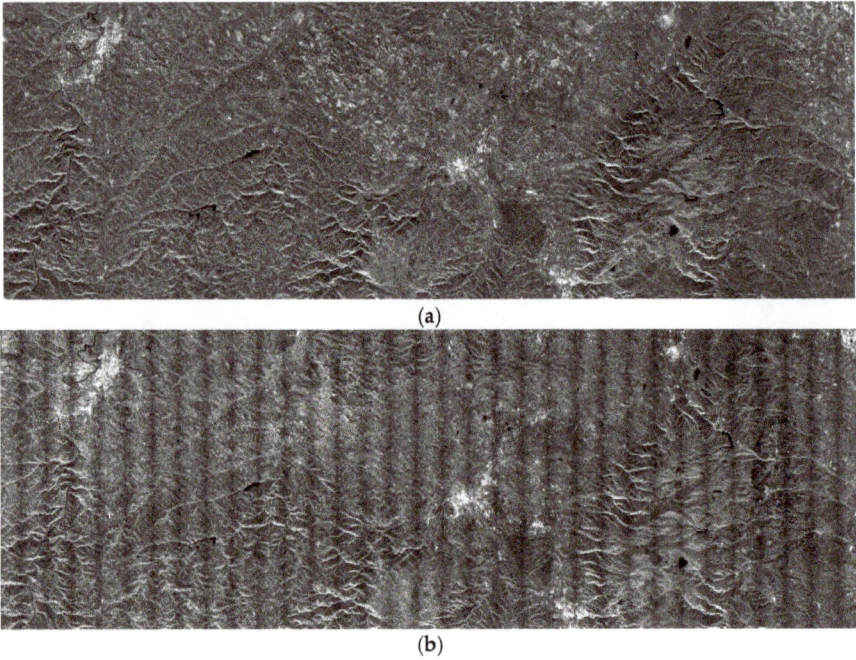

Figure 7. Terrain Observation with Progressive Scans SAR (TOPSAR) (**a**) and ScanSAR (**b**) comparison, both without scalloping correction. One sub-swath is shown. The measured scalloping in the ScanSAR image is around 1.2 dB, as compared to 0.3 dB in the TOPSAR image. Thirty-three ScanSAR bursts were necessary in contrast to only nine required in the TOPSAR mode.

5.4. BiDiSAR

The newly developed bidirectional synthetic aperture radar (BiDi) imaging mode [27] was demonstrated with TerraSAR-X. Bi-directional means the simultaneous imaging of two directions by one antenna into one receiving channel. The BiDi imaging is based on an azimuth pattern with two steered lobes that are pointing into different directions, an increased PRF, and a separation of the simultaneously received images in the Doppler spectral domain. Figure 8 shows the BiDi acquisition geometry for simultaneous forward and backward acquisitions in the directions ψ_{fore} and ψ_{aft}, respectively. The slant range vector in the forward direction is oriented by $90° - \psi_{fore}$ w.r.t. azimuth. In the backward direction, the slant range vector is oriented at $90° + \psi_{aft}$. The same target area is acquired twice as the sensor flies by with an along-track separation in the range of sections in-between the two illuminations.

Figure 9 shows an excerpt of a larger BiDi acquisition in the form of a color composite. The fore image is in red and the aft image in green color. At positions where the backscatter is high and is equal in both images, the colors combine to yellow. There are a number of nonmoving manmade targets that already show a considerable difference in the backscatter due to the acquisition azimuth angle difference of 4.4°. The motion of some ships is clearly visible as they appear partially in red and green, as the ship moved between the two illuminations.

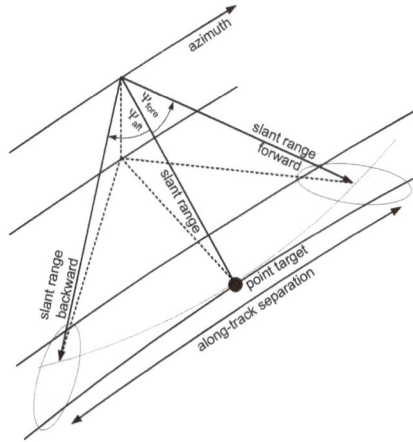

Figure 8. Bidirectional synthetic aperture radar (BiDi) acquisition geometry example with simultaneous fore and aft acquisitions.

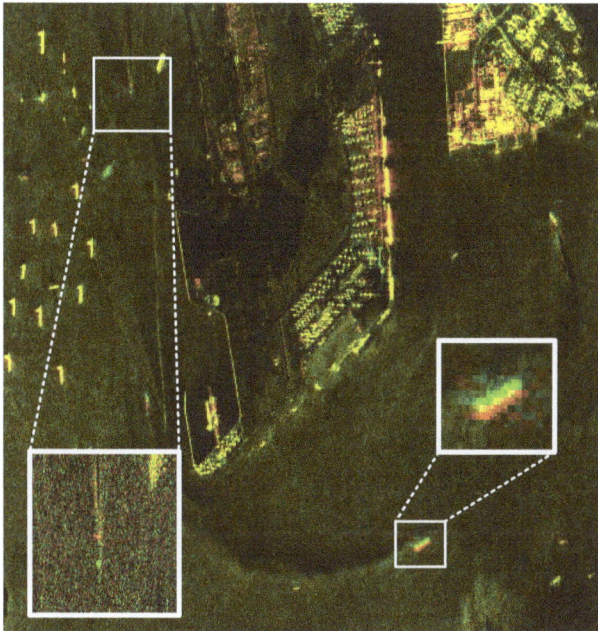

Figure 9. Color composite of fore (red) and aft (green) zoom images of a TerraSAR-X BiDi acquisition of Singapore. The azimuth direction is on the left, from bottom to top. Time lag between fore and aft image is 7.2 s. Further zooms of a small fast and a large ship in the white frames.

5.5. Wrapped Staring Spotlight

Wrapped Staring Spotlight [28] is a new method to extend the azimuth steering capability of phased array SAR antennas in order to achieve improved azimuth geometric and/or radiometric resolution. It extends the steering to directions with very low signal contribution. Point and extended targets in experimental TerraSAR-X acquisitions were evaluated up to ±4.4° steering, i.e., gains up to 45 dB below the grating lobes. It was found that, for TerraSAR-X, an extension of up to about ±3.9° improves the image quality although the cost-benefit ratio decreases with increasing steering angles.

One innovation developed is the Wrapped Commanding of steering angles. Figure 10 shows in green the lobes into desired steering direction and in red the undesired grating lobes. Up to ±2.2°, the on-board available angles can directly be used. Beyond, larger steering angles are obtained—without the need of additional on-board stored azimuth patterns—by re-accessing available angles, and later extraction of the desired steering directions in the SAR processing.

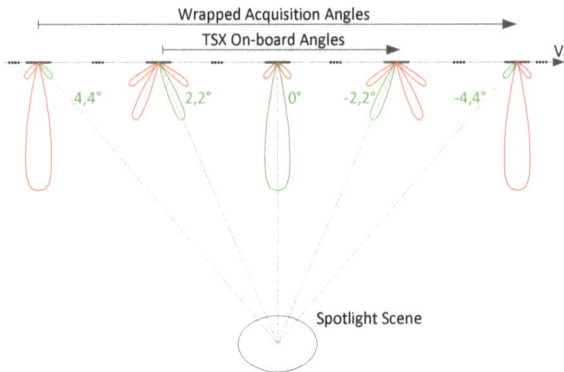

Figure 10. Desired steering lobes (green) and undesired grating lobes (red). TerraSAR-X on-board available angles up to ±2.2°. Steering angles beyond are obtained by re-accessing available steering angles.

Figure 11 shows wrapped corner reflector measurements. The azimuth resolution is steadily improving with increasing processed bandwidth up to 0.14 m, i.e., 68 kHz or ±3.9° in the case of sidelobe suppression weighting. For comparison, the staring spotlight mode of TerraSAR-X exploits only 38 kHz (±2.2°). Two statistical approaches were developed [28] that improve the radiometric performance when integrating signal directions with low SNR, i.e., the space invariant Look-Normalized Pattern Compensation (LNPC) and the space variant Ω-weighting. It was demonstrated that in wrapped TerraSAR-X imaging up to ±4.4°, the contrast for manmade targets was improved by 20%, and for extended targets, an improvement in the equivalent number of looks from two to four looks was measured.

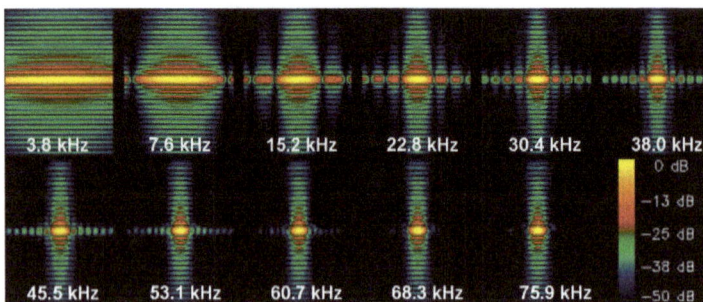

Figure 11. Corner reflector measurements in a TerraSAR-X wrapped staring spotlight image.

6. Satellite Operations

For a period of three years, TSX was operated as a single satellite mission in 514 km altitude. In 2010, a transition to a dual satellite mission needed to be accomplished and even more in October 2010 the two spacecraft entered into a close formation with inter satellite distances in the order of few hundred meters. The close formation flight posed a major challenge for satellite operations. Concepts needed to be developed in order to cope with safety aspects.

One major point of interest was the original safe mode concept of TSX. A satellite safe mode is typically entered by on-board automatic reactions in the case that a malfunction is detected. The original TSX safe mode used thrusters for attitude control in order to profit from the high reactivity of these actuators. Unfortunately, due to the thrusters' orientation attitude controls came along with the side effect of an unintended orbit change. Investigation showed that a safe mode drop might result in a potential collision risk of the spacecraft in orbit, which was not acceptable. As a consequence a multiple stage safe mode concept has been newly developed. The primary safe mode is now based on magnetorquers, which do not alter the relative orbit of TSX and TDX. In the case of an insufficient reactivity, further stages within the safe mode concept are applied, using thrusters in a more controlled way from the perspective of formation flight. The on-board safe mode has been implemented by the satellite manufacturer; however, this major concept change affected satellite operations in a profound way. Many flight procedures needed to be reworked, major on-board software updates became necessary, and in-orbit tests of the safe mode concepts before entering close formation have been performed.

Another major challenge to be mastered is to ensure that the two satellites do not illuminate each other when transmitting a radar beam, as this could result in an irreparable damage of the partner satellite. A well-thought-out concept was established to comply with this requirement during nominal operation, but in particular, during on-ground or on-board caused contingency situations. For the latter one, the so-called Sync-Warnings were implemented. These are isochronal instrument activities on both satellites with a bidirectional exchange of sync pulses via one of six sync horns. In case one of the satellites gets no feedback from the other (i.e., the state/position of the other satellite is different to the predicated one), its instrument is switched to receive-only mode. In order to reduce the mutual illumination risk to a maximum extent during nominal operations, the so-called transmit exclusion window information is delivered to both mission planning and the mission control and monitoring system. This double layer check on ground prevents firstly planning, and secondly, the commanding of possible hazardous activities in forbidden areas. In addition to these ground based measures, the on-board exclusion zone mechanism was implemented, in order to prevent the execution of possible hazardous activities in such areas. See [29] for a detailed description of all TanDEM-X formation flight operation related aspects.

Figure 12 illustrates acquisition activities, the on-board memory fill level, together with the downlink activities for the TerraSAR-X mission and the TanDEM-X mission for about one day. The combined mission planning system enables the concurrent support of the two missions by the two spacecrafts. In order to minimize the impact of the two missions on each other a balanced downlink planning strategy was implemented. The fact that the size of the on-board memory of the two spacecraft differs by the factor of two and the necessity to support the TerraSAR-X near real time feature with its prioritized downlink make the downlink planning even more challenging.

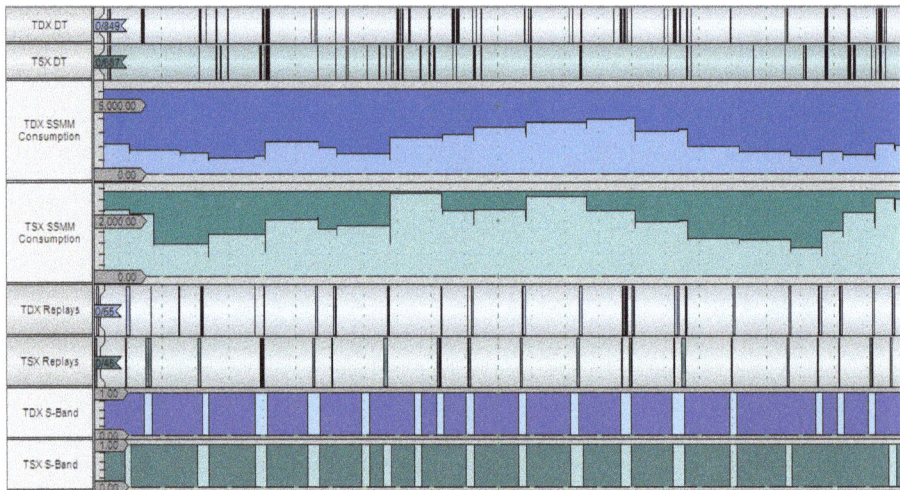

Figure 12. Typical acquisition and downlink activities for TanDEM-X Satellite (TSX) and TanDEM-X Satellite (TDX) for about one day. DT rows: Data acquisition activities (number of tele-commands is given on the left hand side) SSMM rows: The on-board memory utilization (TDX's memory has twice the size of TSX's one, as indicated on the left hand side). Replay rows: The X-Band downlink contacts used (number of replays is given on the left hand side). S-Band rows: TT&C contacts (indicated by light color).

The design of the TSX and TDX spacecraft targeted in both cases for an in-orbit lifetime of 5.5 years. These days, TSX has completed ten years and TDX seven years in orbit. Both spacecraft are still fully functional; SAR operations do not suffer any restrictions due to on-board hardware failures. Furthermore, the state of depletion of consumables gives reason to expect many more years of satellite operations. For example, the measured capacity degradation of the batteries is much less than predicted for the age that the spacecraft have now reached.

Figure 13 details the hydrazine consumption of TSX and TDX. In both cases, approximately half of hydrazine budget is still available for future use. So far, the phases directly after launch (2007 for TSX and 2010 for TDX) were expensive in terms of hydrazine usage for attitude control after separation and for orbit control to adjust the target orbit. In 2008, a safe mode drop on TSX caused an unusually high yearly budget. In those days, the safe mode still used thrusters primarily for attitude control, as described above. Generally, TDX uses more hydrazine because of its helix formation flight relative to TSX. The consumption was remarkably high in the years 2014 and 2015 due to the specific formations that were required in the TanDEM-X science phase.

Beside the primary payload, the SAR instrument, TSX carries secondary payloads, namely a laser communication terminal (LCT) by Tesat-Spacecom GmbH & Co. KG (Backnang, Germany) and a two frequency GPS receiver IGOR, as supplied by the Deutsches GeoForschungsZentrum (GFZ). The latter is also mounted on TDX, and is classified in this case, as primary payload due to the strengthened requirements of the TanDEM-X mission on absolute and relative orbit determination. Both additional payloads showed a very good performance in orbit. On 21 February 2008, the LCT on TSX communicated with a counterpart of the NFIRE satellite, the very first inter-satellite communication with laser technology in Low Earth Orbit.

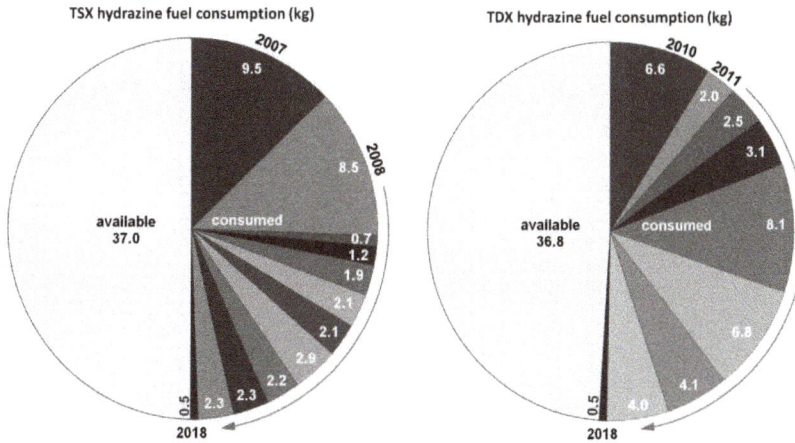

Figure 13. Pie charts of TSX and TDX in orbit hydrazine consumptions. Each slice in the "consumed" region represents an operations year for which the consumed fuel is reported. Approximately half of the hydrazine is still available for both satellites, see region "available".

7. Orbit Control

The TerraSAR-X mission design is relying on the reference orbit, which forms the basis for both orbit control and mission planning, including the scheduling of SAR acquisitions. In contrast to traditional design considerations for Sun-synchronous, frozen eccentricity repeat orbits, the TSX reference orbit must be a closed orbit with matching states at the beginning and end of each 11-day repeat cycle. Therefore, the reference orbit design was formulated as an optimization problem [30]. The implemented TSX reference orbit is expressed in an Earth-fixed frame, and it can be repeated in 11-day intervals throughout the entire mission.

The high requirements on guidance, navigation, and control are mainly driven by SAR interferometry. In order to permit repeat-pass interferometry applications, like subsidence mapping or glacier monitoring, the cross-track distance between radar acquisitions in repeated orbits should be as small as possible. Depending on the availability of digital elevation models, which are used to compensate topographic effects, cross-track distances below 350 m are desirable. Hence, the TSX osculating orbit is controlled within a "tube" that is defined about the Earth-fixed reference orbit. In order to fulfill the requirements, the radius of the "tube" is set to 250 m, which corresponds to the maximum allowed deviation of TSX from the reference orbit in the plane perpendicular to the flight direction. The implemented TSX orbit control concept is described in [31].

Within ten years of TerraSAR-X operations, more than 1400 orbit control maneuvers have been performed. With increasing solar activity strong variations in solar flux and geomagnetic activity occurred, which significantly affected the air density, and hence the orbit decay.

Figure 14 (top) depicts the F10.7 cm solar flux evolution, which serves as an indicator for the solar activity, over the years 2008 to 2017. The diagrams below summarize all of the in-plane maneuvers that were performed in the same period (middle) and the achieved cross-track deviation from the reference orbit (bottom). Obviously, during low solar activity in 2008 and 2009, the maneuvers were relatively small ($\leq$1 cm/s) and the typical period between two successive maneuvers were 10 to 14 days. In contrast, maneuvers with sizes of up to 5 cm/s and maneuver cycles of two to three days became necessary to precisely control the TSX orbit during periods of high solar activity, e.g., at the end of 2011 and in 2014.

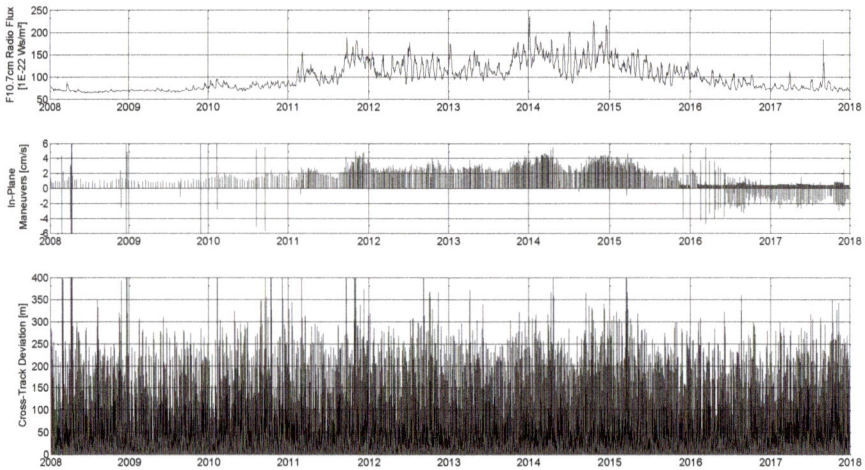

Figure 14. F10.7 cm solar flux (**top**), TSX in-plane maneuver activity (**middle**), and achieved cross-track deviation from the reference orbit (**bottom**) in period 2008–2017.

At the end of 2015 a novel orbit control strategy called "distributed formation flying" was tested and it has been operationally implemented since February 2016. Since then, TSX has contributed to the formation maintenance by means of a small but daily orbit raising maneuver in the order of 3 to 6 mm/s, depending on the formation geometry. In this way, the formation flight for TanDEM-X can be fully supported although the cold-gas, which was previously used by TanDEM-X for formation maintenance, has been almost depleted. Furthermore, the TerraSAR-X orbit control accuracy is not negatively influenced, as can be seen in the bottom of Figure 14. The daily orbit raising partially balances the natural orbit decay. However, additional orbit lowering and eccentricity correction maneuvers are necessary in order to stay within the 250 m control tube. These maneuvers are in anti-flight direction and they can be seen in the middle of Figure 14 as negative spikes in 2016 and 2017.

Note the negative in-plane maneuvers before 2011 have a different cause. The original satellite safe mode implementation made use of the thrusters for coarse attitude and rate control. Safe mode drops, e.g., because of an on-board computer boot, changed the orbit and the maneuvers for the re-acquisition of the reference orbit became necessary. Fortunately, the on-board software was upgraded at the end of 2010 to feature a force-free safe mode based on magnetorquers, as described in Section 6.

Besides the drag make-up maneuvers, three to five inclination control maneuvers are performed per year, with maneuver sizes between 10 and 30 cm/s. Furthermore, every year, one to two critical close approaches with man-made objects require active collision avoidance. In summary, the implemented reference orbit and the orbit control concept have proved to work remarkably well, and more than 99% of the time TSX has been staying inside the 250 m control tube.

8. Performance Monitoring

The quality of the delivered data is the most important benchmark of a SAR mission. Strong effort has to be undertaken, in order to ensure a stable quality of the system and resulting data. For this purpose, a verification and monitoring system was implemented in TerraSAR-X and TanDEM-X. With the short term analysis of instrument telemetry data and all data takes executed during the mission, malfunctions in the satellite systems, and especially in the SAR instruments are detected. Short-term analysis is performed in two different ways: instrument monitoring and data take verification. A long-term monitoring aims on detecting trends and evolutions that may lead to a malfunction in future. These analyses are explained in detail and with examples below.

8.1. Instrument Monitoring

Instrument Monitoring collects and visualizes information about the actual state of the radar instruments and performs automatic checks of measurements against a parameter limit matrix. For that purpose, housekeeping data, which is down-linked via S-band, is checked and analyzed both, automatically and manually, in order to examine instrument health. In critical cases, dedicated procedures are immediately initiated according to the instrument handbook. The instrument-specific housekeeping data is replicated and is stored in a central database, where it is one of the main sources of information for long-term monitoring. The instrument onboard timeline is continuously reviewed w.r.t. data take and command execution state, SAR data downlink history, etc.

8.2. Data Take Verification

The Data Take Verification functionality provides an immediate feedback on data take quality and overall performance of the SAR system after commanding, execution, and SAR screening of the raw data of a TerraSAR-X or TanDEM-X acquisition. It combines information from data take ordering, commanding, as well as the data reception and processing chain, and it therefore provides additional value on top of the regular SAR data screening.

Three main tasks are in the scope of Data Take Verification. The first one is the evaluation of the SAR screening results of an acquisition, where the raw data statistics of in-phase and quadrature channels are checked against predefined quality limits. These checks can identify instrument calibration issues like channel bias or imbalances, but also issues in the commanding like non-optimal receiver gain settings, which would result in insufficient raw data saturation or even overflows in the A/D converters. Another prominent quality parameter is the Doppler centroid that is estimated from the raw data, which is a sensitive parameter for the whole SAR system.

Another check is the completeness of scene coverage w.r.t. the user-requested scene, which depends not only on precise instrument switch-on and switch-off, but also on the correct commanding of the radar echo window. As an example, Figure 15 shows a control plot for an acquisition in Stripmap mode taken over Morocco. The red polygon indicates the scene, as requested by a user, whereas the green polygon shows the boundaries of the recorded and processed scene.

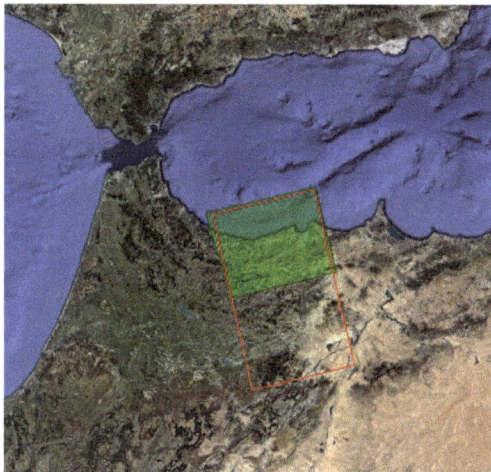

Figure 15. Example for scene coverage verification after data take acquisition, visualized in Google Earth. The red polygon indicates the requested scene, whereas the green polygon shows the acquired scene. In this case, a considerable part of the data take is not available, as the downlink was affected by severe weather conditions.

In the third step of the verification of a data take, or data take pair in the case of a TanDEM-X acquisition, the correct on-board execution of each commanded data take is verified, down to the level of single pulses, via the correlation of the data take command parameters, as generated by the Request-to-Command Converter (R2CC), with the range lines actually being recorded and downlinked by the radar instruments. This check has the potential to identify issues in the complex interleaved commanding of imaging pulses with other instrument-specific pulse types, like warm-up sequences, calibration pulses, and interleaved inter-satellite synchronization in the case of TanDEM-X. It does also provide valuable information in the analysis of issues regarding missing or corrupted range lines in the SAR data.

Throughout the mission, data take verification provided valuable routine monitoring. Due to the reliability of the whole SAR system, major issues were detected in rare cases only. Yet, the system was used extensively during the commissioning phases of the satellites, where it allowed for verifying the SAR system performance and to derive and fine-tune initial calibration parameters.

8.3. Long Term System Monitoring

Long-term system monitoring (LTSM) is an important task to ensure the quality of the SAR products and the appropriate operation of the satellite and the instrument. The LTSM visualizes the most significant parameters of the satellite system.

There are a number of parameters that are checked for each data take in order to monitor the quality of a single acquisition as described in the previous section. In contrast, the LTSM shows the long-term evolution of the system parameters in order to identify trends in the data and to be able to take countermeasures before the occurrence of critical situations.

The most important parameters monitored by LTSM are:

- Instrument parameters:

  - Utilization of the satellites
  - Quality of the Replica amplitude and the Bias of the In-phase and Quadrature Channel
  - Calibration parameters like geometrical accuracy or the absolute calibration factor, including antenna pattern accuracy
  - Status of the Transmit/Receive Modules
  - Drift of the radar frequency
  - Drift of the Doppler of the SAR signal
  - Coverage fulfillment of data takes compared to the user requests

- Satellite parameters:

  - Amount of switches of the traveling wave tube amplifier for X-band downlink
  - Battery performance and utilization
  - Bit error rates of the Solid State Mass Memory

- Mission parameters:

  - Mission phases (for information)
  - SAR outages (for information)
  - Parameters concerning the TanDEM-X Mission like DEM performance

Figure 16 shows, as an example, the evolution of the In-Phase Channel Bias, as visualized by the Long-Term System Monitoring. The In-Phase Channel Bias is a measure for the correct utilization of the dynamic range of the receivers. A drift of 20% over six years was observed with the long-term evaluation. In June 2014, the origin for the In-Phase Channel Bias was readjusted in the system, and the bias is since then again well centered within the dynamic range.

In general, the TerraSAR-X satellite still performs very well. The quality of the data is as good as at the beginning of the mission, both in terms of overall performance and in terms of radiometric accuracy [32].

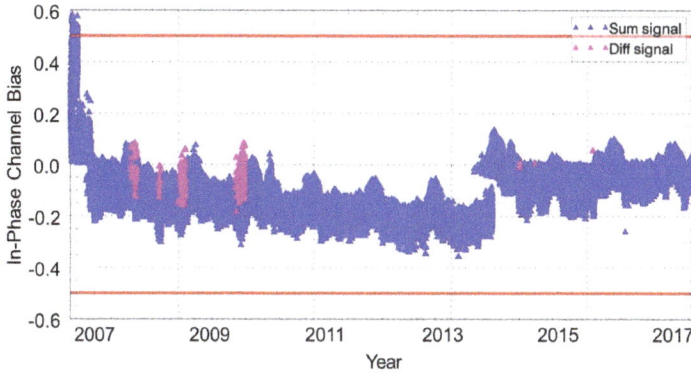

Figure 16. Bias of the In-phase Channel and its readjustment in 2014.

9. TerraSAR-X Mission Utilization

Since the beginning of the operational phase, the number of yearly acquired data takes for basic product generations has been continuously increasing, with only one exception in 2011. When compared to the operations that started in 2008, the acquisitions have almost tripled. Figure 17 visualizes the yearly numbers per imaging mode. Adding the experimental TerraSAR-X-like products we have close to 226,000 acquisitions which are available for user catalog ordering in EOWEB under the headings *TSX-1 Products, TSX-1 Experimental Products, TanDEM-X Pursuit TSX-1 Like,* and *TanDEM-X Pursuit TSX-1 like Experimental.*

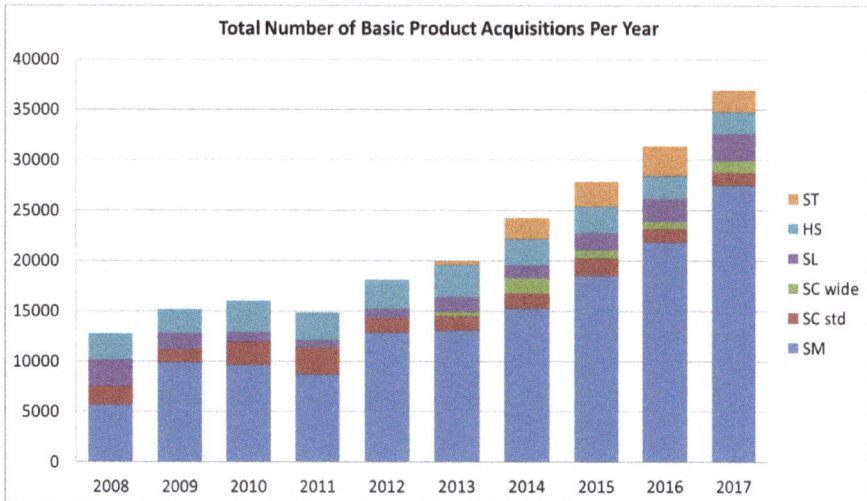

Figure 17. Total number of basic product acquisitions per year.

The TerraSAR-X mission philosophy to let a user decide which acquisitions are performed under what conditions even comprises data takes taken in left-looking mode. Ordering of these acquisitions is consistently integrated into the nominal ordering and the production workflow. In total, over 3000 left-looking data takes covering all imaging modes were acquired. Of particular benefit are those that were taken at latitude south of −80° over Antarctica [33]. Figure 18 shows the left-looking Stripmap acquisitions over Antarctica, as executed upon user request.

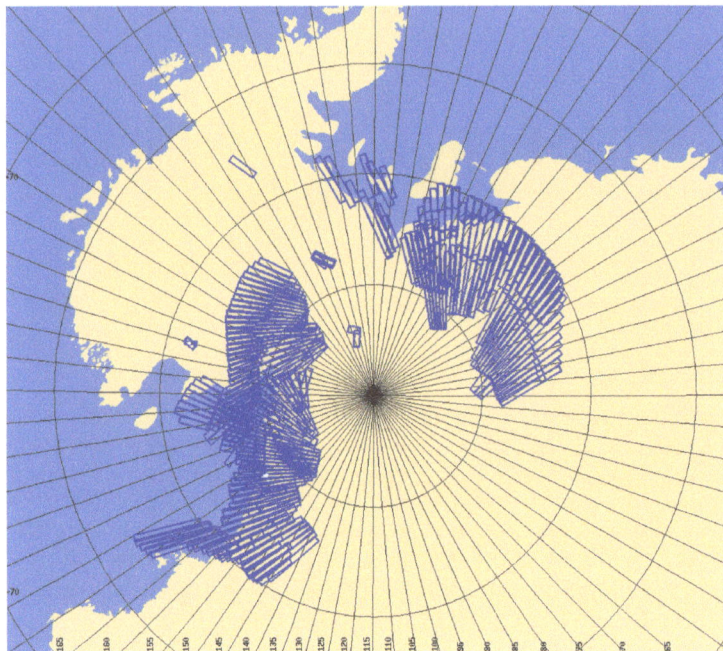

Figure 18. Map visualization of Stripmap left-looking data takes over Antarctica.

10. Conclusions

A decade of successful TerraSAR-X operation has proven that the basic design of the overall system, consisting of space and ground segment, is both robust and flexible enough to meet changing requirements and to fulfill the desires of the scientific and commercial community during such a long period of time. Also, thanks to the great commitment of the ground segment team, all of the modifications could be made without any particular disruption or interruption of operations. The interaction of these factors has enabled ensuring the successful use of the mission by scientists and commercial users for more than ten years. This will continue to be the case since radar performance and calibration of the individual satellites is still well within specification, and no indication of any degradation is noticeable, a fact that is remarkably reflected in the consistently excellent quality of the SAR products. From today's point of view, the satellites and the ground segment are operating without restrictions and the irretrievable resources, specifically the propellant and the battery capacity, allow for operating both satellites for another five years.

Author Contributions: S.B. assumes the role of the Mission Manager performing administrative and organizational tasks, he monitors the system changes and wrote Sections 1 and 10. B.S. is the Ground Segment Systems Engineer. She coordinates and tests the technical system changes and wrote Sections 2, 4, and 9. T.F. is in charge of the SAR product portfolio and user product quality, he wrote Section 3. J.M. managed the commissioning phase and conceived experimental new SAR modes, he wrote Sections 5.2–5.5. R.K. is in charge of all flight dynamics and orbit control aspects, he wrote Section 7. E.M. is the manager of the Mission Operations Segment and in charge of the satellite operations, she wrote Section 6. J.B. is the manager of the System Engineering and Calibration Segment and in charge of the performance monitoring, M.B. is in charge of the overall SAR performance and image quality, they wrote Section 8. F.M. developed the Mission Planning System in order to generate executable acquisition timelines, he contributed to Sections 2, 4, and 6. E.S. manages the Neustrelitz ground station and supervises the SAR operations, he contributed NRT aspects in Section 2. H.B. is in charge of the SAR image processing, respectively the TerraSAR-X Multi-mode SAR Processor (TMSP) and contributed to Section 3. U.S. is in charge of the commanding and monitoring of the radar instrument, he implements command sequences for even the weirdest new SAR modes and wrote Section 5.1.

Acknowledgments: The authors would like to thank the entire TerraSAR-X/TanDEM-X team, which ensures the smooth running of the mission and enables the scientific and commercial exploitation of the results.

Conflicts of Interest: The authors declare no conflict of interest.

Abbreviations

AIS	Automatic Identification System
ATI	Along Track Interferometry
ATIS	Along Track Interferometer Switching
Az.	Azimuth
BiDiSAR	Bidirectional Synthetic Aperture Radar
DEM	Digital Elevation Map
DLR	Deutsches Zentrum für Luft- und Raumfahrt e.V.
DRA	Dual-Receive Antenna
DT	Data Take
DTAR	Distributed Target Ambiguity Ratio
EEC	Enhanced Ellipsoid Corrected
EOWEB	Earth Observation Data Service
Esrange	European Space and Sounding Rocket Range
ESRI	Environmental Systems Research Institute
FTP	File Transfer Protocol
GARS	German Antarctica Receiving Station O'Higgins
GEC	Geocoded Ellipsoid Corrected
GFZ	Deutsches GeoForschungsZentrum
GML	Geography Markup Language
GPS	Global Positioning System
HS	High Resolution Spotlight
IGOR	Integrated GPS and Occultation Receiver
ISSF	Inuvik Satellite Station Facility
KIR	Kiruna Ground Station
KML	Keyhole Markup Language
KMZ	Keyhole Markup Language Zipped
KSAT	Kongsberg Satellite Services
LNPC	Look-Normalized Pattern Compensation
LCT	Laser Communication Terminal
LTSM	Long-term System Monitoring
MGD	Multi-looked Ground-range Detected
NFIRE	Near Field Infrared Experiment
NRT	Near Real Time
NSG	Neustrelitz Ground Station
pol.	Polarization
PRF	Pulse Repetition Frequency
R2CC	Request-to-Command Converter
RE	Radiometrically Enhanced
Rg.	Range
SAINT	SAR AIS Integrated NRT Toolbox
SAR	Synthetic Aperture Radar
SE	Spatially Enhanced

SL	Sliding Spotlight
SM	Stripmap
SNR	Signal-to-Noise Ratio
SRA	Single-Receive Antenna
SSC	Single-look Slant-range Complex
SSMM	Solid State Mass Memory
ST	Staring Spotlight
SvalSat	Svalbard Satellite Station
sync	Synchronization
TDX	TanDEM-X Satellite
TMSP	TerraSAR-X Multi-mode SAR Processor
TOPSAR	Terrain Observation with Progressive Scans SAR
TSX	TerraSAR-X Satellite
TT&C	Telemetry, Tracking and Command

References

1. Werninghaus, R.; Buckreuss, S. The TerraSAR-X Mission and System Design. *IEEE Trans. Geosci. Remote Sens.* **2010**, *48*, 606–614. [CrossRef]
2. Fritz, T.; Eineder, M. *TerraSAR-X Basic Product Specification*; TX-GS-DD-3302, Issue 1.9, 09.10.2013; German Aerospace Center (DLR), Space Agency, National Contact point Space: Bonn, Germany, 2013.
3. Pitz, W.; Miller, D. The TerraSAR-X Satellite. *IEEE Trans. Geosci. Remote Sens.* **2010**, *48*, 615–622. [CrossRef]
4. Buckreuss, S.; Schättler, B. The TerraSAR-X Ground Segment. *IEEE Trans. Geosci. Remote Sens.* **2010**, *48*, 623–632. [CrossRef]
5. Mittermayer, J.; Schättler, B.; Younis, M. TerraSAR-X Commissioning Phase Execution Summary. *IEEE Trans. Geosci. Remote Sens.* **2010**, *48*, 649–659. [CrossRef]
6. Schättler, B.; Kahle, R.; Steinbrecher, U.; Metzig, R.; Balzer, W.; Zink, M. Extending the TerraSAR-X Ground Segment for TanDEM-X. In Proceedings of the 8th EUSAR 2010, Aachen, Germany, 7–10 June 2010; VDE Verlag: Berlin, Germany, 2010.
7. Schättler, B.; Kahle, R.; Metzig, R.; Steinbrecher, U.; Zink, M. The Joint TerraSAR-X/TanDEM-X Ground Segment. In Proceedings of the IEEE Geoscience and Remote Sensing Symposium 2011, Vancouver, BC, Canada, 24–29 July 2011; pp. 2298–2301. [CrossRef]
8. Kahle, R.; Schlepp, B.; Aida, S.; Kirschner, M.; Wermuth, M. Flight Dynamics Operations of the TanDEM-X Formation. In Proceedings of the 12th International Conference on Space Operations, Stockholm, Sweden, 11–15 June 2012.
9. Kahle, R.; Runge, H.; Ardaens, J.-S.; Suchandt, S.; Romeiser, R. Formation flying for along-track interferometric oceanography—First in-flight demonstration with TanDEM-X. *Acta Astronaut.* **2014**, *99*, 130–142. [CrossRef]
10. Kahle, R.; Schlepp, B. Extending the TerraSAR-X Flight Dynamics System for TanDEM-X. In Proceedings of the 4th International Conference on Astrodynamics Tools and Techniques, Madrid, Spain, 3–6 May 2010.
11. Mrowka, F.; Geyer, M.P.; Lenzen, C.; Spörl, A.; Göttfert, T.; Maurer, E.; Wickler, M.; Schättler, B. The Joint TerraSAR-X / TanDEM-X Mission Planning System. In Proceedings of the IEEE Geoscience and Remote Sensing Symposium 2011, Vancouver, BC, Canada, 24–29 July 2011. [CrossRef]
12. Schättler, B.; Mrowka, F.; Schwarz, E.; Lachaise, M. The TerraSAR-X Ground Segment in Service for Nine Years: Current Status and Recent Extensions. In Proceedings of the IEEE Geoscience and Remote Sensing Symposium 2016, Bejing, China, 10–15 July 2016; pp. 1400–1403. [CrossRef]
13. Schwarz, E.; Krause, D.; Berg, M.; Daedelow, H.; Maass, H. Near Real Time Applications for Maritime Situational Awareness. In Proceedings of the 36th International Symposium on Remote Sensing of Enviroment, Berlin, Germany, 11–15 May 2015. [CrossRef]
14. Lehner, S.; Tings, B. Maritime NRT products using TerraSAR-X and Sentinal-1 imagery. In Proceedings of the 36th International Symposium on Remote Sensing of Enviroment, Berlin, Germany, 11–15 May 2015. [CrossRef]
15. Fritz, T.; Eineder, M. TerraSAR-X Experimental Product Specification. TX-GS-DD-3303, Issue 1.3, 06.10.2006. Available online: http://sss.terrasar-x.dlr.de/docs/TX-GS-DD-3303.pdf (accessed on 28 May 2018).

16. Breit, H.; Fischer, M.; Balss, U.; Fritz, T. TerraSAR-X Staring Spotlight Processing and Products. In Proceedings of the 10th EUSAR 2014, Berlin, Germany, 3–5 June 2014; VDE Verlag: Berlin, Germany, 2014; pp. 193–196.

17. Duque, S.; Breit, H.; Balss, U.; Parizzi, A. Absolute Height Estimation Using a Single TerraSAR-X Staring Spotlight Acquisition. *IEEE Geosci. Remote Sens. Lett.* **2015**, *12*, 1735–1739. [CrossRef]

18. Balss, U.; Breit, H.; Fritz, T. Noise-Related Radiometric Correction in the TerraSAR-X Multimode SAR Processor. *IEEE Trans. Geosci. Remote Sens.* **2010**, *48*, 741–750. [CrossRef]

19. Balss, U.; Breit, H.; Fritz, T.; Steinbrecher, U.; Gisinger, C.; Eineder, M. Analysis of internal timings and clock rates of TerraSAR-X. In Proceedings of the IEEE Geoscience and Remote Sensing Symposium 2014, Quebec City, QC, Canada, 13–18 July 2014; pp. 2671–2674. [CrossRef]

20. Balss, U.; Gisinger, C.; Eineder, M. Measurements on the Absolute 2-D and 3-D Localization Accuracy of TerraSAR-X. *Remote Sens.* **2018**, *10*, 656. [CrossRef]

21. Romeiser, R.; Runge, H.; Suchandt, S.; Kahle, R.; Rossi, C.; Bell, P.S. Quality Assessment of Surface Current Fields From TerraSAR-X and TanDEM-X Along-Track Interferometry and Doppler Centroid Analysis. *IEEE Trans. Geosci. Remote Sens.* **2014**, *52*, 2759–2772. [CrossRef]

22. Mrowka, F.; Göttfert, T.; Wörle, M.; Schättler, B.; Stathopoulos, F. TerraSAR-X/TanDEM-X Mission Planning System: Realizing new Customer Visions by Applying new Upgrade Strategies. In Proceedings of the 14th International Conference on Space Operations 2016, Deajeon, Korea, 16–20 May 2016. [CrossRef]

23. Mittermayer, J.; Wollstadt, S.; Prats, P.; Scheiber, R. The TerraSAR-X staring spotlight mode concept. *IEEE Trans. Geosci. Remote Sens.* **2014**, *52*, 3695–3706. [CrossRef]

24. Prats-Iraola, P.; Scheiber, R.; Rodriguez-Cassola, M.; Mittermayer, J.; Wollstadt, S.; De Zan, F.; Bräutigam, B.; Schwerdt, M.; Reigber, A.; Moreira, A. On the processing of very high resolution spaceborne SAR data. *IEEE Trans. Geosci. Remote Sens.* **2014**, *52*, 6003–6016. [CrossRef]

25. Kraus, T.; Bräutigam, B.; Mittermayer, J.; Wollstadt, S.; Grigorov, C. TerraSAR-X Staring Spotlight Mode Optimization and Global Performance Predictions. *IEEE J. Sel. Top. Appl. Earth Observ. Remote Sens.* **2016**, *9*, 1015–1027. [CrossRef]

26. Meta, A.; Mittermayer, J.; Prats, P.; Scheiber, R.; Steinbrecher, U. TOPS Imaging with TerraSAR-X: Mode Design and Performance Analysis. *IEEE Trans. Geosci. Remote Sens.* **2009**, *48*, 759–769. [CrossRef]

27. Mittermayer, J.; Wollstadt, S.; Prats, P.; López-Dekker, P.; Krieger, G.; Moreira, A. Bidirectional SAR imaging mode. *IEEE Trans. Geosci. Remote Sens.* **2013**, *51*, 601–614. [CrossRef]

28. Mittermayer, J.; Kraus, T.; López-Dekker, P.; Prats, P.; Krieger, G.; Moreira, A. Wrapped Staring Spotlight SAR. *IEEE Trans. on Geosc. Remote Sens.* **2016**, *54*, 5745–5764. [CrossRef]

29. Hofmann, H.; Maurer, E. TanDEM-X Formation Flight Operation. In Proceedings of the 4th International Conference on Spacecraft Formation Flying Missions & Technologies, Montreal, QC, Canada, 18–20 May 2011.

30. D'Amico, S.; Arbinger, C.; Kirschner, M.; Campagnola, S. Generation of an Optimum Target Trajectory for the TerraSAR-X Repeat Observation Satellite. In Proceedings of the 18th International Symposium on Space Flight Dynamics, Munich, Germany, 11–15 October 2004; pp. 137–142.

31. Kahle, R.; D'Amico, S. The TerraSAR-X Precise Orbit Control—Concept and Flight Results. In Proceedings of the 24th International Symposium on Space Flight Dynamics, Laurel, MD, USA, 5–9 May 2014.

32. Schwerdt, M.; Schmidt, K.; Klenk, P.; Tous Ramon, N.; Rudolf, D.; Raab, S.; Weidenhaupt, K.; Reimann, J.; Zink, M. Radiometric Performance of TerraSAR-X Mission since Launch—More than Ten Years in Operation. *Remote Sens.* **2018**, *10*, 754. [CrossRef]

33. Floricioiu, D.; Jaber, W.A.; Jezek, K. TerraSAR-X and TanDEM-X Observations of the Recovery Glacier System, Antarctica. In Proceedings of the IEEE Geoscience and Remote Sensing Symposium 2014, Quebec City, QC, Canada, 13–18 July 2014; pp. 4852–4855. [CrossRef]

remote sensing

MDPI

Article

Radiometric Performance of the TerraSAR-X Mission over More Than Ten Years of Operation

Marco Schwerdt *, Kersten Schmidt, Patrick Klenk, Núria Tous Ramon, Daniel Rudolf, Sebastian Raab, Klaus Weidenhaupt, Jens Reimann and Manfred Zink

German Aerospace Center (DLR), Microwaves and Radar Institute, Oberpfaffenhofen,
D-82234 Wessling, Germany; kersten.schmidt@dlr.de (K.S.); patrick.klenk@dlr.de (P.K.);
nuria.tousramon@dlr.de (N.T.R.); daniel.rudolf@dlr.de (D.R.); sebastian.raab@dlr.de (S.R.);
klaus.weidenhaupt@dlr.de (K.W.); jens.reimann@dlr.de (J.R.); manfred.zink@dlr.de (M.Z.)
* Correspondence: marco.schwerdt@dlr.de; Tel.: +49-8153-28-3533

Received: 23 April 2018; Accepted: 13 May 2018; Published: 15 May 2018

Abstract: The TerraSAR-X mission, based on two satellites, has produced SAR data products of high quality for a number of scientific and commercial applications for more than ten years. To guarantee the stability and the reliability of these highly accurate SAR data products, both systems were first accurately calibrated during their respective commissioning phases and have been permanently monitored since then. Based on a short description of the methods applied, this paper focuses on the radiometric performance including the gain and phase properties of the transmit/receiver modules, the antenna pattern checked by evaluating scenes acquired over uniformly distributed targets and the radiometric stability derived from permanently deployed point targets. The outcome demonstrates the remarkable performance of both systems since their respective launch.

Keywords: TerraSAR-X; internal calibration; geometric and radiometric calibration; antenna model verification; antenna pointing determination; radiometric accuracy; calibration targets; long term performance monitoring

1. Introduction

More than ten years ago, on 15 June 2007, the first German synthetic aperture radar (SAR) mission TerraSAR-X for commercial and scientific application was started by launching the first satellite, called TSX. Three years later, in June 2010, the mission was expanded by an additional satellite, TDX. Since then, both satellites have been flying in a close formation in a sun-synchronous dusk-dawn orbit at 514 km altitude to fulfill the tasks of two different missions in parallel:

- the TerraSAR-X mission for providing single multi-mode X- Band SAR data in different operation modes [1] and
- in bistatic operation, the TanDEM-X mission to generate a new global digital elevation model on a 12-m grid and a vertical accuracy better than two meters [2].

The satellites feature an advanced high-resolution X-Band SAR instrument operated at 9.65 GHz and enabling the operation in Spotlight, Stripmap and ScanSAR mode, in different polarizations and over a wide range of incidence angles. For these various acquisition modes, their active phased array antenna electronically steers and shapes the patterns in the azimuth and elevation direction. The antenna consists of 12 panels in azimuth direction with each panel comprising 32 sub-array radiators in elevation direction, whereby each sub-array is fed by its own active transmit/receiver module (TRM). The instrument combines the ability to acquire high resolution images for detailed analysis as well as wide swath images for overview applications. The geometric resolution varies from

0.24 m for Staring Spotlight [3], 1 m for Spotlight, 3 m for nominal Stripmap, 16 m for ScanSAR and 40 m for Wide ScanSAR products. The image width ranges from 4.6 km (Staring Spotlight) to 266 km (Wide ScanSAR). There are over 1000 possible product variations, which result from the combination of different imaging modes, polarizations and elevation angles [4].

This paper focuses on the radiometric performance of the two satellites. A pre-requisite to ensure accurate and reliable SAR data products over the whole mission time is first, an accurate calibration of the system and, then, a permanent monitoring of relevant system parameters. During their commissioning phases, TSX (in 2007) and TDX (in 2010) could be accurately calibrated with outstanding results [5,6]. Since then, both SAR systems have been permanently monitored to detect changes in their performance, like degradation of the satellite hardware and to guarantee reliable and correct operation of the instruments. However, during a second TSX calibration campaign executed for the so called Dual Receive Antenna mode in 2009, a high radiometric stability could be verified two years after launch [7].

The evaluation of long-term system monitoring (LTSM) parameters, like instrument characteristics or the antenna patterns, verify an excellent stability of TSX and TDX. This LTSM has been ensuring a consistent product quality for more than ten years. The success of this performance is based on the innovative design and precise manufacturing of the satellite systems [8,9] on the one hand and on the other on innovative methods, accurate reference targets and the strategy for calibrating and monitoring TSX and TDX over lifetime [5,10–15]. The radiometric performance over lifetime of both systems is analyzed in three steps:

- characterizing the gain and phase stability of TRMs (Section 2) by means of coded calibration pulses,
- monitoring the antenna characteristics by evaluating scenes acquired over distributed targets like the Amazon rain forest (Section 3) and
- analyzing the radiometric stability by means of impulse response functions derived from permanently installed corner reflectors (Section 4).

Beyond that, the quality of SAR data products have also been monitored since launch (Section 5).

2. Stability of Individual Transmit/Receiver Modules

For analyzing the stability of the TRMs within the instrument front end, the pseudo noise (PN) gating method is applied [16]. This method was developed and established in the frame of TerraSAR-X for characterizing the amplitude and phase settings of individual TRMs of an active phased array antenna, while all modules are in operation. Thus, all modules are characterized simultaneously under most realistic conditions—in contrast to switch on/off individual modules resulting in different power loads and consequently to nonrealistic operating conditions. Under the control of the Data and Control Electronics, the phase of each TRM is individually shifted between successive calibration pulses according to a certain code sequence. Based on these specific PN-gating data takes (L0 products), the settings of all TRMs are regularly monitored on transmission and on reception in flight.

Possible drift effects can be found by depicting amplitude and phase trends over time. For example, amplitude and phase deviations with respect to a reference value are plotted in Figure 1 for TSX versus data take execution time for each of the 384 TRMs on transmission. The reference value for each TRM was derived in flight at the beginning of the commissioning phase. The figure shows that all TRMs have been working within the established limits (red lines) so far, and only one outlier was detected in October 2012. To further monitor the TRMs status in detail at that time, several extra PN-gating data takes were ordered, but no further anomalies could be observed, as shown in Figure 1. Hence, no trend can be observed, indicating the stability of the TSX instrument and the TRM settings, respectively.

This characterization of individual TRMs has been performed for TSX and TDX over mission time since launch in 2007 and 2010, respectively. The results, averaged for each TRM over mission elapsed time, are shown in Figure 2 for the measured amplitude settings and in Figure 3 for the phase settings; red lines and black error bars indicate the mean value and standard deviation for each module, respectively.

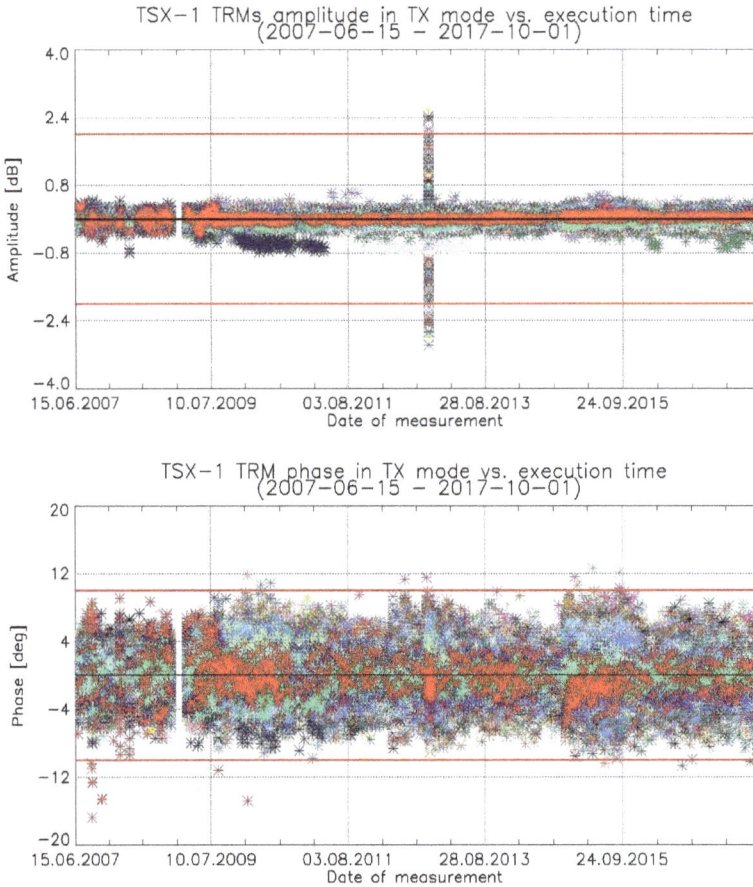

Figure 1. Amplitude and phase deviation versus time of all 384 transmit/receiver modules of TSX measured in flight since launch. Each colored dot along the *y*-axis indicates the measured deviation of one of the 384 TRMs at that time. Red horizontal lines indicate the established limits.

The figures show that the TRM amplitude and phase settings are very stable for transmission and reception: no instrument degradation is detectable for both satellites. Averaging the standard deviations over all modules yields an overview of the instrument quality for each parameter. The results are summarized in Table 1. The amplitude deviation stays below 0.2 dB, the phase deviation below 2.0 °. This is in the order of the accuracy of the PN gating method. It can be summarized that the TRMs have been working in a very stable manner over mission elapsed time, i.e., for more than 10 years for TSX and seven years for TDX.

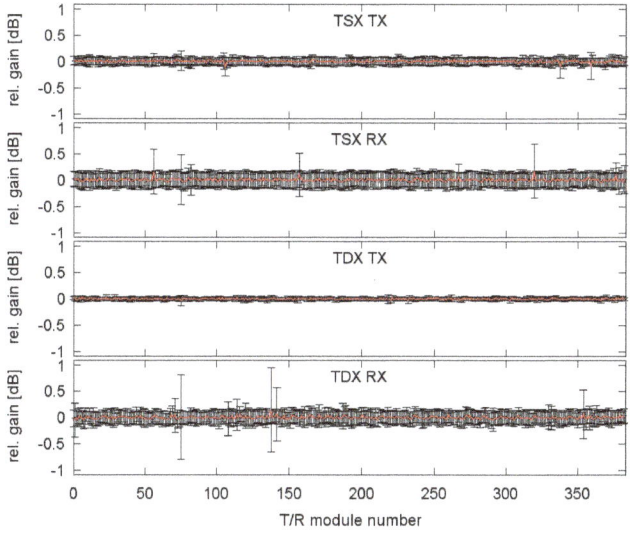

Figure 2. Amplitude deviation versus TRM number on transmission (TX) and reception (RX) for each of the 384 TRMs measured in flight since launch of TSX in 2007 (**top**) and TDX in 2010 (**bottom**). Red curve: the mean value averaged for each TRM over mission elapsed time, black error bars: corresponding standard deviation.

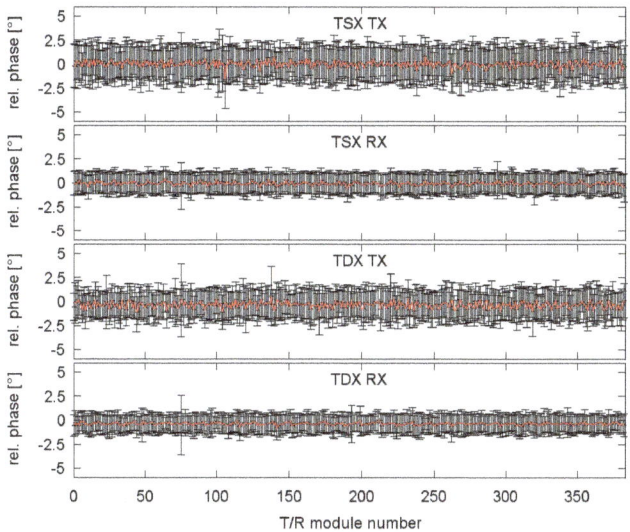

Figure 3. Phase deviation versus TRM number on transmission (TX) and reception (RX) for each of the 384 TRMs measured in flight since launch of TSX in 2007 (**top**) and TDX in 2010 (**bottom**). Red curve: the mean value averaged for each TRM over mission elapsed time, black error bars: corresponding standard deviation.

Table 1. Statistics of amplitude and phase TRM deviations on transmission and reception for TSX and TDX since launch.

Mean of the σ-values of all TRMs	TX		RX	
	Amplitude (dB)	Phase (°)	Amplitude (dB)	Phase (°)
TSX	0.08	1.97	0.16	1.16
TDX	0.03	1.67	0.14	1.05

3. Antenna-Pattern Monitoring

Besides internal calibration, the antenna patterns also have to be monitored to detect any degradation of the front-end panels; this concerns especially potential degradations of the antenna wave guides as these are not covered by internal calibration. For estimating the front-end quality, gamma profiles derived from SAR images acquired over the Amazon rainforest (Figure 4) are compared with antenna reference patterns calculated by means of a precise antenna model [17].

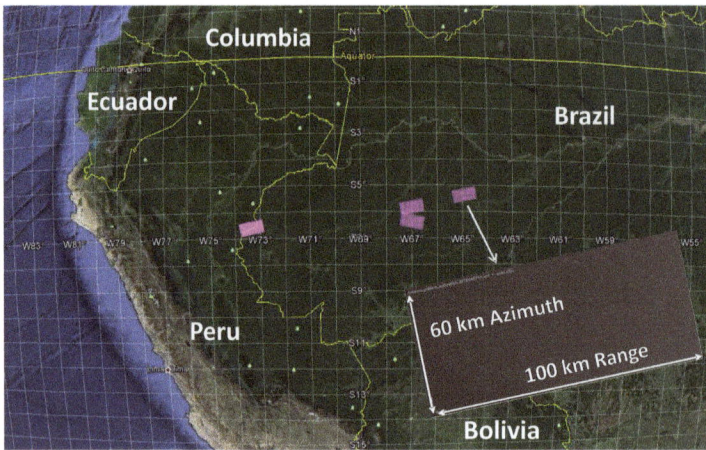

Figure 4. Four acquisition areas (pink) for long term system monitoring of the antenna over the Amazon rainforest serving as quite homogeneous scatterer as seen in the SAR image preview.

According to the strategy described in [5], the instrument is operated in ScanSAR mode for this monitoring task. For ScanSAR acquisitions, the beam is switched sequentially from burst to burst between four neighbouring subswaths to get a broader swath width compared to normal Stripmap acquisitions. By generating an un-normalized gamma profile for each of the four subswaths, the relative gain deviation from subswath to subswath can be determined (in Level1B products, the four subswaths are still separated but combined in a higher level product). Thus, not only the pattern shape of each single beam itself, but also the gain offset between different beams can be verified in flight. For this in-flight verification of the shape and the gain offsets, a strong requirement of ±0.2 dB was defined for the antenna model. The reason for that number was not only driven by the radiometric accuracy budget, but also by the visibility of gain offsets between adjoining beams in ScanSAR images.

An essential assumption for the analysis is a nearly constant gamma profile from rainforest backscattering. This was analyzed for different areas and acquisition periods by TSX and TDX in [18]. In this analysis, higher gamma values were measured for morning acquisitions (descending orbit) compared to afternoon acquisitions (ascending orbit). Hence, an absolute gain comparison over time is infeasible. Nevertheless, the rainforest can still be used to estimate the relative radiometric accuracy and the radiometric stability of the antenna pattern as each ScanSAR acquisition is evaluated separately. A typical measured uncorrected profile together with the antenna reference pattern is depicted on the top panel of Figure 5 for a single ScanSAR acquisition. Note that absolute radiometric calibration was performed by measuring both systems against accurate reference targets [5–7,11].

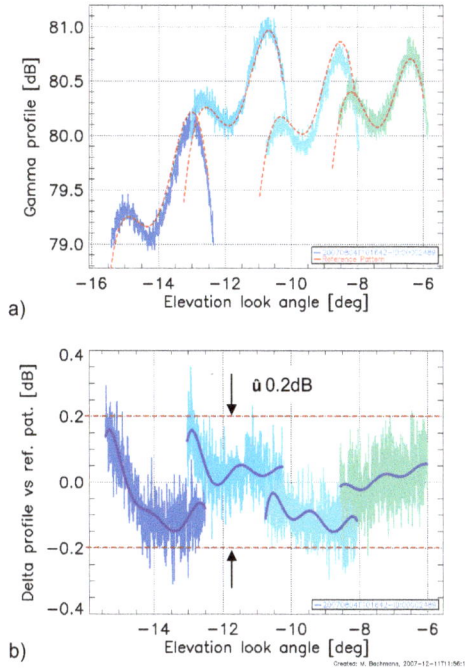

Figure 5. (**a**) antenna reference pattern (dashed red) and gain profiles (blue to green) derived from a ScanSAR image without pattern correction. (**b**) difference between reference and measured pattern to verify the antenna model w.r.t. pattern shape and beam-to-beam gain offset (blue lines are fits of the difference).

In the next step, the minimum and the maximum value as well as the standard deviation are derived from the difference between the reference and the measured antenna patterns, as shown on the bottom panel of Figure 5. Based on these statistic values derived from each set of ScanSAR beams acquired over the Amazon rainforest, the antenna patterns and consequently the front-end panels, especially the antenna wave guides have been monitored since launch of the satellites. This is shown in Figure 6: blue for TSX and green for TDX. Here, the minimum and maximum deviation between the measured and the calculated set of Scan-SAR patterns (four beams) per acquisition are represented by an error bar and the standard deviation by a triangle. Furthermore, the mean values are subtracted to focus on relative radiometric accuracy.

The standard deviation observed from mid-2014 until mid-2017 is slightly higher than during other mission times. This can be traced back to including a larger number of data takes (but acquired over the same test sites shown in Figure 4) in the LTSM analysis during that period, which were affected by weather-related effects such as heavy thunderstorms across the Amazon rainforest. From summer 2017 onwards, a stricter screening policy was adopted again, reducing the observed standard deviation to pre-2014 levels. Nevertheless, the timeline shows a very stable behavior of the antenna pattern. While the standard deviations are almost always inside a limit of +/−0.2 dB (with very few execeptions), the extreme values do not exceed +/−0.3 dB (excluding data takes disturbed by poor weather conditions). Finally, there is no remarkable difference between TSX and TDX.

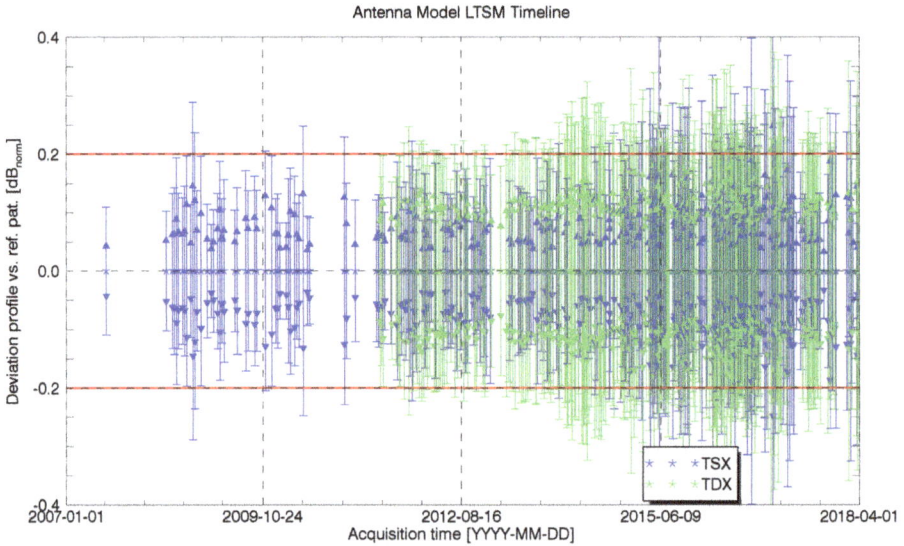

Figure 6. Timeline of antenna pattern statistics derived from SAR scenes acquired over the Amazon rainforest by using ScanSAR mode for TSX (blue) and TDX (green).

4. Radiometric Stability

The radiometric stability is monitored by deriving the radar cross section (RCS) of permanently installed corner reflectors. For this long-term system monitoring, three corner reflectors with a leg length of 1.5 m are permanently installed with fixed alignment near Neustrelitz, Germany (see Figure 7).

The RCS of each corner reflector has been monitored for more than 10 years for TSX and seven years for TDX. The results are shown in Figure 8. First of all, we can see that the theoretical RCS value of 43.42 dBm^2 in X-Band could be metrologically proven for all three corner reflectors by means of TSX and TDX. This confirms both how accurately the two systems were absolutely calibrated and matched to each other during their commissioning phases in 2007 [5] and 2010 [6] (by means of a mix of corner reflectors and transponders), and how accurately the corner reflectors were manufactured—because the mechanical tolerance of a corner reflector defines its absolute radiometric accuracy (1 mm form tolerance corresponds to about 0.2 dB accuracy).

Based upon a dedicated calibration campaign performed for TSX two years after the launch, a precise evaluation of the radiometric stability with a high confidence level could be achieved for TSX [7]. For this purpose, the absolute calibration factor derived from point target measurements executed in 2009 had been compared to that derived during the commissioning phase in 2007. The difference between both factors and consequently the radiometric stability over two years was only 0.15 dB. This stability was more than one magnitude better than the requirement of 0.5 dB over six months. However, a radiometric stability of 0.15 dB over two years would mean that the calibration factor might differ up to 0.75 dB from the initial value after ten years. However, this is not the case. As shown in Figure 8, the RCS values measured for each corner reflector are very stable, confirming as well a very stable radiometric performance of TSX and TDX for the entire duration of their mission time. The standard deviation over this period is below 0.2 dB and indicates that the calibration factor has not been drifting since launch. Hence, the high radiometric stability already explicitly derived in 2009 [7] remains valid over a much longer time period of 10 years.

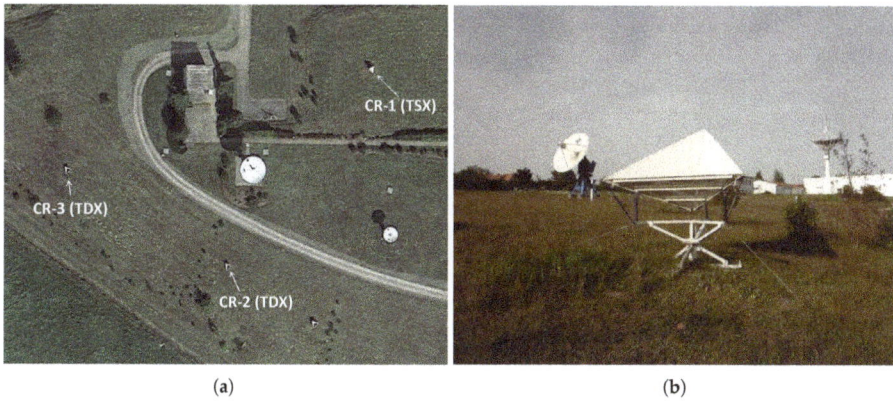

(a)   (b)

Figure 7. DLR calibration site for long-term system monitoring of TSX and TDX near Neustrelitz (Germany) with three permanently installed corner reflectors; (a) seen from the above; (b) closer view of one 1.5 m target showing the fixed alignment.

Moreover, as no trends and no degradation of the two systems can be observed, TSX and TDX still have the same absolute radiometric accuracy as the one derived by a comprehensive calibration campaign executed for both systems during the TDX commissioning phase in 2010 [6]. The radiometric performance of TSX and TDX are summarized in Table 2 and confirms that the satellites are still calibrated to unprecedented quality, 10 and seven years after launch of TSX and TDX, respectively.

Table 2. Radiometric performance parameters for TSX and TDX.

Cal Procedure	Goal	TSX	TDX
Internal Calibration			
Amplitude	0.25 dB	<0.1 dB	<0.1 dB
Phase	1.0 deg	<1.0 deg	<1.0 deg
TRM Setting Characterization			
Amplitude	-	<0.2 dB	<2 deg
Phase	-	<0.2 dB	<2 deg
Antenna Model Verification			
Pattern Shape	±0.2 dB	±0.2 dB	±0.2 dB
Beam-to-BeamGain Offset	±0.2 dB	±0.2 dB	±0.2 dB
Radiometric Calibration			
Radiometric Stability	0.5 dB *	<0.15 dB **	<0.15 dB **
Relative Accuracy	0.68 dB ‡	≤ 0.18 dB ‡	≤0.17 dB ‡
Absolute Accuracy	1.1 dB ‡	<0.34 dB ‡	<0.33 dB ‡

* requirement defined over a period of six months, ** measured by TSX after two years and confirmed by long term system monotoring over 10 years, ‡ StripMap mode.

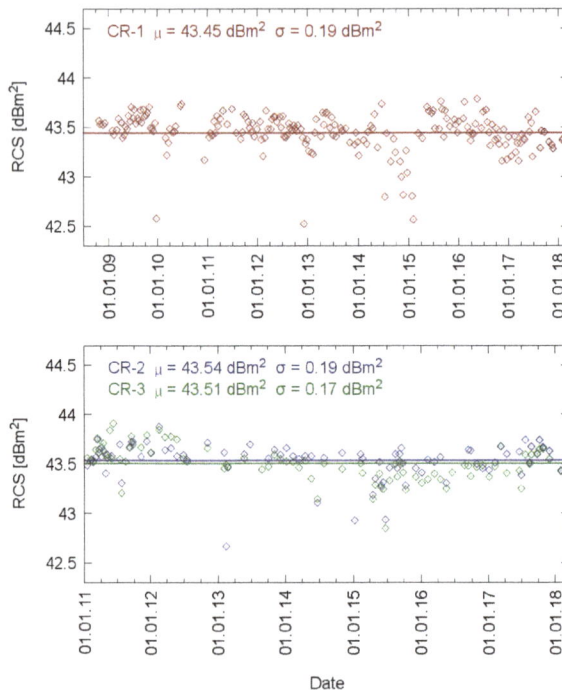

Figure 8. Radar cross section (RCS) of permanently installed corner reflectors derived from SAR images acquired by TSX (**top**) and TDX (**bottom**) over elapsed lifetime.

It should be mentioned that the absolute radiometric accuracy of <0.34 dB in Table 2 is composed of different independent error contributions: the error due to external calibration executed during the commissioning phase (0.16 dB for TSX and 0.14 dB for TDX), the stability of the instrument (<0.15 dB for TSX and TDX), a dynamic range error (0.1 dB for TSX and TDX) as well as an atmospheric loss error (0.24 dB to consider up to moderate rain fall in X-band [19]).

5. Image Quality

In addition to the radiometric performance of the instrument, the quality of SAR images also have been monitored for TSX and TDX since launch. Based on the impulse response function (IRF) derived from the three corner reflectors deployed within DLR's calibration site at Neustrelitz, the following parameters have been derived:

- integrated side lobe ratio (ISLR),
- peak-to-side lobe ratio (PSLR) and
- geometric resolution derived by the main lobe width at −3 dB.

The ISLR and PSLR are separated for azimuth and range direction and depicted in Figure 9 over the mission elapsed time. The statistics show a stable behavior for both satellites, the distributions are all inside the limit and essentially better than the required −18 dB for all parameters.

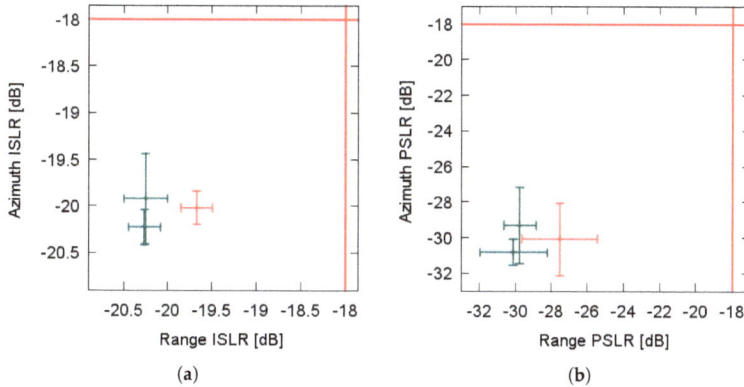

Figure 9. Statistics of image quality parameters derived from the IRF of permanently installed corner reflectors for TSX (red) and TDX (green); (**a**) intergated side lobe ratio (ISLR); (**b**) peak to side lobe ratio (PSLR). The center of the cross indicates the mean and the size of the cross the standard deviation in azimuth and range direction, respectively. All distributions are inside the limits (red lines).

The geometric resolution derived likewise from the corresponding IRF is shown in Figure 10 for TSX and TDX. The resolution of both systems is always inside the limit and essentially better than the required 1.8 m in range and 3.3 m in azimuth. The standard deviation for both systems is better than 1 cm in both direction and confirms likewise the constant product quality over mission time.

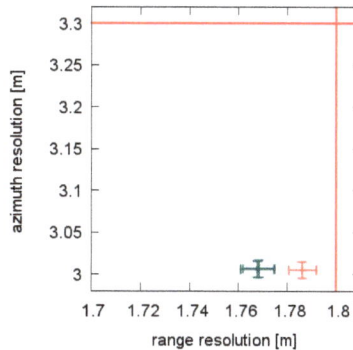

Figure 10. Geometric resolution in StripMap operation derived from the IRF of permanently installed corner reflectors for TSX (red) and TDX (green), the center of the cross indicates the mean and the size of the cross the standard deviation in azimuth and in range direction respectively. All distributions are inside the limits (red lines).

6. Conclusions

The radiometric performance of TSX and TDX has been monitored for the entire mission duration, i.e., more than 10 years for TSX and seven years for TDX. The measurements and extended analyses performed for this long-term system monitoring task show an outstanding stability of the instrument performance. No trends and no degradation have been observed and all parameters show a constant behavior since launch of the respective satellite. Moreover, the measurements against corner reflectors executed for TSX and TDX confirm first a high radiometric stability on the order of one tenth of a dB over a period of 10 years, which is considerably better than that one verified in flight by a dedicated calibration campaign executed in 2009, and second that the absolute radiometric accuracy derived by a comprehensive calibration campaign executed in 2010 is still valid for TSX and TDX.

Never before have two independent spaceborne SAR systems been as accurately calibrated and consequently matched to each other as TSX and TDX. All requirements and goals were not only achieved or succeeded during their individual commissioning phase. They are still valid, 10 and seven years after launch of TSX and TDX, respectively (see Table 2). This confirms not only that the performance of the satellites is still of unprecedented quality, but this also reflects the reliability and the accuracy of sophisticated procedures, modules and reference targets established for calibrating and monitoring both systems over lifetime. Thus, SAR data products of constant high quality have been provided for more than 10 years, and it is not distinguishable whether a TerraSAR-X scene was acquired by TSX or TDX. They are still perfect twins operated as SAR systems in space.

Author Contributions: M.S. had the leading role for preparing this article. By means of point target analysis, K.S. evaluated the radiometric stability and the quality of the SAR data. The work of P.K. was concentrated on the antenna and the verification of the antenna patterns. N.T.R. conducted the characterization of individual transmit/receive modules in the front end. D.R., S.R. and K.W. are responsible for the maintenance and the alignment of the reference targets. J.R. and M.Z. have contributed to the revision of this paper and provided insightful comments and suggestions. The manuscript was written by M.S. All authors have read and approved the final manuscript.

Acknowledgments: The authors would like to thank the whole TerraSAR-X team, particularly the calibration and operation teams of the Microwave and Radar Institute for intensively analyzing and commanding special calibration data takes and initiating automatic long-term system monitoring. But also great thank to Airbus Defence and Space GmbH for developing TerraSAR-X to be such a highly accurate and stable SAR system, a prerequisite for ensuring products of high quality over mission time. And finally the DLR colleagues of the Neustrelitz site for freeing the corner reflectors from snow during the winter.

Conflicts of Interest: The authors declare no conflict of interest.

Abbreviations

The following abbreviations are used in this manuscript:

DLR	German Aerospace Center
IRF	Impulse Response Function
ISLR	Integrated Side Lobe Ratio
LTSM	Long Term System Moniotring
PN	Pseudo Noise
PSLR	Peak to Side Lobe Ratio
RCS	Radar Cross Section
RX	Reception
SAR	Synthetic Aperture Radar
TDX	Second Satellite of the TerraSAR-X Mission
TRM	Transmit/Receiver Module
TSX	First Satellite of the TerraSAR-X Mission
TX	Transmission

References

1. Buckreuss, S.; Werninghaus, R.; Pitz, W. The German Satellite Mission TerraSAR-X. In Proceedings of the 2008 IEEE Radar Conference, Rome, Italy, 26–30 May 2008; pp. 306–310.
2. Zink, M.; Bartusch, M.; Miller, D. TanDEM-X Mission Status. In Proceedings of the 30th International Geoscience And Remote Sensing Symposium, Vancouver, BC, Canada, 24–29 July 2011.
3. Kraus, T.; Bräutigam, B.; Mittermayer, J.; Wollstadt, S.; Grigorov, C. TerraSAR-X Staring Spotlight Mode Optimization and Global Performance Predictions. *IEEE J. Sel. Top. Appl. Earth Observ. Remote Sens.* **2016**, *9*, 1015–1027. [CrossRef]

4. Buckreuss, S.; Schättler, B.; Fritz, T.; Steinbrecher, U.; Böer, J.; Bachmann, M.; Mittermayer, J.; Maurer, E.; Kahle, R.; Mrowka, F.; et al. Ten Years of TerraSAR-X Operations. *Remote Sens.* **2018**, under review.
5. Schwerdt, M.; Bräutigam, B.; Bachmann, M.; Döring, B.; Schrank, D.; Gonzalez, J.H. Final TerraSAR-X Calibration Results Based on Novel Efficient Calibration Methods. *IEEE Trans. Geosci. Remote Sens.* **2010**, *48*, 677–689. [CrossRef]
6. Schwerdt, M.; Gonzalez, J.H.; Bachmann, M.; Schrank, D.; Döring, B.; Ramon, N.T.; Antony, J.W. In-Orbit Calibration of the TanDEM-X System. In Proceedings of the 2011 IEEE International Geoscience and Remote Sensing Symposium, Vancouver, BC, Canada, 24–29 July 2011; pp. 2420–2423.
7. Schwerdt, M.; Schrank, D.; Bachmann, M.; Schulz, C.; Döring, B.; Gonzalez, J.H. TerraSAR-X Re-Calibration and Dual Receive Antenna Campaigns performed in 2009. In Proceedings of the 8th European Conference on Synthetic Aperture Radar, Aachen, Germany, 7–10 June 2010.
8. Grafmüller, B.; Herschlein, A.; Fischer, C. The TerraSAR-X Antenna System. In Proceedings of the 2005 IEEE International Radar Conference, Arlington, VA, USA, 9–12 May 2005; pp. 222–225.
9. Stangl, M.; Werninghaus, R.; Schweizer, B.; Fischer, C.; Brandfass, M.; Mittermayer, J.; Breit, H. TerraSAR-X Technologies and First Results. *IEE Radar Sonar Navig.* **2006**, *153*, 86–95. [CrossRef]
10. Schwerdt, M.; Hounam, D.; Alvarez-Pérez, J.L.; Molkenthin, T. The Calibration Concept of TerraSAR-X, a Multiple Mode High Resolution SAR. *Can. J. Remote Sens.* **2005**, *31*, 30–36. [CrossRef]
11. Döring, B.; Schwerdt, M.; Bauer, R. TerraSAR-X Calibration Ground Equipment. In Proceedings of the European Radar Conference, Munich, Germany, 10–12 October 2007.
12. Bräutigam, B.; Rizzoli, P.; González, C.; Schulze, D.; Schwerdt, M. SAR Performance of TerraSAR-X Mission with Two Satellites. In Proceedings of the 8th European Conference on Synthetic Aperture Radar, Aachen, Germany, 7–10 June 2010.
13. Bräutigam, B.; Schwerdt, M.; Bachmann, M. An Efficient Method for Performance Monitoring of Active Phased Array Antennas. *IEEE Trans. Geosci. Remote Sens.* **2009**, *47*, 1236–1243. [CrossRef]
14. Bachmann, M.; Schwerdt, M.; Brautigam, B. TerraSAR-X Antenna Calibration and Monitoring Based on a Precise Antenna Model. *IEEE Trans. Geosci. Remote Sens.* **2010**, *48*, 690–701. [CrossRef]
15. Tous-Ramon, N.; Schrank, D.; Bachmann, M.; Alfonzo, G.C.; Polimeni, D.; Böer, J.; Schwerdt, M. Long Term System Monitoring Status of the TerraSAR-X and the TanDEM-X Satellites. In Proceedings of the 9th European Conference on Synthetic Aperture Radar, Nuremberg, Germany, 23–26 April 2012; pp. 1–4.
16. Hounam, D.; Schwerdt, M.; Zink, M. Active Antenna Module Characterisation by Pseudo-Noise Gating. In Proceedings of the 25th ESA Antenna Workshop on Satellite Antenna Technology, Noordwijk, The Netherlands, 18–20 September 2002.
17. Bachmann, M.; Schwerdt, M.; Bräutigam, B. Accurate Antenna Pattern Modeling for Phased Array Antennas in SAR Applications—Demonstration on TerraSAR-X. *Int. J. Antennas Propag.* **2009**, *2009*, 492505. [CrossRef]
18. Rizzoli, P.; Bräutigam, B.; Zink, M. TanDEM-X Large-Scale Study of Tropical Rainforest for Spaceborne SAR Calibration in X-Band. In Proceedings of the 10th European Conference on Synthetic Aperture Radar, Berlin, Germany, 3–5 June 2014; pp. 1–4.
19. Danklmayer, A.; Döring, B.; Schwerdt, M.; Chandra, M. Assessment of Atmospheric Propoagation Effects in SAR Images. *IEEE Trans. Geosci. Remote Sens.* **2009**, *47*, 3507–3518. [CrossRef]

remote sensing

MDPI

Article

Long-Term Validation of TerraSAR-X and TanDEM-X Orbit Solutions with Laser and Radar Measurements

Stefan Hackel [1,*], Christoph Gisinger [2], Ulrich Balss [2], Martin Wermuth [1] and Oliver Montenbruck [1]

[1]   German Space Operations Center, Deutsches Zentrum für Luft- und Raumfahrt, 82230 Wessling, Germany;
      martin.wermuth@dlr.de (M.W.); oliver.montenbruck@dlr.de (O.M.)
[2]   Remote Sensing Technology Institute, Deutsches Zentrum für Luft- und Raumfahrt,
      82230 Wessling, Germany; christoph.gisinger@dlr.de (C.G.); ulrich.balss@dlr.de (U.B.)
*   Correspondence: stefan.hackel@dlr.de; Tel.: +49-8153-28-3024

Received: 18 April 2018; Accepted: 13 May 2018; Published: 15 May 2018

Abstract: Precise orbit determination solutions for the two spacecrafts TerraSAR-X (TSX) and TanDEM-X (TDX) are operationally computed at the German Space Operations Center (GSOC/DLR). This publication makes use of 6 years of TSX and TDX orbit solutions for a detailed orbit validation. The validation compares the standard orbit products with newly determined enhanced orbit solutions, which additionally consider GPS ambiguity fixing and utilize a macro model for modeling non-gravitational forces. The technique of satellite laser ranging (SLR) serves as a key measure for validating the derived orbit solutions. In addition, the synthetic aperture radar (SAR) instruments on-board both spacecrafts are for the first time employed for orbit validation. Both the microwave instrument and the optical laser approach are compared and assessed. The average SLR residuals, obtained from the TSX and TDX enhanced orbit solutions within the analysis period, are at 1.6 ± 11.4 mm (1σ) and 1.2 ± 12.5 mm, respectively. Compared to the standard orbit products, this is an improvement of 33 % in standard deviation. The corresponding radar range biases are in the same order and amount to -3.5 ± 12.5 mm and 4.5 ± 14.9 mm. Along with the millimeter level position offsets in radial, along-track and cross-track inferred from the SLR data on a monthly basis, the results confirm the advantage of the enhanced orbit solutions over the standard orbit products.

Keywords: TerraSAR-X; TanDEM-X; LEO; POD; SLR; SAR; Satellite Laser Ranging; radar ranging; satellite orbit; validation

---

1. Introduction

The TerraSAR-X satellite mission [1,2] consists of the two spacecrafts: TSX and TDX, which are equipped with Synthetic Aperture Radar (SAR) instruments for active remote sensing. The ongoing mission is successfully observing the Earth for more than 10 years. Precise information on the satellite positions is of vital importance for many of its SAR applications, e.g., for SAR imaging geodesy [3–6] or the generation of the global digital elevation model from bistatic interferometry [7]. The satellite orbits are determined by the German Space Operations Center (GSOC) in a reduced-dynamic approach, which incorporates measurements from the onboard Global Positioning System (GPS) receivers as well as the modeling of gravitational and non-gravitational forces [8]. For orbit validation, the technique of Satellite Laser Ranging (SLR) is widely used and embedded as state-of-the-art [9,10]. Data obtained from various laser ranging stations for TSX and TDX, and the notably high ranking of the mission in the tracking priority list provide highly reliable SLR data. In addition to the laser reflector, the spacecrafts are equipped with accurate SAR instruments as primary payloads. The SAR enables ranging measurements from the spacecrafts to on-ground trihedral Corner Reflectors (CRs), which provide a sharp reflection of the transmitted chirp signals, and thus can also provide high accuracy in ranging if the data is processed according to geodetic standards [5,6,11]. Therefore, the orbit analysis published in this paper makes use of both SLR and SAR measurements. The derived SAR and SLR measurements are compared in terms of residuals, and estimated spacecraft position offsets. SLR allows estimation of the offsets in radial, along-, and cross-track direction, which is comparable to the range and azimuth residuals, obtained from the SAR analysis.

The remainder of the publication is structured as follows. Sections 2 and 3 describe the TerraSAR-X orbit, the payloads and methods used for precise orbit determination, as well as the orbit solutions investigated in our study. Section 4 provides the details on the SLR and the SAR ranging techniques, including descriptions of the data sets and the contributing geodetic stations. In Section 5, we present and discuss the results for each of the two techniques, and provide a comparative analysis based on monthly estimates for the remaining offsets. The paper concludes with Section 6 with a discussion of the key findings and the final conclusions.

2. Spacecrafts and Orbits

For the present study, two different satellite orbit solutions of the spacecrafts TSX and TDX are analyzed. The spacecrafts were launched on 15 June 2007, and 21 June 2010, respectively [2,12], which enables us to study the long-term behavior of the orbit solutions. The aim of the TerraSAR-X mission is Earth observation, based on individual, or combined SAR observations of the Earth surface. The GSOC at the German Aerospace Center (DLR) in Oberpfaffenhofen, Germany, is in charge of mission control and the operational orbit determination. Both spacecrafts are orbiting the Earth on a Sun-synchronous, 97.44 ° inclined dusk-dawn orbit at an average altitude of 514 km. The orbital period is roughly 90 minutes, and one complete orbital repeat cycle takes 11 days, which provides high repeatability [13,14]. Due to the chosen orbit configuration along with the terminator, the spacecraft is almost constantly illuminated by the Sun. Due to the objective of the interferometric campaigns, the strict orbit requirements for TerraSAR-X are formulated in the form of a 250 m toroidal tube around a pre-flight determined reference trajectory. Star trackers for attitude determination, reaction wheels and magnetorquers for momentum unloading are utilized for determining the spacecrafts attitude with a pointing accuracy smaller than 40 arcsec [12].

The spacecrafts are equipped with identical payloads, of which the SAR payload primarily serving the Earth observation is employed for additional orbit validation within this study. An illustration of the spacecraft TerraSAR-X is shown in Figure 1, indicating the SAR and SLR payloads.

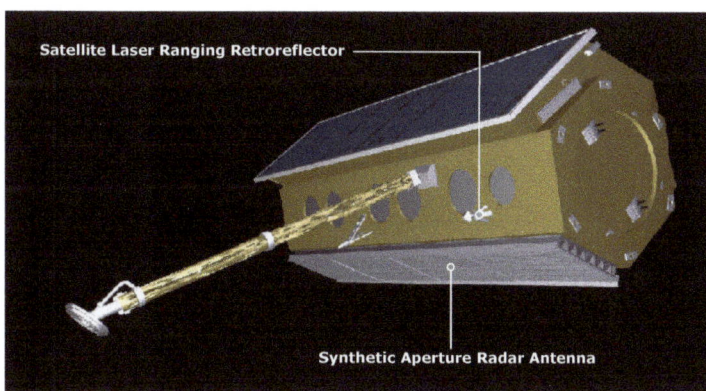

Figure 1. Illustration of TerraSAR-X with selected optical and radar payload for orbit validation.

For the purpose of attitude and orbit determination, each spacecraft is equipped with star trackers for attitude determination, as well as three GPS receivers. The Integrated Geodetic and Occultation

Remote Sens. **2018**, *10*, 762

Receiver (IGOR) serves as the primary receiver for Precise Orbit Determination (POD) and baseline reconstruction, and provides geodetic-grade dual-frequency measurements [15]. The second and third are MosaicGNSS receivers, which mainly serve for on-board timing and basic orbit information for aligning the spacecraft with the ground track and the nadir direction [16]. In addition, both spacecrafts carry a Laser Retro Reflector (LRR), which allows accurate orbit validation with SLR. The IGOR receiver and the LRR unit are contributed by the University of Texas Center for Space Research and the Deutsches GeoForschungsZentrum Potsdam (GFZ).

3. Precise Orbit Determination Concept and Models

Precise orbit determination solutions of TSX and TDX are computed at GSOC, applying the method of reduced-dynamic orbit determination. DLR's GNSS High Precision Orbit Determination Software Tools (GHOST) are utilized for POD and SLR validation [17]. The orbit determination employs a dynamical orbit model considering gravitational and non-gravitational forces, as well as complementary empirical accelerations, which are modeled in addition to the initial daily state vector of the satellite and the scaling factors for individual force model constituents. Estimation parameters with respect to the on-board GPS measurements include the epoch-wise receiver clock offsets and an integer-valued ambiguity of the ionosphere-free carrier phase combination for each continuous tracking pass.

Non-gravitational accelerations for Earth and solar radiation pressure, as well as the aerodynamic accelerations, are considered by a macro model formulation, which approximates the spacecraft as an extruded, equilateral hexagon [18].

GPS observations from the geodetic-grade IGOR receiver are considered by L_1 and L_2 code and phase measurements, which are combined in a ionosphere-free linear combination. In addition, a single-receiver ambiguity resolution concept is employed that builds on dedicated GPS orbit, clock, and wide-lane bias products provided by the CNES/CLS (Centre National d'Études Spatiales/Collecte Localisation Satellites) analysis center of the International GNSS Service (IGS) [19].

Table 1 shows an overview of the employed key models and parameters that have been used to generate the enhanced orbit solutions, which are the main scope of the present study. These orbit solutions are abbreviated and referred to in the following as ENHanced (ENH). Note that the orbit solution described above is not identical to the operationally computed Precise Science Orbits (PSOs), which consider non-gravitational forces combined with a simplified satellite shape (cannon-ball model), and employ float GPS ambiguities [18]. The PSOs are utilized for ease of comparison only. Regarding the temporal coverage, this analysis covers both orbit solutions for TSX and TDX during the period of 1 January 2012 to 31 December 2017.

Table 1. GHOST models and data sets for TerraSAR-X precise orbit determination.

GPS Measurement	
GPS observations	Undifferenced L_1/L_2 pseudorange and carrier phase range, 30-s sampling, daily arcs
GPS orbits and clocks	CNES-CLS grg products [20], 30-s sampling
GPS satellite antenna	IGS igs08.atx/igs14.atx phase center offsets and variation [21]
GPS satellite biases	CNES-CLS wide-lane satellite biases [20], ftp://ftpsedr.cls.fr/pub/igsac/
GPS antenna	PCO + PV corrections from in-flight calibration; center-of-mass variation
Attitude	Quaternions (measured)
Reference frame	IGb08 [21], IGS14 [22], from DOY 30/2017 onwards
Phase windup	Modeled [23]

Table 1. *Cont.*

Orbit	
Earth gravity field	GOCO03S [24] up to order and degree 100; rate terms $\dot{C}_{20}$, $\dot{C}_{21}$, $\dot{S}_{20}$
Luni-solar gravity	Point-mass model; analytical series of luni-solar coordinates
Solid Earth and pole tides	IERS2003
Ocean tides	CSR/Topex3.0 [25]
Relativity	Post-Newtonian correction
Spacecraft parameters	Time-varying mass ($\approx$1319 kg); 8 panel box-wing macro-model
Solar radiation pressure	Macro-model; conical Earth shadow model
Earth radiation pressure	Macro-model; CERES Earth radiation data [26]
Aerodynamics	Macro-model; NRLMSISE-00 density model [27], NOAA/SWPC solar flux and geomagnetic activity data (ftp://tp.swpc.noaa.gov/pub/indices/)
Maneuvers	Constant thrust in RTN direction
Empirical acceleration	Piecewise constant accelerations in RTN direction; 10-min intervals
Reference frame	ICRF
Earth orientation	IERS1996; IGS final EOPs; center-of-mass/center-of-figure offset
Numerical integration	Self-starting variable-order variable step size multistep method [28]
Estimation	
Filter	Batch least squares estimation
Estimation parameters	Epoch state vector, scale factors for SRP and drag/lift, empirical accelerations and maneuvers, clock offsets, phase ambiguities
Stochastic models	White observation noise, elevation-independent

4. Orbit Validation Techniques

In the field of precise orbit determination, SLR is well-established as a state-of-the-art validation technique, since it provides independent measurements of high quality and reliability [10]. Additionally, we present a long-term analysis of SAR range and azimuth residuals derived from the measurements of the TerraSAR-X radar payloads, which for the first time provide a second independent method to validate the impact of the enhanced orbit solution. Both methods rely on global geodetic ground infrastructure, which is visualized in Figure 2. On the SLR side, the measurements are obtained by 11 International Laser Ranging Service (ILRS) [29] stations, while the SAR measurements are supported by five CRs located at three stations. The following sections discuss both methods, and provide further details on the data reduction and the processing strategies.

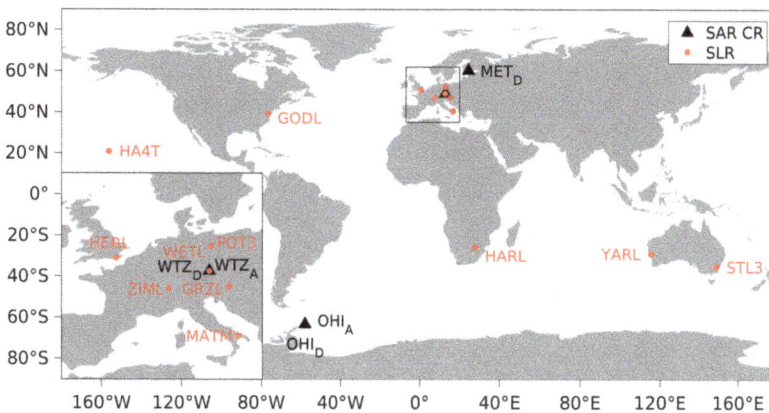

Figure 2. Selected stations of ground network, utilized for orbit validation. Radar corner reflectors, utilized by the SAR analysis, are indicated by triangles; the selected ILRS stations for SLR by dots.

4.1. Satellite Laser Ranging

4.1.1. Technique

Satellite laser ranging observations are optical two-way runtime measurements from ground stations to the LRR on-board the satellite. The LRR of TSX and TDX comprises four individual prisms in a rotationally symmetric arrangement [30], see Figure 3, to ensure reliable reflection properties over a wide range of incidence angles. The individual prisms are arranged in a regular 45° pyramid.

Basically, the ground station actively emits a laser pulse, which is reflected by the prisms on the spacecraft. The runtime measurements translate into the range r_{SLR}, which is compared to the geometric range derived from the orbit solutions r_{POD}:

$$\Delta r = r_{\text{SLR}} - r_{\text{POD}}. \tag{1}$$

The obtained SLR range residuals Δr are a measure of the orbit accuracy. Of vital importance are the well-known positions of the LRR onboard the spacecraft and of the laser tracking stations on ground. The coordinates of the LRR were precisely determined along with the phase patterns of the optical systems during the pre-flight calibration on ground, see Table 2. The SLR station coordinates, on the other hand, are related to the terrestrial frame and its models are discussed in the next section.

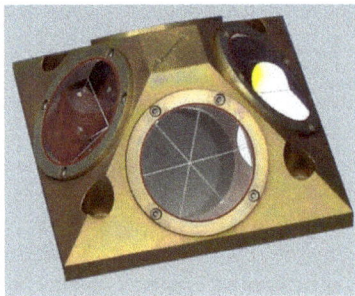

Figure 3. Laser retro reflector array of TSX (TerraSAR-X) and TDX(TanDEM-X) (courtesy GFZ Potsdam).

Satellite laser ranging residuals, which are obtained from the approach as described above, do not contain any information on the 3-D spatial variation of the residuals across individual tracking passes. Therefore, an extended validation approach has been published in Hackel et al. [18], where the obtained SLR residuals are further utilized for estimating SLR-based position offsets of the spacecraft. The extended data interval of 30 days serves as the basis for monthly position estimates in the spacecraft's radial, along-track, and cross-track direction.

Table 2. Selected models for Satellite Laser Ranging orbit validation.

Station coordinates	SLRF2014 [31]
Solid Earth and pole tides	IERS2003
Ocean tide loading	GOT00.2 [32]
Tropospheric refraction	IERS2010 [33]
Relativity	Space-time curvature correction
LRR phase correction	Neubert et al. [30]

4.1.2. Models

The modeling of SLR makes use of station coordinates, which refer to the Satellite Laser Ranging Frame 2014 (SLRF2014) [31]. This ensures best consistency with the IGb08, and IGS14 frames of the GPS orbit products, which are employed in the orbit determination of the TSX and TDX orbit solutions [22,34]. The phase center range correction of the LRR of each spacecraft is considered as a function of azimuth with respect to local spacecraft heading and the nadir angle, and is provided by the GFZ [30]. The selected models for solid Earth tides, ocean tidal loading, and tropospheric refraction are introduced in Table 2.

4.1.3. Tracking Data Analysis

The SLR tracking data are provided by the ILRS, which observes a multitude of satellites according to a mission priority list. The spacecrafts TSX and TDX are highly ranked at positions 3 and 2 [35], which results in a large number of tracking points. The SLR measurements are averaged in time intervals of 5 s, which finally result in one Normal Point (NP) [36]. The number of normal points within one tracking pass, i.e., the passage where the instrument on ground follows and tracks the spacecraft may vary from pass to pass. For TSX, on average 15 NPs are collected during one tracking pass, which corresponds to an average tracking length of 75 s.

Within the six year period, the spacecraft TSX was tracked by 34 different ILRS stations, yielding a wide range from only 5 up to 75,000 normal points per station. For the present study, a subset of 11 reliable and high-performance ILRS stations has been selected, which are shown in Figure 2.

Considering the 11 stations, a total of 206,401 normal points is available for TSX within the 6 years (cf. Figure 4). Applying an elevation cutoff-angle of 10°, and a threshold of 6 cm for the residuals, this leads to a total of 204,667 NPs, which are employed within this study. This corresponds to a screening rate of 0.84%. For the spacecraft TDX, the number of accepted normal points used throughout this analysis amounts to 195,506. Despite the higher ranking of TDX, the number of available normal points during the analysis period is slightly larger for TSX.

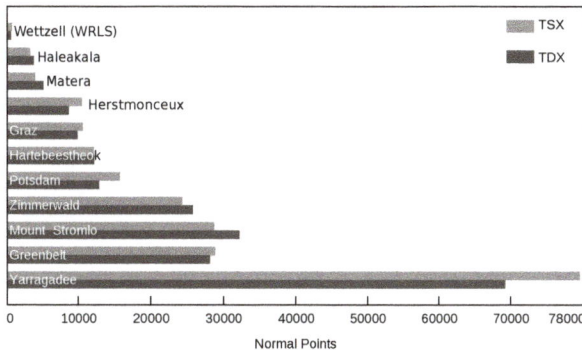

Figure 4. Number of accepted normal points for selected ILRS stations between 2012 and 2017.

4.2. Radar Ranging

4.2.1. Technique

Similar to SLR, a spaceborne SAR payload allows for the measurement of the 2-way signal round trip time between the sensor and the ground. Contrary to the transmission and reception of laser pulses dedicated to the pass of an individual satellite, the radar instrument illuminates a large footprint in side-looking geometry and discriminates the echoes according to the time of flight and the time of reception [37]. Provided that the SAR image processor employs strict geometrical standards, the processing of this raw data matrix for the actual SAR image is able to accurately preserve this underlying timing information. In particular, the often used approximations that reduce computational efforts need to be avoided, for instance the simplification of the platform movement during signal transmission and signal reception, which is also known as 'stop-go' approximation [37].

For the TerraSAR-X mission, all these SAR-related aspects have been carefully resolved in the TerraSAR-X Multimode SAR Processor (TMSP). The processor generates well-defined level 1b radar images with each pixel referring to the time of closest approach (zero-Doppler time) and the corresponding two-way round trip time [38,39]. In order to accurately relate this 2-D timing information to a dedicated location within the image, one may create a bright point response by placing a trihedral CR, see Figure 5. From the SAR sensor perspective, the CR needs to be large enough to offer a strong signal return in the radar image, and it must fulfill tight limits in terms of plate orthogonality, planar surface geometries, and mechanical stability [40]. If such a reflector is permanently installed and the reference coordinates are known in the International Terrestrial Reference Frame (ITRF, for the latest release 2014 see [22]), then the SAR measurements may be verified on a pass by pass basis by analyzing the radar timings. As already discussed for the SLR, the terrestrial CR coordinates have to be consistent with IGb08 and IGS14 frame realizations used in the orbit determination.

Figure 5. Corner reflectors (CRs) at geodetic observatories. CR with 0.7 m inner leg dimension at GARS O'Higgins (**left**), and CRs with 1.5 m inner leg dimension at Wettzell (**middle**) and at Metsähovi (**right**).

In accordance with the TerraSAR-X imaging model, we use the range-Doppler equations [37] in zero-Doppler geometry, which relate the satellite trajectory, given by the time-dependent position vector $\mathbf{X}_s$ and velocity vector $\dot{\mathbf{X}}_s$, and the reflector position vector $\mathbf{X}_r$ with the observed radar times t and τ, also referred to as slow time and fast time or azimuth and range, respectively.

$$\tau = 2/c \cdot |\mathbf{X}_s(t) - \mathbf{X}_r| \tag{2}$$

$$0 = \frac{\dot{\mathbf{X}}_s(t) \cdot (\mathbf{X}_s(t) - \mathbf{X}_r)}{|\dot{\mathbf{X}}_s(t)| \cdot |\mathbf{X}_s(t) - \mathbf{X}_r|} \tag{3}$$

The conversion of geometrical distance for the τ uses the speed of light in vacuum c. The slow time t is linked to the satellite trajectory and can be resolved by interpolating a given orbit solution and performing an iterative search for the instant of Doppler-zero using the Equation (3). Subsequently, the corresponding round trip time τ is derived from Equation (2). From a geometrical point of view, the combination of the Equations (2) and (3) models a circle located at the satellite's zero-Doppler position and oriented according to the zero-Doppler plane, and which intersects with the reflector position on ground, see Figure 6.

In order to compute the reference radar times from the geometric model, the tidal-related solid Earth effects have to be taken into account when defining the $\mathbf{X}_r$ for each pass. This is identical to the modeling of SLR station coordinates or other reference markers in the ITRF, see Section 4.1.2 and Table 2. The details of the ITRF models that have to be applied are given in the conventions issued by the International Earth Rotation and Reference Systems Service (IERS) [41]. We include all the models listed in chapter 7 of the conventions in our SAR analysis, and the reflector position $\mathbf{X}_r$ is corrected for the epoch of the image acquisition before computing the reference timings with the Equations (2) and (3).

The measured radar timings are extracted from the images by performing the so-called point target analysis [11], which detects the center of the reflector's point signature in the image with better than 1/1000 of a pixel. The precision of the extraction in range and azimuth is equivalent to the signal-to-noise ratio, or Signal to Clutter Ratio (SCR) in radar terminology, and may be expressed by [42,43]:

$$\sigma_{R,A} = \frac{\sqrt{3}}{\pi\sqrt{2}} \cdot \frac{\rho_{R,A}}{\sqrt{SCR}} \tag{4}$$

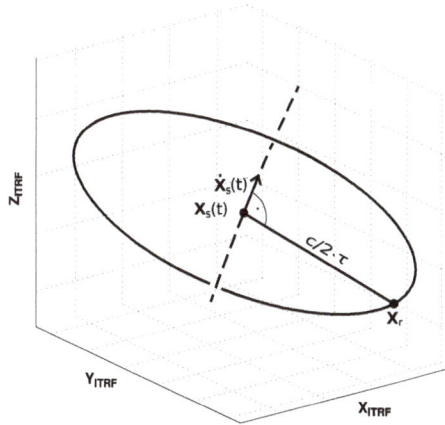

Figure 6. Zero-Doppler geometry of a SAR acquisition in the ITRF for a corner reflector at X_r and a satellite pass given by the dashed line.

The $\rho_{R,A}$ denotes the image resolution in range and azimuth, respectively. For the CRs with 0.7 m and 1.5 m inner leg dimension that we use in our study, see Section 4.2.2, we determined a typical SCR of 31 dB and 46 dB for the TerraSAR-X high resolution spotlight images (0.6 m by 1.1 m resolution, [39]) employed in the analysis. According to Equation (4), these SCR values translate into range and azimuth precisions at the millimeter to centimeter level, see Table 3.

Table 3. Expected range and azimuth localization precision computed from Equation (4) for the TerraSAR-X high resolution spotlight mode and the reflectors used in our study.

CR Inner Leg	SCR [dB]	σ_R [mm]	σ_A [mm]
0.7 m	31	6.6	12.1
1.5 m	46	1.2	2.1

Regarding the TerraSAR-X azimuth measurements, a limitation was found in the link between the GPS time provided by the secondary MosaicGNSS receiver (cf. Section 2) and the SAR payload, but this could be improved to about 1–2 cm by exploiting the accurately known pulse repetition interval of the SAR instrument [44]. For the range measurements, additional corrections are required for the atmospheric path delays, for which we rely on the permanently operated GNSS receivers at the geodetic stations hosting the CRs. The slant range tropospheric delay is computed from the zenith delay product provided on a daily basis by the IGS [45], and the slant range ionospheric delay is inferred from the dual-frequency GNSS observations at the stations. The details of our methods are given in [5]. The tropospheric products are estimated with an accuracy of better than 5 mm [45], whereas the cross-comparison of the ionospheric delays from the two or more GNSS receivers available at each of our test sites indicate a similar 5 mm quality level for the ionospheric correction. Finally, there are the range and azimuth geometrical calibration constants of the SAR instrument, which need to be determined in dedicated experiments, as described in Section 4.2.3.

In summary, the combination of CRs with reference coordinates and the orbit solution allows for the computation of residuals in range and azimuth using the radar payload. The correction of the atmospheric path delay in the measurements and the modeling of the reference coordinates according to the geodetic conventions ensure accurate SAR-based residuals, which are also useful when testing different orbit solutions. Like in the case of SLR, the SAR residuals may be interpreted as orbital errors and can in principle be decomposed into radial, along-track and cross-track residuals [18]. However, it has to be emphasized that the SAR provides only one range per pass, i.e., the range at the instant of zero-Doppler, and the corresponding azimuth is basically the along-track error, whereas the SLR enables the tracking of the entire pass.

4.2.2. Ground Infrastructure

In the course of our long-term monitoring of the geometrical quality of TerraSAR-X, the three geodetic stations marked in Figure 2 have become equipped with permanent CR installations, see Figure 5. The Wettzell geodetic station in Germany hosts two CRs with 1.5 m inner leg dimension, with one CR aligned for the ascending passes (WTZ$_A$) and the other CR aligned for the descending passes (WTZ$_D$). The reflectors became available in July 2011 and October 2013, respectively. The Metsähovi geodetic station in Finland was equipped with a 1.5 m CR in October 2013 that is aligned for descending passes (MET$_D$). Moreover, two CRs with 0.7 m inner leg dimension were permanently installed at the German Antarctic Receiving Station (GARS) O'Higgins in March 2012. They are oriented for ascending passes (OHI$_A$) and descending passes (OHI$_D$). All of the five CRs are linked by local ties to the reference coordinates of their respective station. The local ties have been determined in terrestrial geodetic surveys of the CR phase centers, i.e., the intersection of the three orthogonal plates. The accuracy of the local ties as reported from the surveys is better than 5 mm for each of the reflectors, and the transformation to the global ITRF, release 2008 [46], was carried out with the dedicated transformation parameters of each station. This ensures that the ITRF reference coordinates of the CRs retain the accuracy of the local survey. For a general description of the methods to determine the local ties at geodetic stations, see for instance the example of Wettzell [47,48].

Repeated measurements of the local ties of the WTZ$_A$ reflector in 2011, 2012 and 2014 revealed small changes in the CR position during the first years, which were caused by soil compaction below the concrete pad supporting the reflector mount, see Table 4. The changes occurred mostly in the vertical direction and were considered as a piece-wise linear displacement in addition to the ITRF velocity of the Wettzell station when modeling the CR reference coordinates. Beyond 2014, the reflector was determined to be stable. Because of these early experiences, a more stable ground was chosen to place the concrete pad of the second Wettzell reflector, while for the reflectors at the other stations we do not expect such secondary deformation, because there the reflector mounts are directly attached to stable bedrock.

Table 4. Changes in the position [cm year^{-1}] of WTZ$_A$ derived from the repeated local tie surveys of October 2011, 2012 and 2014.

Period	North	East	Height
Before 10/2012	0.5	−001	−0.9
10/2012–10/2014	0.1	−0.2	−0.5
After 10/2014	0.0	0.0	0.0

On 29 January 2017, the IGS adopted the latest ITRF solution, namely the ITRF2014 [22]. This frame updates results in refined coordinate solutions for all the IGS GNSS sites, which in turn affect the orbit and clock products of the GNSS constellations provided by the IGS after the aforementioned date. Consequently, the orbits of TSX and TDX also refer to the renewed ITRF2014 solution, because

the IGS products are used in the reduced dynamic orbit determination, see Table 1 and the details in [18]. To ensure a consistent SAR data analysis beyond January 2017, the reference coordinates of all the reflectors were transformed from ITRF2008 to ITRF2014 using the official transformation parameters [22]. The station velocities of Wettzell, GARS O'Higgins and Metsähovi required for linearly transforming the reference coordinates of the reflectors to the epoch of the SAR acquisition were also taken from the corresponding ITRF2008 and ITRF2014 solution files, which are available at the IERS [49].

4.2.3. TerraSAR-X Image Acquisitions

From 2012 to 2017, the satellites TSX and TDX acquired in total 1,033 scenes for the five reflectors located at the geodetic stations. The imaging mode was the TerraSAR-X high resolution spotlight mode, which features an average resolution of 0.6 m by 1.1 m in slant range and azimuth, as well as a scene extent of approximately 5 km by 10 km [39]. Out of these acquisitions, 68 scenes had to be eliminated from the processing because they were rendered unusable by snow or water in the reflectors, which significantly reduce the signal backscatter. The degraded measurements are easily detected by computing the SCR from the CR point response in the radar image, and comparing it to the average SCR of the data series. In addition, 10 scenes were eliminated because of non-final alignments of the CRs or because of gross outliers at the decimeter level in the processed SAR residuals. The remaining scenes are distributed across several pass geometries that may be identified by the incidence angle at the CR at Doppler zero, see Table 5. Both satellites TSX and TDX captured data for all of the available reflectors, but in the case of TDX the majority of the data was acquired at GARS O'Higgins, while for TSX the data distribution across the sites is more homogeneous.

Table 5. TerraSAR-X repeat pass geometries for the CRs at the geodetic observatories used in the study. The different geometries are identified by the incidence angles of the side-looking radar beams at zero Doppler.

CR	Pass Geometries [°]
WTZ_A	34, 46
WTZ_D	33, 45, 54
OHI_A	30, 38, 45
OHI_D	35, 43
MET_D	27, 37, 45

In order to center the SAR measurements and account for any unresolved biases caused by internal electronic delays of the SAR instrument, radar payloads are empirically calibrated against stable CRs with known reference coordinates. For the TerraSAR-X mission, the calibration was performed during the commissioning phase according to the initial requirements of 1 m [39,50], but because of our improvement in the TerraSAR-X range and azimuth measurements, we adopted the procedure to determine our own refined calibration constants for TSX and TDX [5,6]. The CR located at the Metsähovi station is used for this task. For the analysis presented in this study, the calibration constants were derived separately for both the standard PSOs and the enhanced orbit products, see Table 6. The changes in the calibration constants caused by the enhanced orbit product become as large as 1 cm. We may conclude that other uncompensated biases, e.g. from the orbit, are included in these constants in addition to the actual electronic instrument delays. This is also the reason why we decided to redetermine these constants after we introduced the refined modeling of the TerraSAR-X radar observations, see Section 4.2.1. Naturally, the remaining offset in the Metsähovi SAR residuals will become very small, but the offsets at the remaining four CRs can be used to judge the quality and the consistency of the SAR data and the two orbit products.

Table 6. Geometrical SAR calibration constants [m] of the TerraSAR-X mission derived from the Metähovi corner reflector for the different orbit solutions.

Solution	TSX		TDX	
	Range	Azimuth	Range	Azimuth
PSO	−0.3017	−0.0747	−0.2805	−0.0539
ENH	−0.2933	−0.0710	−0.2714	−0.0608

5. Validation Results and Discussion

5.1. Laser Ranging

For the 6 year period, and for both spacecrafts, TSX and TDX, the corresponding SLR residuals are shown in Table 7. Figure 7 shows a time series of TSX and TDX satellite laser ranging residuals obtained from the two types of orbit solutions. In 2012, the series exhibits a data gap in January, which is caused by a planned outage of SLR tracking due to a spacecraft campaign. In general, the SLR residuals result in a mean offset of 2 mm over the 6 years of data. The standard deviation reaches up to 17 mm, which is reduced by 6 mm in the case of the enhanced solutions. The TDX series shows an increased amount of residual scatter in spring 2012, whereas both spacecrafts show similar patterns of larger residuals in mid 2014. Overall, the series do not show notable systematic variations with time.

Station-wise residual series help to identify systematic effects related to individual stations. As an example, Figure 8 shows station-wise SLR residuals series of two ILRS stations with totally different characteristics. The ILRS station Graz, Austria, exhibits no bias and a standard deviation of only 0.8 mm, whereas Mount Stromlo, Australia, shows a bias of 10 mm, along with a standard deviation of 12 mmfor both the PSO and the ENH solutions. Outliers, which are most probably related to the station itself, are indicated by the vertically aligned dots stemming from specific passes. They can easily be identified in the series of Mount Stromlo, Australia. Contrary, the station in Graz, Austria, is almost free from such systematics. Offsets in station-wise SLR residuals hint to potential deficiencies in the knowledge of the SLR station coordinates and biases, as demonstrated in [10].

Table 7. Satellite laser ranging residuals ($\bar{x} \pm \sigma_x$ [mm]) and amount of normal points (N_{np}) for selected ILRS stations of the TSX and TDX orbit validation.

Station	TSX			TDX		
	N_{np}	PSO	ENH	N_{np}	PSO	ENH
Graz	9966	−2.7 ± 13.6	−0.7 ± 7.3	9299	−3.9 ± 14.0	−0.9 ± 7.9
Greenbelt	27,060	−11.8 ± 17.2	−7.5 ± 11.0	26,413	−10.4 ± 16.6	−6.8 ± 12.3
Haleakala	2994	0.1 ± 13.5	2.3 ± 10.4	3507	0.6 ± 14.4	3.1 ± 11.0
Hartebeesthoek	11,425	3.5 ± 20.0	6.8 ± 17.5	11,460	1.9 ± 21.0	5.4 ± 18.6
Herstmonceux	9865	−9.1 ± 12.0	−5.7 ± 6.8	8127	−9.7 ± 13.5	−6.2 ± 7.9
Matera	3658	−3.7 ± 14.2	−3.7 ± 9.2	4743	−8.5 ± 14.1	−5.1 ± 11.1
Mount Stromlo	26,925	8.0 ± 16.8	9.6 ± 12.0	30,216	7.7 ± 16.6	8.6 ± 12.7
Potsdam	14,780	−6.1 ± 14.3	−3.2 ± 9.5	12,106	−7.6 ± 14.4	−4.8 ± 10.0
Wettzell	636	4.0 ± 12.9	7.4 ± 10.1	516	−0.6 ± 13.0	3.1 ± 12.3
Yarragadee	74,533	0.8 ± 16.3	4.9 ± 8.3	64,889	1.2 ± 17.4	4.4 ± 9.7
Zimmerwald	22,825	−4.0 ± 12.9	−2.4 ± 8.2	24,230	−5.1 ± 14.2	−2.9 ± 10.7
Total	204,667	−1.5 ± 16.9	1.6 ± 11.4	195,506	−1.6 ± 17.5	1.2 ± 12.5

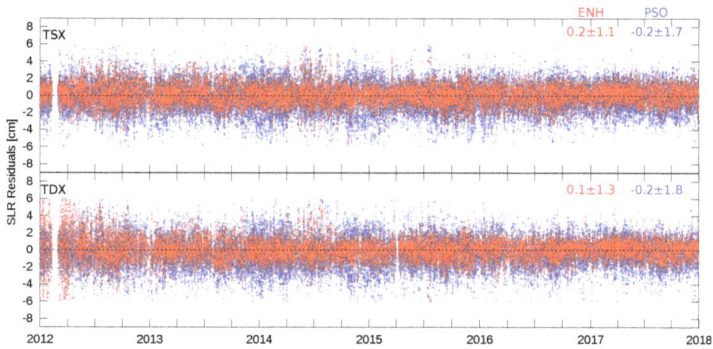

Figure 7. Satellite laser ranging residuals obtained for the TSX and TDX enhanced orbit solutions (ENH, red) and the precise science orbits (PSO, blue).

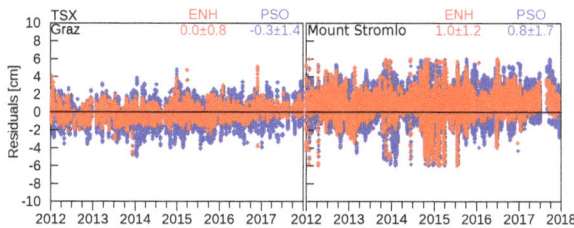

Figure 8. Satellite laser ranging residuals of TSX orbit solutions from the ILRS stations Graz, Austria, and Mount Stromlo, Australia.

The SLR results listed in Table 7 demonstrate a 17.5 mm to 11.4 mm consistency across the selected stations. Basically, the SLR residuals constitute a measure of the one-dimensional orbit error. However, a simplified correlation to the associated 3-D position error is given by $\sigma_{pos} = \sqrt{3}\sigma_{SLR}$ [10], and amounts to 7.2 mm to 5.8 mm for the mentioned orbit products. Overall, the satellite laser ranging residuals in Table 7 result in a mean value of 1.6 mm and a standard deviation of 11.4 mm for TSX, which is similar to the TDX solutions of 1.2 ± 12.5 mm. Compared to the results obtained with the precise science orbit products, the enhanced solutions show a 33 % reduction of the standard deviation, which is clear evidence for the improvements achieved with this solution. Depending on the elevation cut-off angle, the satellite laser ranging residuals are confined to a cone of 50°–80° with respect to the satellite's nadir direction. Therefore, the SLR measurements are most sensitive to radial contributions of the orbit error.

5.2. Radar Ranging

For the SAR-based orbit validation, two independent sets of residuals can be analyzed, namely a set of residuals for the range and a set of residuals for the azimuth. Table 8 lists the results for both types of observations as well as for TSX and TDX. The findings show a very good agreement regarding the standard deviations across the stations and the two orbit solutions tested. With the PSOs, the standard deviation of the SAR ranging is in the order of 15 mm, while for the ENH solution we see an improvement to 10 mm. Note that the standard deviations of the GARS O'Higgins range results are larger when compared to standard deviations of the other two stations hosting CRs. Part of the explanation may be found in the different CR sizes, i.e., 0.7 m versus 1.5 m. The estimates presented earlier in Table 3 indicate a drop in the range localization precision by approximately 5 mm for the smaller reflectors. Moreover, there is a slight tendency towards the positive range for the TDX range data of GARS O'Higgins between mid 2016 to mid 2017, see Figure 9, which is present in both orbit solutions. The effect also has an impact on the mean and standard deviations computed for this site. The much sparser data of TSX for GARS O'Higgins indicate a similar behavior; thus we think it is related to CR installations at GARS O'Higgins, and it is also the reason why there is less improvement in the overall TDX results and the ENH orbit solution, because the data is dominated by the acquisitions at GARS O'Higgins.

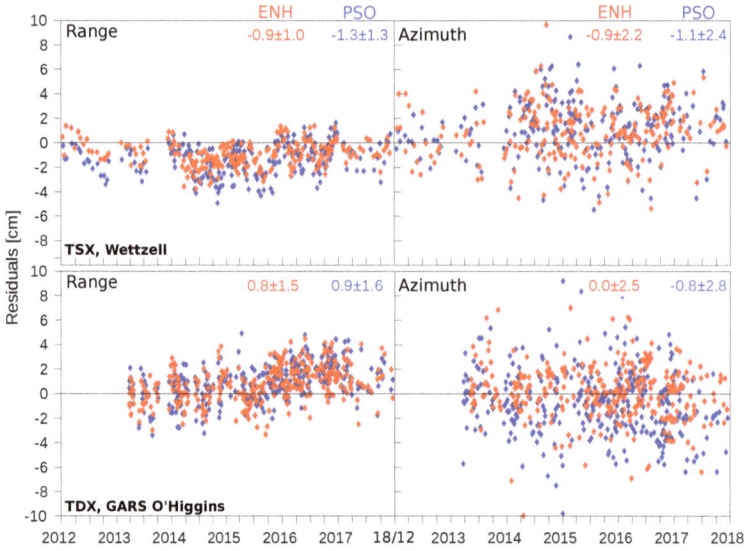

Figure 9. TerraSAR-X range and azimuth residuals derived from the Wettzell corner reflectors (top) and the GARS O'Higgins corner reflectors (bottom).

Table 8. SAR residuals ($\bar{x} \pm \sigma_x$ [mm]) and underlying number of data takes (N_{dt}) for the stations with corner reflectors of the TSX and TDX orbit validation.

Station	TSX			TDX		
	N_{dt}	PSO	ENH	N_{dt}	PSO	ENH
Wettzell Range	235	-13.7 ± 13.1	-9.4 ± 10.3	66	-11.5 ± 12.9	-7.4 ± 8.1
O'Higgins Range	135	1.3 ± 16.3	1.9 ± 14.6	305	9.3 ± 15.7	7.6 ± 15.2
Metsähovi Range	178	-0.1 ± 11.2	0.3 ± 10.0	36	-0.4 ± 12.8	-0.3 ± 10.1
Wettzell Azimuth	235	10.7 ± 23.9	9.2 ± 21.9	66	5.2 ± 29.2	8.6 ± 24.6
O'Higgins Azimuth	135	18.0 ± 25.9	10.0 ± 23.7	305	-7.6 ± 28.2	0.1 ± 25.1
Metsähovi Azimuth	178	0.0 ± 17.6	0.0 ± 15.2	36	0.0 ± 18.2	0.0 ± 14.8
Total Range	548	-5.6 ± 15.1	-3.5 ± 12.5	407	5.1 ± 16.9	4.5 ± 14.9
Total Azimuth	548	9.0 ± 23.6	6.4 ± 20.9	407	-4.8 ± 28.0	1.4 ± 24.4

Regarding the results in azimuth, one immediately notices the increase in the standard deviations by approximately a factor of 1.5 when compared to range. Once more there is an impact of the smaller CRs of GARS O'Higgins, resulting in azimuth standard deviations of 25 mm. However, for Wettzell we would expect the azimuth results to be closer to the 16 mm azimuth standard deviation observed at Metsähovi. Instead, the Wettzell azimuth results are almost identical to the findings of GARS O'Higgins. The main driver for the TerraSAR-X quality in azimuth is the temporal link between the absolute time provided by the MosaicGNSS receiver and the SAR payload. The limit of the link is considered to be approximately 1 microsecond once the significant digits truncated by the on-board data quantization have been restored [44]. An uncertainty of 1 microsecond translates into approximately 7 mm in azimuth when taking into account the average TerraSAR-X orbit velocity of 7683 m/s. However, the azimuth results we obtain from the experiments show a larger noise than we would expect from this consideration, and therefore further investigations on the SAR payload and the processing are required to identify their possible cause.

The detailed station results for Wettzell and GARS O'Higgins are shown in Figure 9. The plots confirm that the larger azimuth standard deviations are indeed due to an increase in the azimuth noise, and that the range is the more sensitive measurement of the SAR. There are slight temporal variations visible in the SAR range results of both orbit solutions, which are similar to the variations found in the temporal series of the SLR residuals, see Figure 8. Most prominent is the aforementioned systematic effect in the range for the TDX data at GARS O'Higgins. However, because of the smaller amount of stations in the SAR analysis, there is no definite conclusion whether they are due to SAR payloads or because of local signals related to the CR sites. As a summary, Figure 10 provides a comprehensive view of all the SAR ranging residuals for the ENH and the PSO orbit solutions. The reason for the small number of TSX data and the lack of data for TDX in the first year lies in the setup of the CR sites. During the first year, only one reflector was available at Wettzell, Germany, which was covered solely by the TDX spacecraft. In spite of the details discussed for the solutions of the individual CR sites, these combined plots confirm the very good overall consistency of both TSX and TDX across the sites, as well as the high quality of the ENH orbit solution.

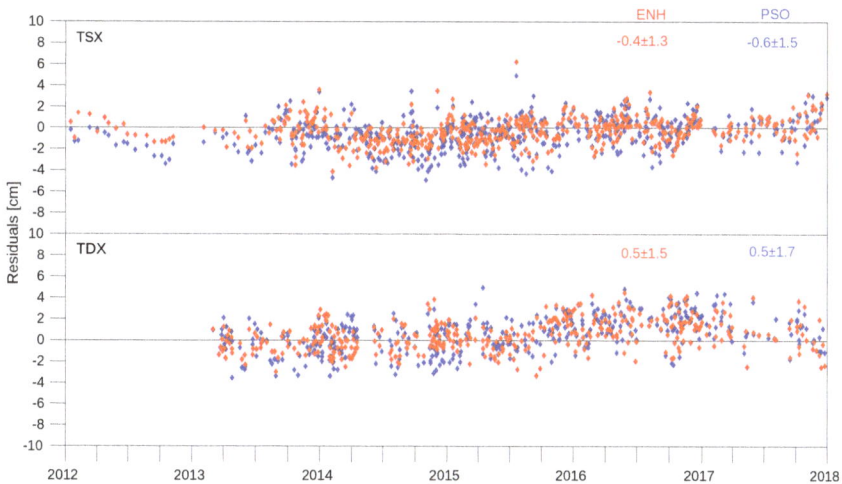

Figure 10. Radar ranging residuals obtained for TSX and TDX enhanced (ENH, red) and precise science orbit (PSO, blue) orbit solutions.

5.3. Position Offset Estimates

The derived SLR range residuals give a measure of the radial accuracy. Whenever the scope is on physical effects or on the accelerations in a certain direction, the SLR-based position offset estimates can improve the interpretation. Therefore, the laser measurements are further utilized for an estimation of spacecraft position offsets in radial, along-, and cross-track directions. The applied setup of SLR preprocessing and data screening is identical to the approach described above, and the analysis period also covers 6 years and both orbit solutions for each of the spacecrafts. The values listed in the results are based on monthly averages.

In principle, the SAR residuals in range and azimuth allow for the same position offset analysis. However, the permanently right-looking nature of the SAR instruments yields radar ranges that basically cover only one cross-direction of the orbit, for which the ranges have an angular separation of approximately 15 to 20 degrees, see the pass geometries given in Table 5. The advantage of measuring all the pass geometries and the considerably larger amount of tracking data enables the SLR to reliably determine the monthly position offsets for all three directions with an estimated accuracy at the sub-millimeter level. In contrast, our tests with the SAR residuals showed estimated accuracies for the cross-track and radial position offsets at the centimeter level, and an accuracy at the millimeter level for the along-track position offsets. Because of these results, we decided to keep the SAR data in the geometry of range and azimuth, but to apply a monthly averaging to preserve the comparison with the SLR-based position offsets. Prior to the monthly averaging, the full rate SAR residuals of range and azimuth were reduced by global 2σ tests. The tests eliminate approximately 4 % of the SAR residuals in the TSX and TDX data series, which exceed the thresholds of two times the estimated standard deviations. They also remove the few remaining outliers in the range and azimuth residuals (see for instance the ranging residuals shown in Figure 10) that would otherwise distort the monthly mean estimates.

Table 9 comprises the estimated offset values that are associated with the time series visualized in Figures 11 and 12. The SLR analysis shows average monthly offsets well below 12 mm in the spacecraft's radial, along-, and cross-track direction. The cross-track component underlines the clear advantage of the ENH solutions, for which the remaining mean values are reduced from one centimeter to less than one millimeter. The induced shift between the PSO and the ENH solutions is similar for both spacecrafts and can bee seen in Figure 11. The along-track component is also slightly improved. The monthly SAR azimuth results are comparable to the SLR along-track estimates and thus we decided to place them side by side in Table 9. The comparison of both results indicate that the SLR along-track offsets are approximately 2–3 times more reliable than the 5–7 mm offsets found for SAR azimuth, but one should also take into account the large differences in the underlying data, namely about 2000 NPs per month for the SLR versus the 5–10 SAR acquisitions per month at the CR sites. Nevertheless, the improvement of the ENH solutions is also clearly visible in the SAR azimuth offsets, especially in the mean value of TDX that is reduced from −7.2 mm to 1.4 mm. The monthly SAR range offsets given in Table 9 confirm the findings for the ENH solutions. Again we observe a reduction of both the mean values and the standard deviations well below 10 mm. Note that the slightly larger numbers for TDX are due to the systematics in the GARS O'Higgins data during the years 2016 and 2017, which was already discussed in the previous section and can be seen once more in Figure 12. In summary, we can conclude that both techniques agree in their monthly assessment of the TerraSAR-X orbit solutions.

Table 9. Monthly position offsets in radial, along-track and cross-track derived from satellite laser ranging, and monthly range and azimuth offsets derived from the SAR measurements. Obtained from the precise science (PSO) and the enhanced orbit solutions (ENH) for TSX and TDX ($\bar{x} \pm \sigma_x$ [mm]).

Type	Orbit	TSX			TDX		
		Radial	Along-Track	Cross-Track	Radial	Along-Track	Cross-Track
SLR	PSO	−1.6 ± 2.8	2.1 ± 4.1	−11.9 ± 8.1	−2.3 ± 3.4	2.7 ± 3.4	−10.0 ± 8.0
SLR	ENH	2.4 ± 2.7	0.4 ± 3.2	0.5 ± 1.9	1.5 ± 3.1	1.8 ± 2.7	−0.1 ± 1.9
			azimuth	range		azimuth	range
SAR	PSO		6.7 ± 8.9	−6.8 ± 8.2		−7.2 ± 12.6	5.1 ± 10.6
SAR	ENH		5.4 ± 7.9	−3.4 ± 6.9		1.4 ± 9.2	3.1 ± 8.3

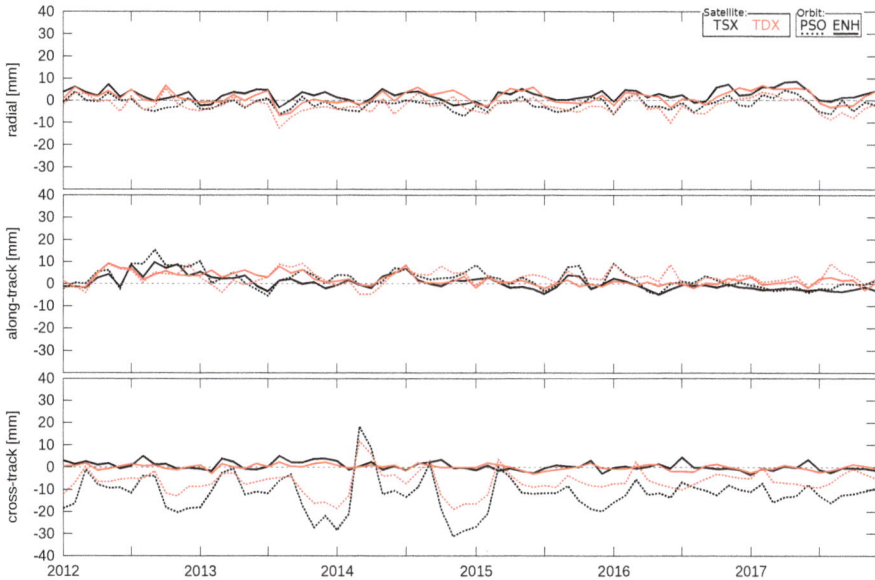

Figure 11. Series of monthly-derived position offsets obtained from satellite laser ranging for the enhanced (ENH) and the precise science orbits (PSOs) of the TerraSAR-X mission.

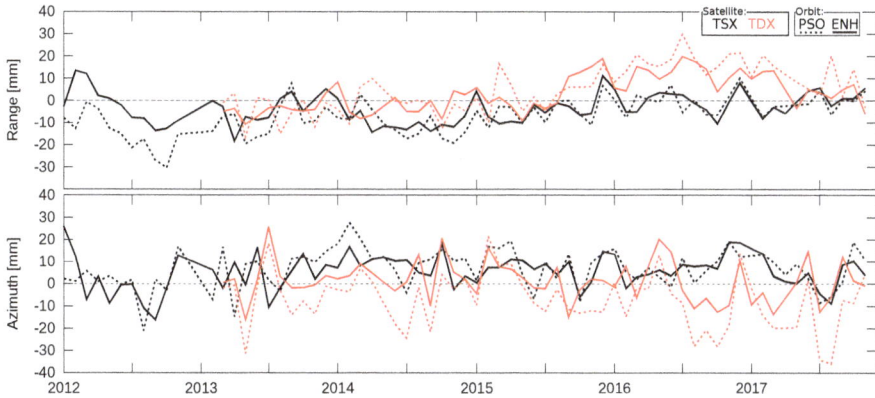

Figure 12. Series of monthly-derived SAR range and azimuth residuals for the enhanced (ENH) and the precise science orbits (PSOs) of the TerraSAR-X mission.

6. Conclusions and Outlook

Independent distance measurements to/from satellites are perfectly suited to orbit validation. SAR and SLR are two active techniques, which allow measuring the distance to a point target, known as a LRRs onboard the satellite in case of laser, or CR on ground in case of radar observations. Both techniques require availability of precise coordinates, which is an important pre-condition. Taking into account 11 stations from the ILRS network for SLR, and 3 stations with trihedral CRs for the SAR ranging, orbit solutions of the spacecrafts TSX and TDX are thoroughly analyzed within the 2012–2017 period. Two types of orbit solutions are computed at GSOC/DLR and employed for comparison: the operational PSO products, and enhanced orbit solutions, which incorporate refined dynamical models, and the GPS integer ambiguity fixing. The validation using the SLR residuals is based on differences between observed and modeled station-to-satellite ranges, and demonstrates a consistency of the TSX orbit products with the optical measurements at a level of 12–17 mm. Corresponding radar ranging residuals are obtained from differences between observed and modeled satellite-to-station ranges, are consistent at a level of 18–26 mm. The results obtained from TDX solutions are similar. Regarding the number of available tracking data, 400,173 SLR normal points are available for the spacecrafts TSX and TDX within the 6 years, whereas 1910 SAR measurements in range and azimuth are available in the same period. The outstanding number of SLR tracking points is because of the high tracking priority of the spacecrafts.

The number of available data limits the capabilities of radar ranging to estimate spacecraft offsets. In case of SLR, a 10-mm shift in cross-track direction can be observed when switching from the PSO to the enhanced orbit solutions. For instance, such offsets can be utilized for identifying center-of-gravity position corrections, or SLR station corrections [10,19]. Similar to the SLR, the SAR range residuum also shows a shift of several millimeters.

Regarding the abilities of both validiation techniques, the following conclusions can be stated. The SLR depends on weather and cloud conditions, whereas SAR is principally free of these limitations. On the other hand, SAR is limited by the corrections for the atmospheric path delays, in particular for the ionosphere, because the range measurements behave like single-frequency microwave observations. The SLR may observe all passes from horizon to horizon, whereas conventional spaceborne SAR payloads observe only in a fixed right-looking (or left-looking) direction and provide only the observations at the instant of zero Doppler. SLR needs dedicated stations while the SAR satellites have global access, provided that passive CRs are available.

Especially for radar satellite missions, the proposed SAR validation paves the way for another, independent orbit validation tool, which is of particular interest for the Sentinel-1 mission, because these spacecrafts are not equipped with LRRs [51]. Since the SAR-based validation technique requires less infrastructure when compared to an SLR station, an extension of the ground corner reflector network is envisaged.

Author Contributions: All authors conceived the study. Oliver Montenbruck developed the POD software package and thoroughly revised and edited the work. Martin Wermuth is in charge of operational TerraSAR-X and TanDEM-X orbit determination. Ulrich Balss modified the TerraSAR-X SAR processor for the refined azimuth timing, and prepared the SAR data. Christoph Gisinger analyzed the orbit solutions via radar ranging. The orbit solution estimation, as well as the SLR-based orbit validation were performed by Stefan Hackel.

Funding: The work was partly funded by the German Helmholtz Association HGF through its DLR@Uni Munich Aerospace project "Hochauflösende geodätische Erdbeobachtung".

Acknowledgments: We thank our cooperation partners—the Federal Agency for Cartography and Geodesy (BKG) and the Finnish Geospatial Research Institute (FGI)—for their kind allowance to install the corner reflectors at their property in Wettzell and Metsähovi, respectively, and for their local support in maintenance. We thank our colleagues from DLR's Remote Sensing Date Center (DFD), who installed and maintain the corner reflectors at GARS O'Higgins.

Conflicts of Interest: The authors declare no conflict of interest.

Abbreviations

The following abbreviations are used in this manuscript:

CR	Corner Reflector
CNES/CLS	Centre National d'Études Spatiales/Collecte Localisation Satellites
DLR	German Aerospace Center
ENH	ENHanced
GNSS	Global Navigation Satellite System
GHOST	High Precision Orbit Determination Software Tools
GPS	Global Positioning System
GFZ	Deutsches GeoForschungsZentrum
GARS	German Antarctic Receiving Station

GSOC	German Space Operations Center
IERS	International Earth Rotation and Reference Systems Service
ILRS	International Laser Ranging Service
IGS	International GNSS Service
IGOR	Integrated Geodetic and Occultation Receiver
ITRF	International Terrestrial Reference Frame
LRR	Laser Retro Reflector
NP	Normal Point
PTA	Point Target Analysis
POD	Precise Orbit Determination
PSO	Precise Science Orbits
SAR	Synthetic Aperture Radar
SCR	Signal to Clutter Ratio
SLR	Satellite Laser Ranging
SLRF	Satellite Laser Ranging Frame
TMSP	TerraSAR-X Multimode SAR Processor

References

1. Werninghaus, R.; Buckreuss, S. The TerraSAR-X Mission and System Design. *IEEE Trans. Geosci. Remote Sens.* **2010**, *48*, 606–614. [CrossRef]
2. Moreira, A.; Krieger, G.; Hajnsek, I.; Hounam, D.; Werner, M.; Riegger, S.; Settelmeyer, E. TanDEM-X: A TerraSAR-X Add-On Satellite for Single-Pass SAR Interferometry. *IEEE Trans. Geosci. Remote Sens.* **2004**, *2*, 1000–1003.
3. Eineder, M.; Minet, C.; Steigenberger, P.; Cong, X.; Fritz, T. Imaging geodesy—Toward centimeter-level ranging accuracy with TerraSAR-X. *IEEE Trans. Geosci. Remote Sens.* **2011**, *49*, 661–671. [CrossRef]
4. Schubert, A.; Jehle, M.; Small, D.; Meier, E. Mitigation of atmospheric perturbations and solid Earth movements in a TerraSAR-X time-series. *J. Geod.* **2012**, *86*, 257–270. [CrossRef]
5. Gisinger, C.; Balss, U.; Pail, R.; Zhu, X.; Montazeri, S.; Gernhardt, S.; Eineder, M. Precise Three-Dimensional Stereo Localization of Corner Reflectors and Persistent Scatterers with TerraSAR-X. *IEEE Trans. Geosci. Remote Sens.* **2015**, *53*, 1782–1802. [CrossRef]
6. Balss, U.; Gisinger, C.; Eineder, M. Measurements on the Absolute 2-D and 3-D Localization Accuracy of TerraSAR-X. *Remote Sens.* **2018**, *10*, 656. [CrossRef]
7. Krieger, G.; Moreira, A.; Fiedler, H.; Hajnsek, I.; Werner, M.; Younis, M.; Zink, M. TanDEM-X: A Satellite Formation for High-Resolution SAR Interferometry. *IEEE Trans. Geosci. Remote Sens.* **2007**, *45*, 3317–3341. [CrossRef]
8. Yoon, Y.; Eineder, M.; Yague-Martinez, N.; Montenbruck, O. TerraSAR-X Precise Trajectory Estimation and Quality Assessment. *IEEE Trans. Geosci. Remote Sens.* **2009**, *47*, 1859–1868. [CrossRef]
9. Tapley, B.D.; Schutz, B.E.; Eanes, R.J. Satellite laser ranging and its applications. *Celest. Mech.* **1985**, *37*, 247–261. [CrossRef]
10. Arnold, D.; Montenbruck, O.; Hackel, S.; Sośnica, K. Satellite Laser Ranging to Low Earth Orbiters—Orbit and Network Validation. *J. Geod.* **2018**, 1–20. [CrossRef]
11. Balss, U.; Cong, X.; Brcic, R.; Rexer, M.; Minet, C.; Breit, H.; Eineder, M.; Fritz, T. High precision measurement on the absolute localization accuracy of TerraSAR-X. In Proceedings of the 2012 IEEE International Geoscience and Remote Sensing Symposium (IGARSS), Munich, Germany, 22–27 July 2012; pp. 1625–1628.
12. Kahle, R.; Kazeminejad, B.; Kirschner, M.; Yoon, Y.; Kiehling, R.; D'Amico, S. First In-Orbit Experience of TerraSAR-X flight dynamic operations. In Proceedings of the 20th International Symposium on Space Flight Dynamics, Annapolis, MD, USA, 24–28 September 2007.
13. Buckreuss, S.; Balzer, W.; Mühlbauer, P.; Werninghaus, R.; Pitz, W. The TerraSAR-X satellite project. In Proceedings of the 2003 IEEE International Geoscience and Remote Sensing Symposium, Toulouse, France, 21–25 July 2003; Volume 5, pp. 3096–3098.
14. Eineder, M.; Runge, H.; Boerner, E.; Bamler, R.; Adam, N.; Schättler, B.; Breit, H.; Suchandt, S. SAR interferometry with TerraSAR-X. In Proceedings of the FRINGE 2003 Workshop, Frascati, Italy, 1–5 December 2000.

15. Montenbruck, O.; Kroes, R. In-flight performance analysis of the CHAMP BlackJack GPS Receiver. *GPS Solut.* **2003**, *7*, 74–86. [CrossRef]

16. Montenbruck, O.; Yoon, Y.; Ardaens, J.S.; Ulrich, D. In-Flight Performance Assessment of the Single Frequency MosaicGNSS Receiver for Satellite Navigation. In Proceedings of the 7th International ESA Conference on Guidance, Navigation and Control Systems, County Kerry, Ireland, 2–5 June 2008.

17. Wermuth, M.; Montenbruck, O.; van Helleputte, T. GPS High Precision Orbit Determination Tools (GHOST). In Proceedings of the 4th International Conference on Astrodynamics Tools and Techniques, Madrid, Spain, 3–6 May 2010.

18. Hackel, S.; Montenbruck, O.; Steigenberger, P.; Balss, U.; Gisinger, C.; Eineder, M. Model improvements and validation of TerraSAR-X precise orbit determination. *J. Geod.* **2017**, *91*, 547–562. [CrossRef]

19. Montenbruck, O.; Hackel, S.; Jäggi, A. Precise orbit determination of the Sentinel-3A altimetry satellite using ambiguity-fixed GPS carrier phase observations. *J. Geod.* **2017**. [CrossRef]

20. Loyer, S.; Perosanz, F.; Mercier, F.; Capdeville, H.; Marty, J.C. Zero-difference GPS ambiguity resolution at CNES-CLS IGS Analysis Center. *J. Geod.* **2012**, *86*, 991–1003. [CrossRef]

21. Schmid, R.; Dach, R.; Collilieux, X.; Jäggi, A.; Schmitz, M.; Dilssner, F. Absolute IGS antenna phase center model igs08.atx: status and potential improvements. *J. Geod.* **2016**, *90*, 343–364. [CrossRef]

22. Altamimi, Z.; Rebischung, P.; Métivier, L.; Collilieux, X. ITRF2014: A new release of the International Terrestrial Reference Frame modeling nonlinear station motions. *J. Geophys. Res. Solid Earth* **2016**, *121*, 6109–6131. [CrossRef]

23. Wu, J.; Wu, S.; Hajj, G.; Bertiger, W.; Lichten, S. Effects of antenna orientation on GPS carrier phase. *Manuscr. Geod.* **1993**, *18*, 91–98.

24. Mayer-Gürr, T.; Pail, R.; Schuh, W.D.; Kusche, J.; Baur, O.; Jäggi, A. The new combined satellite only model GOCO03s. In Proceedings of the International Symposium on Gravity, Geoid and Height Systems GGHS2012, Venice, Italy, 9–12 October 2012.

25. Eanes, R.J.; Bettadpur, S. *The CSR 3.0 Global Ocean Tide Model: Diurnal and Semi-Diurnal Ocean Tides from TOPEX/POSEIDON Altimetry*; Technical Report 6; Center for Space Research, University of Texas at Austin: Austin, TX, USA, 1995.

26. Priestley, K.J.; Smith, G.L.; Thomas, S.; Cooper, D.; Lee, R.B.; Walikainen, D.; Hess, P.; Szewczyk, Z.P.; Wilson, R. Radiometric Performance of the CERES Earth Radiation Budget Climate Record Sensors on the EOS Aqua and Terra Spacecraft through April 2007. *J. Atmos. Ocean. Technol.* **2011**, *28*, 3–21. [CrossRef]

27. Picone, J.M.; Hedin, A.E.; Drob, D.P.; Aikin, A.C. NRLMSISE-00 empirical model of the atmosphere: Statistical comparisons and scientific issues. *J. Geophys. Res.* **2002**, *107*, 1468. [CrossRef]

28. Shampine, L.F.; Gordon, M.K. *Computer Solution of Ordinary Differential Equations: The Initial Value Problem*; W.H. Freeman & Co., Ltd.: San Francisco, CA, USA, 1975.

29. Pearlman, M.R.; Degnan, J.J.; Bosworth, J.M. The International Laser Ranging Service. *Adv. Space Res.* **2002**, *30*, 135–143. [CrossRef]

30. Neubert, R.; Grunwaldt, L.; Neubert, J. The Retro-Reflector for the CHAMP Satellite: Final Design and Realization. In Proceedings of the 11th International Workshop on Laser Ranging, Deggendorf, Germany, 21–25 September 1998.

31. International Laser Ranging Service. SLRF2014 Station Coordinates. 2017. Available online: ftp://ftp. cddis.eosdis.nasa.gov/pub/slr/products/resource/SLRF2014_POS+VEL_2030.0_170605.snx (accessed on 8 June 2017).

32. Ray, R. *A Global Ocean Tide Model from TOPEX/Poseidon Altimeter: GOT99*; Tm-209478; NASA Technical Memorandum; NASA: Greenbelt, MA, USA, 1999.

33. Mendes, V.; Pavlis, E. High-accuracy zenith delay prediction at optical wavelengths. *Geophys. Res. Lett.* **2004**, *31*, L14602. [CrossRef]

34. Pavlis, E. SLRF2008: The ILRS reference frame for SLR POD contributed to ITRF2008. In Proceedings of the 2009 Ocean Surface Topography Science Team Meeting, Buellton, CA, USA, 19 June 2009.

35. International Association of Geodesy. International Laser Ranging Service Mission Priorities as of December 25, 2017. Available online: https://ilrs.cddis.eosdis.nasa.gov/missions/mission_operations/priorities/index.htmll (accessed on 25 December 2017).

36. International Laser Ranging Service. TerraSAR-X. 2015. Available online: http://ilrs.gsfc.nasa.gov/missions/satellite_missions/current_missions/tsar_general.html (accessed on 15 January 2015).

37. Cumming, I.G.; Wong, F.H. *Digital Processing of Synthetic Aperture Radar Data*; Artech House: Norwood, MA, USA, 2005.

38. Breit, H.; Fritz, T.; Balss, U.; Lachaise, M.; Niedermeier, A.; Vonavka, M. TerraSAR-X SAR Processing and Products. *IEEE Trans. Geosci. Remote Sens.* **2010**, *48*, 727–740. [CrossRef]

39. TerraSAR-X Ground Segment Basic Product Specification Document. TX-GS-DD-3302, v1.9, 09.10.2014. 2013. Available online: http://sss.terrasar-x.dlr.de (accessed on 26 March 2018).

40. Döring, B.; Schwerdt, M.R.B. TerraSAR-X Calibration Ground Equipment. In Proceedings of the Wave Propagation in Communication, Microwaves Systems and Navigation (WFMN), Chemnitz, Germany, 4–5 July 2007.

41. Petit, G.; Luzum, B. (Eds.) *IERS Conventions (2010)*; Verlag des Bundesamts für Kartographie und Geodäsie: Frankfurt, German, 2010. Available online: http://tai.bipm.org/iers/conv2010/conv2010.html (accessed on 26 March 2018).

42. Stein, S. Algorithms for Ambiguity Function Processing. *IEEE Trans. Acoust. Speech Signal Process.* **1981**, *ASSP-29*, 588–599. [CrossRef]

43. Bamler, R.; Eineder, M. Accuracy of Differential Shift Estimation by Correlation and Split-Bandwidth Interferometry for Wideband and Delta-k SAR Systems. *IEEE Geosci. Remote Sens. Lett.* **2005**, *2*, 151–155. [CrossRef]

44. Balss, U.; Breit, H.; Fritz, T.; Steinbrecher, U.; Gisinger, C.; Eineder, M. Analysis of Internal Timings and Clock Rates of TerraSAR-X. In Proceedings of the 2014 IEEE International Geoscience and Remote Sensing Symposium (IGARSS), Quebec City, QC, Canada, 13–18 July 2014; pp. 2671–2674.

45. Byram, S.; Hackman, C.; Tracey, J. Computation of a High-Precision GPS-Based Troposphere Product by the USNO. In Proceedings of the 24th international technical meeting of the satellite division of the institute of navigation (ION GNSS 2011), Portland, OR, USA, 20–23 September 2011; pp. 572–578.

46. Altamimi, Z.; Collilieux, X.; Métivier, L. ITRF2008: An improved solution of the international terrestrial reference frame. *J. Geod.* **2011**, *85*, 457–473. [CrossRef]

47. Schlüter, W.; Zernecke, R.; Becker, S.; Klügel, T.; Thaller, D. Local Ties between the Reference Points at the Fundamentalstation Wettzell. In Proceedings of the IERS Workshop on Site Co-Location, Matera, Italy, 23–24 October 2003; IERS Technical Note 33; Richter B., Dick W.R., Schwegmann, W., Eds.; pp. 64–70. Available online: https://www.iers.org/IERS/EN/Publications/publications.html (accessed on 26 March 2018).

48. Klügel, T.; Mähler, S.; Schade, C. Ground Survey and Local Ties at the Geodetic Observatory Wettzell. In Proceedings of the 17th International Workshop on Laser Ranging, Bad Koetzting, Germany, 16–20 May 2011. Available online: https://cddis.nasa.gov/lw17/ (accessed on 26 March 2018).

49. International Earth Rotation and Reference Systems Service (IERS). The International Terrestrial Reference Frame (ITRF). 2018. Available online: https://www.iers.org/IERS/EN/DataProducts/ITRF/itrf.html (accessed on 26 March 2018).

50. Schwerdt, M.; Bräutigam, B.; Bachmann, M.; Döring, B.; Schrank, D.; Gonzalez, J.H. Final TerraSAR-X Calibration Results Based on Novel Efficient Methods. *IEEE Trans. Geosci. Remote Sens.* **2010**, *48*, 677–689. [CrossRef]

51. Peter, H.; Jäggi, A.; Fernández, J.; Escobar, D.; Ayuga, F.; Arnold, D.; Wermuth, M.; Hackel, S.; Otten, M.; Simons, W.; et al. Sentinel-1A—First precise orbit determination results. *Adv. Space Res.* **2017**, *60*, 879–892. [CrossRef]

remote sensing

MDPI

Article

Measurements on the Absolute 2-D and 3-D Localization Accuracy of TerraSAR-X

Ulrich Balss *, Christoph Gisinger and Michael Eineder

DLR, Remote Sensing Technology Institute, Muenchener Str. 20, D-82234 Wessling, Germany;
christoph.gisinger@dlr.de (C.G.); michael.eineder@dlr.de (M.E.)
* Correspondence: ulrich.balss@dlr.de; Tel.: +49-8153-28-2145

Received: 23 March 2018; Accepted: 20 April 2018; Published: 23 April 2018

Abstract: The German TerraSAR-X radar satellites TSX-1 and TDX-1 are well-regarded for their unprecedented geolocation accuracy. However, to access their full potential, Synthetic Aperture Radar (SAR)-based location measurements have to be carefully corrected for effects that are well-known in the area of geodesy but were previously often neglected in the area of SAR, such as wave propagation and Earth dynamics. Our measurements indicate that in this way, when SAR is handled as a geodetic measurement instrument, absolute localization accuracy at better than centimeter level with respect to a given geodetic reference frame is obtained in 2-D and, when using stereo SAR techniques, also in 3-D. The TerraSAR-X measurement results presented in this study are based on a network of three globally distributed geodetic observatories. Each is equipped with one or two trihedral corner reflectors with accurately (<5 mm) known reference coordinates, used as a reference for the verification of the SAR measured coordinates. Because these observatories are located in distant parts of the world, they give us evidence on the worldwide reproducibility of the obtained results. In this paper we report the achieved results of measurements performed over 6 1/2 years (from July 2011 to January 2018) and refer to some first new application areas for geodetic SAR.

Keywords: synthetic aperture radar; TerraSAR-X; geolocation; absolute localization accuracy; stereo sar; imaging geodesy

---

1. Introduction

Space-borne SAR is mainly known for its ability to provide image observations of the Earth's surface and the measurement of relative shifts making use of the carrier phase (i.e., SAR interferometry)—independent from weather and time of day. However, SAR offers several additional capabilities. The objective of this paper is to highlight the ability to provide also absolute localization accuracy of bright, well-detectable radar targets at centimeter level [1–4] with respect to a given geodetic reference frame like International Terrestrial Reference Frame, release 2014 (ITRF2014) [5]. Such a high accuracy—where TerraSAR-X/TanDEM-X lead worldwide among space borne SAR sensors [6]—can be achieved only if the SAR data are processed and calibrated with meticulous care and if they are corrected for well-known effects such as wave propagation and solid Earth dynamics, as it is done in geodesy [7].

Aiming at centimeter level accuracy, some basic aspects of SAR geolocation have to be reassessed. Therefore, the paper will start with a discussion of refinements compared to traditional techniques in the concept of SAR based location measurements. Furthermore, the verification of the localization accuracy requires a standard of comparison at least as good as SAR. Thus, we had to exercise utmost care with the installation and survey of our corner reflector (CR) test sites. If at least two SAR images with different acquisition geometries are involved, Stereo SAR techniques even allow to derive accurate 3-D on-ground coordinates when combining the ensemble of 2-D image coordinates. Equipped with this set of tools, we analyzed the geolocation accuracy of TerraSAR-X in detail, considering many potential influences like incidence angle, acquisition mode or polarization as well as the long-term stability of the SAR instrument or potential dependencies on the geographic location of the target. Our measurement results in 2-D and 3-D and their interpretation are the key aspects of this work. A brief outlook on an upcoming further improvement in geolocation accuracy and on some first applications for precise geolocation in the area of imaging geodesy, an area brought about by the geolocation capabilities of TerraSAR-X, completes the paper.

2. Background

2.1. Location Measurements by SAR

Radar systems indirectly measure geometric distances by means of the two-way travel time of radar pulses. The conversion from travel time to geometric distance, i.e., range, depends on the velocity of light (divided by two in order to convert from *two*-way travel time to *one*-way distance). Usually, the vacuum velocity of light is used in this context which however is larger than the true signal travel velocity if the signal passes through the Earth's atmosphere. Here, electrons in the ionosphere as well as dry air and water vapor, mainly contained in the troposphere, introduce additional signal delays which have to be taken into account. The impact of the tropospheric delay on TerraSAR-X range measurements typically varies between 2.5 and 4 m depending on terrain height and incidence angle. The dispersive ionospheric delay at X-band for TerraSAR-X (9.65 GHz) amounts from several centimeters up to a few decimeters. The indirect annotation of the signal delays as part of the geometrical SAR range bias in units of length is common practice when generating SAR products, because it is convenient when correcting range measurements, but this leads to a mixing of different effects and complicates the usage of alternative atmospheric correction methods.

As atmospheric delays are likewise relevant for the Global Navigation Satellite System (GNSS) [8], which also applies centimeter wavelength radio signals (i.e., L-band, about 1-2 GHz), the International GNSS Service (IGS) [9] provides path delay products inferred from the permanently operated GNSS receivers of the global IGS network. While the correction values for the non-dispersive (mainly tropospheric) delay can be directly applied to the SAR ranges, the correction values for the dispersive (mainly ionospheric) delay need to be adapted to the differences in the radio frequency (9.65 GHz instead of L-band) and the change in orbit height. Because the orbits of TerraSAR-X are significantly lower than the GNSS orbits and lie still within the upper ionosphere, only a portion of the entire Total Electron Content (TEC) in the ionosphere that is seen by GNSS has to be considered in TerraSAR-X datatakes. We found that the usage of a constant TEC scaling factor of 75% provides reasonable results [10].

Similar to the range, the azimuth coordinate of a ground target is indirectly given by a time measurement. As focused TerraSAR-X images correspond to zero-Doppler geometry, the azimuth position of the target in the image represents the time of closest approach between sensor and target. The conversion from azimuth time to a spatial coordinate is given by the zero-Doppler condition [11] and requires precise knowledge of the satellite's orbit and the temporal synchronization between the orbit timeline and the operation of the radar instrument.

When assigning SAR measurements to a geodetic reference frame, the geodynamic effects shifting the true position of a ground target have to be considered in accordance with geodetic conventions [7]. The most prominent of these effects are solid Earth tides and plate tectonics which cause a periodic variation of up to a few decimeters over the course of a day or linearly accumulating shifts on the order of centimeters per year, respectively. The non-tidal atmospheric pressure loading (not part of the conventions) and the ocean tidal loading weigh on the tectonic plates and their variations shift the target position by up to several centimeters each. Pole tides occur due to the dynamics of the Earth's rotational axis and are modeled as deviations with respect to a mean rotational pole. Their amount also varies at the millimeter level. Even weaker effects are caused by ocean pole tides and atmospheric tidal loading. Our correction values for the geodynamic effects applied to TerraSAR-X are based on the International Earth Rotation and Reference Systems Service (IERS) conventions, release 2010 [7].

However, plate tectonics are not covered by the IERS conventions as the geodetic techniques usually estimate them as part of the linear station coordinates, and therefore we have to consider them when modelling the target coordinates or reduce them in the SAR measurements.

In summary, the SAR measurements in range τ_m and azimuth t_m, corresponding to the zero-Doppler geometry of a CR, may be written as:

$$\tau_m - \Delta\tau_{cal} = \| X_s(t_{0D}) - (X_t + \Delta X_{geo}) \| \cdot 2/c + \Delta\tau_{tro} + \Delta\tau_{ion}$$
$$t_m - \Delta t_{cal} \Leftrightarrow \frac{\dot{X}_s(t_{0D}) \cdot (X_s(t_{0D}) - (X_t + \Delta X_{geo}))}{\| \dot{X}_s(t_{0D}) \| \cdot \| X_s(t_{0D}) - (X_t + \Delta X_{geo}) \|} = 0 \tag{1}$$

where X_s and $\dot{X}_s$ are the satellite position and velocity vectors at the instant of closest approach t_{0D}, i.e., the zero-Doppler time which may be expressed in seconds of day UTC; the X_t denotes the ITRF target coordinates that are corrected for the geodynamic effects ΔX_{geo}; the c is the speed of light in vacuum, the $\Delta\tau_{tro}$ and $\Delta\tau_{ion}$ are the path delays of troposphere and ionosphere, and $\Delta\tau_{cal}$ and Δt_{cal} refer to time biases, i.e., the geometrical calibration constants of the SAR sensor. Note that these equations are but an extension of the well-known range-Doppler equation describing the SAR observation geometry [11]. Both equations are coupled through the time dependency of the satellite state vector, and the azimuth is only implicitly defined by the geometric zero-Doppler condition; thus the $\Leftrightarrow$ has to be read as "corresponds to" and the $t_m - \Delta t_{cal}$ should equal to the geometric t_{0D}. Note that strictly speaking, the azimuth also experiences secondary contributions from the atmospheric delays if there is a steep horizontal or temporal gradient in the atmospheric portion traversed by the two-way signal, which is transmitted and received by the radar at separate times at separate locations. The typical integration time of a single target, however, is in order of some seconds; hence we do not consider this in our azimuth measurements.

2.2. SAR Positioning with Stereo SAR

Measuring the position of a target in a radar image determines only its 2-D location in the slant range and azimuth geometry defined by the 3-D to 2-D projection of the SAR imaging process. In order to determine the target's underlying 3-D location, at least two images with sufficient angular separation have to be used, which enable a stereo setup and ultimately the retrieval of the CR coordinates. Expanding this concept to a more general method by employing least-squares parameter estimation, any number of radar images larger than one may be combined in a consistent estimation of the target position. This general approach is the core of our Stereo SAR method which is extensively discussed in [12]. In the following, we provide only a short summary and the reader interested in more details is referred to the paper. By using the range-Doppler equations (Equation (1)) as two condition equations that couple the measurements τ_m, t_m with the target position X_t, one can solve for the coordinates by performing a rigorous linearization and employing an iterative computation scheme. The theory behind this method may be found in [13]. In accordance with Equation (1) the condition equations read:

$$\| X_s(t_m^*) - X_t \| - \tau_m^* = 0 \qquad \frac{\dot{X}_s(t_m^*) \cdot (X_s(t_m^*) - X_t)}{\| \dot{X}_s(t_m^*) \| \cdot \| X_s(t_m^*) - X_t \|} = 0. \tag{2}$$

The τ_m^* and τ_m^* are the refined measurements already corrected for the atmospheric path delay and with the geometrical calibration applied. The least-squares method resolves the remaining inconsistencies of the equations by estimating observation residuals, which are minimized according to the L2-norm, and it estimates the optimal target coordinates at the same time. Be aware that the satellite orbit X_s and $\dot{X}_s$ is provided with the SAR image annotation and it is kept fixed in the processing. The orbit is modelled through short arc polynomials [12], but the along-track relationship of the azimuth observation with the orbital state vector is fully maintained through the linearization. We do not introduce any a priori variance information for the individual measurements, but variance component estimation [14] is part of the solution process to infer a common variance for all the range measurements and a common variance for all the azimuth measurements. This ensures optimal weighting between the two types of SAR measurements.

The outcomes are not only the position coordinates but also the estimation of the positioning accuracy, i.e., the variance-covariance (VC) matrix of the coordinates stemming from the solution of the mathematically overdetermined problem. The VC matrix characterizes the geometry of the underlying stereo setup and the quality of the SAR measurements inferred from the minimized residuals; hence it also enables a reliable characterization of the coordinate solution. It is a fully populated 3 by 3 matrix with the main diagonal holding the variances σ_{xx}^2, σ_{yy}^2 and σ_{zz}^2 referring to the global frame, but the VC matrix may be transformed to express the estimated accuracy for any frame orientation, for instance the local north, east and height frame. A useful way of visualizing the VC matrix is the error ellipsoid, which is the decomposition of the matrix for its eigenvalues [13,15]. However, the confidence level of the VC matrix of 3-D coordinates as computed by the least-squares method is only about 20% [13]; hence for our positioning results we scale all the variance-covariance estimates to a more reliable confidence level of 95% by applying the corresponding scaling factor of the F-distribution [15], which is typically in the order of 2.85 because of the large number of TerraSAR-X images available for our analysis.

3. Materials and Methods

3.1. Verification of the Geolocation Accuracy of SAR

A widely used approach to verify the geolocation accuracy of SAR is based on artificial point-like targets deployed on ground [16]. Their positions in focused SAR images are measured and compared to the expected values that are derived from the knowledge of the target's on-ground coordinates and the determination of the satellite position at closest approach. The conversion from spatial coordinates to expected radar time coordinates is based on the zero-Doppler equations (Equation (1)) and the interpolation of the satellite's orbit, i.e., the reverse approach to the location measurements of SAR, discussed in Section 2.1.

Due to the well-defined shape of their impulse-response and to their high radar cross section (RCS), the exact positions such artificial targets in SAR images are precisely detectable so that the location accuracy mainly depends on the geometric features of the object of research: the end to end SAR system consisting of SAR sensor and SAR processing system. In contrast for weaker "natural" targets, their detectability in the surrounding clutter is the limiting factor for the location accuracy so that such targets give little incidence on the geometric accuracy of the end to end SAR system. Following [17], which includes a modified form of earlier work by Stein [18] and Swerling [19], the effect of the clutter on the obtainable localization accuracy is given by:

$$\sigma = \frac{\sqrt{3}}{\pi\sqrt{2}} \cdot \frac{\rho}{\sqrt{SCR}} \approx 0.39 \cdot \frac{\rho}{\sqrt{SCR}} \tag{3}$$

where σ denotes the expected clutter contribution to the standard deviation of the target location, and ρ is the resolution of the target in terms of the 3 dB width of its impulse response. Different from [17], an additional factor $\sqrt{1/2}$ is introduced here, considering that we examine the localization accuracy of a single SAR measurement, whereas [17] refers to the mutual distance of two SAR measurements.

The most common types of point-like targets are passive corner reflectors (CRs) and active radar transponders. Appreciated features of purely passive elements like CRs are that they neither need any power-supply nor do they introduce any (possibly unknown) additional delays to the signal round trip time. Moreover, the electric phase center position is equal to the inner corner position which can be determined readily. The intersection of the three orthogonal plates is a well-defined point, for which the ITRF2014 coordinates may be accurately determined by terrestrial geodetic survey.

While the verification approach sketched in Figure 1 looks simple on the first glance, its realization requires meticulous care during all processing steps when centimeter or even millimeter accuracy is aspired (In [20], Small et al. give a detailed guideline on the installation of a CR test site). The CRs have to be firmly installed (preferably on stable bedrock or second-best on a firm concrete foundation) and correctly aligned. Their reference coordinates have to be very accurately determined. If possible, the deployment at well-known geodetic reference sites is advised when verifying the SAR geolocation capability. As any motion or re-orientation operation is prone to introduce spurious errors, we advise against any actions that could invalidate the geodetic reference coordinates of the phase center. Instead the reflectors should be kept fixed and their coordinates verified every 1-2 years.

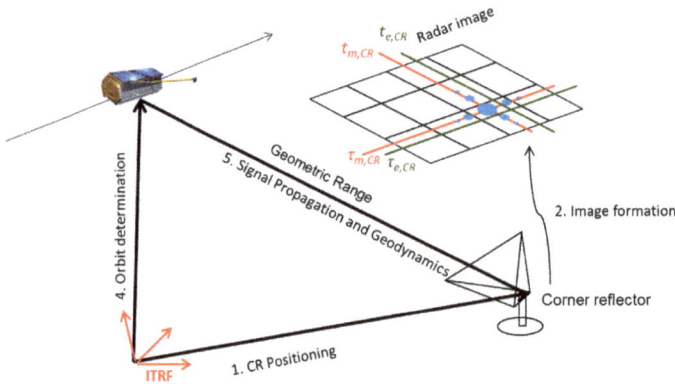

Figure 1. Schematic view of measurement arrangement and procedures.

Once the CRs are installed and SAR datatakes are acquired, the target positions in the focused images have to be measured with sub-pixel resolution using oversampling and interpolation techniques, and the extracted radar times have to be corrected for the estimated propagation delays and geodynamic effects (as detailed in [21]). We recommend to use Single-look Slant-range Complex (SSC) image products for geolocation measurements instead of ground-projected image products (i.e., TerraSAR-X product types MGD, GEC and EEC), as any ground-projection introduces an avoidable additional error source.

By performing many verification measurements at several test sites far apart from each other, and employing SAR images acquired in different SAR modes, at different times, as well as varying the site position within an image and the overall imaging conditions (e.g., features like polarization, incidence angle or orbit direction: ascending or descending), the SAR imaging system can be very accurately geometrically calibrated and verified. Moreover, error estimates can be given and residual systematic biases can be characterized. Regular observations over long time spans (years) enable investigations of seasonal effects and the long-term stability of the measurement results.

3.2. Geometric Recalibration of the Sensor

The operational TerraSAR-X products are already prepared for the consideration of atmospheric effects [22] and their product annotation contains estimates for tropospheric and ionospheric signal propagation delays. These delay values are composed of scene-dependent tropospheric and ionospheric delays derived from a simplified static height-dependent model [23], and a set of associated system calibration constants determined during the mission's initial calibration campaigns. With these annotated parameters systematic biases in the measured target positions are avoided and geolocation accuracy at decimeter level can be achieved [22], which is already beyond the product specification of 1 m [24].

However, these calibration constants need to be refined when we change the measurement concept and correct the SAR geolocation measurements for signal propagation with the more precisely measured atmospheric delay values provided by the IGS, and also take the geodynamic effects into account. Because of different biases of the atmospheric and geodynamic models involved, an adapted set of constants was necessary and we had to determine recalibration constants for both TerraSAR-X satellites, TSX-1 and TDX-1, to match this refined measurement concept. Note that, using our refined geodetic concept, TerraSAR-X is now fully compatible with the well-established standards of the GNSS community. Because the image user has no influence on the operational SAR focusing process, the only manageable way to apply our recalibration constants is that the user explicitly modifies his measured azimuth and range times by them.

The recalibration constants result from the median location offsets in azimuth and range obtained at reference test site(s). We use the median instead of the mean value because it is less sensitive to outliers, as the presence of a few of them is hardly avoidable in long measurement series.

Because most of our datatakes are in the HH polarization, we started to recalibrate this polarization channel first. After that, the calibration of the VV channel is efficiently performed on the base of dual-pol (HH/VV) datatakes. Here, both polarization channels are imaged at the same time, so that almost all influencing factors (like the exact amount of atmospheric delays and solid Earth dynamics, the exact satellite orbit and the exact CR position) are equal in both channels and consequently cancel out in a relative measurement of the target position offset performed in both channels. In consequence, also error contributions from the limited knowledge of these influences cancel out and we needed significantly less dual-pol datatakes to determine only the difference between the calibration constants of both channels. In comparison, the amount of single-pol datatakes required in an independent calibration of the VV channel with comparable accuracy was much larger.

3.3. Our TerraSAR-X Test Sites

The network of our test sites [3] grew step by step over the course of our investigations. In the final state, three test sites with in total five trihedral CRs are involved in our measurement series. The geographic distribution of these test sites spans from Germany and Finland to the Antarctic Peninsula (see Figure 2a-c). All of our CRs are mounted close to a local IGS reference station. Benefitting from this vicinity, the ground positions of the reflectors are known very precisely (< 5 mm) relative to the station reference coordinates from terrestrial geodetic survey. The particulars of the individual TerraSAR-X measurement series and the incidence angles of the datatakes involved are summarized in Table 1.

| (a) | (b) | (c) |

Figure 2. Corner reflectors at our test sites: (**a**) Wettzell, Germany; (**b**) GARS O'Higgins, Antarctic Peninsula; (**c**) Metsähovi, Finland.

Table 1. Key figures of our SAR measurement series at geodetic observatories. Each CR is seen from two or three adjacent orbits though the CR is optimally oriented for only one of them (see column "optimum incidence angle").

Corner Reflector/Measurement Series	Acquisitions Started at	Optimum Incidence Angle [1]	Additional Incidence Angle(s)
Wettzell Ascending	12 Jul. 2011	34°	46° (since 2 Mar. 2013)
Wettzell Descending	11 Dec. 2013	45°	33°, 54°
GARS O'Higgins Ascending	27 Mar. 2013	38°	30°, 45°
GARS O'Higgins Descending	24 Mar. 2013	35°	42°
Metsähovi Descending	4 Nov. 2013	37°	27°, 46°

[1] w.r.t. the orientation of the CR.

In July 2011, we installed our first 1.5 m CR at the Geodetic Observatory Wettzell, Germany (Figure 2a). It is firmly mounted on a concrete foundation and horizontally oriented for acquisitions from ascending orbits and vertically aligned for an incidence angle of 34 degrees. In order to investigate possible angular dependencies in the position offsets measured with TerraSAR-X, we set up a second, parallel acquisition series of this CR using datatakes from an adjacent ascending orbit with a 46 degrees incidence angle. In order to preserve the precisely measured geodetic reference coordinates of the CR, we left its orientation unchanged and accepted the slight reduction in the RCS by about 2 dB. This fixed installation method was also adopted for the other subsequently installed reflectors, which are aligned for either ascending or descending passes, and their vertical orientation was adjusted for the center of the radar beam used at the site, see column "optimum incidence angle" in Table 1.

In December 2013, we installed a second 1.5 m CR at Wettzell. Oriented for descending orbits and regularly imaged with different incidence angles (45 degrees, which is optimum w.r.t. the CR orientation, and additionally 33 and 54 degrees) from three adjacent orbits, it enlarges the number of acquisition geometries at this test site.

Since March 2013, there are two 0.7 m CRs at our second test site, close to the German receiving and research station GARS O'Higgins located at the Antarctic Peninsula (Figure 2b). One of the CRs is oriented for ascending, the other for descending orbit passes. Because of the demanding local weather conditions, each CR is mounted above the usually experienced winter snow level using an additional 1 m base frame (see Figure 2b), and both are covered by a high-frequency transparent Gore-Tex™ (Gore-Tex is a trademark of W.L. Gore & Associates, Newark, DE, USA) canvas. The latter significantly eases the local maintenance due to snow. The CRs are still not maintenance-free, but the canvas avoids snow and ice within the reflector where they are more difficult and more cumbersome to remove than at the covering canvas.

Finally, there is our third test site at the geodetic observatory in Metsähovi, Finland, where one 1.5 m CR oriented for descending orbits was installed in October 2013 (Figure 2c). Similar to GARS O'Higgins, the CR is mounted above the expected snow level on a 1 m frame base. Benefiting from local ground conditions, the frame base is anchored by rods and rock epoxy glue residing in deep holes drilled into stable bedrock.

3.4. TerraSAR-X Datatakes

As of 1 February 2018, both TerraSAR-X sensors—TSX-1 and TDX-1—acquired in total 1060 datatakes for our test sites. The datatake acquisition is still ongoing in order to evaluate possible long-term trends. High resolution imaging modes are preferable when aiming for centimeter or millimeter measurement accuracy level with SAR, because the mode directly affects the image resolution and the signal to clutter ratio (SCR) and thus the extraction accuracy of the range and azimuth coordinates from the SAR image [17–19]. Consequently, we opted for HS300 mode, i.e., the high resolution sliding spotlight mode with an average slant range and azimuth resolution of 0.6 m by 1.1 m [24], for the majority of our TerraSAR-X acquisitions, because the even better Staring Spotlight mode ST300 was not available at the time when our measurement series started. Thus, only a few ST300 datatakes are included in our acquisition series. Almost all of our datatakes are in HH polarization. The only exception is a recently started HS dual-pol (HH/VV) measurement campaign at Wettzell where we investigate polarization channel dependent offsets. In order to avoid saturation of our 1.5 m CRs in the focused SAR images, the L1B products of the datatakes were ordered to be processed with gain attenuation (10 dB for HS300 and even 20 dB for ST300). The nominal backscatter values for our CRs are 30.2 dBm^2 for the 0.7 m and 43.4 dBm^2 for the 1.5 m reflectors.

TerraSAR-X datatakes are subject to DLR's copyright and to security regulations of German law, which prohibit any unauthorized redistribution. Thus, we have to refer scientists interested in reproducing our measurement results to the EOWEB catalogue [25], where TerraSAR-X L1B products can be ordered by registered users (scientific proposal submission under [26]). The image center coordinates, required to select datatakes of our test sites from catalogue, are listed in Table 2.

Table 2. Image center coordinates of the datatakes underlying our measurements.

Test Site	Latitude [°N]	Longitude [°E]
Wettzell	+49.145	+12.876
GARS O'Higgins	−63.321	−57.902
Metsähovi	+60.217	+24.395

4. Results and Analysis

4.1. Monitoring the Operational Readiness of a CR

A lesson learned from our long-term measurement series, is to be aware of weather influences affecting the performance of a CR [21]. Snow within the CR changes the backscatter geometry and in consequence firstly lowers the CR's RCS and secondly—much more problematic for our measurements—it changes the actual position of the CR phase center. Because the CR models we installed have drainage holes at their converging corners, rain is in general not a problem, except when leaves fallen or blown into the CR may clog the drainage hole.

As the significantly lowered RCS is a good indicator for such disturbances, we routinely monitor the RCS and compare its value against the expected value derived from theory or experience. Based on an empirically derived threshold of 3 dB RCS loss relative to the expected value, we are able to identify the TerraSAR-X measurements affected by such irregular conditions and exclude them from further analyses, where they would otherwise occur as coarse (i.e., decimeter-level) outliers. During our analysis we did not observe any significant correlation of moderate RCS losses of less than 3 dB with the shifts in the SAR measured CR locations.

Over the course of our investigations and based on this assumption, we identified in total 79 disturbed location measurements, reducing the number of datatakes in the analysis from 1060 to 981. In contrast, there still remain 18 measurements (approximately 1.8%) with a conspicuously high location error of 7-11 cm where no obvious external cause could be identified. With the exception of one, these values are found in azimuth. Because we have no indication for an external cause, these measurements must not be excluded from analysis.

4.2. Geometric Recalibration Constants

We have chosen the CR at the Metsähovi test site as the reference for our TerraSAR-X recalibration. Here, the mounting of the CR on stable bedrock provides confidence that the geodetic coordinates of its phase center are even more long-term stable than the ones from our other test sites. Because the measurement series are in HH polarization mode, we originally focused our recalibration activities on this mode and just recently complemented them by deriving recalibration constants for VV polarization too. Table 3 shows our new calibration constants to be applied by the user. To remain consistent with the sign convention of the operational calibration constants, our recalibration constants have to be subtracted from the measured radar times at all test sites (i.e., the absolute value of the negative constants has to be added!). For convenience, Table 3 also shows the values converted to distances. Note that we use an average ground track velocity as conversion factor for azimuth here, whereas the precise value in a given SAR image depends on geometric factors like the incidence angle and the topographic height.

In addition to the operational range calibration constant that is already implicitly applied during TerraSAR-X image focusing, we had to introduce a range recalibration constant that the user has to apply explicitly. The amount of our range recalibration constant corresponds to about 30 cm spatial distance, see Table 3. The cause for its necessity can be found mainly in the different handling of the wet part of the tropospheric delay in both measurement concepts. Being the most volatile contribution to the tropospheric delay, its consideration is beyond the capabilities of the simplified model underlying the annotated delay value and in consequence, it contributes in average to the operational calibration constant. When we make use of the measured tropospheric delay of the IGS, the wet delay is already included and must not be considered twice. Thus, our recalibration constant figures as a readjustment of the wet delay's contribution in the operationally applied calibration constant.

The likely explanation for the small difference of 0.27 nanoseconds (equivalent to 4 cm) discovered in the range calibration constants of both polarization channels (cf. Table 3) might be found in the staggered mounting of H and V antenna elements on the satellite's surface and in the slightly different signal routing inside the sensor electronics.

Table 3. Applied geometric calibration constants in seconds and converted to spatial distances.

Satellite	Polarization	Azimuth [s] [1]	Range [s]	Azimuth [m] [2]	Range [m] [3]
TSX-1	HH	$-9.7 \cdot 10^{-6}$	$-2.01 \cdot 10^{-9}$	-0.068	-0.301
TSX-1	VV	$-9.7 \cdot 10^{-6}$	$-2.28 \cdot 10^{-9}$	-0.068	-0.342
TDX-1	HH	$-7.7 \cdot 10^{-6}$	$-1.82 \cdot 10^{-9}$	-0.054	-0.273
TDX-1	VV	$-7.7 \cdot 10^{-6}$	$-2.09 \cdot 10^{-9}$	-0.054	-0.313

[1] For comparison purposes: The operational calibration constant w.r.t. azimuth amounts to $-11.2 \cdot 10^{-6}$ s for TSX-1 and $-6.8 \cdot 10^{-6}$ s for TDX-1. [2] Converted with average TerraSAR-X ground track velocity of 7050 m/s. [3] Converted with speed of light in vacuum and divided by 2 for one-way.

In contrast to the range, the operational calibration constant in azimuth is not applied in the processor; hence a user is able to replace it with our new constants. The differences between the results of the operational calibration and our recalibration are rather small (corresponding to about 11 mm spatial distance for TSX-1 and 6 mm for TDX-1). However, we benefit from the long duration of our measurement series, the more precisely determinable CR reference coordinates in the vicinity to the IGS reference stations, and from the firm orientation of the CRs that avoids any undesirable change of their phase centers. Due to these factors, we were able to slightly refine the azimuth calibration constants of both sensors.

Finally, we want to give some remarks on the cross-polar channels. Because trihedral CRs do not change the plane of polarization, our CRs are invisible in cross-polar image channels, as they are employed in polarimetry. Consequently, a geometric recalibration of the cross-polar channels HV and VH is out of the capabilities of our on-ground measurement equipment, but we can provide assumptions about adequate recalibration constants for the cross-pol channels based on theoretical considerations. Assuming that most of the differences between the examined co-polar channels HH and VV results from construction particulars of the antenna and the sensor electronics, we might conclude that the recalibration constants for HV and VH are approximately in the mean of the HH and VV constants, because HV shares the transmit path with the HH channel and the receive path with the VV channel, while the opposite is true for VH.

4.3. Temporal Stability of SAR Geolocation Results

Figures 3–5 show the temporal progression of the azimuth and range offsets between measured and expected radar coordinates at our three test sites. A slight trend is perceivable from all plots. In all measurement series, there is a tendency of the measured values toward early azimuth and toward far range. In order to estimate the magnitude of this trend, we approximated the temporal progressions of the offsets by a linear function and investigated the obtained gradients. Table 4 and Figure 6 show that the gradients are in the order of millimeters per year and exceed the standard deviation (1σ) expected from error propagation. However, the 1σ just represents a 68% confidence interval whereas for a 95% confidence interval 2σ is required [15]. The gradients in range even exceed 3σ so that a confidence level of more than 99.7% results. Therefore we consider the range effect to be significant. For the azimuth, the results from the different test sites are much less distinct and some of the estimated gradients (Wettzell Ascending and Metsähovi Descending) obviously remain below the 2σ, thus the continuation of the ongoing measurement series is required to reduce the remaining uncertainty.

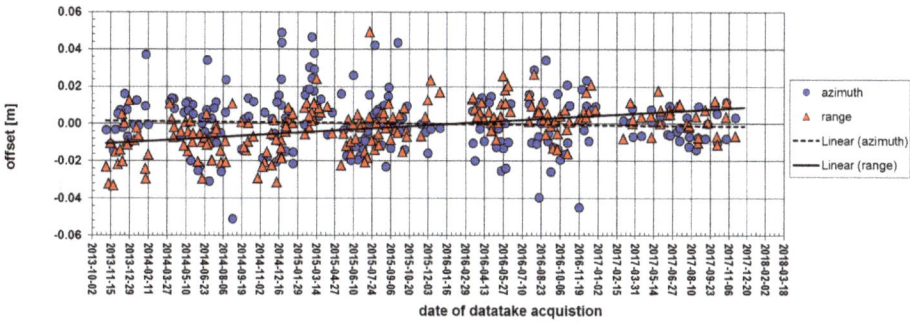

Figure 3. Temporal progression of the azimuth (blue) and range (red) offset obtained at the Metsähovi test site.

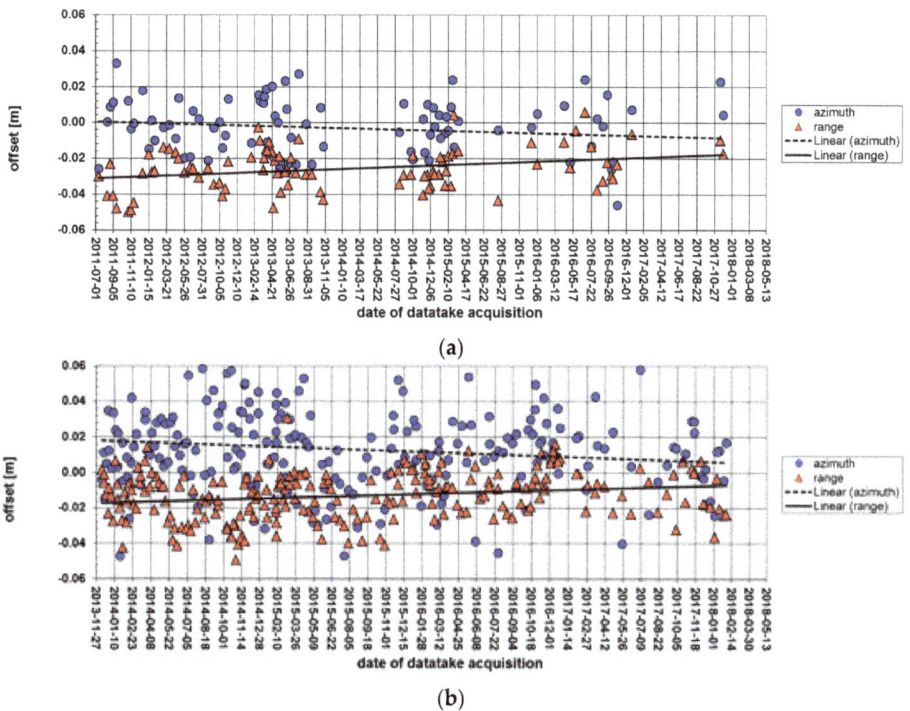

(a)

(b)

Figure 4. Temporal progression of the azimuth (blue) and range (red) offset obtained at the Wettzell test site: (**a**) Wettzell Ascending measurement series; (**b**) Wettzell Descending.

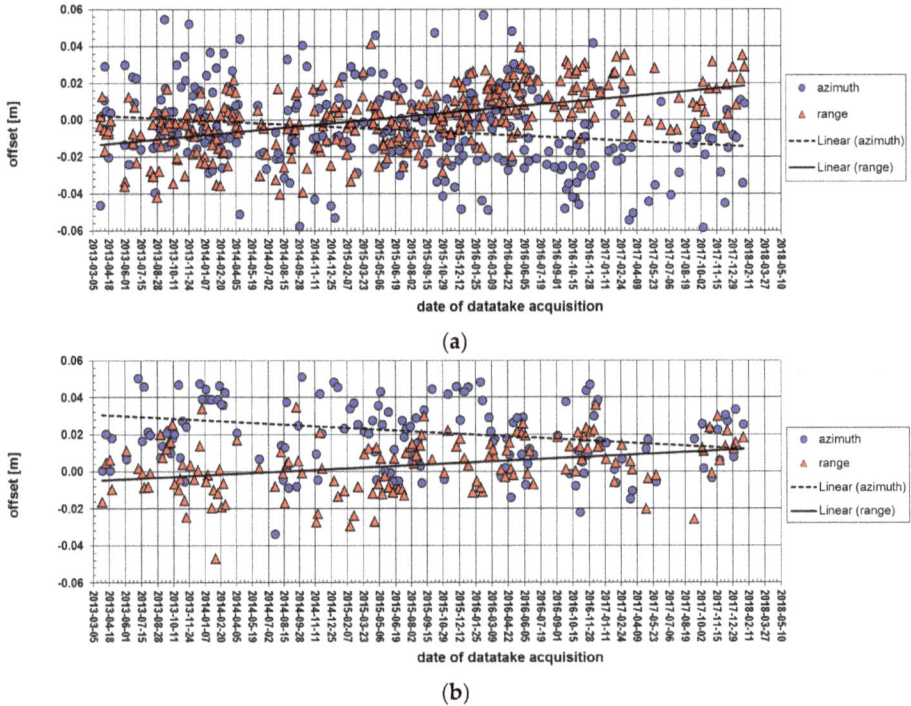

Figure 5. Temporal progression of the azimuth (blue) and range (red) offset obtained at the GARS O'Higgins test site: (**a**) GARS O'Higgins Ascending measurement series; (**b**) GARS O'Higgins Descending.

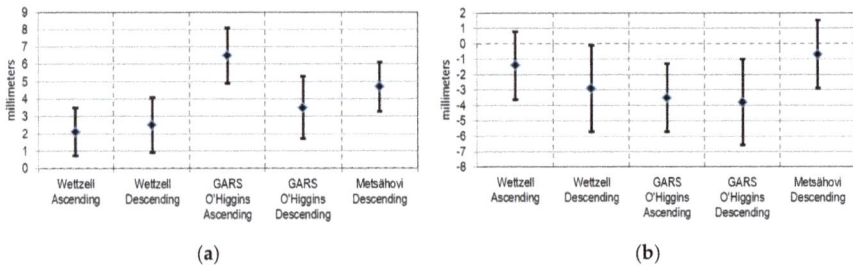

Figure 6. Gradients of the interpolated linear trends at the different test sites: (**a**) range; (**b**) azimuth. The plotted error bars represent a 95% confidence interval (2σ).

Table 4. Interpolated linear trend of the observed offsets: Mean value and standard deviation (1σ) of the gradient.

Measurement Series	Azimuth Gradient [mm/y]	Range Gradient [mm/y]
Wettzell Ascending	-1.4 ± 1.1	$+2.1 \pm 0.7$
Wettzell Descending	-2.9 ± 1.4	$+2.5 \pm 0.8$
GARS O'Higgins Ascending	-3.5 ± 1.1	$+6.5 \pm 0.8$
GARS O'Higgins Descending	-3.8 ± 1.4	$+3.5 \pm 0.9$
Metsähovi Descending	-0.7 ± 1.1	$+4.7 \pm 0.7$

Assuming that the trend in the range measurements is real, finding a possible cause for it is subject of ongoing investigations. We may already excluded some potential causes: The stability of the sensor internal clock rate governing all timings in the SAR payload is routinely monitored in the TerraSAR-X mission and in this way known with nine digit accuracy (the technique is described in [27]). In consequence, this error contribution cannot exceed 3 mm and is therefore below the amount of the observed trends. Also the satellite orbit determination is not considered to be a cause because we have independent measurements of the TerraSAR-X orbits based on Satellite Laser Ranging (SLR) confirming the long-term stability—see [28] for more. Thus, the focus of our further investigations has to be on the SAR sensor itself. We might investigate e.g., whether slight aging effects in electronic components are supposable. The challenge is that to our knowledge no prior space-borne SAR sensor was examined after insert to orbit at a comparable level of detail.

4.4. Analysis for Angular Dependencies

Figure 7a–e show the scatter plots of the azimuth and range offset obtained between measured and expected radar coordinates for the different measurement series. In each plot, different colors represent acquisitions from different incidence angles.

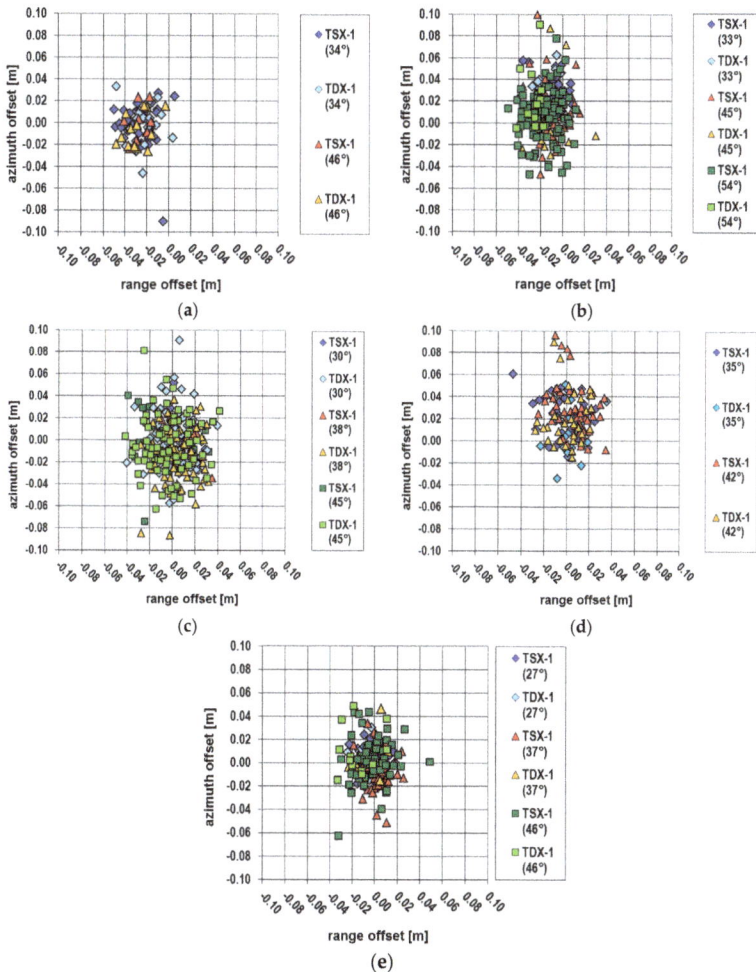

Figure 7. Scatter plots of the obtained azimuth and range offsets: (**a**) Wettzell Ascending; (**b**) Wettzell Descending; (**c**) GARS O'Higgins Ascending; (**d**) GARS O'Higgins Descending; (**e**) Metsähovi Descending.

By visual inspection, the single distributions coincide for the different incidence angles and no significant dependency between incidence angle and location offset is perceivable. In order to quantitatively verify this hypothesis, we performed a statistical analysis of the location offsets. The resulting mean values and standard deviations are listed in Table 5, sorted by incidence angle. The variations across the mean values of one site are usually lower than the standard deviation of the single measurement series, and no particular systematic dependency on the incidence angle can be found. We conclude from these results that the individual beams are consistent.

Table 5. Incidence angle wise mean values and standard deviations (1σ) of the obtained range and azimuth error.

Measurement Series	Incid. Angle	# of Meas.	Azimuth Offset [mm]	Range Offset [mm]
Wettzell Ascending	34	65	−2.9 ± 18.6	−25.1 ± 12.0
	46	23	−4.1 ± 15.5	−27.4 ± 10.5
Wettzell Descending	33	42	+29.5 ± 15.7	−10.4 ± 11.2
	45	75	+11.5 ± 25.3	−8.3 ± 13.2
	54	112	+8.0 ± 23.7	−17.2 ± 13.1
GARS O'Higgins Asc.	30	112	−0.7 ± 23.5	+3.9 ± 17.0
	38	86	−11.6 ± 22.9	+5.0 ± 14.8
	45	104	−5.3 ± 25.7	−5.2 ± 18.8
GARS O'Higgins Desc.	35	62	+17.6 ± 19.8	+0.9 ± 14.9
	42	85	+24.9 ± 23.2	+4.9 ± 14.0
Metsähovi Descending	27	61	+3.5 ± 10.7	−2.5 ± 9.3
	37	75	−4.0 ± 15.6	+1.2 ± 9.7
	46	79	+1.6 ± 22.3	−4.4 ± 15.0

4.5. Comparison of Different Imaging Modes

Figure 8 shows the temporal progression of the azimuth and the range offset at the Wettzell test site during the Staring Spotlight (ST300) campaign from February 2015 until March 2016.

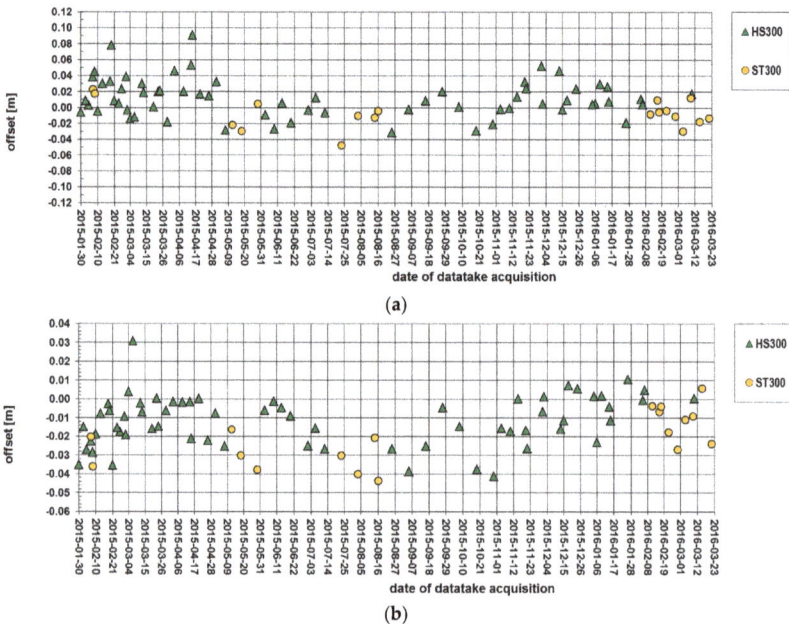

(a)

(b)

Figure 8. Close-up of the temporal progression of the obtained azimuth (**a**) and range (**b**) offset at the Wettzell test site during the ST300 campaign (yellow) interleaving regular HS300 acquisitions (green).

The Staring Spotlight datatakes interleave the series of regular Sliding Spotlight (HS300) acquisitions. The plots indicate the consistency of the location results from both imaging modes as the short term patterns in the azimuth and range offsets progress across the varying imaging modes.

4.6. Location Independency of Results

Table 6 shows the overall mean values and standard deviations of the localization error determined at our different test sites. As Metsähovi is used as our recalibration reference, the mean location offset is consequently close to zero and does not contain any useful information. However, the standard deviation determined at this test site is not influenced by the recalibration and therefore it is still an independent measurement result.

Table 6. CR wise mean values and deviations (1σ) of the obtained range and azimuth error.

Measurement Series	# of Measurements	Azimuth Offset [mm]	Range Offset [mm]
Wettzell Ascending	88	-3.2 ± 17.7	-25.7 ± 11.6
Wettzell Descending	229	$+13.1 \pm 24.3$	-13.0 ± 13.4
GARS O'Higgins Ascending	302	-5.4 ± 24.4	$+1.1 \pm 17.6$
GARS O'Higgins Descending	147	$+21.8 \pm 22.1$	$+3.2 \pm 14.5$
Metsähovi Descending [1]	215	$+0.2 \pm 17.5$	-1.9 ± 12.0

[1] The reason, why the average position offset in this measurement series is almost zero, is that we used the Metsähovi CR as calibration reference for all of our test sites, see Section 4.2.

The obtained mean values at the other test sites indicate the consistency between SAR location measurements and the calibration derived for Metsähovi. They amount from few millimeters up to 2.6 cm. The maximum range bias results for the Wettzell Ascending CR. However, a terrestrial resurvey of this reflector disclosed the cause for the deviating behavior of this single measurement series: The concrete foundation of the CR subsided by about 2 cm during the first 2 years after the CR was installed and surveyed. While we chose the location for our first CR with regard to a minimum disturbance in the radar image from the installations of the IGS reference station, we did not take enough care of the ground conditions and put the foundation on a small mound rising above the local surroundings, see Figure 2a. As lesson learned, we more carefully considered this aspect when installing the other CRs.

The standard deviations of the single measurement series in range vary from 12 to 18 mm. The standard deviation in azimuth is somewhat higher and ranges from 18 to 25 mm, which is still nearly two orders of magnitude below the upper limit defined in the TerraSAR-X product specification [24]. The lower performance of the azimuth localization accuracy is limited by the annotation precision (18.6 microseconds step width) of the raw data acquisition time. Even if it is derived from very accurate pulse per seconds (PPS) information of the on-board GPS receivers [27], the information is stored in the SAR payload with an insufficient number of digits. We strongly recommend future SAR missions to provide a finer time annotation. For TerraSAR-X we found a way to overcome the truncated timing by applying a post processing of the azimuth timing [27]. Since May 2014, this post processing is part of the operational TerraSAR-X SAR processor and lowers the quantization error of the azimuth timing by about one order of magnitude. The remaining error amounts to about 1 microsecond, equivalent to about 7 mm azimuth location error.

Comparing the standard deviations obtained at the different test sites, the 0.7 m CRs at GARS O'Higgins are inferior to the 1.5 m CRs at Wettzell and Metsähovi as it is predicted from theoretical considerations because of the lower RCS resulting in a lower SCR. Using Equation (3) and inserting the parameters of our (HS300) measurement series at GARS O'Higgins or Wettzell and Metsähovi, respectively, the expected clutter contribution to the overall location error in case of a 0.7 m reflector amounts to 6 mm in range and 12 mm in azimuth. In case of a 1.5 m CR the values read 1.0 and 1.8 mm, respectively. Thus, a good deal of the difference in the standard deviations is explainable by the different CR sizes.

4.7. 3-D Coordinates

The solution of the independent SAR positioning with TerraSAR-X at the test sites can be directly compared to the coordinates from the surveys. To ease interpretation, the differences in global X, Y and Z have been rotated to the local north, east, and height frame of the respective station. Since we removed the average ITRF station velocity in the processing (see Section 2.1), the SAR coordinates are reduced to the epoch of the first SAR datatake in the series. Therefore, we transformed the reference ITRF2014 coordinates of the CRs to this epoch to ensure a consistent comparison.

The coordinate differences listed in Table 7 confirm the very high accuracy of the TerraSAR-X positioning ability, if all the known contributions have been compensated for in the measurements and if the sensors are accurately calibrated. Naturally, the differences for the Metsähovi CR are very small because this reflector was used to derive the calibration constants (see Section 4.2). At the other test sites, the remaining differences are usually in the order of 1-2 cm, while the largest difference of 4 cm is found for the east component of the GARS O'Higgins Descending CR. However, when examining the estimated standard deviations, the reason for this becomes clear. Like the Wettzell Ascending reflector, this reflector is only captured by two passes (see Table 5), and the smaller baseline between the two adjacent passes results in a larger uncertainty for the east component. In addition, one would also expect an impact of the reflector size (0.7 m versus 1.5 m), but this is mostly compensated by the larger number of acquisitions at GARS O'Higgins.

Table 7. Differences of the TerraSAR-X coordinates and the reference coordinates in the global ITRF2014 expressed in local north, east and height. The variances of the TerraSAR solution mark the 95% confidence interval of the estimated coordinates.

Measurement Series	ΔN [mm]	ΔE [mm]	ΔH [mm]	σN [mm]	σE [mm]	σH [mm]
Wettzell Ascending	−5.2	−12.7	+17.2	+8.0	+36.6	+27.5
Wettzell Descending	−16.0	+18.3	−7.5	+4.8	+11.5	+11.7
GARS O'Higgins Ascending	−15.1	−10.1	−23.1	+9.1	+17.8	+14.6
GARS O'Higgins Descending	−4.8	−39.9	+16.7	+19.1	+38.3	+34.4
Metsähovi Descending [1]	−1.1	+2.4	−4.0	+4.0	+13.2	+9.8

[1] The reason, why the differences are almost zero, is that we used the Metsähovi CR as calibration reference for all of our test sites, see Section 4.2.

The comparison of the coordinate differences with the 95% confidence interval also listed in Table 7 shows that only a few of the differences are actually significant. The overall error behavior of a higher quality in the north component and a reduced quality for the east and height components is widely consistent with these differences. The explanation for this error behavior lies in the almost polar orbit and the SAR zero-Doppler geometry, for which the local north component is mainly driven by the azimuth measurements, whereas the range has to resolve both east and height. As for the remaining biases, we consider them to be the results of the slightly different behavior of the individual beams (see Table 5), as well as the small biases of the orbit not captured by the single site calibration at Metsähovi. Regarding the orbit, new solutions for TSX-1 and TDX-1 are presented in [28] and a first look on the impact on our TerraSAR-X results is provided in the discussion, see Section 5.2.

Another interesting aspect is the overall error distribution of the coordinate solution, namely the error ellipsoid, which is visualized in Figure 9. Because of the same scaling for a confidence level of 95%, the horizontal and vertical cross sections shown in the graphics directly correspond to the confidence values listed in Table 7. The σN, σE, σH are simply the bounding boxes of the ellipses, while the smallest errors are found to be in the range direction, see Figure 9a. All the ellipses are inclined to about 40 degrees, which is the average incidence angles of the passes for the individual CRs, see the angles listed in Table 5. As expected, the underlying intersection geometries are most sensitive in the SAR range direction, and least sensitive in the perpendicular cross-range direction. The horizontal view, see Figure 9b, has a similar behavior, but here it is the azimuth that dominates the orientation. The smaller axes coincide with the local sensor heading, which is nearly north-south due to the sun-synchronous orbit used by TerraSAR-X [29]. Only towards the polar regions, the tracks start to converge, leading to the slightly more tilted horizontal error ellipses of the CRs at GARS O'Higgins. Therefore, the azimuth is almost only sensitive for the north-south component. These error patterns also explain the different confidence levels derived for local north, east and height (Table 5), and they make it immediately clear, why the ideal setup for the application of SAR positioning is a reflector with a common phase center for ascending and descending passes. This concept was already studied with TerraSAR-X in a small experiment at Wettzell [30].

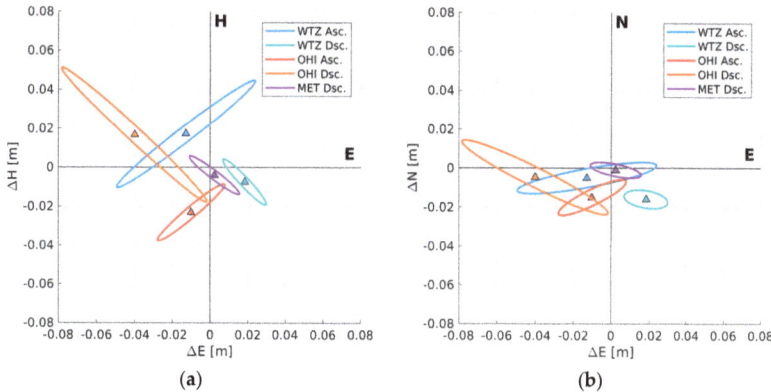

Figure 9. Differences of the TerraSAR-X stereo coordinates to the reference coordinates, and the error ellipsoid of the SAR solution scaled to 95% confidence level: Horizontal cross section in local east/height (**a**); vertical cross section in local north/east (**b**). WTZ = Wettzell, OHI = GARS O'Higgins and MET = Metsähovi.

In summary, the 3-D localization experiments underline the very high absolute accuracy that can be achieved with absolute SAR positioning under ideal circumstances. Moreover, they confirm the performance of our positioning method, the high quality TerraSAR-X radar payloads, the fidelity of the SAR processing, as well as the accuracy of the annotated orbit product.

5. Discussion

In the previous section, we analyzed the geolocation performance of TerraSAR-X with respect to a multitude of contributing parameters: polarization, incidence angle, the pass direction (ascending, descending), the occurrence of long-term trends, and geographic target location. Due to the large amount of topics, the straightforward way of presentation was to discuss each of these aspects in the immediate context of the corresponding measurement results. In order to synthesize these discussions, two comprehensive aspects shall be addressed here. Firstly, we compare the determined a posteriori location error with the a priori error computed from the expected amount of the major error contributions. Secondly, we want to give an outlook on an upcoming further improvement in the attainable geolocation accuracy which is enabled by a new approach in precise orbit determination.

5.1. Analysis of Error Contributions

Table 8 shows a priori estimates for the major contributions to the location error. The estimate for the clutter contribution results from Equation (3), while the estimates for the measurement accuracy of tropospheric and ionospheric delay are taken from the respective data products. Considering the methods discussed in [27], the assumed uncertainty in the azimuth timing, taken from L1B annotation of operational TerraSAR-X products, amounts to about 1 microsecond, corresponding to about 7 mm azimuth error. The Analog to Digital Converter (ADC) sample rate of both TerraSAR-X sensors (about

330 MHz) is operationally monitored to ascertain that the difference between its actual and annotated value does not exceed a margin of 1 Hz. In this way, the residual uncertainty in the ADC sample rate contributes at most 3 mm to the overall ranging error. A conservative estimate for the actual orbit error of TerraSAR-X Science orbits is given in [31].

Comparing the error contributions listed in Table 8, the residual uncertainty in the knowledge of the precise orbit is the dominant contribution in our CRs measurement series (whereas for weaker targets the clutter contribution would be higher and might dominate there). However, as long as we only know an upper limit for the actual orbit error, there remains room for interpretation what could be its true contribution. Regarding the localization accuracies we obtained in our measurement series, we conclude that the actual orbit error must be significantly below the estimated upper limit of 5 cm and that it may amount to about 1 cm.

Table 8. Estimations for the major error contributions in the geolocation result.

Parameter	Azimuth Contribution [mm]	Range Contribution [mm]
Clutter	0.3–12 [1]	0.7–6 [1]
Tropospheric delay	-	0.5–1.3 [2]
Ionospheric delay	-	1.5–5.3 [2]
Timing and ADC sample rate	7	<3
(Science) Orbit	<50 [3]	<50 [3]

[1] Mainly depending on CR size (1.5 or 0.7 m) and imaging mode (ST300 or HS300). [2] Depending on test site and incidence angle. [3] Conservative assumption.

5.2. Geolocation Improvements from More Precise Orbit Determination

Beyond the currently used TerraSAR-X Science orbits, [28] discusses methods to further improve the accuracy of the orbit determination. Once operationally applied, the geolocation accuracy of SAR immediately should benefit from these improvements. We tested the new experimental orbit solutions by repeating our location accuracy measurements in a modified way where we computed the expected radar time coordinates on base of these orbits instead of using the operational TerraSAR-X Science orbit products. Table 9 and Figure 10 show the detailed CR-wise mean values and standard deviations (1σ) of the obtained measurement results and oppose them to the results previously obtained with the operational Science orbits. As a finding of this experiment, we observed that for the 1.5 m CRs the usage of the new orbit solutions TerraSAR-X significantly lowers the standard deviations. They amount to about 1.5 cm in azimuth while in range sub-centimeter level is reached. This represents an accuracy gain of 15% in azimuth and 28% in range compared to the usage of the operational Science orbits. In case of the 0.7 m CRs, where also the clutter significantly contributes to the total location error, the accuracy gain from the improved orbit quality is naturally less distinct (1.5% in azimuth and 10% in range).

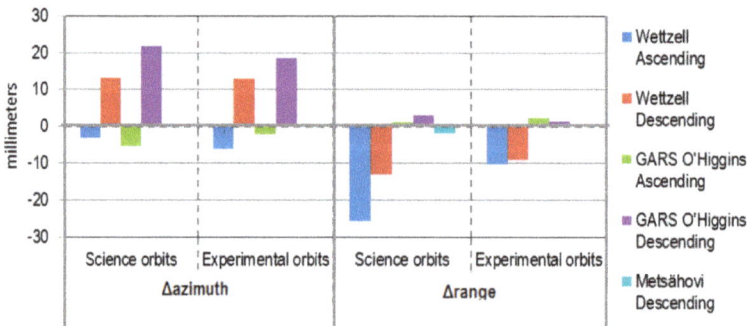

(a)

Figure 10. *Cont.*

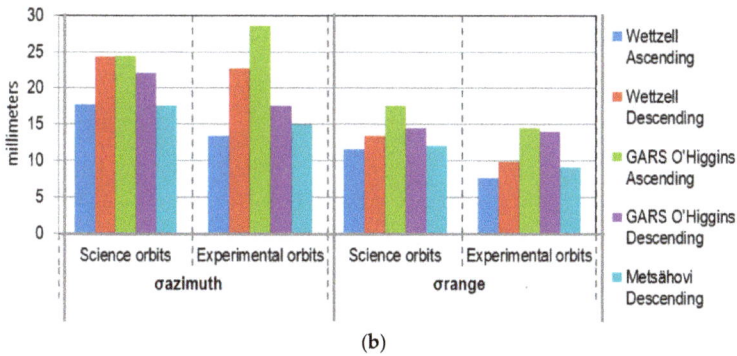

(b)

Figure 10. Comparison of the mean values (**a**) and standard deviations (**b**) in azimuth and range for operational Science orbits (values of Table 6) and new experimental orbits (values of Table 9). The different colors represent the different measurement series.

Table 9. Comparison of the mean values and standard deviations in azimuth and range for operational Science orbits (as they are already shown in Table 6) and for the new experimental TerraSAR-X orbit solutions discussed in [28] (operationally not accessible as yet).

Measurement Series	Azimuth Offset [mm]		Range Offset [mm]	
	Science Orbits	Experimental Orbits	Science Orbits	Experimental Orbits
Wettzell Asc.	-3.2 ± 17.7	-6.1 ± 13.4	-25.7 ± 11.6	-10.2 ± 7.6
Wettzell Desc.	$+13.1 \pm 24.3$	$+12.9 \pm 22.7$	-13.0 ± 13.4	-9.2 ± 9.8
GARS O'Higgins Asc.	-5.4 ± 24.4	-2.2 ± 28.5	$+1.1 \pm 17.6$	$+2.4 \pm 14.5$
GARS O'Higgins Desc.	$+21.8 \pm 22.1$	$+18.5 \pm 17.6$	$+3.2 \pm 14.5$	$+1.4 \pm 14.0$
Metsähovi Desc.	$+0.2 \pm 17.5$	-0.4 ± 15.0	-1.9 ± 12.0	0.0 ± 9.1

Consequently, also the stereo SAR solutions for the CR coordinates profit from the increased orbit quality. Table 10 shows the differences of the SAR measured 3-D coordinates using the new orbits and the GNSS reference coordinates. Figure 11 visualizes a comparison between the results obtained with operational science orbits (which were shown in Table 7) and the results for the new orbits. A closer look on the diagram reveals that the new orbits lead to improvements in almost all the estimated accuracies and the coordinate differences become smaller, in particular the height is now more accurately determined; only at Wettzell there remain significant millimeter level differences for the horizontal positions.

Table 10. Differences of the TerraSAR-X coordinates using the updated orbits and the reference coordinates in the global ITRF2014 expressed in local north, east and height. The variances of the TerraSAR-X solution mark the 95% confidence interval of the estimated coordinates.

Measurement Series	ΔN [mm]	ΔE [mm]	ΔH [mm]	σN [mm]	σE [mm]	σH [mm]
Wettzell Ascending	-9.0	-18.8	$+0.1$	$+6.0$	$+23.5$	$+17.9$
Wettzell Descending	-14.3	$+11.9$	-1.1	$+4.2$	$+9.1$	$+9.2$
GARS O'Higgins Ascending	-6.3	-4.6	-14.1	$+9.0$	$+17.3$	$+14.1$
GARS O'Higgins Descending	-8.4	-24.4	$+10.9$	$+17.0$	$+34.4$	$+30.9$
Metsähovi Descending [1]	$+1.8$	-7.8	$+5.6$	$+3.5$	$+11.4$	$+8.5$

[1] The reason, why the differences are almost zero, is that we used the Metsähovi CR as calibration reference for all of our test sites, see Section 4.2.

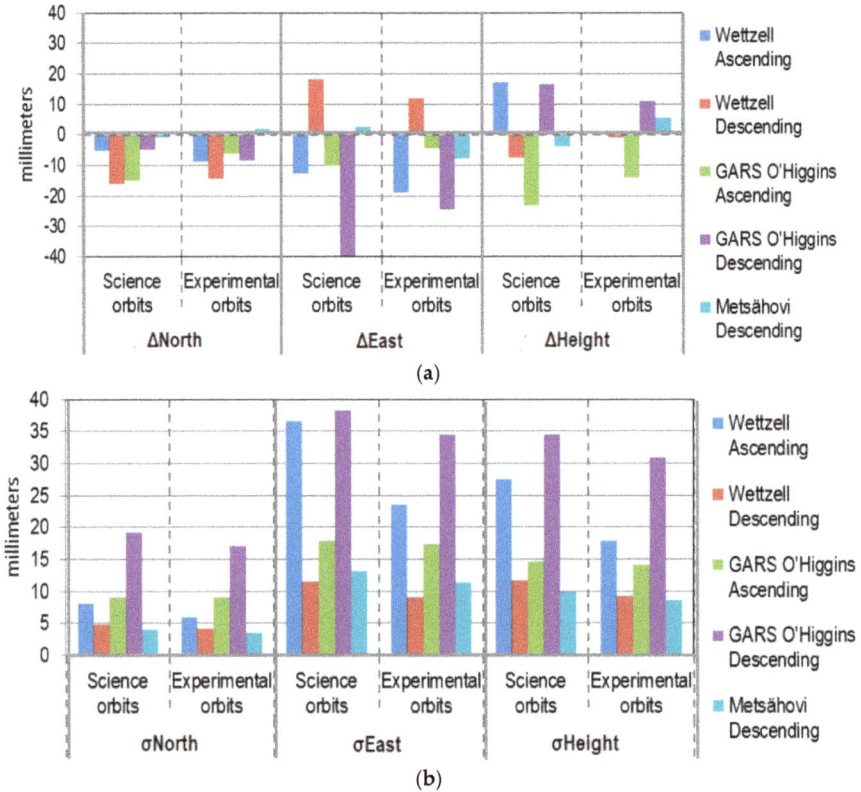

Figure 11. Comparison of the biases (a) and 95% confidence intervals (b) in north, east and height for operational Science orbits (values of Table 7) and new experimental orbits (values of Table 10). The different colors represent the different measurement series.

6. Conclusions

With the setup of three CR test sites as far apart from each other as Germany, Finland and the Antarctic Peninsula, we established a far-distributed test network to verify the absolute localization accuracy of space borne SAR sensors and the worldwide reproducibility of the obtained results. The vicinity of all of our test sites to local IGS reference stations allows very precise determination of the CRs' reference coordinates relative to the respective station ties by terrestrial survey. These precise reference coordinates render a new class of SAR calibration and validation accuracy possible.

Based on our test network, absolute geolocation accuracy at better than centimeter level is proven for both TerraSAR-X sensors (TSX-1 and TDX-1). Prerequisite for such high localization accuracy is meticulous care in processing and calibration of the SAR data and the thorough correction of location measurements from a SAR image for wave propagation effects and for Earth dynamics, consistent with the methods applied in geodesy. In contrast to the "out of the box" usage of TerraSAR-X data, additional external data sources, which are well established in the area of GNSS, have to be accessed in order to obtain precise correction values for atmospheric signal propagation delays. In the process, the geometric instrument calibration constants had to be adapted to the altered measurement approach.

Handling SAR as a geodetic measurement instrument opens completely new application areas for SAR. First examples were already discussed in the literature, e.g., the efficient survey of landmarks along road networks [32] or the stereo measurement of building heights [33]. The synergetic usage of geodetic SAR and SAR tomography is discussed in [34].

Author Contributions: Michael Eineder conceived and designed the basic concept of the project. Ulrich Balss and Christoph Gisinger contribute by refining the concept details w.r.t. their particular area of expertise (SAR signal processing or geodesy, respectively). Ulrich Balss and Christoph Gisinger also cooperate in performing the experiments and analyzing the data, each at his area of expertise. Ulrich Balss contributes analysis tools for point target analysis while Christoph Gisinger contributes tools for estimating signal propagation delays and geodynamic effects. Ulrich Balss wrote substantial parts the paper. Christoph Gisinger contributes to the text by writing Sections 2.2 and 4.7. Abstract, Introduction, Discussion and Conclusions are joint work of the whole author team.

Acknowledgments: The work was partially funded by the German Helmholtz Association HGF through its DLR@Uni Munich Aerospace project "Hochauflösende geodätische Erdbeobachtung." We thank our cooperation partners—Federal Agency for Cartography and Geodesy (BKG) and the Finnish Geodetic Institute (FGI)—for their kind allowance to install the corner reflectors at their property in Wettzell and Metsähovi, respectively, and for their local support. We thank our colleagues from DLR's Remote Sensing Data Center (DFD) who installed and maintain the corner reflectors at GARS O'Higgins.

Conflicts of Interest: The authors declare no conflict of interest. The founding sponsor had no role in the design of the study; in the collection, analyses, or interpretation of data; in the writing of the manuscript, and in the decision to publish the results.

References

1. Schubert, A.; Jehle, M.; Small, D.; Meier, E. Mitigation of Atmosphere Perturbations and Solid Earth Movements in a TerraSAR-X Time Series. *J. Geod.* **2011**, *86*, 257–270. [CrossRef]

2. Balss, U.; Cong, X.Y.; Brcic, R.; Rexer, M.; Minet, C.; Breit, H.; Eineder, M.; Fritz, T. High Precision Measurement on the Absolute Localization Accuracy of TerraSAR-X. In Proceedings of the 2012 IEEE International Geoscience & Remote Sensing Symposium (IGARSS), Munich, Germany, 22–27 July 2012; pp. 1625–1628.

3. Balss, U.; Gisinger, C.; Cong, X.Y.; Brcic, R.; Hackel, S.; Eineder, M. Precise Measurements on the Absolute Localization Accuracy of TerraSAR-X on the Base of Far-Distributed Test Sites. In Proceedings of the 10th European Conference on Synthetic Aperture Radar (EUSAR), Berlin, Germany, 3–5 June 2014; pp. 993–996.

4. Capaldo, P.; Fratarcangeli, F.; Nascetti, A.; Mazzoni, A.; Porfiri, M.; Crespi, M. Centimeter Range Measurement Using Amplitude Data of TerraSAR-X Imagery. In Proceedings of the ISPRS Technical Commission VII Symposium, Istanbul, Turkey, 29 September–2 October 2014; pp. 55–61.

5. Altamimi, Z.; Rebischung, P.; Métivier, L.; Collilieux, X. ITRF2014: A New Release of the International Terrestrial Reference Frame Modeling Nonlinear Station Motions. *J. Geophys. Res. Solid Earth* **2016**, *121*, 6109–6131. [CrossRef]

6. Schubert, A.; Small, D.; Jehle, M.; Meier, E. COSMO-SkyMed, TerraSAR-X, and RADARSAT-2 Geolocation Accuracy after Compensation for Earth-System Effects. In Proceedings of the 2012 IEEE International Geoscience & Remote Sensing Symposium (IGARSS), Munich, Germany, 22–27 July 2012; pp. 3301–3304.

7. Petit, G.; Luzum, B. *IERS Conventions (2010)*; IERS Technical Note 36; Verlag des Bundesamtes für Kartographie und Geodäsie: Frankfurt, Germany, 2010.

8. Hofmann-Wellenhof, B.; Lichtenegger, H.; Wasle, E. *GNSS Global Navigation Satellite Systems*; Springer: Vienna, Austria, 2008; ISBN 978-3-211-73012-6.

9. Meindl, M.; Dach, R.; Jean, Y. *International GNSS Service Technical Report 2011*; Astronomical Institute University of Bern: Bern, Switzerland, 2012.

10. Balss, U.; Gisinger, C.; Cong, X.Y.; Brcic, R.; Steigenberger, P.; Eineder, M.; Pail, R.; Hugentobler, U. High Resolution Geodetic Earth Observation with TerraSAR-X: Correction Schemes and Validation. In Proceedings of the 2013 IEEE International Geoscience & Remote Sensing Symposium (IGARSS), Melbourne, Australia, 21–26 July 2013; pp. 4499–4502.

11. Cumming, I.G.; Wong, F.H. *Digital Processing of Synthetic Aperture Radar: Algorithms and Implementations*; Artech House: Boston, MA, USA, 2005; ISBN 978-1-580-53058-3.

12. Gisinger, C.; Balss, U.; Pail, R.; Zhu, X.X.; Montazeri, S.; Gernhardt, S.; Eineder, M. Precise Three-Dimensional Stereo Localization of Corner Reflectors and Persistent Scatterers With TerraSAR-X. *IEEE Trans. Geosci. Remote Sens.* **2015**, *53*, 1782–1802. [CrossRef]

13. Mikhail, E.M.; Ackermann, F. *Observations and Least Squares*; IEP-Dun-Donnelly, Harper and Row: New York, NY, USA, 1976.

14. Koch, K.R.; Kusche, J. Regularization of geopotential determination from satellite data by variance components. *J. Geod.* **2012**, *76*, 259–268. [CrossRef]

15. Koch, K.R. *Parameter Estimation and Hypotheses Testing in Linear Models*; Springer: Berlin/Heidelberg, Germany, 1999; ISBN 978-3-642-08461-4.
16. Gray, A.L.; Vachon, P.W.; Livingstone, C.E.; Lukowski, T.I. Synthetic Aperture Radar Calibration Using Reference Reflectors. *IEEE Trans. Geosci. Remote Sens.* **1990**, *28*, 374383. [CrossRef]
17. Bamler, R.; Eineder, M. Accuracy of Differential Shift Estimation by Correlation and Split-Bandwidth Interferometry for Wideband and Delta-k SAR Systems. *Geosci. Remote Sens. Lett.* **2005**, *2*, 151–155. [CrossRef]
18. Stein, S. Algorithms for Ambiguity Function Processing. *IEEE Trans. Acoust. Speech Signal Process.* **1981**, *ASSP-29*, 588–599. [CrossRef]
19. Swerling, P. Radar Measurement Accuracy. In *Radar Handbook*; Skolnik, M., Ed.; McGraw-Hill: New York, NY, USA, 1970; pp. 4:1–4:14.
20. Schubert, A.; Small, D.; Gisinger, C.; Balss, U.; Eineder, M. *Corner Reflector Deployment for SAR Geometric Calibration and Performance Assessment*; ESRIN: Frascati, Italy, 2018; in press.
21. Balss, U.; Gisinger, C.; Eineder, M.; Breit, H.; Schubert, A.; Small, D. *Survey Protocol for Geodetic SAR Sensor Analysis*; ESRIN: Frascati, Italy, 2018; in press.
22. Schwerdt, M.; Bräutigam, B.; Bachmann, M.; Döring, B.; Schrank, D.; Gonzalez, J.H. Final TerraSAR-X Calibration Results Based on Novel Efficient Methods. *IEEE Trans. Geosci. Remote Sens.* **2010**, *48*, 677–689. [CrossRef]
23. Breit, H.; Fritz, T.; Balss, U.; Lachaise, M.; Niedermeier, A.; Vonavka, M. TerraSAR-X Processing and Products. *IEEE Trans. Geosci. Remote Sens.* **2010**, *48*, 727–739. [CrossRef]
24. Fritz, T.; Eineder, M. TerraSAR-X Ground Segment Basic Product Specification Document. Available online: http://sss.terrasar-x.dlr.de/docs/TX-GS-DD-3302.pdf (accessed on 15 February 2018).
25. EOWEB Earth Observation on the WEB. Available online: https://centaurus.caf.dlr.de:8443 (accessed on 16 February 2018).
26. TerraSAR-X Science Service System. Available online: http://sss.terrasar-x.dlr.de (accessed on 26 February 2018).
27. Balss, U.; Breit, H.; Fritz, T.; Steinbrecher, U.; Gisinger, C.; Eineder, M. Analysis of Internal Timings and Clock Rates of TerraSAR-X. In Proceedings of the 2014 IEEE International Geoscience & Remote Sensing Symposium (IGARSS), Quebec, QC, Canada, 13–18 July 2014; pp. 2671–2674.
28. Hackel, S.; Gisinger, C.; Balss, U.; Wermuth, M.; Montenbruck, O. Long-Term Validation of TerraSAR-X Orbit Solutions with Laser and Radar Measurements. *Remote Sens. Spec.* **2018**, in press.
29. Werninghaus, R.; Buckreuss, S. The TerraSAR-X Mission and System Design. *IEEE Trans. Geosci. Remote Sens.* **2010**, *48*, 606–614. [CrossRef]
30. Gisinger, C.; Willberg, M.; Balss, U.; Klügel, T.; Mähler, S.; Pail, R.; Eineder, M. Differential geodetic stereo SAR with TerraSAR-X by exploiting small multi-directional radar reflectors. *J. Geod.* **2017**, *91*, 53–67. [CrossRef]
31. Yoon, Y.; Eineder, M.; Yague-Martinez, N.; Montenbruck, O. TerraSAR-X Precise Trajectory Estimation and Quality Assessment. *IEEE Trans. Geosci. Remote Sens.* **2009**, *47*, 1859–1868. [CrossRef]
32. Runge, H.; Balss, U.; Suchandt, S.; Klarner, R.; Cong, X.Y. DriveMark—Generation of High Resolution Road Maps with Radar Satellites. In Proceedings of the 11th ITS European Congress, Glasgow, UK, 6–9 June 2016; pp. 1–6.
33. Eldhuset, K.; Weydahl, D.J. Geolocation and Stereo Height Estimation Using TerraSAR-X Spotlight Image Data. *IEEE Trans. Geosci. Remote Sens.* **2011**, *49*, 3574–3581. [CrossRef]
34. Zhu, X.X.; Montazeri, S.; Gisinger, C.; Hanssen, R.F.; Bamler, R. Geodetic SAR Tomography. *IEEE Trans. Geosci. Remote Sens.* **2016**, *54*, 18–35. [CrossRef]

remote sensing

MDPI

Article

Development of Operational Applications for TerraSAR-X

Oliver Lang [1,*], Parivash Lumsdon [2], Diana Walter [1], Jan Anderssohn [1], Wolfgang Koppe [2], Jüergen Janoth [2], Tamer Koban [3] and Christoph Stahl [3]

[1] Airbus Defence and Space GmbH, Platz der Einheit 14, 14467 Potsdam, Germany; diana.d.walter@airbus.com (D.W.); jan.anderssohn@airbus.com (J.A.)
[2] Airbus Defence and Space GmbH, Claude-Dornier Straße, 88090 Immenstaad, Germany; parivash.lumsdon@airbus.com (P.L.); wolfgang.koppe@airbus.com (W.K.); juergen.janoth@airbus.com (J.J.)
[3] Airbus Defence and Space GmbH, Rechliner Str., 85077 Manching, Germany; tamer.koban@airbus.com (T.K.); christoph.stahl@airbus.com (C.S.)
* Correspondence: oliver.ol.lang@airbus.com; Tel.: +49-331-2374-8424

Received: 30 June 2018; Accepted: 11 September 2018; Published: 25 September 2018

Abstract: In the course of the TerraSAR-X mission, various new applications based on X-Band Synthetic Aperture Radar (SAR) data have been developed and made available as operational products or services. In this article, we elaborate on proven characteristics of TerraSAR-X that are responsible for development of operational applications. This article is written from the perspective of a commercial data and service provider and the focus is on the following applications with high commercial relevance, and varying operational maturity levels: Surface Movement Monitoring (SMM), Ground Control Point (GCP) extraction and Automatic Target Recognition (ATR). Based on these applications, the article highlights the successful transition of innovative research into sustainable and operational use within various market segments. TerraSAR-X's high orbit accuracy, its precise radar beam tracing, the high-resolution modes, and high-quality radiometric performance have proven to be the instrument's advanced characteristics, through, which reliable ground control points and surface movement measurements are obtained. Moreover, TerraSAR-X high-resolution data has been widely exploited for the clarity of its target signatures in the fields of target intelligence and identification. TerraSAR-X's multi temporal interferometry applications are non-invasive and are now fully standardised autonomous tools to measure surface deformation. In particular, multi-baseline interferometric techniques, such as Persistent Scatter Interferometry (PSI) and Small Baseline Subsets (SBAS) benefit from TerraSAR-X's highly precise orbit information and phase stability. Similarly, the instrument's precise orbit information is responsible for sub-metre accuracy of Ground Control Points (GCPs), which are essential inputs for orthorectification of remote sensing imagery, to locate targets, and to precisely georeference a variety of datasets. While geolocation accuracy is an essential ingredient in the intelligence field, high-resolution TerraSAR-X data, particularly in Staring SpotLight mode has been widely used in surveillance, security and reconnaissance applications in real-time and also by automatic or assisted target recognition software.

Keywords: interferometry; surface movement monitoring; ground control points; radargrammetry; automated target recognition; convolutional neural networks (CNN); deep CNN; support vector machine; SVM

1. Introduction

Designed for a five years operation, the German Synthetic Aperture Radar (SAR) satellite TerraSAR-X has already achieved ten years of flawless operation in orbit providing high-resolution radar images in all weather conditions 24 h per day. Developed and constructed by Airbus Defence and Space in Friedrichshafen, Germany for the German Aerospace Center (DLR), the satellite orbits at a height of 514 km above Earth and provides radar imagery to a wide variety of scientific and commercial users. To date, TerraSAR-X has not only doubled its service lifetime, but also boasts 99.9% availability, at an outstanding performance. TerraSAR-X is in full health so that a current assessment indicates that it can be operated for a few more years in space until a follow-on system is in place.

With a user driven end-to-end system design, TerraSAR-X has successfully carried out its scientific and commercial data exploitation commitments through DLR's large science phase outreach and Airbus's industrial partnership. TerraSAR-X has paved the way for its national focus on a sustainable SAR technology strategy, in addition to disseminating know-how and exploring new paths for the next generation of high performance SAR Systems based on advanced features (High Resolution Wide Swath (HRWS), digital beamforming).

As a parallel mission, TerraSAR-X has been a source of data sharing continuity of earth observation projects through the European Space Agency's Copernicus programme, and other international space organisations. In this respect, TerraSAR-X data has been extensively applied to five main thematic areas of land, marine, atmosphere, emergency, and security.

Since the launch of its almost identical twin TanDEM-X (TerraSAR-X add-on for Digital Elevation Measurement) in 2010, both satellites have been flying in a formation at a distance only a few hundred metres apart. Together, they have acquired a large amount of data, which provides the basis for the new global elevation model, WorldDEM. TerraSAR-X and TanDEM-X offer a high repeat rate, irrespective of the area of interest (AOI) or weather conditions. This is crucial in the case of natural or manmade disasters, where reactive mapping is needed to support situation awareness and rescue planning. Following the launch of the PAZ (peace in Spanish), satellite in February 2018, into the same orbit, the three satellites are now operating as a constellation to deliver even higher revisit times, increased spatial coverage and thus improved services. Airbus Defence and Space is now working on next generation Synthetic Aperture Radar satellites as the follow-on to the TerraSAR-X family triplets from 2022.

TerraSAR-X features a unique geometric accuracy and offers flexible area coverage and spatial resolution ranging from 0.25 m to 40.0 m. This answers the needs of a wide range of application domains, such as engineering companies who ensure safe operation of large construction projects, oil and gas enterprises that monitor remote production sites, or intelligence and security agencies that carry out targeted surveillance and detailed change detection of site activities.

The article aims to highlight the successful transition of innovative research into a sustainable and operational use within a commercial framework. The qualities of the TerraSAR-X mission and its setup as dual-use (scientific and commercial) mission perfectly demonstrates that basic research and methodological evolution can lead to relevant information for various industry segments and administrative bodies. In the following sections the methods employed for implementation of three operational applications are outlined. The applications are selected, since they all share a high commercial relevance, but they vary with respect to their operational maturity level. Thus, they reflect the typical transition steps from a scientific into a commercial context.

These operational services include Surface Movement Monitoring, an operational and mature application that is relevant, for example, for civil engineering projects and transport maintenance. Secondly, we discuss precise Ground Control Points, which are useful within the automotive and aeronautics sector. The application is well established, but shall be lifted to the next evolution level by leveraging cutting-edge research findings. Finally, we present an Automatic Target Recognition approach, which is at a medium operational maturity level. Research in machine learning is used to optimise the application's performance. The achievable performance of each application in accordance with the satellite's systematic and processing characteristics is described.

2. Applications and Methods

2.1. Surface Movement Monitoring (SMM)

2.1.1. Overview

SAR-interferometric (InSAR) surface deformation monitoring has gained an increasing interest and acceptance in the commercial market. Several operational aspects are addressed by this technique complementing and even replacing successively terrestrial surveillance approaches. Surface movements induced by manmade activities, such as infrastructure constructions, excavations and underground engineering, or natural disasters (e.g., landslides) can be unexpected and far-reaching, and may endanger infrastructure and even human lives. Satellites monitor wide-area surface movements, as well as single structure displacements [1,2] at regular intervals. They can help to understand the nature of ground instabilities and formulate an adequate response. They may also provide input data into simulation models to better evaluate risks of future surface movements.

The start of the TerraSAR-X mission in 2007 was a significant milestone, which helped to establish products and geo-information services based on Earth remote sensing sources on a global scale. In particular, Multi Baseline Interferometric Techniques (such as Persistent Scatter Interferometry (PSI) [3], and Small Baseline Subsets (SBAS) [4]) profit from a precisely defined satellite orbit tube and precise absolute orbit information. The sensor's high-resolution of up to 0.25 m × 1.0 m allows for monitoring of infrastructure elements (e.g., single buildings) on a large scale. In urban regions, a density of 100,000 valid measurement points per square kilometre is typically exceeded. This high level of detail in combination with complementary geospatial information, e.g., municipal building footprints, provides added value to a broad range of applications. The aggregation of a significant number of individual measurements at respective points and the link to relevant object dimensions (e.g., buildings) does not only reduce the number of single measurement points, but also improves data handling and validity levels for specific problems, in particular for non-expert users. The value-added geospatial information can be disseminated in a user friendly way, e.g., on a web platform. The way in which information is provided is crucial for a widespread acceptance of the technique and consequently the use by decision makers in industry and public authorities. High-resolution surface movement products based on TerraSAR-X synergistically complement small scale information layers, such as the German Surface Movement Service (Bodenbewegungsdienst Deutschland BBD [5]).

The following case study demonstrates the capacity of TerraSAR-X showing a relevant operational use case in an active mining area. The use of TerraSAR-X helps determine relatively dynamic movements, but also identifies phenomena of a slower movement rate; quick and frequent situation updates are achieved in an 11-day repeat cycle, cloudless long-term observation allow for the assessment of slowly ongoing or historic situations. The end-user profits from a high density of information.

2.1.2. TerraSAR-X Surface Movement Monitoring Case Study

Surface movement detection based on InSAR measurements is demonstrated in an opencast mining region in western Germany near the city of Cologne with known tectonic faults, which is also affected by groundwater abstraction. The method is based on interferometric SBAS processing of 34 TerraSAR-X repeat-pass scenes, acquired between February 2015 and March 2016 in StripMap (SM) mode at a spatial resolution of 3 m and a repeat cycle of 11 days. The acquisitions were recorded from a descending orbit (rel. orbit 63, strip009R) with HH (Horizontal transmit and Horizontal receive) polarisation and an incidence angle of approximately 34°. We used a semi-automated processing chain for interferometric SAR analysis and post-processing with the objective of providing user relevant deformation information (Figure 1).

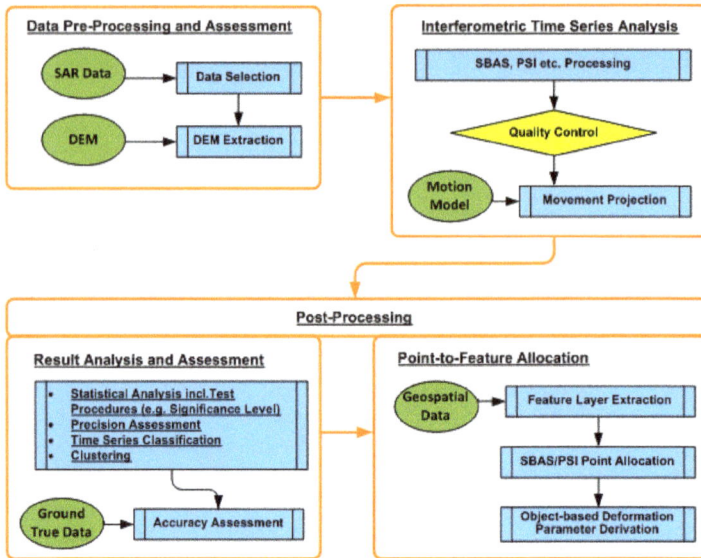

Figure 1. Semi-automated interferometric time series processing and post-processing steps.

The interferometric SBAS processing of the TerraSAR-X data stack was carried out using the SARscape software (Version 5.2) from sarmap S.A. [6]. For the creation of the interferometric-pairs network, geometric baselines have been limited by a threshold of about 183 m to avoid geometric decorrelation. The maximum temporal baseline has been limited to 176 days. The Time-Position Plot in Figure 2 shows the final interferometric network that was used and the differential interferograms for estimation of the deformation time series with measured displacements in line-of-sight (LOS). SBAS processing also includes atmospheric phase screen (APS) for estimation of the atmospheric effects (second inversion). A reliable estimation of these atmospheric components could be achieved by using default APS SARscape setting parameters. This means we used an atmospheric low pass size of 1200 m and atmospheric high pass size of 365 days for atmospheric correction.

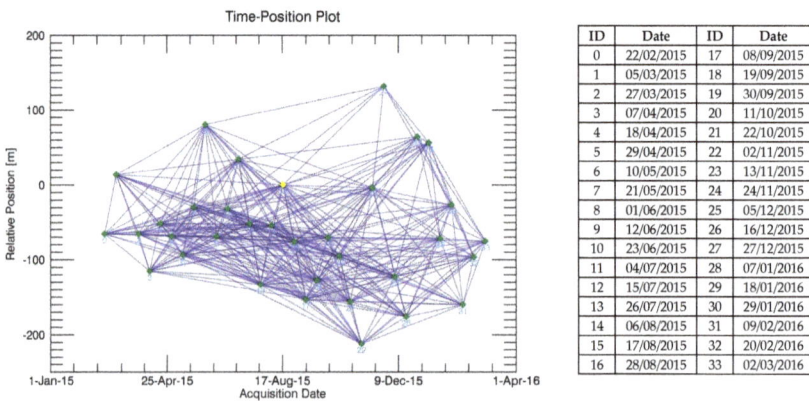

ID	Date	ID	Date
0	22/02/2015	17	08/09/2015
1	05/03/2015	18	19/09/2015
2	27/03/2015	19	30/09/2015
3	07/04/2015	20	11/10/2015
4	18/04/2015	21	22/10/2015
5	29/04/2015	22	02/11/2015
6	10/05/2015	23	13/11/2015
7	21/05/2015	24	24/11/2015
8	01/06/2015	25	05/12/2015
9	12/06/2015	26	16/12/2015
10	23/06/2015	27	27/12/2015
11	04/07/2015	28	07/01/2016
12	15/07/2015	29	18/01/2016
13	26/07/2015	30	29/01/2016
14	06/08/2015	31	09/02/2016
15	17/08/2015	32	20/02/2016
16	28/08/2015	33	02/03/2016

Figure 2. Time-position plot of SBAS connection graph for the used scenes in the case study (left figure). TerraSAR-X scenes were acquired on the listed dates in the table on the right.

After an extensive quality control of SBAS results and under the assumption of no horizontal displacements, the LOS results could be transformed into vertical displacements (see step 'Movement Projection' in Figure 1).

Figure 3a provides the area extent and a detailed view of vertical movement velocity in the area. The dimension of the large-scale subsidence areas is closely connected to the known location of tectonic faults (black hatching lines in Figure 3). Locally, further subsidence areas are detectable, e.g., on waste disposal sites. Figure 4 shows the time series of estimated vertical displacements for three selected measurement positions in Figure 3b. On the west side of the tectonic fault, we can observe a high subsidence velocity of about −17 mm/year (P1), while further east the velocities are clearly smaller (P2—approx. −3 mm/year). Furthermore, the time series of P1 and P2 show a significant linear behaviour in time depending on groundwater withdrawal activities of opencast mining. By contrast on the waste disposal site, the time series of P3 shows a higher velocity, but also a deceleration. This is due to the varying causes of movement.

For a more detailed analysis of objects of interest in a movement area, high-resolution TerraSAR-X data is particularly recommended, in order to fully exploit the high sampling of the surface showing local and small scale specific differences in the movement behaviour. Therefore, high-resolution TerraSAR-X acquisitions are used for enhanced and precise localisation of active tectonic faults (see Figure 3b).

In general, urban structures provide a sufficiently high number of valid measurement pixels. In rural areas with dense vegetation on agricultural land, grasslands, forests and plantations, however, no surface valid movement results can be derived, due to vegetation growth and movement and the resultant radar backscatter changes over time. Infrastructure objects are stable radar backscatter targets, thereby allowing the allocation of a large number of measurement pixels to create feature layers, as shown by railways in Figure 3c, roads (Figure 3d) and buildings (Figure 3e). In the region of interest, approximately 75% of railway networks, and about 40% of road networks have been covered by surface movement results retrieved from TerraSAR-X StripMap data. Therefore, a railway operator has the possibility for high-resolution, large-scale deformation monitoring of his rail network to identify critical sections. Municipalities and district authorities can be supported in their management and fulfilment of duties to detect and maintain traffic lines affected by deformation. Experience shows that insurance companies, for example, are highly interested in building-level movement information in the context of loss adjustment in areas affected by mining-induced surface deformation. For each building, a time series of surface movement measurement pixels is available including, for example, movement velocity and acceleration/deceleration. An extensive analysis allows the extraction of damage-relevant information, such as the maximum tilt of a house, the date when the tilt threshold is surpassed and the potential risk of instability and break. In Figure 3f, houses and railway segments with maximum tilts greater than 0.3 mm/meter are identified. These houses are potentially at risk of damage.

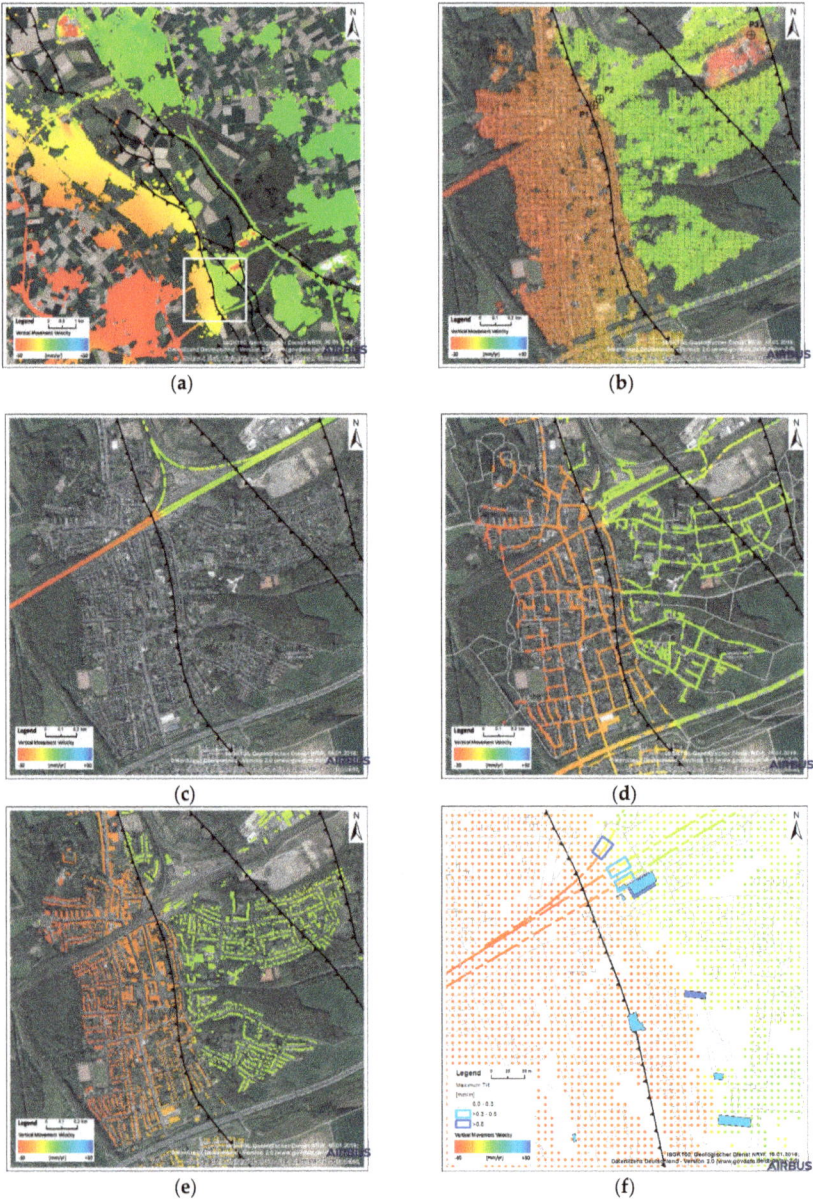

Figure 3. Selection of operational TerraSAR-X surface movement products for an opencast mining region with known tectonic faults (black hatching lines of ISGK100 © Geologischer Dienst North Rhine-Westphalia 2018): (**a**) Vertical movement velocities in the AOI of 13×13 km^2; (**b**) Subset area (white rectangle in (**a**) with Surface Movement Monitoring (SMM) measurement pixels in Horrem (Kerpen, Germany); (**c**) SMM railway allocation product; (**d**) SMM road allocation product; (**e**) SMM building allocation product; (**f**) SMM enhanced product with marking of buildings and railway sections with maximum tilts > 0.3 mm/m in a detailed area. Background: World Imagery (Source: Esri, Digital Globe, GeoEye, Earthstar Geographics) and OSM data (© OpenStreetMap).

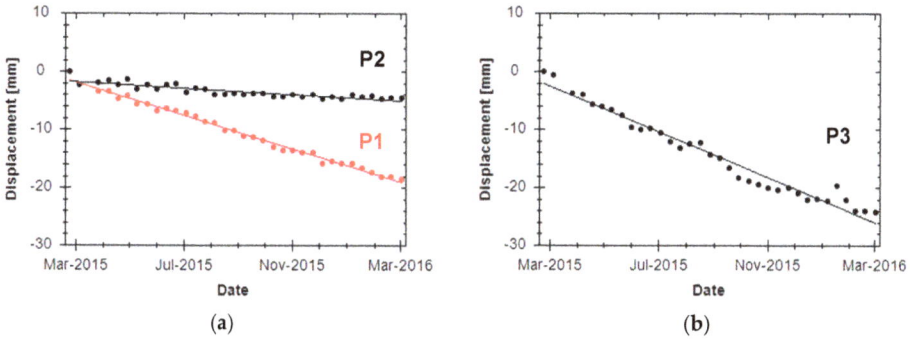

Figure 4. Time series of vertical displacements (plotted points) and linear regression line (solid line) on selected measurements positions in Figure 3b. (**a**) refers to the points P1 and P2, shown in Figure 3b. Though spatially adjacent the temporal displacement of both points significantly differs. (**b**) refers to point P3 and represents the subsiding tendency of a waste disposal site.

2.2. Ground Control Points (GCP)

2.2.1. TerraSAR-X GCP Background

Reliable Ground Control Points (GCPs), i.e., points of known geographical coordinates, are an essential input for the precise orthorectification of remote sensing imagery, the exact location of targets or the accurate georeferencing of a variety of geo-datasets. Although GCPs collected by terrestrial means typically offer a high accuracy, the TerraSAR-X based space-borne approach is of significant interest in such areas where access can be hazardous or may not be authorised. Thanks to TerraSAR-X's precise orbit determination accuracy (within the range of centimetres), the precise radar beam tracing, its high spatial resolution and the resulting high positional accuracy of the imagery, the satellite proves to be highly suitable for obtaining 3D ground information. Based on stereo imagery or multiple image datasets acquired at defined geometrical conditions, GCPs can be obtained at a high accuracy in East (E), North (N) coordinates and in Height (H). The subsequent use of the retrieved points is to establish a control point database for commercial use, particularly in poorly mapped areas, where such information is not available or insufficiently accurate.

TerraSAR-X and TanDEM-X are capable of high-resolution and multi-beam image acquisition. Along with the image data, detailed and very precise orbit data allow for highly accurate 3D information extraction based on stereo or multi-angle image data sets. The image geolocation positioning error is proved to be less than 10 cm in azimuth and range [7,8].

For automated radargrammetric ground control point extraction, a minimum of two images acquired with the same orbit direction or, ideally, from ascending and descending orbits are acquired over the area of interest. The disparity angle α (compare Figure 5) between the acquisition geometry of the 2 images is of great importance for the establishment of a stable geometric model. Measurements of corresponding points in two or more images are used to determine the ground coordinates (Easting, Northing, and Height). The general mathematical background of this intersection procedure is given in [9]. Detailed analysis of image stereo constellations for high performance can be found in [10,11].

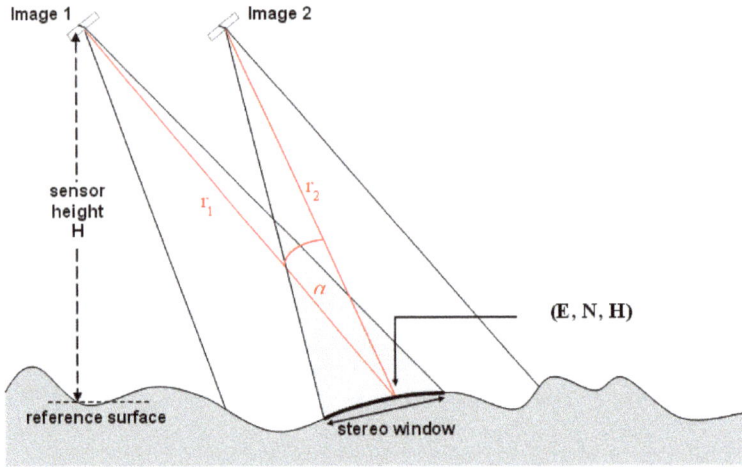

Figure 5. Stereo image configuration for Ground Control Points (GCP) calculation.

2.2.2. TerraSAR-X GCP Case Study

The test site in the South-West of Denver, USA, mainly covers urban/suburban and rural areas. According to the image data acquisition scenario, shown in Figure 6, TerraSAR-X Staring SpotLight (ST) images are acquired as stereo image pairs in HH polarisation. For each acquisition period, three images taken from ascending and descending orbits were used. The image parameters are summarised in Table 1.

Table 1. TerraSAR-X data over south west Denver, USA acquired with Staring SpotLight (ST) mode (rg—range, az—azimuth).

ID	Acquisition Mode	Date	Orbit Direction	Incidence Angle (°)	Resolution rg/az (m)
1	ST	09.05.2014	Descending	42.5	0.88/0.38
2	ST	15.05.2014	Descending	53.5	0.74/0.32
3	ST	08.06.2014	Ascending	35.8	1.03/0.44

In Figure 6, an overview of the Denver test site and the measured points is shown. Points are selected in urban and suburban areas to achieve a mix of diverse surface features, different elements, such as point scatterers (lamp posts and utility poles), road crossings and roundabouts.

Figure 6. Overview of the area of interest (Denver, USA) and selected points for ground coordinate measurements. The white box indicates the subset in Figure 7.

Figure 7 depicts the backscatter response of a point target localised in the TerraSAR-X ST image. Due to the high spatial resolution and radiometric performance of TerraSAR-X ST, the point target is clearly pronounced.

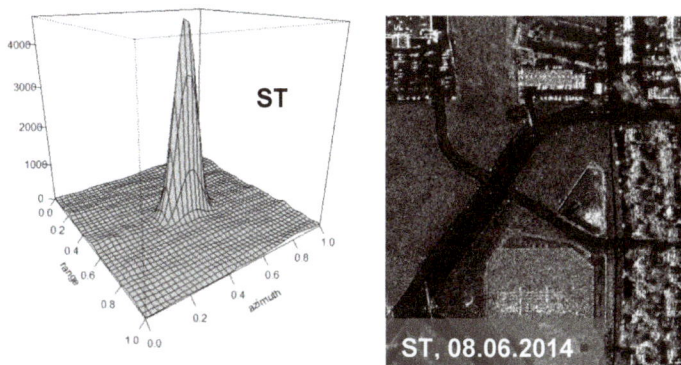

Figure 7. Signal response of a point target (centre peak) in the TerraSAR-X ST image as marked with white box in Figure 6.

For the validation of the TerraSAR-X retrieved Ground Control Points, highly accurate reference coordinates of similar features obtained from Differential Global Positioning System (DGPS) observations, i.e., ITRF 2008 with the Epoch of 2012.0, are employed [12]. The advantage of the DGPS measurements is the very high positional accuracy of the measurements and the global availability with a consistent quality. This reference data was measured with survey grade GPS equipment and processed to an absolute and global accuracy better than 10 cm with CE95 and LE95 confidence levels [12].

The RMSE values of the point residuals in Easting (ΔE), Northing (ΔN) and height (ΔH), as well as for horizontal and spatial displacements are shown in Table 2. The original TerraSAR-X ST pixel spacing is approximately 20 cm [13].

Table 2. Absolute and relative 3D geolocation accuracies for retrieved GCPs from TerraSAR-X ST data (n = 12) [13].

	ΔE (m)	ΔN (m)	ΔH (m)	Δ Horizontal (m)	Δ Spatial (m)
Mean	−0.28	−0.39	−0.07	0.68	0.83
RMSE	0.48	0.54	0.44	0.73	0.85

Difference in planimetric East (ΔE) and North (ΔN), as well as Height of measured GCP compared to DGPS measurements (ΔH).

2.3. Automatic Target Recognition (ATR)

2.3.1. Method

For surveillance, security and reconnaissance applications, high-resolution TerraSAR-X data has been exploited with respect to the detection of relevant ground targets in real-time by automatic and assisted target recognition software. Target detection and recognition is one of Airbus's main technology development products, which are utilised in a variety of applications on both green and blue borders, e.g., for target tracking and recognition, change detection and recognition, anti-piracy or immigration-control, and ship detection of search and rescue operations.

The TerraSAR-X satellite acquires high-quality, high-resolution commercial radar data well suited for purposes of earth observation, target detection and recognition. TerraSAR-X operates in three main acquisition modes: ScanSAR (16 m resolution), StripMap (3 m resolution) and SpotLight modes (0.25 to 0.5 m). In this section, machine learning methods applied to specific target signatures for detection and recognition applications with high-resolution TerraSAR-X images are investigated.

With the increase in processing capabilities through fast multiple parallel graphical units (GPUs) [14], the availability of large-scale annotated datasets, the accessibility of high-resolution SAR imagery and the intense demand from civil applications, the push to develop automated target recognition with transfer learning has grown. Machine learning techniques are often engaged in autonomous detection of the targets based on appropriate training data sets. Convolutional Neural Networks (CNN) include a specific architecture of deep learning and are particularly effective in object and feature classification [15,16]. The CNNs algorithm convolves the input data with a successive number of pre-learned kernels to derive and encode target features prepared for classification. Deep Convolution Neural Network is the stepping block within ATR that iteratively and adaptively customises a variety of relevant features to classify and identify targets. Deep learning does not require any manual pre-built feature extractor, but rather excerpts features automatically. Therefore, an increase in the number of high-quality training datasets enhances feature recognition. In order to minimise experimental risks, the training data set parameters must be adjusted for their quality, quantity and target feature diversity. A hybrid system comprising Deep CNN (DCNN) and Support Vector Machines (SVM) algorithms has been designed at Airbus specifically for TerraSAR-X. This hybrid design has shown to decrease generalisation error producing complex non-linear constraints that give the highest possible differentiation between trained target classes and the surrounding environment, as well as reducing total throughput for ATR applications [17]. The hybrid system efficiently combines convolutional networks effective learning of invariant features, with support vector machines fast and powerful decision making to reduce the number of parameters and thus increases the runtime performance of the network Figure 8 shows the steps employed in the ATR processor.

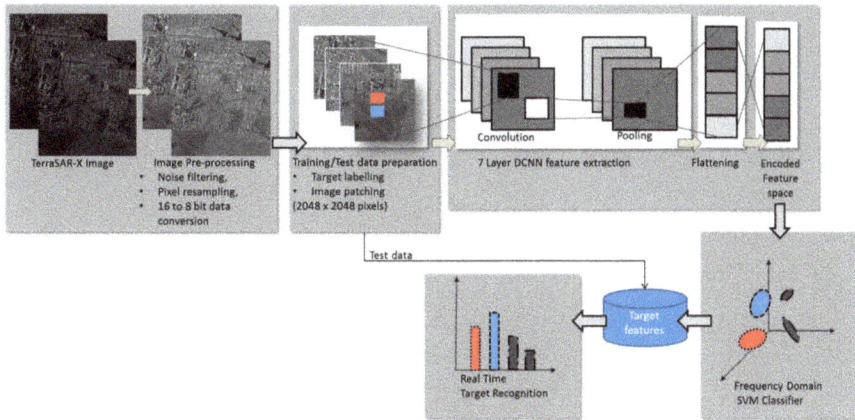

Figure 8. Automatic Target Recognition (ATR) real-time processing steps emulated on a stand-alone computer using Synthetic Aperture Radar (SAR)-specific feature extractor and frequency domain high-speed Support Vector Machine classifier.

Input data are stacks of high-resolution TerraSAR-X level1B images in SpotLight mode. Images are acquired at different incidence angles to account for flexibility in recognition. TerraSAR-X images are processed to Enhanced Ellipsoid Corrected geometric projection format.

Each target of interest's position is manually identified and labelled as a training image patch. Following image preparation, the prepared image patches are fed into the DCNN processor. Therein distinctive features, such as target brightness, texture, edges, shadows and corners are automatically obtained. Each patch covers the labelled target and its surrounding pixels where target backscattering characteristic are described by the scatter cluster rather than the scatter point extraction. These features are combined layer by layer to achieve complex non-linear class features that fully encode target characteristics and are superior to any manual operation. The pre-trained layers are obtained by training the stacked convolutional auto-encoder on labelled SAR scene images.

As depicted in Figure 9, the classification architecture consists of several (here, five) convolutional layers, followed by max pooling with rectified linear units (ReLU) activation functions and fully connected layers. In this example, convolutional layers have adopted a pyramid structure, which means as the convolution layers increase, the outputs of each layer are down sampled by maxpooling. The fully-connected layer preserves the 512-neurons. The output of the last fully connected layer is fed to the Softmax probability activation function, before going into the SVM classifier. The Frequency-Domain Support Vector Machine is an Airbus proprietary disruptive technology, which enables high-speed machine learning applications [18].

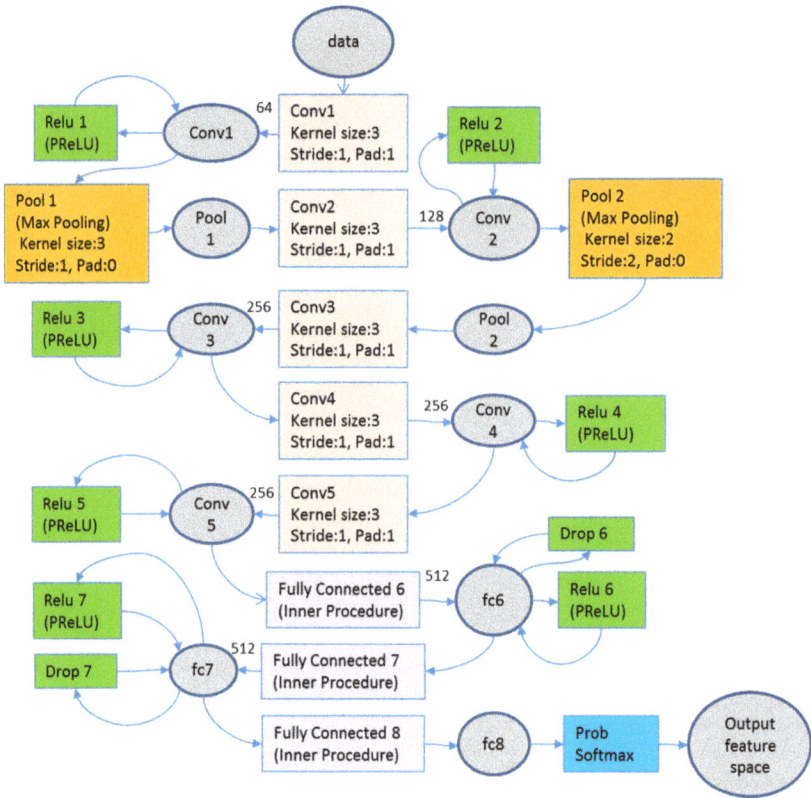

Figure 9. Adaptation and processing steps of Deep Convolutional Neural Network for SAR ATR.

The application of automatic target recognition is performed in two steps: Detection and identification. Detection is the pure localisation of the target. Identification is achieved in two ways; a target can be identified as a member of a class (e.g., class of airplanes) or a specific type within that class (e.g., military aircraft or civilian passenger aircraft).

The performance capabilities of an ATR application depend on appropriate feature detection, characterised by its spatial and radiometric resolutions, and the noise level. The training phase enables "learning by examples" and data-driven generalisation, but performance correlates to availability and quality of training imagery. A variety of pre-training tasks can be utilised to enhance image-signal-to-clutter ratio, for example noise reduction with spatially variant apodization [19], polarisation whitening filters [20], or radiometrically/spatially enhanced image products [21]. The SAR image acquisition geometry and environmental conditions are other important factors for the ATR process. Here, ATR performance depends not only on sensor acquisition geometry, but also on the target geometry, its shadow, layover, foreshortening or occlusion by neighbouring objects. These limitations may be partially overcome by exploitation of target libraries created from simulated SAR data for a specific target size, shape, orientation, and revetment. The accuracy of labelled training targets and classes and the validity of ground-truth data in all-weather conditions are less compromised with simulated SAR libraries. In addition to the above, considerations are made for user requirements, where ATR applications are required to perform within minimum processing time (near-real-time), under limited or reduced data dimensions dealing with computational complexity, achieving a reasonably high (>90%) average precision.

2.3.2. Automatic Target Recognition Case Study

The presented ATR experiment is executed on TerraSAR-X High Resolution SpotLight images (with a resolution of 0.5 m) processed respectively with Spatially Enhanced (SE) and Radiometrically Enhanced (RE) variants and TerraSAR-X Staring SpotLight images (with a ground resolution of approximately 0.25 m × 1.0 m) with the Spatially Enhanced (SE) processing variant. In total, 93 images, for three sites, and 26 images per area of interest in different modes containing positively labelled targets of interest and negatively excluding targets are used for training and testing. The areas of interest are selected to contain a large number of targets suitable for detection and recognition. Targets of interest are four models of aircrafts with lengths ranging from 12 m to 46 m. Targets include a variety of ground parking orientations, due to interim use of the aircraft. Table 3 provides the aircrafts' primary dimensions.

Table 3. Aircraft dimensions for four targets considered in the ATR experiment.

Aircraft Model	Length (m)	Height (m)	Wing Span (m)
Bear (Tupolev Tu-95)	46.2	12.1	50.1
Backfire (Tupolev Tu-22)	42.4	11.0	34.3 or 23.3
Midas/Candid (Ilyushin IL-76M)	46.6	14.7	50.6
Colt (Antonov An-2)	12.4	4.1	14.2 lower, 18.2 upper wing

The areas of interest selected are Ryazan, Ukrainka and Engels airbases in Russia. Training data sets are acquired at different incidence angles and at different orbit orientations to accommodate for different acquisition geometries. The time series of the images accounts for the possibility of seasonal variation throughout the year (i.e., snow and rain). Figures 10 and 11 provide distribution of variety of images per site per acquisition for each site.

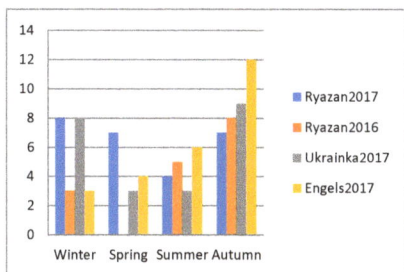

Figure 10. Number of TerraSAR-X images per season for each location.

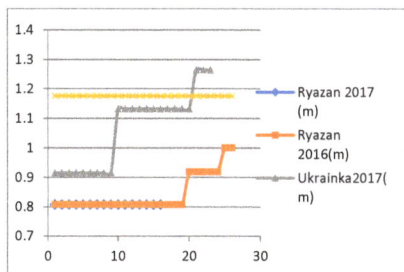

Figure 11. Ground range resolution of TerraSAR-X images for each location.

The image pre-processing step of the experiment started with conversion of TerraSAR-X images from 16 to 8 bit by histogram truncation and rescaling. The experiment was performed on image patches of 2048×2048 pixels. All training and test data patches were resampled to 0.4 m pixel spacing. The positively labelled target chip size in the training depended on the object class size. Data chips were utilised to train the DCNN processor. The frequency domain SVM [22] was trained on the output of the DCNN. In this experiment, the quantitative evaluation of the ATR processor gave very promising results. It showed especially very low false alarm rates on the additional TerraSAR-X test images. Figure 12 provides the performance of the processor for each target on tested data. The quantitative average precision of ATR processor on TerraSAR-X images is given in Table 4. It can be deduced that, since the pose or ground parking geometry of the targets in this experiment did not vary significantly, the tested target signature remained similar, and was mainly affected by signal to noise variation, due to the sensor system and the local weather conditions. Examples of SAR image signature, and the corresponding computer graphics models of each labelled aircraft, are shown in Figure 13.

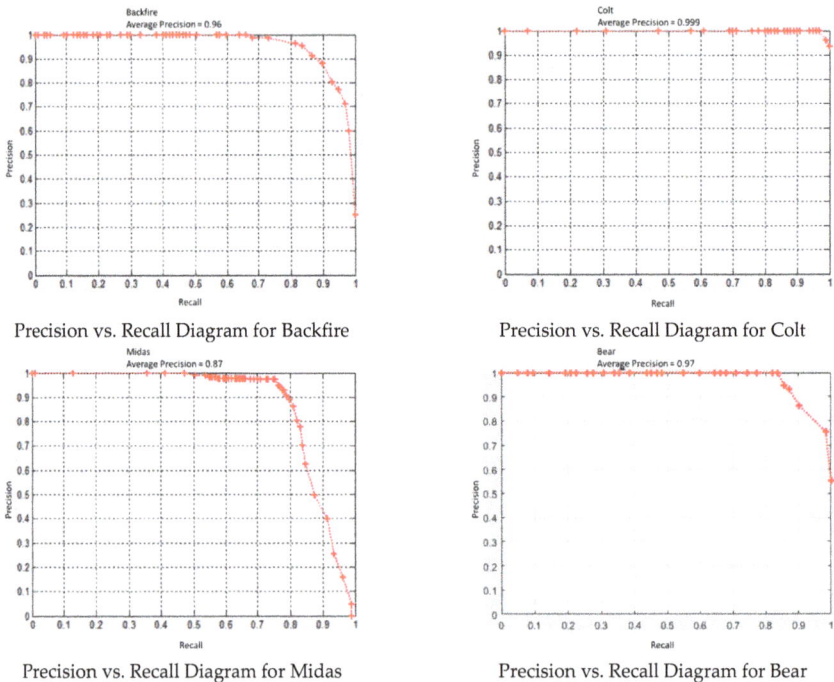

Precision vs. Recall Diagram for Backfire

Precision vs. Recall Diagram for Colt

Precision vs. Recall Diagram for Midas

Precision vs. Recall Diagram for Bear

Figure 12. Performance of the processor for each target on tested data.

Table 4. Average precision (AP) results of Automatic Target Recognition based on DCNN and frequency domain SVM classifier.

Aircraft Model	AP
Bear (Tupolev Tu-95)	97%
Backfire (Tupolev Tu-22)	96%
Midas/Candid (Ilyushin IL-76M)	88%
Colt (Antonov An-2)	100%

Remote Sens. **2018**, *10*, 1535

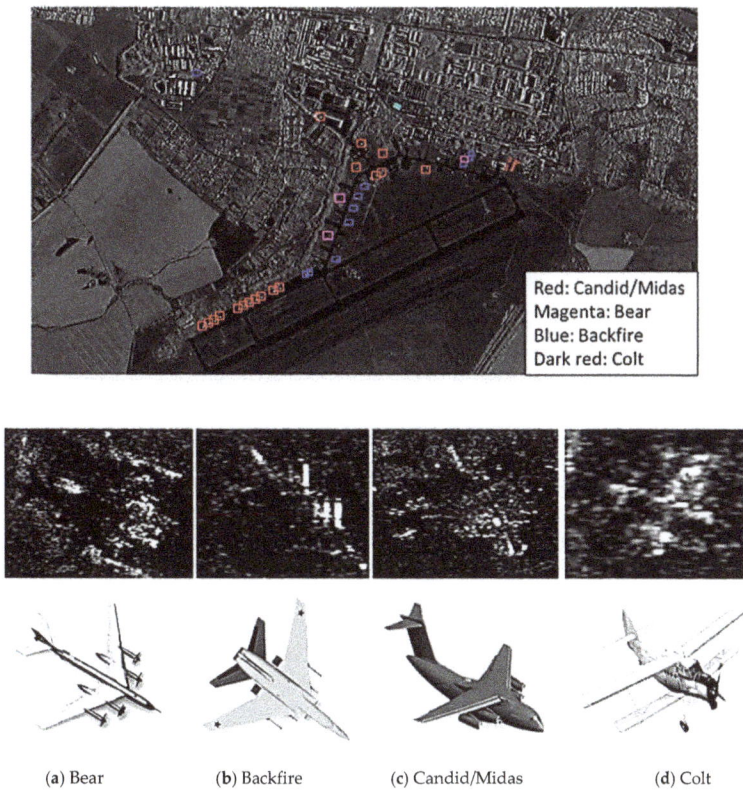

(a) Bear	(b) Backfire	(c) Candid/Midas	(d) Colt

Figure 13. Target detection and classification of TerraSAR-X Staring SpotLight images for Ryazan airbase.

In preparation for the ATR experiment, every aspect of SAR imagery was considered. Products derived from ATR applications ultimately will be utilised to support SAR image analysts where timeliness and precision are of highest importance.

The fusion of pre-processing steps, employment of deep convolution neural network and feature enhancement methodologies in the frequency domain to extract complex and non-linear target attributes minimised efforts to generate optimised target features and reduced the required processing time.

The analysis showed that the classifier's performance is limited significantly by the sensor's acquisition geometry, target shape, size, orientation, revetment, and environmental variations. The target's geometric signature sensitivity indicates the need for feature annotation libraries at a range of incidence angles with small intervals and limited squint variation. Target ground aspects and orientation angles greater than 10 degrees reduce the probability of correct recognition. Therefore, provision for additional training data with different parking positions must be made. In addition, significant differences between signatures of the same target were observed from different imaging modes of TerraSAR-X. The ATR classifier cannot be reused and applied to High Resolution SpotLight and Staring SpotLight images without re-training the system. This confirmed the importance of SAR image resolution and noise level in ATR processing applications.

During this experiment, it was found that the hybrid DCNN design had a much superior generalisation performance, due to abstract feature learning and reacted better at target scale variation. It also performed in an acceptable run-time.

Future experiments are planned to analyse simulated SAR signatures of targets within the same target class category. Simulated SAR data will be utilised to determine specific target attributes that differentiate them from each other. In addition, all scenarios will be tested with different ground parking poses and at different signal-to-clutter levels.

3. Conclusions

Since the launch of TerraSAR-X in 2007 a steady flow of high-resolution SAR data in X-band frequency has been processed and disseminated. The raw SAR data is processed by the TerraSAR-X multi-mode SAR processor at DLR to achieve an optimised geometric and radiometric resolution. In addition to the TerraSAR-X basic products, Airbus Defence and Space provides TerraSAR-X value added geo-information products to a variety of applications. In this paper, three key operational applications were presented. Characterised by high radiometric focusing quality and phase stability, TerraSAR-X surface movement products based on StripMap or High Resolution SpotLight mode imagery have demonstrated an average InSAR phase velocity measurement at a resolution along the line of sight of the sensor, which is equivalent to 1 mm/year.

Similarly, with the advantage of multiple radiometric looks in azimuth and an orbital stability better than 10 cm, the application of ground control points from Staring SpotLight images, provides absolute geolocation accuracy RMSE of < 0.85 m.

In the field of automated target detection and recognition, armed with a large number of training images and objects and advances in deep machine learning techniques, new challenges are tackled, and a fresh wave of products are currently being developed. Thanks to both radiometric stability, absolute geometric focusing and driven by high-resolution training signatures, the operational machine learning application has achieved an average precision of 88% to 97% for selected targets.

The advances in development of operational applications fuel the design requirements for the next generation of synthetic aperture radar satellites as follow-on missions to the current constellation. Based on current achievements, the new era of high-resolution wide swath sensors has started. The importance of high-resolution imagery with greater orbital accuracy and highly focused and stable radiometric quality are vital needs of the follow-on missions.

Author Contributions: O.L. and J.J. supervised and administrated the contribution of all authors, reviewed and edited the manuscript. W.K. wrote the total manuscript for Ground Control Points experiment. D.W., J.A. and O.L. designed analyzed, validated results and wrote the manuscript for Surface Movement Monitoring. P.L. contributed to literature review and assembly of all results for Automatic Target Recognition. C.S. and T.K. designed Deep Neural Network architecture and performed all ATR analysis.

Funding: This research received no external funding.

Conflicts of Interest: The authors declare no conflict of interest.

References

1. Perissin, D.; Ferretti, A. Urban-Target Recognition by Means of Repeated Spaceborne SAR Images. *IEEE Trans. Geosci. Remote Sens.* **2007**, *45*, 4043–4058. [CrossRef]
2. Airbus Defence and Space Case Studies. Available online: http://www.intelligence-airbusds.com/en/ 6988-case-study-gallery-details?item=41559&products_services=\protect\T1\textbraceleftCase_Study: Products__Services:value\protect\T1\textbraceright&search=&market=2397&keyword= (accessed on 12 September 2018).
3. Ferretti, A.; Prati, C.; Rocca, F. Permanent Scatterers in SAR Interferometry. *IEEE Trans. Geosci. Remote Sens.* **2001**, *39*, 8–20. [CrossRef]
4. Berardino, P.; Fornaro, G.; Lanari, R.; Sansosti, E. A New Algorithm for Surface Deformation Monitoring Based on Small Baseline Differential SAR Interferograms. *IEEE Trans. Geosci. Remote Sens.* **2002**, *40*, 2375–2383. [CrossRef]

5.   Bodenbewegungsdienst Deutschland (BBD). Available online: https://www.bgr.bund.de/DE/Themen/Erdbeben-Gefaehrdungsanalysen/Fernerkundliche_Gefaehrdungsanalysen/fernerkundliche_gefaehrdungsanalysen_node.html (accessed on 12 Sep 2018).

6.   Harris Geospatial Solutions. SARscape Help Manual. Available online: https://www.harrisgeospatial.com/docs/pdf/sarscape_5.1_help.pdf (accessed on 31 July 2018).

7.   Eineder, M.; Minet, M.C.; Steigenberger, P.; Cong, X.; Fritz, T. Imaging Geodesy—Towards Centimeter-Level Ranging Accuracy with TerraSAR-X. *IEEE Trans. Geosci. Remote Sens.* **2011**, *49*, 661–671. [CrossRef]

8.   Fritz, T.; Eineder, M. TerraSAR-X Basic Product Specification Document. TX-GS-DD-3302. Airbus DS 2014. Available online: http://www.intelligence-airbusds.com/en/228-terrasar-x-technical-documents (accessed on 12 December 2014).

9.   Raggam, H.; Almer, A. Mathematical Aspects of Multi-Sensor Stereo Mapping. In Proceedings of the 1990 IEEE International Geoscience and Remote Sensing Symposium (IGRASS): Remote Sensing—Science for the Nineties, Washington, DC, USA, 20–24 May 1990.

10.   Koppe, W.; Wenzel, R.; Hennig, S.; Janoth, J.; Hummel, P.; Raggam, H. Quality assessment of TerraSAR-X derived ground control points. In Proceedings of the 2012 IEEE International Geoscience and Remote Sensing Symposium (IGRASS), Munich, Germany, 22–27 July 2012.

11.   Raggam, H.; Perko, R.; Gutjahr, K.H.; Koppe, W.; Kiefl, N.; Hennig, S. Accuracy Assessment of 3D Point Retrieval from TerraSAR-X Data Sets. In Proceedings of the 2010 EUSAR European Conference on Synthetic Aperture Radar, Aachen, Germany, 7–10 June 2010.

12.   Hummel, P. *Remotely Sensed Ground Control Points*; Compass Data Inc.: Centennial, CO, USA, 2014; Available online: http://www.compassdatainc.com (accessed on 12 December 2014).

13.   Koppe, W.; Hennig, S.; Henrichs, L. 3D Point Measurement from Space Using TerraSAR-X HS and ST Stereo Imagery. In Proceedings of the DGPF (German Society for Photogrammetry, Remote Sensing and Geoinformation) Conference, Köln, Germany, 16–18 March 2015.

14.   Mizell, E.; Biery, R. *Introduction to GPUs for Data Analytics Advances and Applications for Accelerated Computing*; O'Reilly Media, Inc.: Sebastopol, CA, USA, 2017; Available online: https://www.network.co.jp/files/9615/0846/8069/GPUs_Data_Analytics_Book.pdf (accessed on 4 June 2018).

15.   Dertat, A. Applied Deep Learning—Part 4: Convolutional Neural Networks, @ Pinterest Nov 8, 2017. Available online: https://towardsdatascience.com/applied-deep-learning-part-4-convolutional-neural-networks-584bc134c1e2 (accessed on 4 June 2018).

16.   Huang, Z.; Pan, Z.; Lei, B. Transfer Learning with Deep Convolutional Neural Network for SAR Target Classification with Limited Labeled Data. *Remote Sens.* **2017**, *9*, 907. [CrossRef]

17.   Kroll, C.; von der Werth, M.; Leuck, H.; Stahl, C.; Schertler, K. Combining high-speed SVM learning with CNN feature encoding for real-time target recognition in high-definition video for ISR missions. In Proceedings of the SPIE, Automatic Target Recognition XXVII, Anaheim, CA, USA, 1 May 2017; Volume 10202, p. 1020208.

18.   Schertler, K.; Liebelt, J. Automatic Learning Method for the Automatic Learning of Forms of Appearance of Objects in Images. U.S. Patent No. 9361543B2, 7 June 2016.

19.   Evers, C.M. Novel Techniques for Enhancing SAR Imaging Using Spatially Variant Apodization. Master's Thesis, The Graduate School of The Ohio State University, Columbus, OH, USA, 2011.

20.   Novak, L.M.; Hesse, S.R. Optimal polarizations for radar detection and recognition of targets in clutter. In Proceedings of the Automatic Object Recognition II, Aerospace Sensing, Orlando, FL, USA, 16 September 1992; Volume 1700. [CrossRef]

21.   *TerraSAR-X Image Product Guide, Basic and Enhanced Radar Satellite Imagery*; Issue 2.1, Airbus Defence and Space: Ottobrunn, Germany, 2015.

22.   Huang, F.J.; LeCun, Y. Large-scale learning with SVM and convolutional nets for generic object categorization. In Proceedings of the IEEE Conference on Computer Vision and Pattern Recognition, New York, NY, USA, 17–22 June 2006; pp. 284–291.

remote sensing

MDPI

Article

Staring Spotlight TerraSAR-X SAR Interferometry for Identification and Monitoring of Small-Scale Landslide Deformation

Farnoush Hosseini *, Manuele Pichierri, Jayson Eppler and Bernhard Rabus

School of Engineering Sciences, Simon Fraser University, 8888 University Drive, Burnaby, BC V5A 1S6, Canada; manuele_pichierri@sfu.ca (M.P.); jaysone@sfu.ca (J.E.); btrabus@sfu.ca (B.R.)
* Correspondence: farnoush_hosseini@sfu.ca; Tel.: +1-778-782-4846

Received: 1 May 2018; Accepted: 22 May 2018; Published: 28 May 2018

Abstract: We discuss enhanced processing methods for high resolution Synthetic Aperture Radar (SAR) interferometry (InSAR) to monitor small landslides with difficult spatial characteristics, such as very steep and rugged terrain, strong spatially heterogeneous surface motion, and coherence-compromising factors, including vegetation and seasonal snow cover. The enhanced methods mitigate phase bias induced by atmospheric effects, as well as topographic phase errors in coherent regions of layover, and due to inaccurate blending of high resolution discontinuous with lower resolution background Digital Surface Models (DSM). We demonstrate the proposed methods using TerraSAR-X (TSX) Staring Spotlight InSAR data for three test sites reflecting diverse challenging landslide-prone mountain terrains in British Columbia, Canada. Comparisons with corresponding standard processing methods show significant improvements with resulting displacement residuals that reveal additional movement hotspots and unprecedented spatial detail for active landslides/rockfalls at the investigated sites.

Keywords: TSX Staring spotlight; high resolution InSAR; small-scale movements; atmospheric phase; layover; DSM blending

1. Introduction

Recent years have witnessed a widespread increase in landslide activity due to climate warning (deglaciation, thaw of alpine permafrost). This affects urban areas and anthropogenic activities in the proximity of landslide-prone areas by posing a serious threat to human life, infrastructure and transportation networks (roads, pipelines, etc.) [1]. In order to reduce the human and economic impact of landslide disasters and define accurate risk scenarios, the use of instruments and techniques capable of providing timely and effective information on mass movements is crucial [2].

Synthetic aperture radar interferometry (InSAR) techniques are attractive for identification and reconnaissance monitoring of landslide hazards, given their capability to detect surface displacements on the scale of the radar wavelength in almost all weather conditions [3,4]. TerraSAR-X (TSX) is ideal for regional-scale landslide investigations, as it can provide high spatial resolution repeat-pass InSAR measurements with wide-area coverage and short repeat intervals [5]. In particular, the TSX Staring Spotlight (TSX-ST) mode is currently the best data source on the market for monitoring small-scale landslides; compared to classic (sliding) spotlight imaging, which has a rotation center located below the scene, TSX-ST improves resolution in azimuth from 1 m to 0.25 m by means of an azimuth steering the antenna to a rotation center within the imaged scene, which also allows for a significantly lower InSAR phase noise compared to other TSX modes [6,7]. TSX-ST allows detection of small precursor landslides, and resolves the details of fault line discontinuities at the meter scale. The short repeat interval of 11 days allows capturing of temporal details in the mass movement time series, while also, the magnitude of spatial displacement gradients for faster slides, preventing aliasing in the corresponding interferograms. Given that most landslides occur in heterogeneous areas (e.g., close to forests or urban settlements) and are affected by pronounced seasonal weather changes, TSX-ST data further provide the added benefit of increased spatial and temporal coherences.

The goal of our study is to investigate the suitability of TSX-ST data for the identification and monitoring of small-scale landslides using enhanced InSAR processing methods. A number of representative case studies of landslide-prone areas in British Columbia (BC), characterized by a diversity of geological phenomena, were observed with TSX-ST InSAR. The sites are used to demonstrate the enhanced InSAR methods used for processing. These methods include a novel Digital Surface Model (DSM) blending scheme to improve topographic phase compensation, a technique for phase recovery in layover-affected areas, and an atmospheric treatment optimization algorithm. The study provides the groundwork for future investigations aimed at developing accurate, robust, and automated time series analysis methods for a comprehensive landslide InSAR solution.

The paper is organized as follows. In Section 2, for three of the case studies (i.e., Hope Slide, Mt. Currie, and Boundary Range Slide; addressed as "test sites" in the remainder of the text) selected to illustrate the enhanced high-resolution processing methods, we summarize the relevant background on the ground movement in more detail, including choice of the TSX-ST InSAR datasets. In Section 3, we describe the enhanced methods for InSAR processing of high-resolution surface displacement maps. In Section 4, we present results from applying these methods to the three selected test sites.

2. Data and Test Sites

Three test sites (see Figure 1) that provide representative diversity for demonstrating the method enhancements are presented in the next section. Site characteristics for these case studies are shown in Table 1. The TSX-ST datasets for these three sites are summarized in Table 2. Figure 2 shows the network of generated interferograms using the datasets of Table 2. The surface deformation maps of Section 4 are obtained, for each case study, using the interferogram in the network of Figure 2 with the least atmospheric contamination and the largest temporal baseline. If more than one interferogram in the network exhibits, comparable levels of atmospheric contamination, and temporal baselines, we selected the one with the smallest perpendicular baseline.

In the following, we describe the location and morphology of the sites, placing emphasis on the geological characteristics that make them suitable examples for the individual method enhancements.

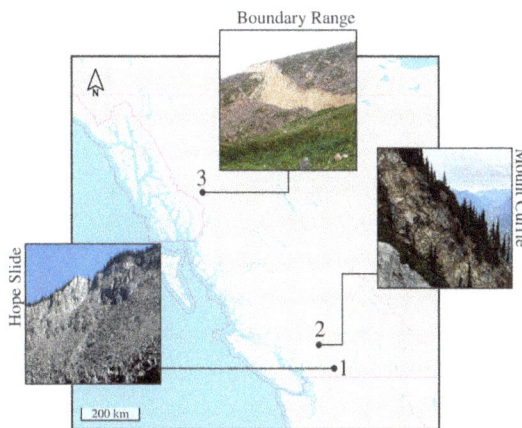

Figure 1. Case studies processed with TerraSAR-X Staring Spotlight Interferometric Synthetic Aperture Radar (InSAR) and their locations. (1) Hope Slide, (2) Mount Currie, (3) Boundary Range.

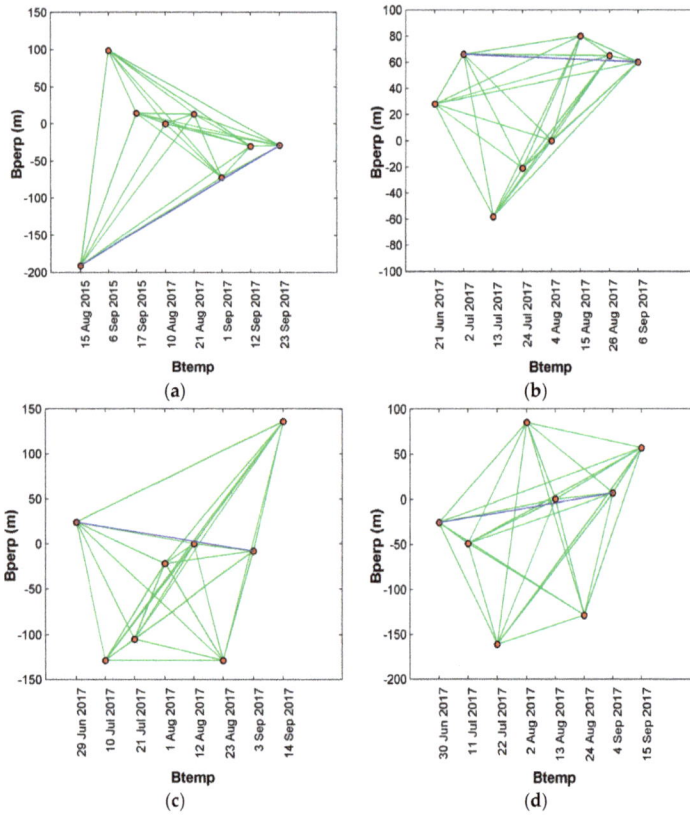

Figure 2. Network of the generated interferograms based on temporal (Btemp) and perpendicular (Bperp) baselines for (**a**) Hope Slide; (**b**) Mount Currie; (**c**) Boundary Range Slide (ascending); (**d**) Boundary Range Slide (descending). For each test site, the deformation maps of Section 4 are generated using the interferograms highlighted in blue.

Table 1. Case studies and their characteristics.

Test Site	Location	Description	Type of Risk	InSAR Observed Phenomena
1. Hope Slide	121.26°W 49.30°N	Slide scar (1965 landslide)	Unstable steep scar, major highway nearby	Strong deformation gradient near the headscarp. Evidence of rockfalls near sheer zone. Subtle movement in the accumulation zone. Deforming patch near the highway.
2. Mount Currie	122.78°W 50.25°N	Steep mountain ridge	Rockfall events, precursors to catastrophic landslide	Deformation near main tension cracks.
3. Boundary Range Slide	132°W 57.5°N	Compound rockslide	Slope instability, large scale movements recorded	Large scale movement at the bottom of the landslide (toppling zone). Sliding and transition zones are also detectable.

Table 2. Test sites and TerraSAR-X Staring Spotlight dataset (ascending (Asc.)/descending (Desc.)). (Incidence angles are expressed in degrees (deg), HH polarization refers to Horizontal transmit and Horizontal receive, and DSM is Digital Surface Model.).

Test Site	Time	No. Scenes	Orbit	Incidence (deg)	Polarization	Methods Demo
Hope Slide	August 2015–September 2017	8	Desc.	36	HH	Enhanced atmospheric correction
Mount Currie	June 2017–September 2017	6/6	Asc./Desc.	22/34	HH	Coherent layover topographic correction
Boundary Range Slide	June 2017–September 2017	8/8	Asc./Desc.	41/38	HH	Enhanced DSM blending

2.1. Hope Slide

One of the largest landslides recorded in Canada occurred at this test side. The historic Hope Slide occurred in January 1965, in the Nicolum Valley in the Cascade Mountains near the municipality of Hope, British Columbia [8]. The landslide buried Highway 3 under 47 million cubic meters of rock [9]. Seven faults and three dominantly brittle shear zones, along with tension cracks and trench-like features, have been recognized as the main deformation features in the Hope Slide area [10,11]. The aforementioned geological features, along with the historical landslide in the Hope Slide, draws attention to using the TSX-ST images to perform a comprehensive analysis of slope stability on and around Hope Slide, to better understand the triggers and dynamics of mass movements in the area. Due to the topography of the area, the Hope Slide is exposed to strong turbulent mixing processes, atmospheric stratification, and changes in atmospheric pressure in time, which introduce significant atmospheric errors to these high resolution TSX-ST images [12]. Addressing and modelling the atmospheric effect (e.g., by means of the methodology described in Sections 3.1 and 3.2) plays a key role in retrieving the deformation phase from the TSX-ST acquisitions for the Hope Slide.

2.2. Mount Currie

This test site comprises the northeast trending ridge above the village of Pemberton and Hamlet of Mount Currie, BC, Canada [13]. This site experienced a series of rockfall events near the ridge crest in 2015 and 2016. The rockfall activities, which may be precursors to a more catastrophic event associated with tectonic structures close to the Mount Currie ridgeline, raised concerns about the safety of human lives and infrastructure in the nearby communities. As Mount Currie's deformation features (including rock slides, rockfalls, lineaments, and deep-seated gravitational slope deformation [14]) are also bringing about small-scale movements, the unprecedented spatial details of the TSX-ST can shed light on these movements and their mechanisms in Mount Currie. The challenge we need to face for using InSAR in such a rugged topography is shadow/layover effect, which decreases the quality of derived deformation maps from TSX-ST images. Even though the TSX-ST images have a high resolution, in a mountainous area with rugged topography, as with Mount Currie, the intractable geometrical distortions (shadow/layover) restrict the potential of repeat-pass InSAR. As a result, the methodology detailed in Section 3.3 is crucial to extracting additional information from the affected areas.

2.3. Boundary Range Slide

The Boundary Range Slide, located in the Coast Mountains of northwestern BC, Canada, has developed since 1956, when the bedrock started moved extensively and rapidly [15]. The retreat and downwasting of the Boundary Range Glacier appeared to have played a key role in the evolution of such a landslide. The investigations in 2008 showed that this landslide is approximately 1 km wide, 500 m long, and is actively deforming [16]. Clayton et al. [17], identified three geomorphic zones within the main landslide, including a lower zone that is toppling, an upper zone of sliding, and a transition zone of extension in between. As the area is changing very rapidly (0.8, 0.33, and 0.19 m/year for the toppling, transition, and sliding zones, respectively), an up-to-date high-resolution DSM is required to subtract the topographic and atmospheric contributions of the phase from the interferograms. To do so, we use a 2017 high resolution photogrammetric DSM (1 m) covering a small patch within the TSX-ST footprint. This DSM is merged with a low resolution Shuttle Radar Topography Mission (SRTM) extending over the whole area (30 m, generated in 2000) using the enhanced blending method described in Section 3.3.

3. Methods

In this section, we describe the developed individual enhanced methods for high-resolution InSAR processing, and how they integrate with the standard InSAR processing steps. Due to the current shortness of our InSAR time series, advanced stack-based InSAR methods cannot be used reliably, and "standard processing" here refers to basic pairwise InSAR (D-InSAR).

The D-InSAR phase, as the phase difference between two complex valued SAR images, can be decomposed as

$$\phi_{int} = \phi_{def} + \phi_{topo} + \phi_{atm} + \phi_{ref} + \phi_n. \tag{1}$$

The deformation term ϕ_{def} accounts for temporal changes in phase occurring between the two acquisitions, e.g., due to surface movement or dielectric changes. The term ϕ_{topo} is the topographic phase component due to difference in satellite orbit (spatial baseline) for the two acquisitions. ϕ_{atm} is the phase component from propagation delays due to atmospheric water vapor variations in time and space. This quantity can be modelled as the sum of a "static" atmospheric term, which is a function of elevation, and a "dynamic" atmospheric term (based on the dynamic nature of the troposphere governed by minute-scale temporal variations of atmospheric water vapor largely at spatial scales > 3 km). The "flat earth" phase ϕ_{ref} is proportional to the spatial baseline and the bald Earth ellipsoid, as seen by the side-looking SAR geometry. Finally, ϕ_n accounts for any residual phase noise in the sensor electronics and due to temporal decorrelation of the imaged surface.

To obtain accurate surface displacement maps, ϕ_{def} is estimated from the observed interferogram by appropriately modeling/estimating and compensating the other phase components in Equation (1). Standard methods to carry out the phase corrections to ϕ_{def} are implemented in an automated python-based InSAR stack processing chain built on top of the Gamma software (e.g., [18]). Except for reference phase ϕ_{ref}, whose removal is uncontentious, we have developed enhancements for high spatial resolution data to the corresponding standard methods. These methods enhancements are detailed in the following subsections.

3.1. Atmospheric Phase Correction

Any changes of the troposphere (e.g., water vapor distribution and air pressure) between the two acquisitions of an interferogram lead to a phase error component ϕ_{atm}. If the atmosphere is at rest at the time of a SAR acquisition, the equations of motion become [8]

$$0 = \frac{-1}{\rho} \frac{\partial P}{\partial x},$$

$$0 = \frac{-1}{\rho} \frac{\partial P}{\partial y}, \tag{2}$$

$$g = \frac{-1}{\rho} \frac{\partial P}{\partial z},$$

with P atmospheric pressure, ρ air density, g gravity, and x, y, z being easting, northing, and elevation, respectively. Equation (2) shows that for a static atmosphere, P depends on z only. As a consequence, the static atmospheric contribution of ϕ_{atm}, which is proportional to the pressure difference between the SAR acquisitions forming the interferogram, will also be a function solely of z [9]. For most atmospheric conditions, this function is an exponential that is well approximated by a linear or quadratic polynomial, pol(a,z), with coefficient vector a. In contrast to the static contribution, which depends on elevation, the dynamic contribution to ϕ_{atm} is proportional to the difference in the random turbulent patterns of moving water vapor at the SAR acquisition times. While these patterns are hard to predict or model, they have little spatial energy at scales below 5 km [19], which separates them from most of the landslide displacement patterns we want to measure, which occur at scales of 1 km or less.

According to [20,21], a linear or a power-law relationship between the phase and topography can be assumed to correct for the atmospheric correction from the high-resolution interferometric phase. Therefore, we implemented a method to estimate both the static and dynamic part of ϕ_{atm} using correlation analysis between the phase of atmospheric-contaminated interferograms and ground elevation from an accurate high-resolution DSM via the iterative algorithm detailed in Figure 3. Areas of suspected movement are masked from the interferogram, which is segmented into overlapping tiles of ~0.5 km × 0.5 km. The algorithm in Figure 3 is carried out for each tile. In a first iteration, the coefficients are found by least squares fitting over an area of the interferogram tile. The support area for the fit must have a small enough elevation diversity to contain the atmospheric phase component safely within one 2π interval, to prevent incorrect fitting of wrapped phase values. In a second iteration, the coefficient vector, a, can be refined by adding a correction (Δa) via fitting the phase residual after removing pol(a,z) found in the first step. As the dynamic range of the phase residual is diminished compared to the original phase, the support area and elevation diversity can be increased in the second step, which increases the accuracy of the polynomial fit. Iterations are continued until Δa falls below a predefined threshold. For each tile, the constant coefficient of the polynomial (phase offset) is interpreted as the average dynamic phase of the tile, while the higher order coefficients are interpreted as describing the static phase contribution. The phase offsets of these overlapping tiles are then interpolated to generate a smooth phase screen over the scene.

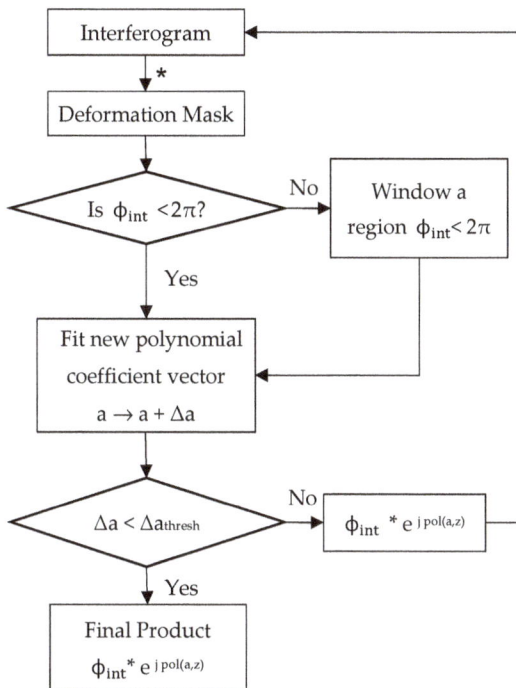

Figure 3. Static atmospheric phase removal: flowchart of the adopted iterative method for each tile.

3.2. Layover Topographic Correction Method

Due to the SAR side-looking geometry, whenever the terrain slope exceeds the incidence angle, more than one area on the ground contributes to the backscatter of a single resolution cell in the SAR image; this is the well-known layover effect [22] (Figure 4a). In layover areas, InSAR produces a superimposed phase whose decorrelation properties are governed by volume coherence criteria even if the scattering mechanisms are all purely from surface scattering. In particular, there is a much faster decorrelation with spatial baseline than for non-layover regions. However, for most current space-borne sensors, the orbital tubes, and thus maximum spatial baselines, are typically kept small, and layover areas, particularly in high resolution images, are often sufficiently coherent to produce an InSAR phase that is at least theoretically usable. Steep and rugged mountainous terrain in high-resolution SAR generally exhibits a significant number of disjointed layover areas as incidence angle cannot be chosen too large, to avoid shadowing. In order to exploit coherent layover phase for detecting ground movement, one needs to model the superposition of the phase from its different ground contributions, and to carry out the topographic phase corrections correspondingly [23–27]. Ground motion can be detected and attributed to a unique area on the ground only when there is a dominant (as revealed by the superposition modeling) phase contribution in the layover. For this study, we have introduced a simple new technique to topographically correct the phase in layover areas for this case based on an accurate SAR geometry and a high-resolution DSM (such as from photogrammetric DSM). The ray-tracing concept of the technique is shown in Figure 4b.

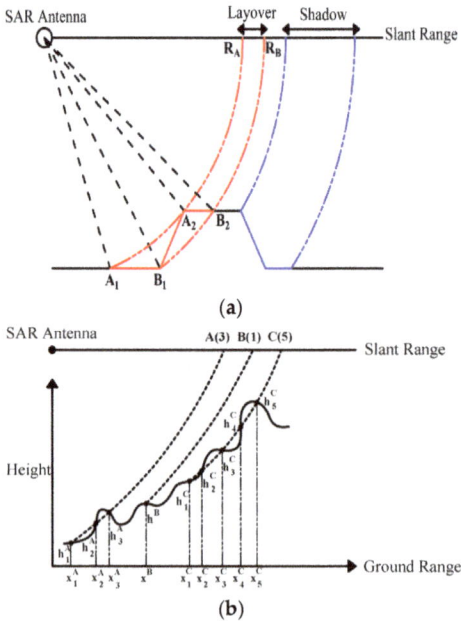

Figure 4. Concept of (**a**) layover and shadow; (**b**) layover topographic correction method.

The technique consists of two steps. First, we construct a series of line-of-sight (LOS) vectors from the sensor to the DSM for every (zero-Doppler) range line of the SAR image as a function of look angle, which is measured at the sensor with respect to its state vector. This delivers both elevation and slant range, which are calculated as the length of the look vector, and series as a finally-sampled function of the look angles. Sections of the range line that are not illuminated by the SAR, due to shadow, are first found and marked via ray tracing as a binary illumination mask matching the series. The slant range series is then converted to a non-integer slant range pixel, and together with the illumination mask, is used to find all elevations, their corresponding ground locations, and the local incidence angles that share the same integer slant range pixel. The result is stored in a suitable numerical structure. Then, the second step analyzes this result in terms of a simple relative contribution, making up the backscatter of the non-unique layover pixels. This uses the local incidence angles of the contributions assuming a cosine scattering law [28] and constant surface roughness. If one of the ground areas dominate the backscatter (intensity > 75 percent of total), we assign the corresponding elevation and ground location to the slant range DSM and geocoding look-up table, respectively. A further sophistication currently being implemented is to use actual geocoded backscatter interpolated from only the uniquely locatable slant range pixels to estimate the individual backscatter contributions making up the non-unique pixels. By reducing the initial topographic phase error, our method also will improve performance of future InSAR time series analysis in coherent layover areas. The extra height error correction in layover found through time series decomposition [29,30] will be usable also as an a posteriori measure of validity for the backscatter superposition model of the just-presented topographic correction method, and consequently, also of the geolocation of the residual deformation phase.

3.3. Digital Surface Model Blending Method

The quality of the topographic correction of the D-InSAR phase ϕ_{topo} depends highly on the accuracy of the DSM used. Baseline-dependent errors in the deformation phase stemming from DSM inaccuracies can be removed via time series analysis [31], but for shorter series, particularly the simple D-InSAR case, it is crucial to use as good a DSM surface as can be constructed. A standard problem is the merging of continuous existing DSMs with lower resolution and accuracy (such as SRTM) with newer higher resolution DSMs (often from photogrammetry and Light Detection and Ranging (LiDAR)) to generate the DSM surface used by the InSAR processing chain [32]. To avoid artifacts in the topographically-corrected phase across where DSMs were joined together, smooth/differentiable transitions are desirable. For the standard InSAR case of merging two or more DSMs, there is usually a clear hierarchy of trustworthiness, which suggests that the older coarse resolution DSM should be altered and matched to the newer high-resolution DSM at their boundary, while the latter should remain unaltered. By contrast, the state-of-the-art algorithms we are aware of (e.g., [33]) only allow merging of DSMs through symmetric blending. We, therefore, developed a simple new method based on thin-plate spline fitting, that "molds" the coarse DSM to the unaltered high-resolution DSM, while also minimizing any distortions and stitching errors along the boundaries during the blending procedure. The concept of the new method is shown in Figure 5.

The method consists of two steps that are described below:

- Step one is to co-register the high- and low-resolution DSM in the overlapping area to find and fix any potential datum issues, elevation offsets, horizontal shifts, or rotations, etc. To do so, for each DSM, we derived feature points along the ridgelines by calculating and thresholding the ratio between the large and the small eigenvalue of the Hessian matrix. Mutual information method [34,35] is used to calculate the initial translation and rotation components of the affine transformation between the two DSMs. To refine, first correspondence between the two sets of feature points is established using the scale-invariant feature transform (SIFT [36]). Based on this feature pair list, the Gauss–Newton method [37] is used to refine the affine transformation parameters iteratively.
- Step two defines a belt-shaped region of width between 100 to 150 postings of the coarse DSM (Figure 6) surrounding the high-resolution DSM(s). A thin plate spline is then fitted across this belt, with clamped boundary conditions of $\Delta h = h_{highres} - h_{lowres}$ at the inner edge (=boundary of the high-resolution DSM) and 0 at the outer edge of the belt. The result of this fit is a correction surface that is subtracted from the coarse DSM. The so-corrected coarse DSM can then be seamlessly merged with the high-resolution DSM by simply inserting the latter into the former.

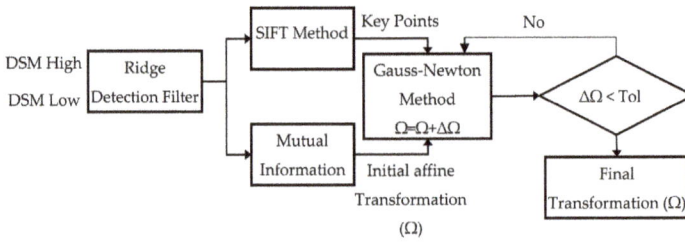

Figure 5. Digital surface model (DSM) co-registration: flowchart of the adopted method (SIFT is the Scale Invariant Feature Transform method and Tol is tolerance.).

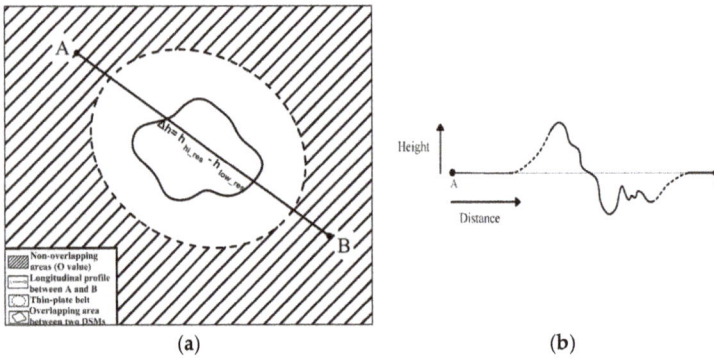

Figure 6. DSM blending: (**a**) Concept of asymmetric DSM blending method; (**b**) longitudinal profile between A and B.

4. Results and Discussion

The methods detailed in Section 3 are now applied to the three test sites presented in Section 2 to generate accurate surface deformation maps. The results of this study are illustrated and discussed in the following.

4.1. Hope Slide

4.1.1. Atmospheric Phase Correction

As highlighted in Section 2.1, we expect the Hope Slide interferograms to be influenced considerably by atmospheric changes due to the topography of the area (see Table 3). For instance, the phase pattern of the Hope Slide interferogram shown in Figure 7a (June to August 2015) is indicative of pronounced atmospheric effects that, if not corrected, would significantly affect the accuracy of the estimated surface deformation. The procedure detailed in Sections 3.1 and 3.2 is applied to mitigate the phase induced from both static atmosphere (Figure 7b) and dynamic atmosphere Figure 7c). A remarkable reduction of the atmospheric-induced effects can be observed by comparing the interferograms before (Figure 7a) and after (Figure 7d) atmospheric correction.

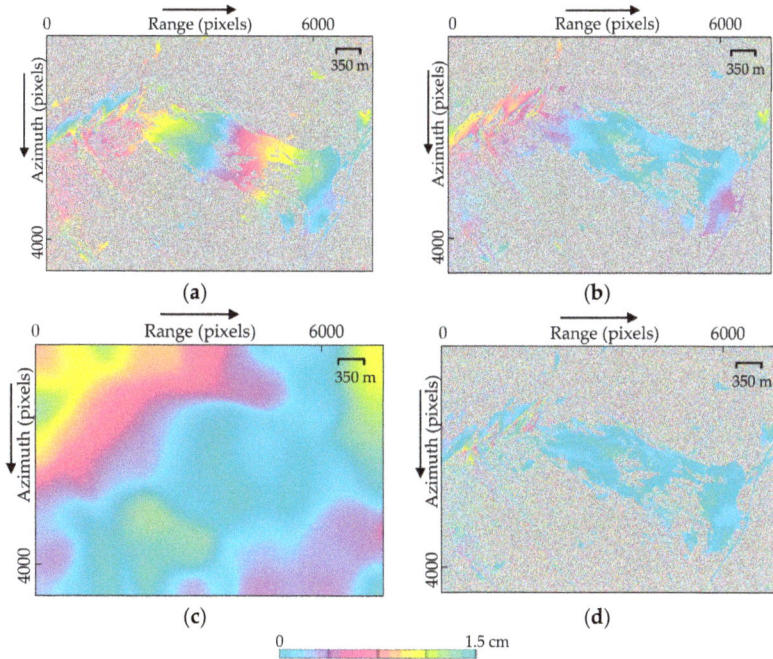

Figure 7. Hope Slide: TSX interferogram (21 June 2015–4 August 2015). (**a**) Original; (**b**) static atmospheric corrected; (**c**) dynamic atmospheric screen; (**d**) static and dynamic atmospheric removed.

Table 3. Hope Slide presented interferograms.

Section Demo	Pair	Orbit	Incidence (deg)	Polarization	Perp. Baseline (m)	Weather
Atmospheric phase correction	21 June 2015 4 August 2015	Desc.	36	HH	45	Cloudy, 10 °C Foggy, 14 °C
Surface deformation map	15 August 2015 23 September 2017	Desc.	36	HH	160	Clear, 13 °C Partly Cloudy, 16 °C

4.1.2. Surface Deformation Map

Figure 8 shows an atmospherically corrected interferogram, August 2015 to September 2017, spanning close to the maximum observed time period (see Table 3). The spatial baseline of this interferogram is approximately 160 m. Despite having a relatively large spatial baseline, this interferogram has the least dynamic atmospheric contamination and the largest temporal baseline. In the same figure, four areas with small-scale deformation are highlighted in red, with the corresponding zoom insets displayed on either side.

The first inset shows a newly identified area with rockfall potential near the headscarp of the Hope Slide. The most evident deformation feature is a small region containing a full interferometric color fringe indicative of a strong deformation gradient. The second inset shows another similar active part of the landslide. The concentration of the regional tectonic features, such as faults and shear zones (Figure 9), which reduce the quality of rock mass [11], can cause onset of rockfalls or slides compatible with the deformation pattern observed in the interferogram. The third inset shows subtle movement in the accumulation zone of the Hope Slide, where the debris of the historic landslide has accumulated and created a bulge. The interferogram suggests subtle movement away from the SAR sensor and along the LOS direction for this region. This movement may indicate secondary slides in the deposits of the historic landslide, or alternatively, subsidence/settling of the bulge.

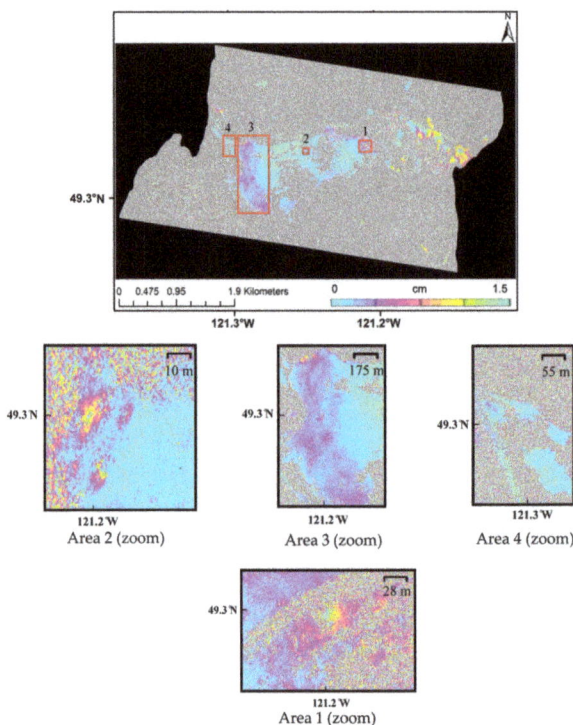

Figure 8. Hope Slide: Geocoded TSX interferogram (15 August 2015–23 September 2017). The four areas where deformation is observed are highlighted in red, and the zoom insets and the scales are also presented.

Figure 9. Hope Slide: geology map (Source: Google Earth). Trench (brown), fault (blue), and shear (red) are presented.

The fourth inset shows a small deforming patch located near the highway. This area corresponds to the distal edge of the coarse portion of the 1965 Hope Slide deposit. As the movement is observed in the LOS direction, there are two possible deformation scenarios: either the landslide deposit might be subject to subsidence on its eastern side, or the eastern side of the hill is experiencing a slow landslide. Although the deformation is subtle in this zone, at its current rate, it may lead to minor damage to road prism and the need for maintenance.

4.2. Mount Currie

4.2.1. Layover Topographic Correction

As discussed in Section 2.2, all Mount Currie interferograms are expected to be affected by small-scale dissected layover, owing to rugged topography and steep incidence angle. The method of Section 3.2 is therefore applied to all interferograms in order to topographically correct the phase in layover areas. The algorithm allows for pinpointing of areas whose non-zero residual phase in the original interferogram is caused by erroneous topographic correction inside layover/shadow areas, rather than actual deformation.

In Figure 10, we demonstrate the performance of the proposed layover topographic correction method for a layover-affected subregion of an 11-day interferogram with a spatial baseline of 21 m (see Table 4). A comparison of the interferograms before and after correction (Figure 10a,b), respectively) shows a number of layover artifacts in the original interferogram (highlighted with red arrows in Figure 10a) as well as shadow-affected areas that are now corrected and masked out.

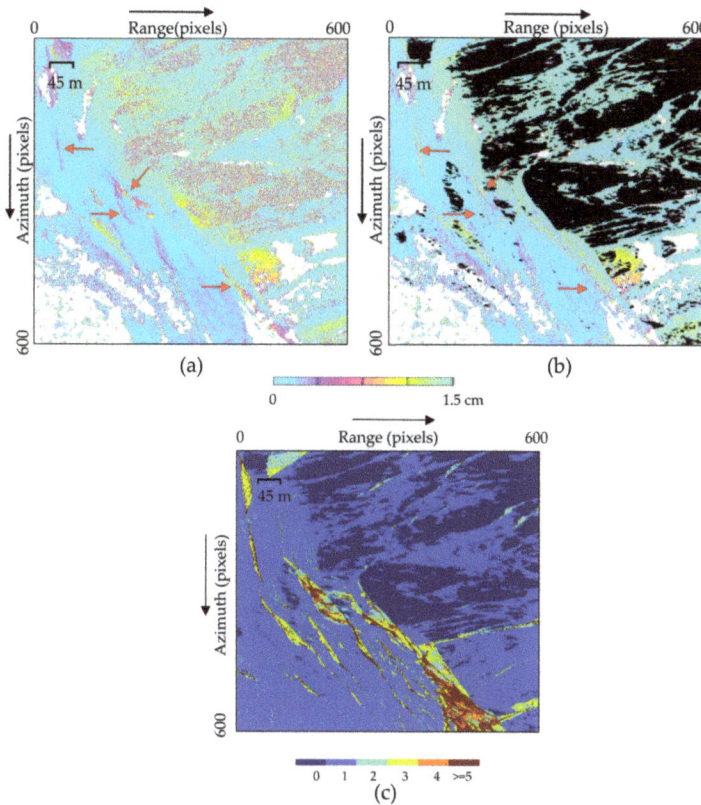

Figure 10. Layover topographic correction: Mount Currie, TSX staring spotlight interferogram (24 July–4 August 2017). (**a**) Conventional method; (**b**) layover corrected with shadow mask; (**c**) number of topographic solutions. Red arrows show the false motion areas, which have been removed after layover correction.

Table 4. Mount Currie presented interferograms.

Section Demo	Pair	Orbit	Incidence (deg)	Polarization	Perp. Baseline (m)	Weather
Layover topographic correction	24 July 2017 4 August 2017	Desc.	34	HH	21	Clear, 28 °C Clear, 26 °C
Surface deformation map	2 July 2017 6 September 2017	Desc.	34	HH	5	Clear, 27 °C Windy, 18 °C

4.2.2. Surface Deformation Map

Figure 11 shows a 66-day interferogram spanning the period 2 July to 6 September 2017. Besides a negligible residual atmospheric phase for both acquisitions of the pair, this interferogram also has the smallest spatial baseline (5 m) and longest possible temporal baseline achievable with our data to date. Details about this pair are available in Table 4.

Surface movements are observed in two areas on the northern face of Mount Currie. The first region, which is situated along the ridge east of Currie, is a source for potential rock avalanches according to BGC Engineering report [14]. The interferogram indicates ongoing deformation in this area, which encompasses the main tension cracks. The deformation observed in the interferogram is consistent with the deformation pattern previously documented in this area [13]. The high quality information of this interferogram suggests a rock avalanche precursor motion, which, if it happens, might dam Green and Lillooet rivers and damage the airport [14]. The second area shows another noticeable block movement occurring near a zone characterized by a deep-seated gravitational slope deformation delimited by tension cracks. The observed movement suggests the potential of rockfalls in this zone.

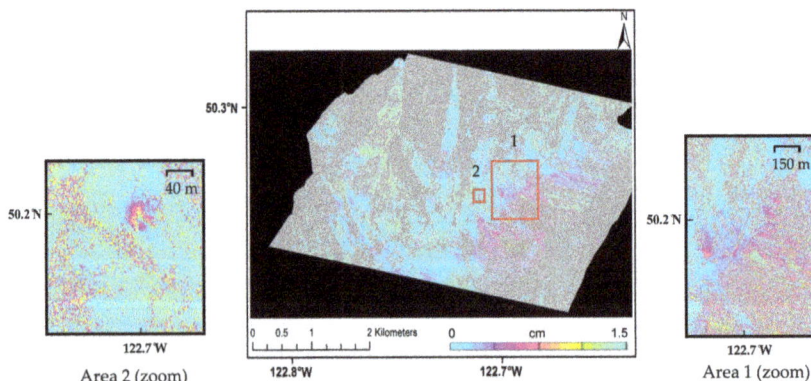

Figure 11. Mount Currie: Geocoded TSX interferogram (2 July 2017–6 September 2017). The two areas where deformation is observed are highlighted in red, and the zoom insets and the scales are also presented.

4.3. Boundary Range Slide

4.3.1. Digital Surface Model Blending

With respect to high deformation rate and ongoing glacial retreat in the Boundary Range Slide, it is essential to use an updated high-resolution DSM for the InSAR processing. To update the existing DSM of the area (SRTM, 30 m resolution), we merged the latter with a more recent high-resolution photogrammetric DSM (1 m) covering a small patch within the radar footprint using the thin-plate blending method detailed in Section 3.3.

The shaded relief of the generated DSM is displayed in Figure 12a. For comparison, we also computed a second DSM by blending photogrammetric and SRTM DSMs via a conventional amalgamation approach implemented in ArcGIS software. This approach takes the average of the two DSMs in the overlapping areas and along the boundaries to generate a smoother blended DSM. A small patch of each DSM has been selected over the highlighted area in Figure 12a, and is presented in Figure 12b,c to visually compare them.

The elevation profile for a transect of both blended DSMs (i.e., between A and B, see Figure 12a), plotted in Figure 12d, indicates good agreement between photogrammetric and thin-plate blended DSM, confirming that the proposed method is only altering the lower resolution DSM, leaving the high-resolution DSM intact. When using the conventional approach, on the other hand, one can observe pronounced differences between photogrammetric and blended DSM (error up to 62 m, see Figure 12e) owing to the symmetric smoothing of the high- and low-resolution DSMs, regardless of their substantial differences in acquisition dates and resolutions. In addition to that, the conventional approach generates noticeable discontinuities/wrong elevations at the boundaries between low- and high-resolution DSMs (Figure 12c).

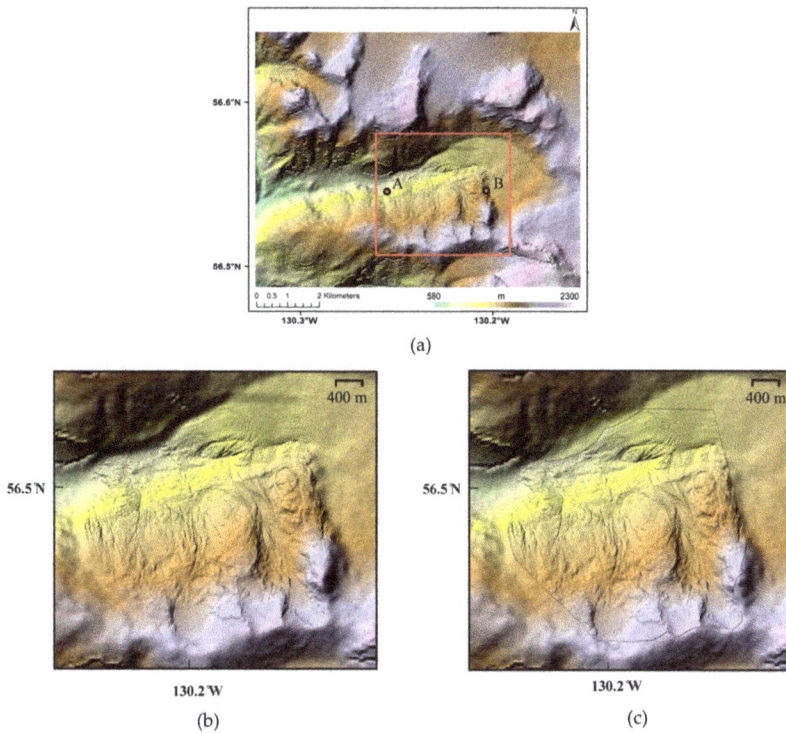

(a)

(b)

(c)

Figure 12. *Cont.*

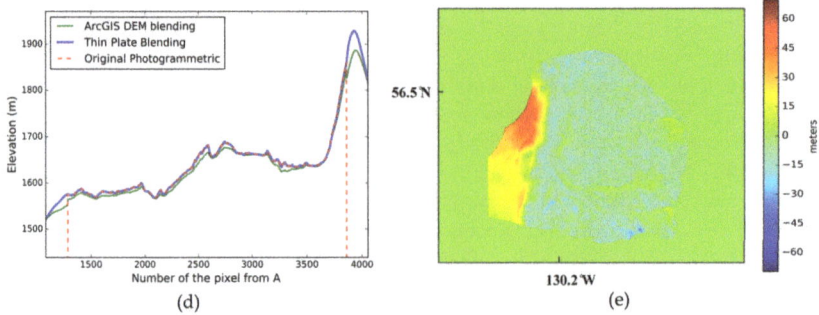

Figure 12. DSM blending, Boundary Range Slide. (**a**) Thin-plate blended DSM; (**b,c**) zoom of the red box in (**a**) for thin-plate blended and ArcGIS blended DSMs, respectively; (**d**) height profile along the A–B transect shown in (**a**); (**e**) difference between photogrammetric and ArcGIS blended DSM over the red box in (**a**).

4.3.2. Surface Deformation Map

The ascending and descending interferograms in Figure 13 are characterized by a very small spatial baseline (i.e., 27 m and 33 m respectively), and a temporal baseline that is close to the maximum that the study allows (i.e., 66 days, from June to September 2017). These interferograms have also a negligible identified residual atmospheric phase for both acquisitions of the two pairs (details in Table 5).

Surface movement is observed in one major area (highlighted in red in Figure 13). Owing to the high resolution of ST-TSX data, the approximate boundaries of the sliding, transition, and toppling zones of the Boundary Range Slide are recognizable based on their rates of deformation (see the zoom insets in Figure 13). The location of the Boundary Range Slide main scarp, which forms the upper boundary of the sliding zone, is clearly detectable in the interferograms. The interferometric phase further shows large scale movement at the bottom of the landslide area corresponding to the toppling zone as the fastest moving part of the slide identified by previous studies [16]. The transition zone is also detectable in the interferograms, separating the two abovementioned areas.

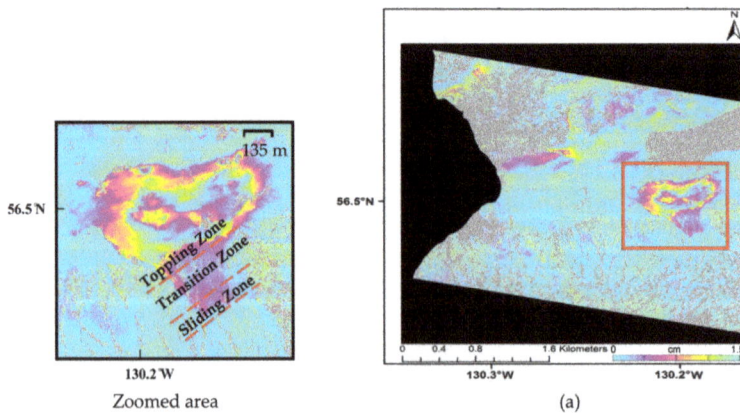

Figure 13. *Cont.*

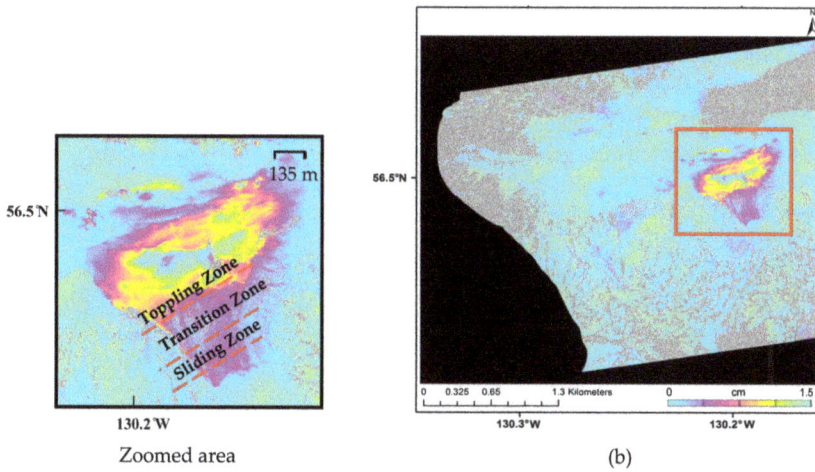

Zoomed area

(b)

Figure 13. Boundary Range Slide: Geocoded TSX. (**a**) Descending orbit interferogram (30 June 2017–4 September 2017); (**b**) ascending orbit interferogram (29 June 2017–3 September 2017). The two areas where deformation is observed are highlighted in red, and the zoom insets and the scales are also presented.

Table 5. Boundary Range Slide presented interferograms.

Section Demo	Pair	Orbit	Incidence (deg)	Polarization	Perp. Baseline (m)	Weather
Surface deformation map	30 June 2017 4 September 2017	Desc.	38	HH	33	Cloudy, 15 °C Mainly Clear, 21 °C
Surface deformation map	29 June 2017 3 September 2017	Asc.	41	HH	27	Cloudy, 17 °C Clear, 18 °C

5. Conclusions

This study presents three enhanced processing methods developed for high resolution InSAR data to improve the estimation of surface deformation. The first method aims to compensate for the InSAR phase induced by static and dynamic atmosphere by fitting a linear model to phase and relative height from a DSM for each pixel of the interferogram. The second method aims to correct the InSAR phase over layover-affected areas based on SAR geometry and high-resolution DSMs. The third method blends low- and high-resolution DSMs by molding the former to match the latter at the boundaries using thin plate spline fitting.

The proposed algorithms demonstrate using TerraSAR-X Staring Spotlight datasets over three sites in BC having geological characteristics that make them representative end-members for this study. Performance comparisons between original and corrected interferograms show that the proposed methods significantly improve the final interferometric deformation maps by removing the non-deformation phase components. High coherence and unprecedented spatial detail of the results indicate excellent suitability of TerraSAR-X Staring Spotlight data for identification and monitoring small-scale landslides in challenging mountainous areas.

The observed movements in line-of-sight direction of the satellite reveal that all the test sites are susceptible to future rockfalls at small and large scales. Therefore, continued assessment of the sites with high resolution InSAR over longer time periods is expected to quantify the likelihood of larger events, such as rock avalanches or landslides, to help avoid future losses.

Author Contributions: F.H. and B.R. conceived of the presented idea. F.H., J.E. and B.R. developed the methods and F.H. performed the computations and analyzed the results. Both F.H. and M.P. contributed to the final version of the manuscript. The project was supervised by B.R.

Remote Sens. **2018**, *10*, 844

Funding: This research was funded through a combination of an NSERC Engage project with BGC, and the NSERC MDA-CSA Industrial Research Chair program in Synthetic Aperture Radar Technologies, Methods, and Applications at Simon Fraser University.

Acknowledgments: We would like to thank German Aerospace Agency (DLR) for acquiring and providing TerraSAR-X images under the proposals LAN3540_rabus and LAN2465_rabus, respectively. We also like to thank BGC Engineering Inc. for providing photogrammetric data and sharing geological background information with us. But also great thank to the two anonymous reviewers for helping with improving the clarity and structure of the manuscript.

Conflicts of Interest: The authors declare no conflicts of interest.

References

1. Jakob, M.; Lambert, S. Climate change effects on landslides along the southwest coast of British Columbia. *Geomorphology* **2009**, *107*, 275–284. [CrossRef]
2. Soldati, M.; Corsini, A.; Pasuto, A. Landslides and climate change in the Italian Dolomites since the late glacial. *Catena* **2004**, *55*, 141–161. [CrossRef]
3. Curlander, J.C.; McDonough, R.N. *Synthetic Aperture Radar*; John Wiley & Sons: New York, NY, USA, 1991; Volume 396.
4. Bamler, R.; Hartl, P. Synthetic aperture radar interferometry. *Inverse Probl.* **1998**, *14*, R1. [CrossRef]
5. Krieger, G.; Moreira, A.; Fiedler, H.; Hajnsek, I.; Werner, M.; Younis, M.; Zink, M. Tandem-x: A satellite formation for high-resolution sar interferometry. *IEEE Trans. Geosci. Remote Sens.* **2007**, *45*, 3317–3341. [CrossRef]
6. Carrara, W.G.; Goodman, R.S.; Majewski, R.M. *Spotlight Synthetic Aperture Radar*; Artech House: Norwood, MA, USA, 2007.
7. Mittermayer, J.; Wollstadt, S.; Prats-Iraola, P.; Scheiber, R. The TerraSAR-X staring spotlight mode concept. *IEEE Trans. Geosci. Remote Sens.* **2014**, *52*, 3695–3706. [CrossRef]
8. Donati, D.; Stead, D.; Brideau, M.; Ghirotti, M. A remote sensing approach for the derivation of numerical modelling input data: Insights from the Hope Slide, Canada. In *ISRM AfriRock—Rock Mechanics for Africa*; Internation Society for Rock Mechanics and and Rocke Engineering: Cape Town, South Africa, 2017.
9. Mathews, W.; McTaggart, K. Hope rockslides, British Columbia, Canada. In *Developments in Geotechnical Engineering*; Elsevier: New York, NY, USA, 1978; Volume 14, pp. 259–275.
10. Von Sacken, R.S. New Data and Re-Evaluation of the 1965 Hope Slide, British Columbia. Ph.D. Thesis, University of British Columbia, Vancouver, BC, Canada, 1991.
11. Brideau, M.-A.; Stead, D.; Kinakin, D.; Fecova, K. Influence of tectonic structures on the Hope Slide, British Columbia, Canada. *Eng. Geol.* **2005**, *80*, 242–259. [CrossRef]
12. Ding, X.-L.; Li, Z.-W.; Zhu, J.-J.; Feng, G.-C.; Long, J.-P. Atmospheric effects on InSAR measurements and their mitigation. *Sensors* **2008**, *8*, 5426–5448. [CrossRef] [PubMed]
13. Bovis, M.J.; Evans, S.G. Rock slope movements along the Mount Currie "fault scarp," southern coast mountains, British Columbia. *Can. J. Earth Sci.* **1995**, *32*, 2015–2020. [CrossRef]
14. BGC Enginnering Inc. *Mount Currie Landslide Risk Assessment*; BGC Enginnering Inc.: Vancouver, BC, Canada, 2018.
15. Clayton, M.; Stead, D.; Kinakin, D. The Mitchell Creek landslide, BC, Canada: Investigation using remote sensing and numerical modeling. In Proceedings of the 47th US Rock Mechanics/Geomechanics Symposium, San Fransisco, CA, USA, 23–26 June 2013.
16. Clayton, M.A. Characterization and analysis of the Mitchell Creek Landslide: A Large-Scale Rock Slope Instability in Northwestern British Columbia. Ph.D. Thesis, Science: Department of Earth Sciences, Cambridge, UK, 2014.
17. Clayton, A.; Stead, D.; Kinakin, D.; Wolter, A. Engineering geomorphological interpretation of the Mitchell Creek landslide, British Columbia, Canada. *Landslides* **2017**, *14*, 1655–1675. [CrossRef]
18. Werner, C.; Wegmüller, U.; Strozzi, T.; Wiesmann, A. Gamma SAR and interferometric processing software. In Proceedings of the ERS-Envisat Symposium, Gothenburg, Sweden, 16–20 October 2000; p. 1620.
19. Rabus, B.T.; Ghuman, P.S. A simple robust two-scale phase component inversion scheme for persistent scatterer interferometry (dual-scale psi). *Can. J. Remote Sens.* **2009**, *35*, 399–410. [CrossRef]

20. Bekaert, D.; Hooper, A.; Wright, T. A spatially variable power law tropospheric correction technique for insar data. *J. Geophys. Res. Solid Earth* **2015**, *120*, 1345–1356. [CrossRef]

21. Bekaert, D.; Walters, R.; Wright, T.; Hooper, A.; Parker, D. Statistical comparison of InSAR tropospheric correction techniques. *Remote Sens. Environ.* **2015**, *170*, 40–47. [CrossRef]

22. Kwok, R.; Curlander, J.C.; Pang, S.S. Rectification of terrain induced distortions in radar Imagery. *Photogramm. Eng. Remote Sens* **1987**, *53*, 507–513.

23. Jakowatz, C.V.; Thompson, P. A new look at spotlight mode synthetic aperture radar as tomography: Imaging 3-d targets. *IEEE Trans. Image Process.* **1995**, *4*, 699–703. [CrossRef] [PubMed]

24. Munson, D.C.; O'Brien, J.D.; Jenkins, W.K. A tomographic formulation of spotlight-mode synthetic aperture radar. *Proc. IEEE* **1983**, *71*, 917–925. [CrossRef]

25. Ausherman, D.A.; Kozma, A.; Walker, J.L.; Jones, H.M.; Poggio, E.C. Developments in radar imaging. *IEEE Trans. Aerosp. Electron. Syst.* **1984**, *4*, 363–400. [CrossRef]

26. Mersereau, R.M.; Oppenheim, A.V. Digital reconstruction of multidimensional signals from their projections. *Proc. IEEE* **1974**, *62*, 1319–1338. [CrossRef]

27. Walker, J.L. Range-doppler imaging of rotating objects. *IEEE Trans. Aerosp. Electron. Syst.* **1980**, *AES-16*, 23–52. [CrossRef]

28. Bamler, R. Principles of synthetic aperture radar. *Surv. Geophys.* **2000**, *21*, 147–157. [CrossRef]

29. Rabus, B.; Eppler, J.; Sharma, J.; Busler, J. Tunnel monitoring with an advanced insar technique. In Proceedings of the Radar Sensor Technology XVI, International Society for Optics and Photonics, Baltimore, MD, USA, 23–27 April 2012; p. 83611F.

30. Eppler, J.; Rabus, B. Monitoring urban infrastructure with an adaptive multilooking insar technique. In Proceedings of the Fringe 2011, Frascati, Italy, 19–23 September 2012; p. 68.

31. Fattahi, H.; Amelung, F. DEM error correction in InSAR time series. *IEEE Trans. Geosci. Remote Sens.* **2013**, *51*, 4249–4259. [CrossRef]

32. Petrasova, A.; Mitasova, H.; Petras, V.; Jeziorska, J. Fusion of high-resolution DEMs for water flow modeling. *Open Geospat. Data Softw. Stand.* **2017**, *2*, 6. [CrossRef]

33. Ormsby, T. *Getting to Know Arcgis Desktop: Basics of Arcview, Arceditor, and Arcinfo*; ESRI, Inc.: Hong Kong, China, 2004.

34. Viola, P.; Wells, W.M., III. Alignment by maximization of mutual information. *Int. J. Comput. Vis.* **1997**, *24*, 137–154. [CrossRef]

35. Chen, H.-M.; Arora, M.K.; Varshney, P.K. Mutual information-based image registration for remote sensing data. *Int. J. Remote Sens.* **2003**, *24*, 3701–3706. [CrossRef]

36. Lowe, D.G. Distinctive image features from scale-invariant keypoints. *Int. J. Comput. Vis.* **2004**, *60*, 91–110. [CrossRef]

37. Wang, K.; Zhang, T. Gauss-Newton method for DEM co-registration. In Proceedings of theInternational Conference on Intelligent Earth Observing and Applications 2015, Guilin, China, 23–24 October 2015; International Society for Optics and Photonics: Bellingham, WA, USA, 2015; p. 98080M.

remote sensing

MDPI

Article

A New InSAR Phase Demodulation Technique Developed for a Typical Example of a Complex, Multi-Lobed Landslide Displacement Field, Fels Glacier Slide, Alaska

Bernhard Rabus * and Manuele Pichierri

School of Engineering Science, Simon Fraser University, Burnaby, BC V5A 1S6, Canada; mpichier@sfu.ca
* Correspondence: btrabus@sfu.ca; Tel.: +1-778-782-4846

Received: 9 May 2018; Accepted: 20 June 2018; Published: 22 June 2018

Abstract: Landslides can have complex, spatially strongly inhomogeneous surface displacement fields with discontinuities from multiple active lobes that are deforming while failing on nested slip surfaces at different depths. For synthetic aperture radar interferometry (InSAR), particularly at lower resolutions, these characteristics can cause significant aliasing of the wrapped phase. In combination with steep terrain and seasonal snow cover, causing layover and temporal decorrelation, respectively, traditional phase unwrapping can become unfeasible, even after topographic phase contributions have been removed with an external high-resolution digital surface model (DSM). We present a novel method: warp demodulation that reduces the complexity of the phase unwrapping problem for noisy and/or aliased, low-resolution interferograms of discontinuous landslide displacement. The key input to our warp demodulation method is a single (or several) reference interferogram(s) from a high-resolution sensor mode such as TerraSAR-X Staring Spotlight with short temporal baseline and good coherence to allow localization of phase discontinuities and accurate unwrapping. The task of constructing suitable phase surfaces to approximate individual to-be-demodulated interferograms from the reference interferogram is made difficult by strong and spatially inhomogeneous temporal, seasonal, and interannual variations of the landslide with individual lobes accelerating or decelerating at different rates. This prevents using simple global scaling of the reference. Instead, our method uses an irregular grid of small patches straddling strong spatial gradients and phase discontinuities in the reference to find optimum local scaling factors that minimize the residual phase gradients across the discontinuities after demodulation. Next, for each to-be-demodulated interferogram, from these measurements we interpolate a spatially smooth global scaling function, which is then used to scale the (discontinuous) reference. Demodulation with the scaled reference leads to a residual phase that is also spatially smooth, allowing it to be unwrapped robustly after low-pass filtering. A key assumption of warp demodulation is that the locations of the phase discontinuities can be mapped in the reference and that they are stationary in time at the scale of the image resolution. We carry out extensive tests with simulated data to establish the accuracy, robustness, and limitations of the new method with respect to relevant parameters, such as decorrelation noise and aliasing along phase discontinuities. A realistic parameterization of the method is demonstrated for the example of the Fels Glacier Slide in Alaska using a recent late-summer high-resolution staring spotlight interferometric image pair from TerraSAR-X to demodulate. We show warp demodulation results for also recent but early-summer, partially incoherent interferograms of the same sensor, as well as for older and coarser aliased interferograms from RADARSAT-2, ALOS-1, and ERS.

Keywords: InSAR; landslide; phase unwrapping; phase demodulation; TerraSAR-X; RADARSAT-2; ALOS-1; ERS

1. Introduction

Interferometric synthetic aperture radar [1,2] (InSAR) processing techniques for measuring the Earth's surface deformation have evolved dramatically over the last decades. First generation techniques to extract the deformation phase, based on the simple removal of the topographic phase components in pairwise interferograms [3–5] have been complemented by second generation time series analysis techniques, which exploit temporal networks of interferograms for the simultaneous statistical modeling of the topographic, atmospheric, and deformation phase components, with [6–9] or without prior phase unwrapping [10–12]. Lastly, corresponding third generation techniques have been developed for the wrapped [13] and unwrapped case [14,15], respectively, which use additionally spatially adaptive filtering to improve coherence, while preserving spatial resolution. Also, for the phase unwrapping of interferometric networks, more powerful spatio-temporal analogues [16] have been developed beyond the more basic pairwise algorithms [17]. The problem of unwrapping the InSAR phase has also been simplified already via partial demodulation of the wrapped phase, both through the availability of higher resolution global digital elevation models (DEMs) (e.g., [18,19]) and the successful introduction of deformation models for specific types of phenomena causing surface displacement. Examples where additional deformation modeling and demodulation can lead to small phase residuals that ease phase unwrapping considerably include volcanoes, earthquakes, and glaciers [20,21].

In contrast, despite the technical sophistication in processing techniques and the availability of high-resolution DEMs, recovering from InSAR data the highly temporally variable and spatially discontinuous deformation fields of large complex landslides have remained challenging to both time series methods and phase unwrapping algorithms [22]. The inability to achieve a small phase residual via modeling of the landslide deformation is mainly due to the lack of accurate geophysical models in combination with other compromising factors, such as poor or patchwise-variable phase quality from decorrelation (e.g., due to precipitation, seasonal snow melt, and vegetation) or from surface disintegration caused by the deformation itself, as well as a multitude of small layover and shadow areas caused by rugged, steep terrain.

We have developed a simple and robust phase demodulation method—*warp demodulation*—that uses high-resolution interferometric data to unwrap the phase of interferograms that are partially incoherent and/or aliased due to coarser spatial resolution. The method relies crucially on the availability of at least one high-resolution coherent template where active faults can be delineated. The parameterization of warp demodulation has been specifically tailored to the example of the Fels Glacier Slide, Alaska, and consequently, the applicability of the method is restricted currently to the diversity provided by this example. The deformation field of the Fels Glacier Slide has pronounced discontinuities caused by multiple active lobes failing on nested slip surfaces at various depths. There also is strong and spatially inhomogeneous temporal variation of the deformation field, as well as significant spatio-temporal coherence variations caused by precipitation and melting of seasonal snow cover. However, the vegetation cover of the Fels Glacier Slide is low (alpine tundra), and the terrain ruggedness is only moderate. To provide a wider context for the current version of the warp demodulation method, we summarize in Table 1 the general complicating factors to the successful phase unwrapping of interferograms of landslide motion and how far these are realized by our case study on the Fels Glacier Slide. We believe our method to be transferable with some adaptation to other landslide scenarios.

The scope of this paper is also limited to only the spatial demodulation of the discontinuous landslide displacement field of the Fels Glacier Slide. Extending our warp demodulation method to include a direct coupling with the strong temporal non-linearity originating from the pronounced temporal variations in the seasonal and inter-annual landslide drivers observed for our example and most other landslides is the subject of follow-up work. The presented demodulation solution can be regarded as a first step towards a robust and automated InSAR time series analysis solution optimized for complex landslide deformation.

The paper is organized as follows. Section 2 outlines the novel phase demodulation method. In Section 3, we first present results from a sensitivity analysis of the new method using simulated data. We then demonstrate the method for the Fels Glacier Slide using recent high-resolution Staring Spotlight [23,24] TerraSAR-X (TSX-ST) data to demodulate also recent but seasonally less coherent data from TerraSAR-X, as well as earlier, coarser resolution aliased data from RADARSAT-1, ALOS-1, and ERS. The results are discussed in terms of capabilities and limitations, as well as in the context of planned extensions of the new method in Section 4.

Table 1. Complicating factors to the successful phase unwrapping/demodulation of landslide interferograms and the relevance of the Fels Glacier Slide example.

Factor	Fels Glacier Slide
Strong spatial deformation gradients	YES
Presence of deformation discontinuities (active faults) producing phase aliasing	YES
Temporal variability of deformation (spatially inhomogeneous—multiple nested active lobes with contrasting deformation driver sensitivity)	YES
Decorrelation due to heavy vegetation cover	NO
Extremely rugged topography with extensive layover/shadow-affected areas	NO
Decorrelation due to rain and seasonal snow cover	YES

2. Method

Our initial idea to construct appropriate demodulation surfaces that reduce the complexity of the phase unwrapping problems for single interferograms of spatially discontinuous landslide motion was to use a wireframe model similar to [25]. The wireframe model consists of an adjustable parametric surface constructed over a triangular mesh. The mesh boundary includes cuts at the active fault line locations of the slide, which enable the model to fit the displacement field first-order differentiable except at its discontinuities. In this study, we use a related but much simpler *warp demodulation* model that directly matches an unwrapped reference interferogram to the to-be-demodulated interferograms at discrete locations along the active faults. The locally optimized match factors are then interpolated to fit the reference to the wrapped interferograms with a global multiplicative warp function. Compared to the full wireframe model, our warp demodulation model has the key advantage of being straightforward to parametrize robustly. The disadvantage of the warp demodulation model of leaving a larger demodulation residual that can still be wrapped is largely neutralized by this residual being continuous and slowly varying spatially. As a result, the residual—in contrast to the original discontinuous interferogram—can be unwrapped easily and accurately after applying a simple boxcar low pass (multi-looking). The steps of our method are schematically depicted at the top of Figure 1 and are explained in more detail as follows:

Step 1 Detect Active Fault Locations. Active fault locations can be mapped robustly and accurately by using a curvature criterion (second-order derivative; minimum threshold on the largest eigenvalue and contrast of eigenvalues of the two-dimensional (2D) Hessian matrix of factor >10) applied to a smoothed version of the reference interferogram. This, or a similar discontinuity detection approach, must be used by the wireframe model, which requires continuous parameterization of each active fault line in its entirety. For the warp demodulation model, it is sufficient for each fault line to be represented by a (generally coarser and irregularly spaced) representative set of points. Such a set can be found by subsampling the output from the described curvature criterion discontinuity detector. This approach was used for the Fels Glacier Slide example (see Section 3.2). Alternatively, it is possible to digitize the point set representing the faults manually on the reference interferogram (as was done for the sensitivity analysis in Section 3.1); this approach may be generally more robust, particularly if a less coherent and high-resolution reference is used.

Step 2 Find Local Match Factors. The reference is matched to the to-be-demodulated interferograms within circular patches straddling the active faults at the pointset locations found in the first step. The radius of the patches is an important parameter of the model whose influence on the accuracy and robustness of the demodulation is discussed later. Optimum multiplicative factors to scale the reference before demodulation are obtained now for each patch by maximizing the (signal-biased) coherence estimate of the residual inside the patch. While this type of match (coherence optimization) essentially minimizes the dynamic range of the coherent phase residual of the patch (incoherent phase noise inside the patch does not change during the optimization and thus does not affect it) it is insensitive to the phase offset between the interferogram and the scaled reference. This is a key feature of our warp demodulation method, which favors spatial smoothness over smallness of the demodulation residual.

Step 3 Thin Plate Spline Fit. The vector of optimum match factors at the point set locations is now interpolated to form a continuous function of the interferogram coordinates using a thin plate spline fit. This function is then multiplied with the reference to warp the latter into an optimum demodulation phase surface for the interferogram. This demodulation phase is then wrapped, and its complex conjugate is multiplied with the to-be-demodulated interferogram to form the residual phase.

Step 4 Unwrap Residual. The residual phase, optimally smooth and slowly varying spatially, is multi-looked and unwrapped using standard Minimum Cost Flow (MCF) techniques [17]. The unwrapped residual is added to the constructed (unwrapped) demodulation phase of step 3 to form the final unwrapped interferogram.

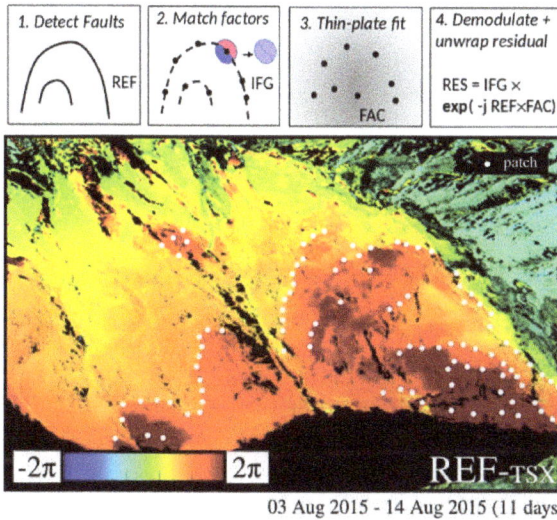

Figure 1. Top: the four steps of the warp demodulation method; IFG: to-be-demodulated wrapped interferogram; REF: high-resolution unwrapped reference interferogram; FAC: interpolated factor map; and RES: wrapped residual. Bottom: real data example of reference interferogram (TerraSAR-X Staring Spotlight, 0.6 m ground resolution), Fels Glacier Slide, Alaska.

The warp demodulation method has several unique characteristics: (i) as already mentioned, the method is designed to produce a smooth demodulation residual; the mean phase offset for each patch between the reference and the to-be-demodulated interferogram deliberately is not considered in finding the optimum local matching factors; (ii) phase structures inside the patch instead are used for the optimization; primarily, the phase discontinuity along the fault line dissecting the patch is used, but secondarily, the gradients and curvature features of the displacement phase of the landslide lobes are also used. The ability to use this secondary information allows the method to succeed in finding the correct demodulation surface, even if phase aliasing occurs across the fault due to the acceleration of a lobe along its downslope axis. The chosen patch radius plays an important role in how robustly the method delivers on this ability, and this is discussed in Section 3.1, where we present a full parametric sensitivity analysis of the method; (iii) our method relies on approximate stationarity of the phase discontinuities between the reference and the interferogram: significant shifting of active faults leads to artifacts also discussed in Section 3.1. The method biases the local scale factor to zero for a "flat" coherence optimum, which handles the case correctly that when an active fault in the reference is inactive in the interferogram, the resulting residual is smooth. However, for the opposite case where a fault in the interferogram is not present in the reference, the coherence optimization will be insensitive and cannot diminish the signal (residual will remain discontinuous). The wireframe model has the same problem; if new active faults are generated, then the reference needs to be updated to include the new faults as cuts. As there, (iv) the warp demodulation method can be generalized straightforwardly by including several references appropriately spaced temporally and selecting the one close enough in time for each to-be-demodulated interferogram to capture adequately the formation (and cessation) of active faults.

3. Results

In this section, we present results from a sensitivity analysis of the new method, as well as demonstration results using actual data. Table 2 contains a list of symbols and acronyms used in Section 3. We want to point out explicitly that our warp demodulation method assumes the existence of a reference interferogram of "good quality", quantified as high coherence and unobscured by layover with an MCF unwrapping result that is correct except for perhaps localized small errors (restricted to the vicinity of a few pixels around minor incoherent areas). Therefore, within the scope of the present study, the effects arising from shortcomings of the reference have not been included in the sensitivity analysis.

Table 2. List of parameters used in Section 3.

Symbol	Description
IFG	To-be-demodulated wrapped interferogram
REF	High-resolution unwrapped reference interferogram
FAC	Thin-plate interpolated warp demodulation factor
RES	Wrapped warp demodulation residual
RES-UNW	Unwrapped warp demodulation residual
UNW	IFG unwrapped through warp demodulation
Δh	Downward shift of the IFG active lobe (with respect to REF); units: pixels
f_{REF}	Along-lobe deformation rate in the REF interferogram; units: fringes/lobe
f_{INF}	Along-lobe deformation rate in the INF interferogram. units: fringes/lobe
Δ_f	Difference (between REF and IFG) in the along-lobe deformation rate; units: fringes/lobe
r	Demodulation patch radius; units: fraction of the main lobe's diameter
κ	Steepness of logistic curve
GAU	Atmospheric phase error surface
σ	Standard deviation of GAU

3.1. Sensitivity Analysis with Simulated Data

The warp demodulation method is applied to a dataset of simulated InSAR observations for two different landslide scenarios:

- Single-lobe: the deforming active lobe is modeled by a semi-ellipse (vertical semi-axis of 80 pixels, horizontal axis of 100 pixels, Figure 2a).
- Multi-lobe: three nested semi-elliptical lobes are simulated (see Figure 2b for dimensions).

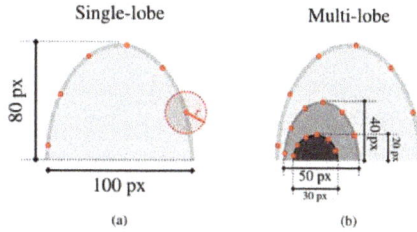

Figure 2. Sensitivity analysis with simulated data. Schematic depiction of the two simulated landslide scenarios: (**a**) single-lobe (semi-ellipse) and (**b**) multi-lobe (three nested semi-ellipses). The red dots identify the location of the demodulation patches (of radius *r*).

The goal of this analysis is to assess the robustness and limitations of warp demodulation with respect to the following landslide deformation and/or demodulation parameters (see also Figure 3):

- Downward shift of the to-be-demodulated wrapped interferogram (IFG) active lobe (with respect to high-resolution unwrapped reference interferogram (REF)).
- Along-lobe deformation gradient types (linear or logistic ramp) and deformation rates.
- IFG "disturbances" (i.e., atmospheric and random phase noise (related to speckle)).
- Radius of the demodulation patches *r* (expressed in a fraction of the main lobe's diameter).

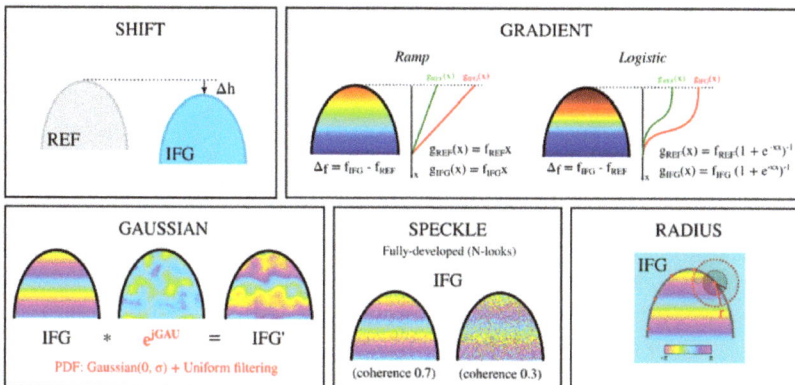

Figure 3. Sensitivity analysis with simulated data. Schematic depiction of the simulation experiments. SHIFT: robustness of the warp demodulation method against downward shifts Δh of the IFG active lobe; GRADIENT: algorithm sensitivity to along-lobe deformation differences Δ_f between REF and IFG for ramp/logistic deformation gradients; GAUSSIAN: IFG affected by Gaussian noise; SPECKLE: IFG affected by fully developed speckle; RADIUS: Impact of patch radius *r* on demodulation performance.

Table 3 provides an overview of the experiments and a summary of the simulation results. The most relevant findings of the simulation study are detailed in Sections 3.1.1–3.1.3

Table 3. Sensitivity analysis with simulated data. Overview of the simulation experiments and results.

Experiment	Description		Model Parameters	Observations
SHIFT	Robustness against downward shift Δh of the IFG active lobe (with respect to REF) at different REF along-lobe deformation rates f_{REF}	VAR	Δh = [0, 30] px f_{REF} = 1, 1.5, 2, 4 fringes/lobe	• RMSE monotonically increasing with Δh • RMSE positively associated with f_{REF}
		FIXED	Δ_f = 0 fringes/lobe r = 0.25 6 patches Gradient: ramp	
GRADIENT	Robustness against along-lobe deformation differences Δ_f between REF and IFG for ramp/logistic along-lobe deformation gradients	VAR	Gradient: ramp/logistic Δ_f = [0, 2] fringes/lobe f_{REF} = 1, 1.5, 2, 4 fringes/lobe κ = 2, 4 (logistic)	• RMSE constant until Δ_f = 1 fringe/lobe, then abruptly increasing • Ramp (Δ_f > 1 fringe/lobe): RMSE negatively associated with f_{REF}
		FIXED	Δh = 0 px r = 0.25 6 patches	
SPECKLE	IFG affected by fully developed speckle (distribution as in [26])	VAR	Gradient: ramp/logistic Δ_f = [0, 2] fringes/lobe f_{REF} = 1, 1.5, 2, 4 fringes/lobe Coherence = 0.3, 0.5, 0.7	• RMSE trend in agreement with GRADIENT • RMSE negatively associated with coherence
		FIXED	Δh = 0 px r = 0.25 6 patches Number of looks = 5 κ = 4 (logistic)	
GAUSSIAN	IFG affected by Gaussian noise (atmosphere-like disturbance)	VAR	Gradient: ramp/logistic Δ_f = [0, 2] fringes/lobe f_{REF} = 1, 1.5, 2, 4 fringes/lobe σ = 3, 7, 11	• RMSE trend in agreement with GRADIENT • RMSE and σ are positively associated
		FIXED	Δh = 0 px r = 0.25 6 patches κ = 4 (logistic)	
RADIUS	Impact of patch radius r on demodulation performance	VAR	Gradient: ramp/logistic f_{REF} = 1, 1.5, 2, 4 fringes/lobe r = [0.1, 0.7]	*Single-lobe:* • RMSE positively associated with r when r > 0.35 • Minimum RMSE when 0.3 < r < 0.5 *Multi-lobe:* Phase artifacts within the secondary lobes as r increases
		FIXED	Δh = 0 px 6 patches Δ_f = 2 fringes/lobe (ramp, single-lobe) Δ_f = 0.75 fringes/lobe (ramp, multi-lobe) Δ_f = 0.75 fringes/lobe (logistic) κ = 4 (logistic)	

3.1.1. Along-Lobe Deformation Gradients

In Figure 4, the Root-Mean-Square Error (RMSE) of the demodulated unwrapped phase is displayed as a function of the difference (i.e., between REF and IFG) in the along-lobe deformation rate $\Delta_f = f_{IFG} - f_{REF}$ for four different REF along-lobe deformation rates f_{REF}. The warp demodulation method is run using six equally spaced demodulation patches of $r = 0.25$ located on the main lobe discontinuity. The statistics are obtained by assuming either a ramp or a logistic along-lobe deformation gradient. In the latter, a curve steepness $\kappa = 4$ is imposed.

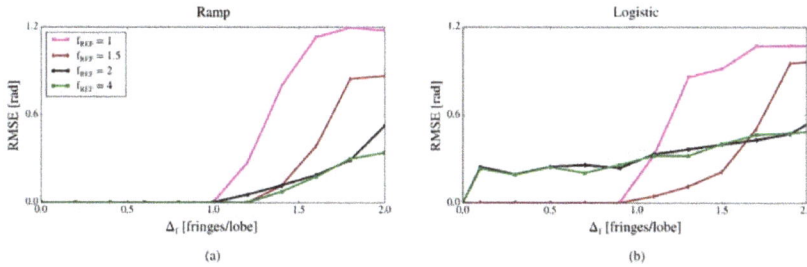

Figure 4. Sensitivity analysis with simulated data, algorithm robustness against along-lobe deformation differences between REF and IFG. RMSE of the demodulated unwrapped phase as a function of the difference in the along-lobe deformation rate $\Delta_f = f_{IFG} - f_{REF}$ for four different REFs along-lobe deformation rates f_{REF}. The statistics are obtained by assuming (**a**) a ramp along-lobe deformation gradient or (**b**) a logistic gradient (curve steepness $\kappa = 4$).

Ramp gradient: The RMSE remains substantially insensitive to f_{REF} and is approximately equal to zero if $\Delta_f < 1$ fringe/lobe. Above this threshold, the RMSE tends to increase with Δ_f until the latter reaches approximately 1.6–1.8 fringes/lobe, then the two quantities become again uncorrelated. Such an increase is markedly abrupt for $f_{REF} = 1$ fringe/lobe (from 0 to 0.96 rad as Δ_f increases from 1 to 1.5 fringes/lobe).

For Δ_f values bigger than 1 fringe/lobe, the RMSE is also negatively associated with f_{REF}. For instance, RMSE at $\Delta_f = 1.5$ fringes/lobe is seen to increase by 0.85 rad in response to a f_{REF} reduction by 50%.

Logistic gradient: If $f_{REF} = 1$ fringes/lobe, the RMSE remains approximately equal to zero until approximately $\Delta_f = 1$ fringe/lobe, when it abruptly increases, reaching a maximum of 1.05 rad at $\Delta_f = 1.7$ fringes/lobe. The same behavior with Δ_f, although with a less abrupt increase, is also found for $f_{REF} = 1.5$ fringes/lobe. If $2 < f_{REF} < 4$ fringes/lobe, the RMSE shows a positive association with Δ_f (Pearson's correlation coefficient $R = 0.93$); in this case, the statistics appear substantially insensitive to f_{REF} and range from a minimum of 0.2 rad (at $\Delta_f = 0.3$ fringes/lobe) to a maximum of 0.5 rad (at $\Delta_f = 1.9$ fringes/lobe).

3.1.2. Effect of Atmosphere-Like Disturbances

We simulated a two-dimensional atmospheric phase error surface (GAU) that shares the spatial characteristics of the dynamic atmosphere. GAU is assumed to be zero-mean Gaussian-distributed (with standard deviation σ) and is smoothed via an 11-pixel uniform filter. The to-be-demodulated interferogram IFG′ now has the form:

$$IFG' = IFG \cdot e^{j \, GAU} \tag{1}$$

In Tables 4 and 5, we reported the mean RMSE of the demodulated unwrapped phase computed for a population of 50 randomly generated IFG′ observations at three different σ values. The statistics are obtained by assuming either a ramp or a logistic ($\kappa = 4$) along-lobe deformation gradient.

Table 4. Sensitivity analysis with simulated data, effect of atmosphere-like disturbances. RMSE (in radians) of the demodulated unwrapped phase at three different σ values (ramp gradient).

		$\sigma = 3$	$\sigma = 7$	$\sigma = 11$
$\Delta_f = 0.5$	$f_{REF} = 1$	0.181	0.417	0.670
	$f_{REF} = 4$	0.180	0.433	0.675
$\Delta_f = 1.0$	$f_{REF} = 1$	0.179	0.423	0.674
	$f_{REF} = 4$	0.178	0.432	0.683
$\Delta_f = 1.5$	$f_{REF} = 1$	1.087	1.185	1.345
	$f_{REF} = 4$	0.260	0.467	0.719

Table 5. Sensitivity analysis with simulated data, effect of atmosphere-like disturbances. RMSE (in radians) of the demodulated unwrapped phase at three different σ values (logistic gradient).

		$\sigma = 3$	$\sigma = 7$	$\sigma = 11$
$\Delta_f = 0.5$	$f_{REF} = 1$	0.180	0.431	0.667
	$f_{REF} = 4$	0.308	0.624	0.832
$\Delta_f = 1.0$	$f_{REF} = 1$	0.181	0.451	0.705
	$f_{REF} = 4$	0.477	0.651	0.866
$\Delta_f = 1.5$	$f_{REF} = 1$	1.161	1.257	1.414
	$f_{REF} = 4$	0.528	0.692	0.905

The mean RMSE shows a trend with Δ_f that is in agreement with the results of Section 3.1.1 (constant until $\Delta_f = 1$ fringe/lobe, then abruptly increasing). Compared to Section 3.1.1, all RMSE measurements of Tables 4 and 5 are now affected by a positive offset that is positively associated with σ (e.g., compared to the case where $\sigma = 3$, the RMSE at $\sigma = 11$ is on average 0.44 rad greater).

3.1.3. Impact of the Demodulation Patch Radius

To assess the performance of the warp demodulation approach with respect to the radius of the demodulation patches r, the RMSE of the demodulated unwrapped phase is plotted against r when either a ramp or a logistic deformation gradient is imposed (Figure 5a,b, respectively). All simulations are performed using six patches located on the main lobe discontinuity.

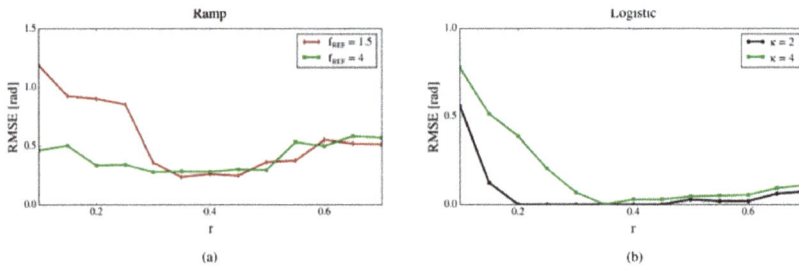

(a)   (b)

Figure 5. Sensitivity analysis with simulated data, impact of the demodulation patch radius r on the algorithm performance. RMSE of the demodulated unwrapped phase as a function of r. (**a**) Ramp gradient, $\Delta_f = 2$ fringes/lobe (in red: $f_{REF} = 1.5$ fringes/lobe; in green: $f_{REF} = 4$ fringes/lobe); (**b**) Logistic gradient, $\Delta_f = 0.75$ fringes/lobe and $f_{REF} = 4$ fringes/lobe (in black: $\kappa = 2$; in green: $\kappa = 4$).

Ramp gradient ($\Delta_f = 2$): When $r < 0.35$, the RMSE is consistently larger at $f_{REF} = 1.5$ fringes/lobe than at $f_{REF} = 4$ fringes/lobe (maximum difference of 0.72 rad at $r = 0.1$). Above this threshold, the RMSE becomes insensitive to f_{REF} and positively associated with r ($R = 0.93$ at $f_{REF} = 1.5$ fringes/lobe; $R = 0.9$ at $f_{REF} = 4$ fringes/lobe). The smallest RMSE (0.25 rad) is observed when $0.3 < r < 0.5$.

Logistic gradient ($\Delta_f = 0.75$, $f_{REF} = 4$): The RMSE is consistently smaller at $\kappa = 2$ than $\kappa = 4$ and is seen to decrease with increasing r when the latter is below 0.2 ($\kappa = 2$) or 0.35 ($\kappa = 4$). Above these thresholds, the RMSE remains consistently below 0.1 rad and positively associated with r ($R = 0.84$ at $\kappa = 2$; $R = 0.96$ at $\kappa = 4$).

The warp demodulation is also applied to a multi-lobe scenario with $\Delta_f = 0.75$ fringes/lobe and $f_{REF} = 1.5$ fringes/lobe (ramp gradient). As shown in Figure 6, a small radius $r = 0.1$ results in localized phase errors within the main lobe (RMSE = 0.206 rad). When $0.3 \leq r \leq 0.5$, the phase gradient along the main lobe is accurately reconstructed. However, a number of phase artifacts within the secondary lobes can now be observed. The number of artifacts, and hence the RMSE, appear to increase with increasing r ($r = 0.3$: RMSE = 0.181 rad, $r = 0.5$: RMSE = 1.318 rad).

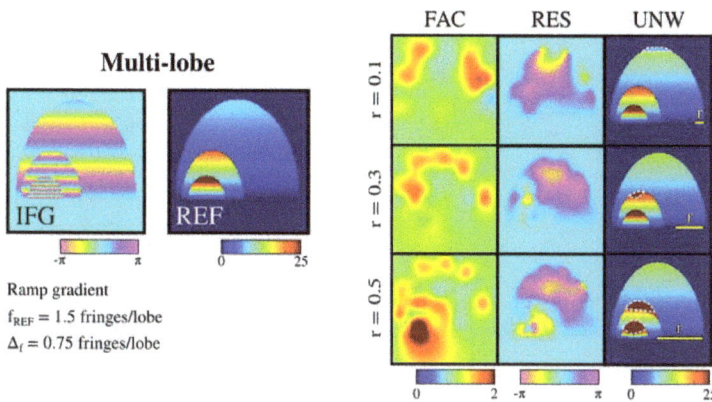

Figure 6. Sensitivity analysis with simulated data, impact of the demodulation patch radius r on the algorithm performance (multi-lobe scenario). IFG: To-be-demodulated interferogram (wrapped); REF: reference interferogram (unwrapped); FAC: demodulation factor; RES: demodulation residual (unwrapped); UNW: demodulated interferogram. FAC, RES, and UNW are obtained for three different r values. The phase errors in UNW are highlighted with white dashed lines.

3.2. Real Data Example: Fels Glacier Slide

In the following, we give a demonstration of the warp demodulation method for a representative large, complex landslide test site, the Fels Glacier Slide in Alaska. The Fels Glacier Slide [27] occupies an instable, south-facing slope above and immediately adjacent to the lower Fels Glacier. The Fels Slide is about 3 km from the Alaska Pipeline corridor to the west and about 2 km from the Denali Fault to the south; earthquakes generated by the Denali Fault, such as the M7.9 one in November 2002, significantly affect the slide dynamics. Other obvious drivers of the dynamics of the slide are meteorology (mostly snow melt and rain) and the ongoing recession of the Fels Glacier, which additionally destabilizes the toe of the slide. On the surface, the deforming area of the slide is about 3 km wide and has approximate fall line and elevation extents of 2 km and 600 m, respectively. The mechanism of the deformation is a deep-seated (depth and shape of the failure plane unknown) gravitational mass deformation forming two large lobes separated east–west by a central dissecting gully. Overlain onto these two deep-seated main lobes are several smaller more surficial deforming lobes that cause strong discontinuities in the deformation field where the surface ruptures along extensive downward curving arcuate fissures. While the deformation is classified as creep-type, movement amounts can easily exceed 30 cm per year for the fastest areas of the slide.

As inputs to our demodulation method, to construct the high-resolution reference "model" we use a TSX-ST image pair 3–14 August 2015 (0.6 m spatial resolution; shown in Figure 1), and for the to-be-demodulated lower resolution interferograms, we have selected three representative examples, one each from ERS, ALOS-1, and RADARSAT-2 capturing different resolution (30 m, 15 m, and 3 m, respectively), seasonal and annual separation, coherence conditions, and interferogram periods with respect to the reference (Table 6). The TSX-ST reference is unwrapped using MCF and re-projected to the line-of-sights of the respective other sensors. The re-projection assumes an approximate motion direction following the average fall line as derived from a spatially smoothed version (500 m scale) of the external DSM used for the topographic corrections and is generally robust due to the similarity of acquisition geometries of the different sensors. Track angles in Table 6 are very similar, as expected for space-borne data of the same orbit direction; look angles are more different but still only vary by less than 20 degrees, which for only moderately rugged topography (as is the case for the Fels Glacier Slide) does not lead to stark layover/shadow contrasts for the different incidence angles, guaranteeing a smooth re-projection map. After re-projection of the template, the warp demodulation method is then applied as introduced in Section 2 and analyzed in Section 3.1.

Table 6. Synthetic aperture radar (SAR) datasets used to test the warp demodulation method.

Sensor	Resolution (m)	Temp. Baseline	Incidence Angle (deg)	Track Angle (deg)
TSX-ST	0.6	11 days	41.3	−169.0
RADARSAT-2	3	48 days	44.0	−171.0
ALOS-1	15	368 days	38.8	−168.7
ERS	30	35 days	23.3	−164.3

Standard MCF unwrapping (using coherence costs and masking areas with coherence below 0.3) results of four selected test interferograms over the Fels Glacier Slide are shown in Figure 7, subfigures (a)–(d). For easy comparison, subfigures (e)–(h) of Figure 7 show the final corresponding warp demodulation results presented in detail in Figures 8–11 (lower right panels). Problems from erroneous branch cuts made by the MCF algorithm are immediately apparent for both the ERS case (within the main lobe, Figure 7a) and most strikingly for the TSX case (Figure 7d). For the ALOS-1 and RSAT-2 cases (Figure 7b,c, respectively), MCF performs better; however, more localized branchcut problems are still easily identifiable.

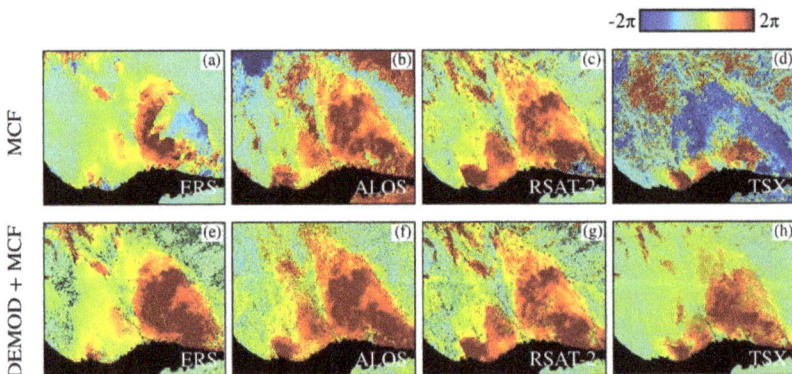

Figure 7. (**a–d**) Standard MCF unwrapping results of four test interferograms ((**a**) ERS, (**b**) ALOS-1, (**c**) RADARSAT-2, and (**d**) TerraSAR-X Staring Spotlight) over the Fels Glacier Slide. (**e–h**) Warp demodulation results for the same test sites for direct comparison (same as subfigures (**f**) in Figures 8–11). The unwrapped phase is clipped to ±2π.

15 May 2016–26 May 2016

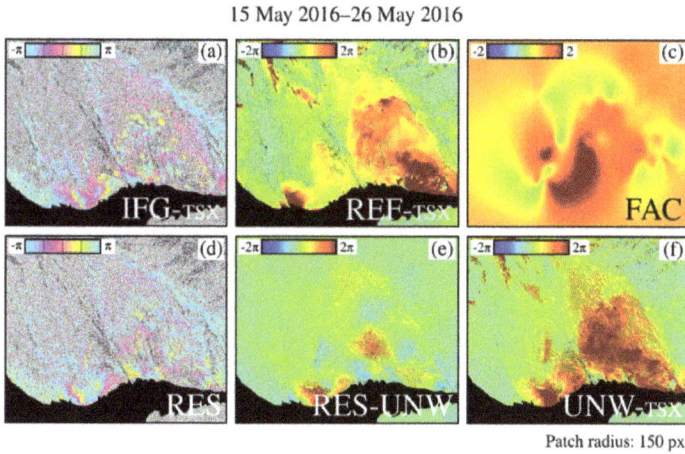

Patch radius: 150 px

Figure 8. Warp demodulation result for a representative TerraSAR-X Staring Spotlight interferogram (X-Band, 0.6 m ground range resolution, 11-day repeat cycle) covering 1 cycle. (**a**) IFG-TSX, to-be-demodulated interferogram (wrapped); (**b**) REF-TSX, reference interferogram (unwrapped); (**c**) FAC, demodulation factor; (**d**) RES: demodulation residual (wrapped); (**e**) RES-UNW: demodulation residual (unwrapped); and (**f**) UNW-TSX, demodulated interferogram. The unwrapped phase in (**b,e,f**) is clipped to $\pm 2\pi$. The demodulation patch radius is 150 pixels.

20 Jun 2010–7 Aug 2010

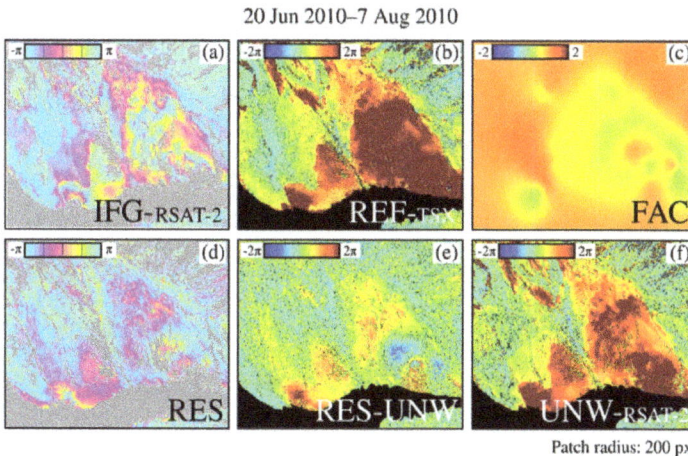

Patch radius: 200 px

Figure 9. Warp demodulation result for a representative RADARSAT-2 interferogram (C-Band, 3 m ground range resolution, 24-day repeat cycle) covering 2 cycles. (**a**) IFG-RSAT-2, to-be-demodulated interferogram (wrapped); (**b**) REF-TSX, reference interferogram (unwrapped); (**c**) FAC, demodulation factor; (**d**) RES: demodulation residual (wrapped); (**e**) RES-UNW: demodulation residual (unwrapped); and (**f**) UNW-RSAT-2, demodulated interferogram. The unwrapped phase in (**b,e,f**) is clipped to $\pm 2\pi$. The demodulation patch radius is 200 pixels.

7 Aug 2007–9 Aug 2008

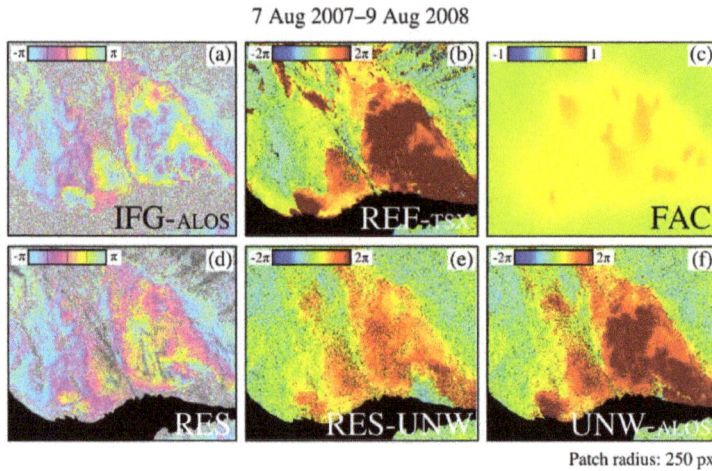

Patch radius: 250 px

Figure 10. Warp demodulation result for a representative ALOS-1 interferogram (L-Band, 15 m ground range resolution, 46-day repeat cycle) covering 8 cycles. (**a**) IFG-ALOS, to-be-demodulated interferogram (wrapped); (**b**) REF-TSX, reference interferogram (unwrapped); (**c**) FAC, demodulation factor; (**d**) RES: demodulation residual (wrapped); (**e**) RES-UNW: demodulation residual (unwrapped); and (**f**) UNW-ALOS, demodulated interferogram. The unwrapped phase in (**b,e,f**) is clipped to $\pm 2\pi$. The demodulation patch radius is 250 pixels.

27 Jul 1995–31 Aug 1995

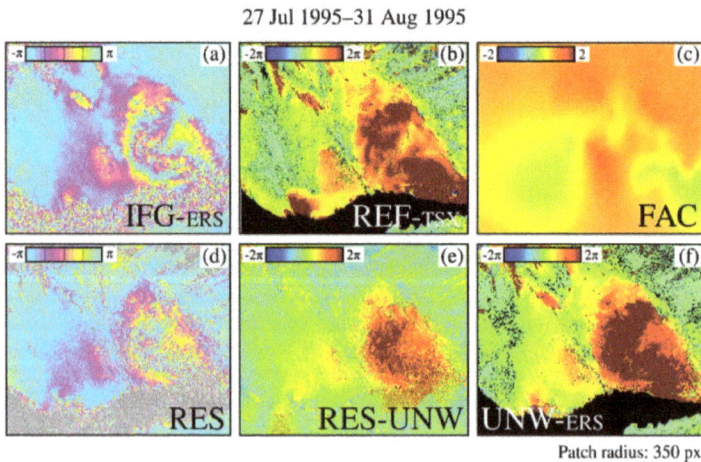

Patch radius: 350 px

Figure 11. Warp demodulation result for a representative ERS interferogram (C-Band, 30 m ground range resolution, 35-day repeat cycle) covering 1 cycle. (**a**) IFG-ERS, to-be-demodulated interferogram (wrapped); (**b**) REF-TSX, reference interferogram (unwrapped); (**c**) FAC, demodulation factor; (**d**) RES: demodulation residual (wrapped); (**e**) RES-UNW: demodulation residual (unwrapped); and (**f**) UNW-ERS, demodulated interferogram. The unwrapped phase in (**b,e,f**) is clipped to $\pm 2\pi$. The demodulation patch radius is 350 pixels.

Respective warp demodulation results for the same test sites are presented in Figures 8–11 (as just stated the unwrapped interferogram (UNW) of the lower right sub-figures are also shown in Figure 7e–h). The test case of Figure 8 (TSX-ST 2016) selects a partially incoherent interferogram with strong seasonal deformation contrast but with the same resolution and close in absolute time to the reference template (TSX-ST 2015). The figure order of the remaining three test cases (RADARSAT-2 2010, ALOS-1 2007/2008, and ERS 1995) aligns with going further into the past of the slide deformation, as well as towards coarser resolution. Each figure has six panels, which starting in the upper-left and going clockwise represent the following: (a) IFG, the to-be-demodulated interferogram; (b) REF, the line-of-sight re-projected unwrapped reference; (c) FAC, the warp demodulation factor map; (d) RES, the wrapped demodulated interferogram; (e) RES-UNW, the unwrapped demodulated interferogram, and (f) UNW, the final interferogram unwrapped through warp demodulation. Note that REF appears different in each figure, as creating the reference template involves line-of-sight re-projections of the (same) TerraSAR-X interferometric pair into the (all different) viewing geometries of the other sensors. The choice of demodulation patch sizes reflects our reasoning on the different resolutions and atmospheric and random noise phase error levels of the different sensors. For example, ERS has the largest radius (350 pixels); ERS has comparable atmospheric error to RADARSAT-2, but its random noise is higher due to factor 10 lower resolution. ALOS-1 (patch radius 250 pixels), covering one entire year, has random noise only slightly lower than ERS over one month, but tropospheric atmospheric noise levels in L-band are expected to be much lower than for ERS. Due to its considerably high-resolution, the RADARSAT-2 patch radius is the smallest (200 pixels).

The findings from the real data experiments indicate that, unlike standard MCF unwrapping (Figure 7d), the warp demodulation method performs creditably over the TSX-ST interferogram of May 2016 (Figure 8) despite the extensive (around 50 percent of area) low coherence areas of the original interferogram. Further, the findings confirm (most obvious in the 48-day C-band interferogram of Figure 9) that our method can handle significant aliasing in IFG that "hides" active discontinuities through a near integer 2-pi phase ambiguity; warp demodulation still finds the correct FAC values across these discontinuities by matching between IFG and REF the spatial texture of the phase surfaces on both sides of the fault. RES in Figure 9d is smooth and within a single fringe, which indicates a successful warp demodulation.

Also, for the other two cases (Figures 10 and 11) involving sensors with considerably coarser resolution, our method still performs satisfactorily in producing phase residuals that are both spatially smooth and sufficiently small to reconstruct the unwrapped phase of the test interferograms. Reconstruction is not perfect, firstly due to the error sources identified in Section 3.1 (e.g., migration and introduction of deformation discontinuities), plus a few additional ones (such as small unwrapping errors in the reference in some small layover regions but particularly in the fast-moving area where the eastern lobe meets the Fels Glacier). Errors in the reconstruction are by and large limited to this fastest moving part of the slide (most visible in Figure 10e, the unwrapped residual of the ALOS-1 case), where the reference likely would also not be representative due to separation in time even in the absence of any phase unwrapping errors.

4. Discussion

The sensitivity analysis of the warp demodulation method in Section 3.1 using synthetic data identifies a number of error sources assuming a perfect reference. Our method, for obvious reasons, cannot reconstruct areas where the reference does not exist. However, in areas where the high-resolution interferogram used to generate the demodulation reference is of high quality and has been correctly unwrapped, our method is robust with respect to almost all potential error sources; the ones analyzed in Section 3.1 and others associated with the described processing steps, such as the re-projection of the unwrapped reference to the line-of-sight of the to-be-demodulated interferogram. Of particular interest is the ability of our method to recover the unwrapped phase correctly across a lobe discontinuity, even if the phase has several wraps when followed along the discontinuity delineating the lobe. This ability depends on the presence of a significantly strong spatial texture (local gradient or curvature) in the phase that can be matched between reference and to-be-demodulated interferogram to break the two-pi ambiguity of the phase jump across the discontinuity. To match spatial textures robustly, a large enough patch size is critical. However, a very large size can also cause patches to overlap with multiple discontinuities, where they are closely spaced, which prevents a smooth residual should differential deformation changes occur for the different discontinuities with respect to the reference. The analysis in Section 3.1.3 revealed quantitatively how patch size trades off quantitatively for realistic lobe configurations. Our warp demodulation method is generally robust for a range of patch sizes between a half and a fifth of the main lobe diameter. An enhanced future version of our method will use an adaptive patch size that takes the variable spacing of phase discontinuities (slide faults) into account to combine the increase in robustness with patch size with preserved high-resolution where several discontinuities are nested close together.

A noteworthy exception to the generally robust behavior of the warp demodulation method is when the reference gets "out of date" (i.e., the number of discontinuities of the deformation field identified in the reference does not match that in the to-be-demodulated interferogram (this can be considered a more extreme case of the subtle shifting of discontinuities included in the analysis of Section 3.1)). For the case of a qualitative mismatch of discontinuities, where either (i) an extra discontinuity or (ii) a discontinuity identified in the reference is not active in the interferogram, there is a pronounced asymmetry. For case (ii), the warp factor will down-scale the reference locally at the inactive discontinuity; the warp modulation method is robust to this case (=smooth residual). However, for case (i), the residual will be insensitive to the warp factor and the residual will be locally discontinuous. Thus, if too many discontinuities are missing in the reference and there are additional high-resolution datasets available closer in time, then a new reference should be constructed and used instead.

As mentioned before, our method is easily generalized to several references and can also be used iteratively (i.e., successfully warp-demodulated interferograms can themselves be used as reference). A full generalization of our method to a wireframe representation in space and time is considered as a next step. For a slide with a deep, historic InSAR dataset, this would then include an explicit parameterization of the spatio-temporal pattern of the active faults of the landslide, including formation, migration, and healing of faults. The patch measurements carried out with our original warp demodulation concept presented in Section 2 could be used to make the explicit fault parametrization converge. Once found, the explicit fault parametrization can then be used to demodulate/unwrap the full time series of interferograms (as discontinuous aliased phase surfaces) in a single optimization step (as per the original wireframe concept).

Besides the robustness of the method, the interpretation of the two results per interferogram (i) warp factor map and (ii) residual phase map is also of interest for discussion. As our method targets smoothness but not necessarily the overall smallest size of the phase residual after demodulation, neither the residual, nor the warp factor maps can be easily interpreted quantitatively as a measure of bulk slide activity relative to that of the reference. Qualitatively, a smooth residual shows persistence of the fault pattern between the interferogram and the reference. Also, a factor map in the 1-range plus a small residual indicates that the deformation field of interferogram and the reference are very similar. However, beyond that—despite obviously containing important information about the (temporal) changes of the slide dynamics—the interpretation of the pair residual/warp factor map is complex. The studying of a variety of real data examples, such as the ones in Figures 8–11, has revealed that the warp factor difference forms one measured change in the shallow (discontinues) deformation field of the slide, while the size of the residual indicates a change in the deep deformation pattern between the interferogram and the reference. More experiments with simulated data are planned to understand this connection more quantitatively.

5. Conclusions

We have developed a new demodulation method (warp demodulation) capable of producing smooth phase residuals to improve the phase unwrapping performance for older, lower resolution, aliased interferograms of complex, multi-lobed, discontinuous landslide deformation. Warp demodulation hinges on the identification of the discontinuity (active fault) pattern on the slide from a high-resolution interferogram, as well as a reference template constructed from the same high-resolution interferogram to match locally the older, coarser interferograms—TerraSAR-X staring spotlight mode (TSX-ST) was found to be exceptionally well suited for this purpose.

A comprehensive sensitivity analysis found the warp demodulation method to be accurate and reasonably robust for all investigated sources. To demonstrate our method, we presented results from three data sets of increasingly coarser resolution and distant in time from the TSX-ST reference. Performance was convincing in all three cases, showing that current high-resolution InSAR and our method can retrieve value from older, low-resolution data that cannot be processed on their own due to low coherence and aliasing, which is useful for the historic analysis of landslide dynamics.

Author Contributions: Conceptualization, B.R.; Investigation, B.R. and M.P.; Methodology, B.R.; Software, B.R. and M.P.; Validation, B.R. and M.P.; and Writing–Review and Editing, B.R. and M.P.

Funding: This work was funded through the NSERC-MDA-CSA SAR Chair research program at Simon Fraser University (SFU).

Acknowledgments: The authors would like to thank the German Aerospace Center (DLR) for providing the TerraSAR-X data (staring spotlight mode) in the framework of the DLR project LAN2435. They would also like to thank MDA Inc. (through their in-kind support for the SAR Chair) for providing access to RADARSAT-2 data and the Alaska SAR Facility (ASF) for providing access to ERS and ALOS-1 data.

Conflicts of Interest: The authors declare no conflict of interest.

References

1. Rosen, P.A.; Hensley, S.; Joughin, I.R.; Li, F.K.; Madsen, S.N.; Rodriguez, E.; Goldstein, R. Synthetic aperture radar interferometry. *Proc. IEEE* **2000**, *88*, 333–382. [CrossRef]

2. Bamler, R.; Hartl, P. Synthetic aperture radar interferometry. *Inverse Probl.* **1998**, *14*, R1. [CrossRef]

3. Strozzi, T.; Farina, P.; Corsini, A.; Ambrosi, C.; Thüring, M.; Zilger, J.; Wiesmann, A.; Wegmüller, U.; Werner, C. Survey and monitoring of landslide displacements by means of L-band satellite SAR interferometry. *Landslides* **2005**, *2*, 193–201. [CrossRef]

4. Kimura, H.; Yamaguchi, Y. Detection of landslide areas using satellite radar interferometry. *Photogramm. Eng. Remote Sens.* **2000**, *66*, 337–344.

5. Rott, H.; Scheuchl, B.; Siegel, A.; Grasemann, B. Monitoring very slow slope movements by means of SAR interferometry: A case study from a mass waste above a reservoir in the Otztal Alps, Austria. *Geophys. Res. Lett.* **1999**, *26*, 1629–1632. [CrossRef]

6. Berardino, P.; Fornaro, G.; Lanari, R.; Sansosti, E. A new algorithm for surface deformation monitoring based on small baseline differential SAR interferograms. *IEEE Trans. Geosci. Remote Sens.* **2002**, *40*, 2375–2383. [CrossRef]

7. Berardino, P.; Costantini, M.; Franceschetti, G.; Iodice, A.; Pietranera, L.; Rizzo, V. Use of differential SAR interferometry in monitoring and modeling large slope instability at Maratea (Basilicata, Italy). *Eng. Geol.* **2003**, *68*, 31–51. [CrossRef]

8. Lanari, R.; Casu, F.; Manzo, M.; Zeni, G.; Berardino, P.; Manunta, M.; Pepe, A. An overview of the Small Baseline Subset Algorithm: A DInSAR technique for surface deformation analysis. *Pure Appl. Geophys.* **2007**, *164*, 637–661. [CrossRef]

9. Rabus, B.T.; Ghuman, P.S. A simple robust two-scale phase component inversion scheme for persistent scatterer interferometry (dual-scale PSI). *Can. J. Remote Sens.* **2009**, *35*, 399–410. [CrossRef]

10. Ferretti, A.; Prati, C.; Rocca, F. Permanent scatterers in SAR interferometry. *IEEE Trans. Geosci. Remote Sens.* **2001**, *39*, 8–20. [CrossRef]

11. Colesanti, C.; Ferretti, A.; Novali, F.; Prati, C.; Rocca, F. SAR monitoring of progressive and seasonal ground deformation using the permanent scatterers technique. *IEEE Trans. Geosci. Remote Sens.* **2003**, *41*, 1685–1701. [CrossRef]

12. Hilley, G.E.; Bürgmann, R.; Ferretti, A.; Novali, F.; Rocca, F. Dynamics of slow-moving landslides from permanent scatterer analysis. *Science* **2004**, *304*, 1952–1955. [CrossRef] [PubMed]

13. Ferretti, A.; Fumagalli, A.; Novali, F.; Prati, C.; Rocca, F.; Rucci, A. A new algorithm for processing interferometric data-stacks: SqueeSAR. *IEEE Trans. Geosci. Remote Sens.* **2011**, *49*, 3460–3470. [CrossRef]

14. Eppler, J.; Rabus, B. Monitoring urban infrastructure with an adaptive multilooking InSAR technique. In Proceedings of the FRINGE 2011, Frascati, Italy, 19–23 September 2011.

15. Rabus, B.; Eppler, J.; Sharma, J.; Busler, J. Tunnel Monitoring with an Advanced InSAR Technique. *Proc. SPIE* **2012**, *8361*, 83611F.

16. Costantini, M.; Malvarosa, F.; Minati, F.; Pietranera, L.; Milillo, G. A three-dimensional phase unwrapping algorithm for processing of multitemporal SAR interferometric measurements. In Proceedings of the IEEE International Geoscience and Remote Sensing Symposium, Toronto, ON, Canada, 24–28 June 2002; Volume 3, pp. 1741–1743.

17. Costantini, M. A novel phase unwrapping method based on network programming. *IEEE Trans. Geosci. Remote Sens.* **1998**, *36*, 813–821. [CrossRef]

18. Van Zyl, J.J. The Shuttle Radar Topography Mission (SRTM): A breakthrough in remote sensing of topography. *Acta Astronaut.* **2001**, *48*, 559–565. [CrossRef]

19. Krieger, G.; Moreira, A.; Fiedler, H.; Hajnsek, I.; Werner, M.; Younis, M.; Zink, M. TanDEM-X: A satellite formation for high-resolution SAR interferometry. *IEEE Trans. Geosci. Remote Sens.* **2007**, *45*, 3317–3341. [CrossRef]

20. Stramondo, S.; Moro, M.; Tolomei, C.; Cinti, F.R.; Doumaz, F. InSAR surface displacement field and fault modelling for the 2003 Bam earthquake (southeastern Iran). *J. Geodyn.* **2005**, *40*, 347–353. [CrossRef]

21. Lu, Z.; Dzurisin, D.; Biggs, J.; Wicks, C.; McNutt, S. Ground surface deformation patterns, magma supply, and magma storage at Okmok volcano, Alaska, from InSAR analysis: 1. Intereruption deformation, 1997–2008. *J. Geophys. Res. Solid Earth* **2010**, *115*. [CrossRef]

22. Woo, K.S.; Eberhardt, E.; Rabus, B.; Stead, D.; Vyazmensky, A. Integration of field characterisation, mine production and InSAR monitoring data to constrain and calibrate 3-D numerical modelling of block caving-induced subsidence. *Int. J. Rock Mech. Min. Sci.* **2012**, *53*, 166–178. [CrossRef]

23. Carrara, W.G.; Goodman, R.S.; Majewski, R.M. *Spotlight Synthetic Aperture Radar: Signal Processing Algorithms*; Artech House: Norwood, MA, USA, 1995.

24. Mittermayer, J.; Wollstadt, S.; Prats-Iraola, P.; Scheiber, R. The TerraSAR-X staring spotlight mode concept. *IEEE Trans. Geosci. Remote Sens.* **2014**, *52*, 3695–3706. [CrossRef]

25. Bozdagi, G.; Tekalp, A.M.; Onural, L. 3-D motion estimation and wireframe adaptation including photometric effects for model-based coding of facial image sequences. *IEEE Trans. Circuits Syst. Video Technol.* **1994**, *4*, 246–256. [CrossRef]

26. Lee, J.S.; Pottier, E. *Polarimetric Radar Imaging: From Basics to Applications*; CRC Press: Boca Raton, FL, USA, 2009.

27. Newman, S.D. Deep-Seated Gravitational Slope Deformations near the Trans-Alaska Pipeline, East-Central Alaska Range. Master's Thesis, Simon Fraser University, Burnaby, BC, Canada, 2013.

remote sensing

MDPI

Article

Land Subsidence in Coastal Environments: Knowledge Advance in the Venice Coastland by TerraSAR-X PSI

Luigi Tosi [1,*], Cristina Da Lio [1], Pietro Teatini [1,2] and Tazio Strozzi [3]

[1] Institute of Marine Sciences, National Research Council, Arsenale Tesa 104, Castello 2737/F, 30122 Venezia, Italy; cristina.dalio@ve.ismar.cnr.it (C.D.L.); pietro.teatini@unipd.it (P.T.)

[2] Department of Civil, Environmental and Architectural Engineering, University of Padova, Via Loredan 20, 35131 Padova, Italy

[3] Gamma Remote Sensing, Worbstrasse 225, 3073 Gümligen, Switzerland; strozzi@gamma-rs.ch

* Correspondence: luigi.tosi@ismar.cnr.it; Tel.: +39-041-240-7949

Received: 24 June 2018; Accepted: 26 July 2018; Published: 29 July 2018

Abstract: The use of satellite SAR interferometric methods has significantly improved the monitoring of ground movements over the last decades, thus opening new possibilities for a more accurate interpretation of land subsidence and its driving mechanisms. TerraSAR-X has been extensively used to study land subsidence in the Venice Lagoon, Italy, with the aim of quantifying the natural and anthropogenic causes. In this paper, we review and update the main results achieved by three research projects supported by DLR AOs (German Aerospace Center Announcement of Opportunity) and conducted to test the capability of TerraSAR-X PSI (Persistent Scatterer Interferometry) to detect ground movements in the complex physiographic setting of the Venice transitional coastal environment. The investigations have been focused on the historical center of Venice, the lagoon inlets where the MoSE is under construction, salt marshes, and newly built-up areas in the littoral. PSI on stacks of stripmap TerraSAR-X images covering short- to long-time periods (i.e., the years 2008–2009, 2008–2011 and 2008–2013) has proven particularly effective to measure land subsidence in the Venice coastland. The very high spatial resolution (3 m) and the short repeat time interval (11 days) of the TerraSAR-X acquisitions make it possible to investigate ground movements with a detail unavailable in the past. The interferometric products, properly calibrated, allowed for a millimetric vertical accuracy of the land movements at both the regional and local scales, even for short-term analyses, i.e., spanning one year only. The new picture of the land movement resulted from processing TerraSAR-X images has significantly contributed to update the knowledge on the subsidence process at the Venice coast.

Keywords: land subsidence; TerraSAR-X; SAR interferometry; coastal environments; Venice lagoon

1. Introduction

The city of Venice and its surrounding lagoon is presently one of the sites most sensitive to land subsidence worldwide. Even a few mm loss of ground elevations with respect to the mean sea level can significantly change the natural lagoon environments and threaten the city's survival. Over the last century, land subsidence and climate change effects concurred to a relative sea level rise (RSLR) in Venice equal to 26 cm [1], a situation that is expected to get dramatically worst if we consider the 53 cm sea level rise projected by the Intergovernmental Panel on Climate Change (IPCC) in 2100 according to the A1B scenario, which accounts for a future energy system balanced between fossil and non-fossil sources [2]. RSLR has contributed to a seven-fold increase of flooding events in Venice over the last decades, causing large inconvenience for the population and enormous damage to the cultural heritage, with significant ecological and environmental impacts in the lagoon and coastland [3]. To protect the city and the lagoon from increased flooding, a series of mobile barriers (MoSE) are under construction at the three inlets (Lido, Malamocco and Chioggia) connecting the Adriatic Sea to the inner water body.

The monitoring of land subsidence in the Venice area began in the 1960s using spirit leveling and over the last decades has been significantly improved using space-borne observation techniques based on synthetic aperture radar (SAR) interferometry. The subsidence monitoring network of the Venice coastland increased from a few hundreds of benchmarks measured by leveling [3] to a few hundred thousand reflectors detected by SAR-based interferometry using ERS and ENVISAT satellites [4]. This reduced the use of in-situ measurements to calibrate and validate interferometric products [5]. The increased spatial coverage of the measured points by the C-band sensors provided a new image of the land subsidence in the Venice coastland. In particular, the high variability of land movements at both local and regional scales was unexpectedly revealed, with areas severely affected by subsidence formerly unknown.

The launch of the new generation X-band SAR sensors opened the challenge to further advance in the knowledge on land subsidence, with particular advantages for coastal areas where the loss of land elevation with respect of the mean sea level is one of the main environmental and geological hazards. Many scientific papers have been published over the last decade concerning the use of TerraSAR-X for measuring displacements in coastal zones, both at the regional scale and for specific infrastructures. For example, in Tianjin and Shanghai, which are two coastal areas in China where land subsidence has seriously threatened the infrastructure efficiency, environment conservation, and population safety, TerraSAR-X-based interferometry was used to quantify land movements at the city scale [6,7] or along the tract of high-speed trains [8,9]. Other coastal cities where land subsidence has been recently mapped by TerraSAR-X images are Jakarta, Indonesia [10], Hanoi, Vietnam [11], and New Orleans, Louisiana [12]. SAR interferometry on TerraSAR-X acquisitions has also been applied in natural coastlands. In one of the most recent contribution, nine Single-Look Slant Range Complex (SSC) images taken between 7 June and 14 September 2013 have been processed Differential SAR interferometry to detect summer thaw subsidence over a yedoma (an organic-rich Pleistocene-age permafrost with ice content of 50–90% by volume) region of the Lena River Delta, Siberia [13]. Concerning the Venice Lagoon, three projects have been carried out over the last decade under the umbrella of the DLR AOs: LAN0242 "Monitoring land subsidence in Venice"; COA0612 "Assessing vertical movements of natural tidal landforms and anthropogenic structures at the Venice Lagoon inlets"; and COA1800 "Ground surface dynamics in the Venice Lagoon: five years of monitoring of natural tidal landforms and anthropogenic structures by TerraSAR-X". These projects have allowed testing the effectiveness of TerraSAR-X in quantifying land subsidence in the Venice coastland with the aim of improving the knowledge on natural and anthropogenic causes. The results of these projects have been published in various papers (e.g., [14–20]).

The goal of this paper is to provide a review of the main results and advancements in the knowledge of land subsidence at the Venice coastland achieved by TerraSAR-X PSI over the last decade, updating the main outcomes obtained in the past years with the analyses recently carried out over the six-year period between 2008 and 2013.

2. TerraSAR-X Image Processing and Output Calibration

Stacks of TerraSAR-X stripmap HH polarization images characterized by a 28° incidence angle acquired with a regular 11-day revisiting time have been processed by IPTA (Interferometric Point Target Analysis) [21,22], one of the available PSI processing chains. A number of 30, 90 and 143 scenes acquired over the periods 2008–2009, 2008–2011 and 2008–2013, respectively, have been used.

Particular attention was paid to the calibration of the interferometric products as they are referred to an arbitrary reference and are affected by the so-called flattening problem, i.e., the slight phase tilt resulting from the inaccuracy in estimating the orbital baseline due to the imperfect knowledge of satellite positions. This latter can be important in regional-scale analyses. To reduce these biases, calibration and de-flattening were previously performed using correction planes modeled through ground-based data [19] by: (i) defining a local reference frame based on a reference point located outside the study area and the subsiding coastland; and (ii) projecting the vertical velocities of the ground-truth data along the line of sight (LOS) of the satellites [20]. Continuous GPS stations (CGPS) are used for calibration and de-flattening of the TerraSAR-X interferometric product: three stations are located inside the area and one, located just outside the SAR frames, is used to refer the horizontal movements. The three-dimensional CGPS velocity vectors were projected onto the SAR LOS using Equation (1):

$$CGPS_{LOS} = \sin(\theta)\cos(\phi)\Delta E + \sin(\theta)\sin(\phi)\Delta N + \cos(\theta)UP \tag{1}$$

where ΔE and ΔN are the local easting and northing components of the CGPS movement, computed by removing the E and N average velocities of the CGPS station located outside the SAR image frame, and UP is the vertical component. θ and ϕ are the incidence angle and the ground track angle, respectively, at the CGPS locations. Notice that the Venice coastland is moving in the horizontal direction almost uniformly (18–20 mm/year northward and 15–17 mm/year eastward with reference to ITRF 2005) and therefore the errors in considering null the relative horizontal movements due to the tectonics at regional scale can be safely ignored. However, since we only used descending orbits, the decomposition of LOS velocity into vertical and East-West components is not possible and LOS velocities still include East-West motions due to local causes. Finally, the differences between the CGPS-LOS and the velocities of the PTs (Point Targets) averaged over a 20–200-m radius centered on the three CGPS stations were used to define the proper equation of tilting plane implemented to correct the interferometric products. Because of the relatively small area of interest and the proper distribution of the CGPS stations within the TerraSAR-X frame, a plane is assumed the most appropriate surface to de-flatten the SAR solution [19,23]. An example of calibration plane is shown in Figure 1a.

A comparison between the CGPS displacements along the LOS direction and the calibrated TerraSAR-X dataset corresponding to the PT closest to the CGPS station is provided in Figure 1b–d. Table 1 provides a quantitative summary in term of LOS velocity. Notice that SFEL CGPS is characterized by a significantly nonlinear trend because of the MoSE-related works started in 2007. If the shorter period (2007–2011) is considered, the average CGPS LOS velocity amounts to -7.46 ± 0.28 mm/year, a value very close to the average SAR calibrated time series.

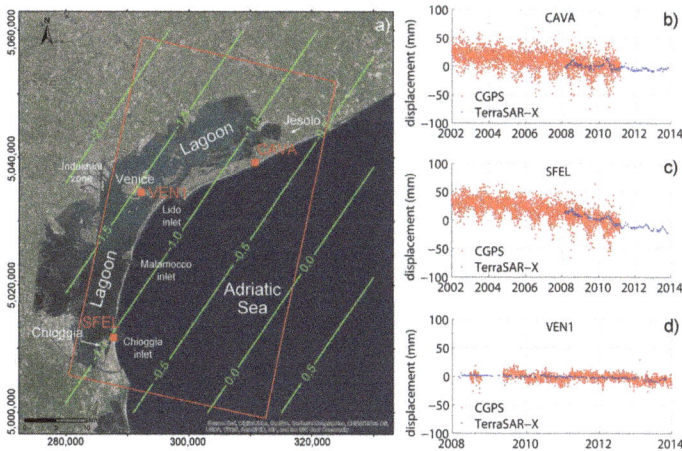

Figure 1. Example of PSI outcome calibration: (**a**) plane model (green isolines, mm/year) based on the Cavallino (CAVA), Chioggia (SFEL) and Venice (VEN1) CGPS displacements; and (**b–d**) comparison between the CGPS and the calibrated PSI time series. Red squares are the locations of the CGPS stations and the footprint of the TerraSAR-X frame (beam strip_006R) is represented by a red box.

Table 1. Comparison between CGPS and representative calibrated PSI average velocities obtained by TerraSAR-X (see Figure 1b–d).

CGPS Station	Time Span	CGPS LOS (mm/year)	PSI (mm/year)
CAVA	2002–2011	-2.83 ± 0.09	-2.20 ± 0.20
SFEL	2000–2011	-4.77 ± 0.09	-6.58 ± 0.20
VEN1	2000–2015	-2.02 ± 0.07	-2.10 ± 0.09

3. Ground Displacements at the Regional Scale

The ground displacements at the regional scale for the 2008–2011 period [19] has been updated extending the time span to 2013. The 2008–2013 displacement map consists of more than 1,900,000 PTs (Figure 2) and shows ground movement velocities ranging from localized gentle uplifts to large subsidence (up to 30 mm/year). The map highlights that this portion of the Northern Adriatic coastland is characterized by a large variability in terms of land subsidence.

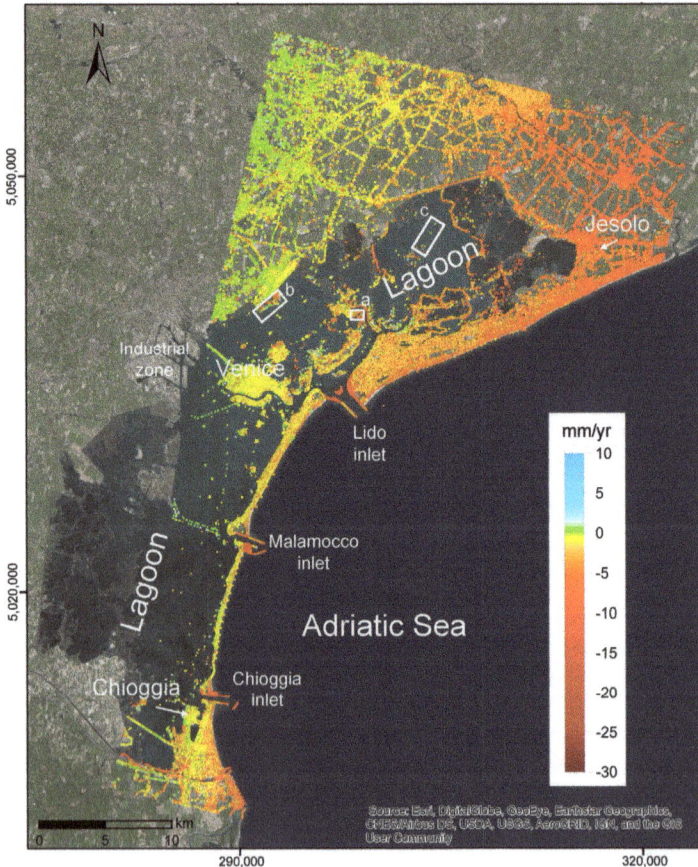

Figure 2. Ground movements at the regional scale. Average land displacements (mm/year) for the Venice coastland obtained by PSI on TerraSAR-X images acquired between March 2008 and November 2013. Positive values indicate uplift, negative values land subsidence. Base map source: Esri, DigitalGlobe, GeoEye, Earthstar Geographics, CNES/Airbus DS, USDA, USGS, AEX, Getmapping, Aerogrid, IGN, IGP, swisstopo, and the GIS User Community.

Investigation of Figure 2 reveals that the northwestern zone is characterized by general stability (average subsidence less than 1 mm/year). Relatively large displacements caused by human activities (e.g., groundwater pumping and development of new urban and industrial zones) affect only small areas. Conversely, the northern coastland is characterized by a quite uniform land subsidence on the order of 4–6 mm/year. The superposition of aquifer exploitation and land reclamation mainly contribute to the observed settlements. Concerning the lagoon basin, it shows the highest variability of the ground displacements, with rates ranging from -1 to -30 mm/year. The largely heterogeneous nature of the shallow Holocene deposits is strongly responsible for the high heterogeneity of the land movements: the lagoon sectors with the thicker unconsolidated deposits generally correspond to the areas experiencing the largest subsidence rates. In the northern lagoon, high sinking rates are often linked to the presence of stone embankments bounding the fish farms. The historical center of Venice and the central lagoon are relatively stable with an average land subsidence of 1–2 mm/year. Here, the compressible Holocene deposits are relatively thin and groundwater withdrawals have been precluded since the early 1970s. Southward, the city of Chioggia and its surroundings are almost stable (subsidence about 1 mm/year), whereas a high subsidence affects the mouths of the main rivers (the Brenta and Adige rivers) crossing the southernmost portion of the study area.

4. Ground Displacement at the Local-Scale

Specific analyses were carried out for peculiar portions of the Venice coastland. We report in the following the results obtained by PSI on the historical center of Venice, the lagoon salt marshes, the network of artificial reflectors established within the lagoon in 2007, the lagoon inlets where the MoSE-related infrastructures have been recently constructed, and new urbanizations developed in the tourist villages on the Adriatic coast.

4.1. Induced Ground Movements in the Historical Center of Venice

Understanding if Venice is affected also today by land subsidence due to anthropogenic activities, i.e., by components that can be removed or reduced, is of paramount importance for the city survival. However, land subsidence measurements generally provide the total movement of the land surface, including both the natural and man-induced components (e.g., [24,25]).

The anthropogenic land subsidence due to activities characterized by large scale and long term effects, e.g., groundwater pumping, ended in Venice a few decades ago [26]. Today, the anthropogenic component of the land subsidence should be related to local and short-time interventions, such as restoration works and inherent deformations of historical structures.

Based on these considerations, it was tested to distinguish between the natural and anthropogenic components of the present land subsidence in the historical center of Venice by properly combining different PSI results [18]. Specifically, a long-term analysis using the ERS/ENVISAT C-band dataset can be reasonably used to quantify the natural component and a short-term analysis of X-band dataset to highlight the anthropogenic displacements. The very high spatial resolution of the X-band satellites and their short revisiting time makes it possible to investigate urban settlements with sufficient measurement accuracy and a high level of space and time detail. Indeed, in Venice, the X-band PSI outcome showed a scatterer density one order of magnitude larger than that obtained by the C-band sensors. To this aim, the long-time 1992–2010 ERS/ENVISAT and short-time 2008–2009 TerraSAR-X images have been processed. Since the differences between the ERS/ENVISAT and TerraSAR-X incidence angles amounts to 5° only, the direct comparison of the two LOS solutions can be carried out introducing a negligible error. The statistical analysis of the displacement distributions pointed out that the average displacement rates detected in Venice by the two satellites are similar, i.e., about -1.0 mm/year. The frequency distribution of the measured displacements shows that 50% of the TerraSAR-X radar reflectors are characterized by movements between 0 and -1.5 mm/year and 25% range from -1.5 and -3.0 mm/year. Regarding the C-band, about 80–85% of the city displacements range between 0 and -1.5 mm/year and 15% from -1.5 to -3 mm/year. As C- and X-band analyses showed similar rates of the average displacement, i.e., mainly the natural component of the subsidence, it is reasonable to assume that the difference between the movements provided by ERS/ENVISAT and TerraSAR-X is likely representative of the effects caused by anthropogenic activities.

Therefore, the two datasets were interpolated by the Kriging method on the same regular grid covering the whole city [18]. The comparison points out the uniformity (in the range between 0 and −1 mm/year) of the long-term displacement rates and the large variability of the short-term movements (Figure 3b) that are superposed to a background velocity similar to that given by ERS/ENVISAT. By removing the C-band interpolated map from the X-band interpolated solution, an estimate of the man-induced displacements was obtained (Figure 3). The map of the differences shows that most of the city is subsiding only due to natural components. However, about 25% of the city has experienced in 2008 some movements due to anthropogenic causes. They developed at local scale and their distribution correspond well with the sites affected by activities linked to the restoration of ancient buildings, consolidations, jet grouting, use of well-points (see the photos in Figure 3), and likely to the ship and boat wave impact on the embankment walls. Generally, the man-induced activities contribute to a subsidence larger than the natural (15%) but there are some areas (10%) where the short-term sinking is smaller than the long-term subsidence.

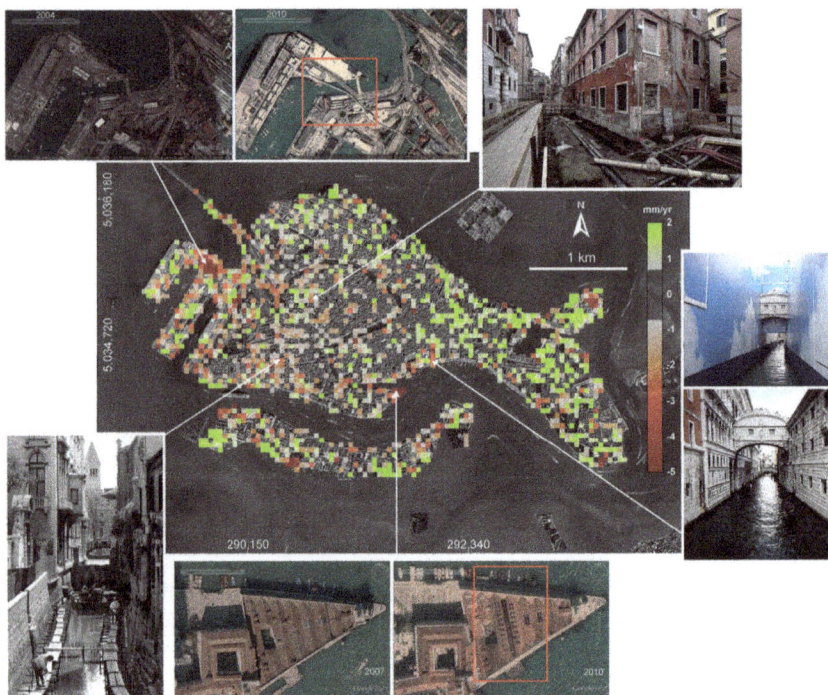

Figure 3. Map of the recent anthropogenic land movements at Venice as obtained by subtracting the ERS-ENVISAT long-term ground movements from the TerraSAR-X short-term ground movements, resampling the PSI solutions on a regular 50 m grid [18]. Negative and positive rates indicate areas where human activities are responsible for land settlements or reduce the natural subsidence, respectively. A few examples of human activities possibly influencing the subsidence values are shown in the photograph insets (available from http://www.insula.it/andGoogleEarthImages).

Finally, notice that because of the intrinsic properties of the TerraSAR-X images and satellite revisiting time, a one-year image stack is sufficient to provide an interferometric product characterized by a sufficient (i.e., millimeter) accuracy. Indeed, the error of a single measurement on a radar scatterer can be computed as follows [27]:

$$r_{err} = \sqrt{\left(r_{err}^{SCR}\right)^2 + \left(r_{err}^{ATM}\right)^2} \tag{2}$$

where r_{err}^{SCR} is the error related to the signal noise and r_{err}^{ATM} the atmospheric displacement error. The r_{err}^{SCR} is related to SCR (Signal-to-Clutter Ratio) and the sensor wavelength λ by [28]:

$$r_{err}^{SCR} = \frac{\lambda}{4} \cdot \frac{1}{\sqrt{2 \cdot SCR}} \tag{3}$$

By considering a SCR value of 10 dB, which is typical for TerraSAR-X scene, $r_{err}^{SCR} = 1.7$ mm. As it can be assumed in the case of TerraSAR-X a variance of the atmospheric residual on the order of $\pi/4$ [19], this corresponds to an atmospheric displacement error $r_{err}^{ATM} = 2.0$ mm. From Equation (2), it results $r_{err} \approx 2.2$ mm. The error of the mean velocity estimated with $n = 30$ TerraSAR-X images over a period of $t = 1$ year is then $r_{err}/(t\sqrt{n}) \approx 0.4$ mm/year.

4.2. Salt Marshes Ground Dynamics

The existence of salt marshes and tidal morphologies is strictly connected to their elevation with respect to the mean sea level (e.g., [29]). In view of the expected climate changes, quantifying land subsidence of these transitional environments is crucial to investigate their long-term possible survival.

However, monitoring with a certain accuracy their movements has always been challenging due to the peculiar features of these morphological forms [30]. They have never been linked to traditional leveling and GPS networks, and also standard Interferometric SAR applications returned very poor results in terms of spatial and temporal coverage. In fact, they are environments difficult to access, submerged by the sea water twice a day, made of largely unconsolidated deposits, without anthropogenic structures, and relatively far from anthropogenic facilities.

IPTA was applied on a stack of 143 TerraSAR-X stripmap images acquired between 2008 and 2013 with a regular 11-day revisiting time [20]. The regularity of the acquisitions, the short satellite revisiting time, the high image resolution (~3 m × 3 m), and the strategies used in the PSI application have allowed us to detect thousands of measurable PTs in the Venice Lagoon salt marshes. The results show that both natural and man-made salt marshes are characterized by a quite wide range of average velocities [20]. Displacements range from small uplifts to subsidence rates of more than 20 mm/year. Generally, land subsidence is much larger on man-made than natural salt marshes, with a significant negative correlation with the marsh age (Figure 4).

A subsidence trend that increases from the littoral strips toward the western lagoon margin characterized the natural portions of the salt marshes. This trend reflects the different geomorphological setting of the lagoon and salt marsh deposits, which are richer in sandy soils along the ancient beach ridges, whereas clayey silts, often rich in organic matter, fill the inter-distributary lowlands and back-barrier zones [31,32]. Land displacements at the man-made salt marshes are characterized by patches with even subsidence and significantly different values among the various zones (Figure 4). This peculiar behavior is related to the type of construction methodology of the artificial salt marshes: coarse sediments are used initially along the marsh bound and then fine materials are used to fill the inner portions.

In addition, it has been observed that land subsidence in vegetated and bare parts of the marshes is significantly different. The comparison between the surface displacements with the presence/absence of halophytic vegetation species clearly reveals that the higher sinking rates occur on the unvegetated salt marshes [20].

Figure 4d–f shows five examples of displacement time series of scatterers located on natural and man-made parts of salt marshes. The analysis of about six years of data regularly acquired every 11 days allows us to exclude significant effect of the tidal regime on the recorded displacements. On the other hand, this is an expected outcome as the tidal fluctuation in Venice is in the order of 1 m only, which is much smaller than the ocean tides to which significant land movements have been associated (e.g., [33]).

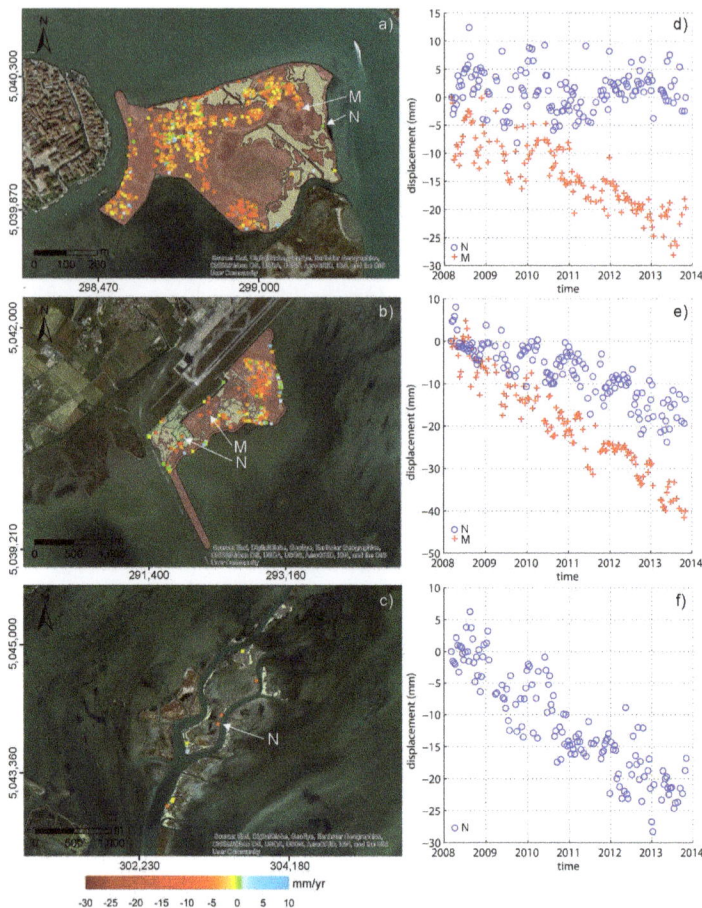

Figure 4. Examples of average land displacement over the years 2008–2013 (mm/year) detected by PSI on the salt marshes: (**a**) Burano; (**b**) Tessera; and (**c**) Cenesa. Salt marsh locations are indicated by white polygons in Figure 2. Positive values indicate uplift, while negative values indicate land subsidence. In (**a-c**) the natural portions of the salt marshes are shadowed in green, while man-made parts are shadowed in red. N and M refer to selected PTs on natural and man-made salt marsh, respectively, for which the displacement time series are provided in (**d–f**). Base map source: Esri, DigitalGlobe, GeoEye, Earthstar Geographics, CNES/Airbus DS, USDA, USGS, AEX, Getmapping, Aerogrid, IGN, IGP, swisstopo, and the GIS User Community.

Moreover, the scattered distribution of the ground movements for some PTs suggests that at least 3–4 years are required to derive a reliable quantification of the ground movement in salt marshes.

The formation age of the salt marshes also affects the subsidence values. The man-made salt marshes dated 2007–2008, 2002–2003 and 1992–1993 shows median subsidence rate amounting to 4.0, 2.8 and 1.3 mm/year, respectively. Natural salt marshes that are approximately 500–1000 year old are characterized by a median subsidence equal to 0.4 mm/year [20].

4.3. Satellite Radar Interferometry on Artificial Reflectors

The key to the PSI approach is the identification and exploitation of time coherent radar reflectors. These scatterers are typically man-made structures within the landscape, such as buildings, utility poles, roadways, or natural features, such as rocks, deserts with little shifting sand, or saline soils. A potentially severe limitation of PSI use in wetlands is the difficulty of identifying stable targets (e.g., [34]). In situ survey revealed that these features in the Venice Lagoon usually correspond to rich-shell silty deposits, construction remnants, woods, and wood posts together with stone-filled rolls (Figure 4); these latter usually place along the marsh bounds to protect them from wave erosion

To improve the coverage of satellite SAR interferometry in salt marshes within the Venice Lagoon, TerraSAR-X and ENVISAT have been tested on a network of 57 Trihedral Corner Reflectors (TCRs) established in 2007 [17]. The TCRs, placed in areas without any other strong scatterer, are characterized by a 60-cm long edge and installed with foundations ranging different depths and at the same height above ground in order to study the possible differences in their relative settlement. Because high tides flood salt marshes, the TCRs were installed at a height of 1 m above the mean sea level, i.e., constantly outside the water.

The experiment provided new information to improve the knowledge of the processes acting in the Venice Lagoon. We found that the northern basin of the lagoon is subsiding at a rate of about 3–4 mm/year, while the central and the southern lagoon regions are more stable [17].

The comparison between the TCR displacements retrieved by TerraSAR-X and ENVISAT revealed that the noise in the time series is smaller with TerraSAR-X than with ENVISAT, because of the better SCR as a consequence of the shorter wavelength and smaller resolution cell in relationship to the size of the TCR [17].

The TerraSAR-X analysis on the TCRs for 2008–2011 provided by Strozzi et al., 2013 [17] has been here extended including till to 2013; an example is shown in Figure 5. The longer period of analysis improved the quantification of the TCRs velocity and confirmed the general previous displacement behavior. The long monitoring interval has revealed an almost yearly fluctuation in the displacement behavior, whose occurrence is under investigation.

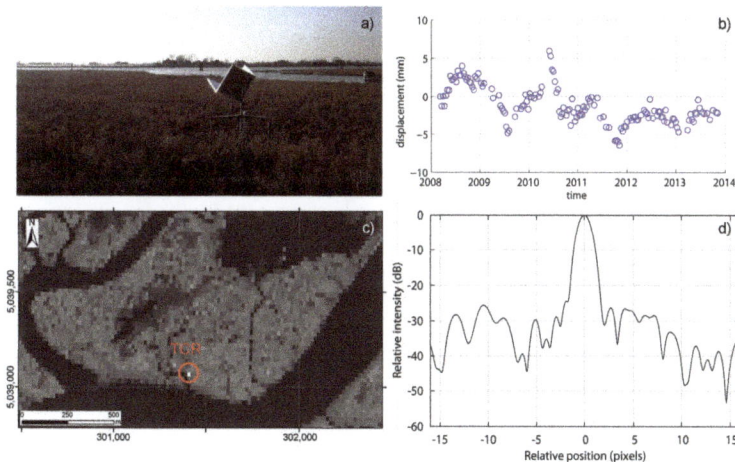

Figure 5. Example of displacement analysis of a TCR using TerraSAR-X images: (**a**) photograph of the TCR installed in a salt marsh located in the northern Venice lagoon; (**b**) time series of displacements obtained using TerraSAR-X data over the 2008–2013 period; (**c**) backscattering intensity image of the salt marsh highlighting the larger intensity (white pixel) in correspondence of the TCR; and (**d**) normalized intensity as a function of range for one for TerraSAR-X acquisition.

Such observations indicate that SAR interferometry on TerraSAR-X images using a large network of artificial reflectors is an effective and powerful methodology to monitor land subsidence in transitional and natural environments where the loss of coherence is a major problem is the use of PSI.

4.4. The MoSE Constructions at the Lagoon Inlets

The geomechanical characterization of coastal soils is of considerable interest to geotechnical and geo-environmental researchers in relation to the stability of large coastal structures, such as breakwaters, jetties, and wharfs [35].

Starting from the mid-1990s, the jetties at the three inlets of Lido, Malamocco, and Chioggia were reinforced, and, since 2005, in the framework of the MoSE works (i.e., the project of mobile barriers for the temporarily closure of the lagoon to the sea), they have been strongly reshaped and supplemented by offshore breakwaters. The construction of the mobile barriers at the three inlets raised concerns about possible important settlements caused by: (i) the load of the complementary structures (e.g., jetties, breakwaters, locks, and an artificial island) on the Quaternary deposits; and (ii) the groundwater drainage required to developed the structures.

Presently, the works at the inlets are still ongoing and monitoring land displacements in these sectors has a double aim: (i) the evaluation of possible effects of the works on the littoral environment; and (ii) the quantification of the consolidation of the new coastal structures. The former issue can support the public authorities in monitoring the environmental impacts of these giant works, while the latter can help engineering companies building the MoSE to check the absolute and differential displacements that can threaten the integrity and efficiency of the structures. In the usual civil engineering practice, leveling, GPS and laser are used to measure the settlement of some selected benchmarks established at fixed intervals along the structure. These traditional methods give straightforward and accurate results, but their applications are restricted to discrete point measurements and sometimes impractical in the rugged ground condition.

The effectiveness of TerraSAR-X to detect ground movements induced by the activities in the MoSE yards has been tested by IPTA on 30 satellite radar images acquired between March 2008 and January 2009 [15]. Significant local settlements up to 70 mm/year have been detected at the three inlets. Sinking rates less than 3 mm/year were measured in the parts of the jetties not affected by the restoration works; conversely, sinking rates up to 30 mm/year are detected in the newer structures. Similar rates have been measured also over the period August 2012–November 2013 (Figure 6a).

The study revealed for the first time that only one year of TerraSAR-X acquisitions suffices to accurately and reliably detect the consolidation trend of large coastal structures. These results have been confirmed by the analyses carried out over the two longer periods, i.e., between 2008 and 2011 [19], and from 2008 to 2013 (Figure 6b,c).

An example of the updated analysis of the displacement history of four PTs detected at the Malamocco inlet is reported in Figure 7. For two PTs, i.e., A and B in Figure 7a, at the tip of the old jetties and in the new breakwater, respectively, the analysis also includes the displacements measured by ENVISAT (Figure 7a). Notice the quite uniform trend of the settlement of the breakwater, which construction started in 2003. The displacements at the place where the mobile gates of the MoSE will be connected are shown in Figure 7b.

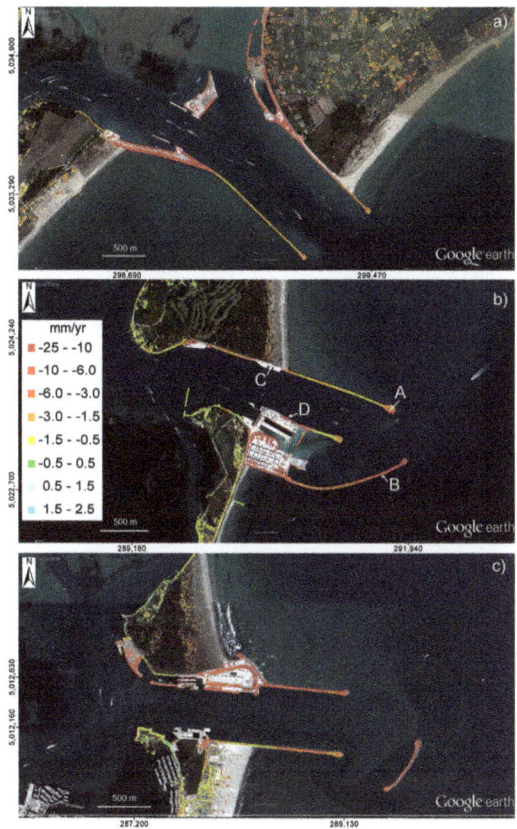

Figure 6. Mean displacement rates from TerraSAR-X interferometry: between August 2012 and November 2013 at the Lido inlet (**a**); and between March 2008 and November 2013 at the Malamocco (**b**) and Chioggia (**c**) inlets. The backgrounds are Google earth images acquired in 2012. Movements are in the satellite line-of-sight direction, negative values indicate settlement and positive indicate uplift.

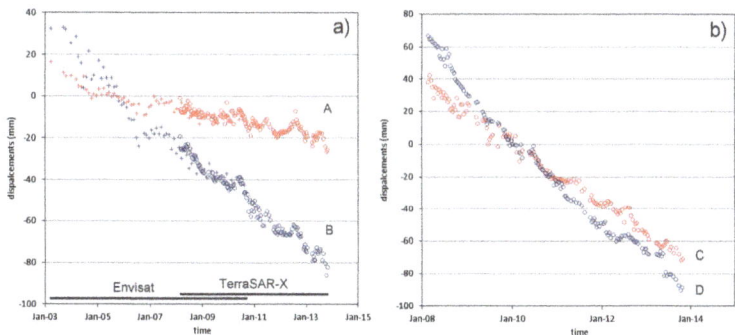

Figure 7. Examples of displacement history in the LOS direction for four PTs (A-D), at the Malamocco inlet (see positions in Figure 6b): (**a**) ENVISAT PSI between April 2003 and November 2007 and TerraSAR-X between March 2008 and November 2013 (A and B); and (**b**) TerraSAR-X PSI between March 2008 and November 2013 (C and D).

4.5. The Newly Built-Up Areas

Over the last decades, newly built-up areas have been developed on the Venice coastland with residential, touristic and industrial purposes. The new urbanization is responsible for subsidence bowls spread all over the Venice coastland, wherein several urban suburbs and industrial/commercial sites have grown rapidly by changing farmland use. The new buildings exert heavy loads on relatively recent coastal deposits and lead to consolidation of the shallow subsoil below their foundations [11,36].

This process has been clearly revealed by TerraSAR-X over the 2008–2011 time span, both at the scale of single buildings (Figure 8) as well as at the scale of villages or portion of small urban centers [19]. Figure 8 shows an updated analysis using the 2008–2013 dataset of the displacement behavior recorded in the surroundings of the 73-m high Aquileia Tower, downtown Jesolo. The tower construction was completed in 2008 and significantly settled (totaling 4 cm) during the following years because of its load. The displacement profile of the PT termed as C in Figure 8c exhibits the log behavior typical of an ongoing consolidation process, with a sinking rate that decreased from more than 10 mm/year over 2009 and 2010 to about 3–4 mm/year in 2012 and 2013, i.e., the subsidence value characterizing the whole area (Figure 8a). Other significant movements due to the built-up of new districts in the tourist villages to the north of the Venice Lagoon are clearly visible in Figure 9a. Local ground movements up to −8 mm/year are detected due to the load exerted by the new constructions, which were established over the period covered by the TerraSAR-X acquisitions (Figure 9b–f) in areas recently reclaimed and therefore characterized by compressible shallow soils.

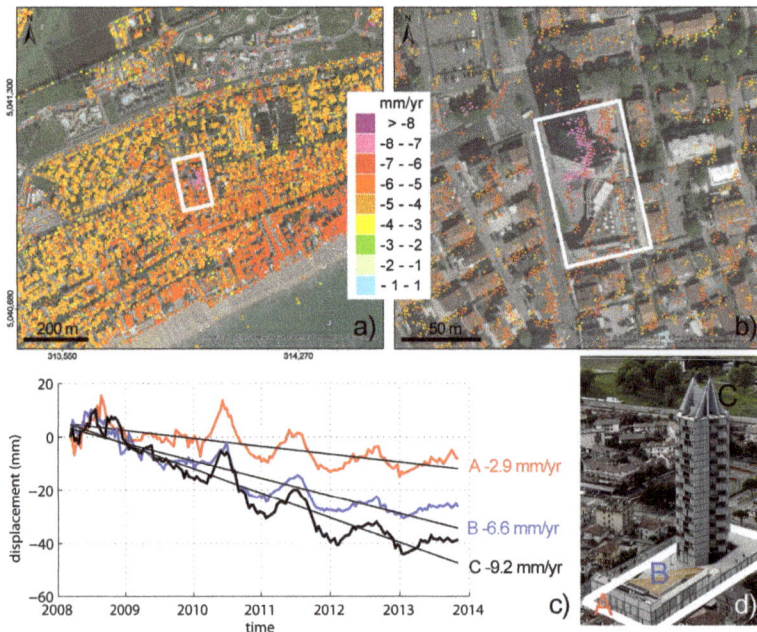

Figure 8. Example of displacements induced by new urbanization at local scale. (**a**) Mean displacement rates between March 2008 and November 2013 from TerraSAR-X interferometry at Jesolo. Negative values indicate settlement and positive indicate uplift. (**b**) Enlargement of the displacement map in the Aquileia Tower area. (**c**) Displacement history in the satellite line-of-sight direction for three scatterers (A, B, C). (**d**) Photograph of Aquileia Tower, Jesolo, which is the cause of the largest movements. Base map source: Esri, DigitalGlobe, GeoEye, Earthstar Geographics, CNES/Airbus DS, USDA, USGS, AEX, Getmapping, Aerogrid, IGN, IGP, swisstopo, and the GIS User Community.

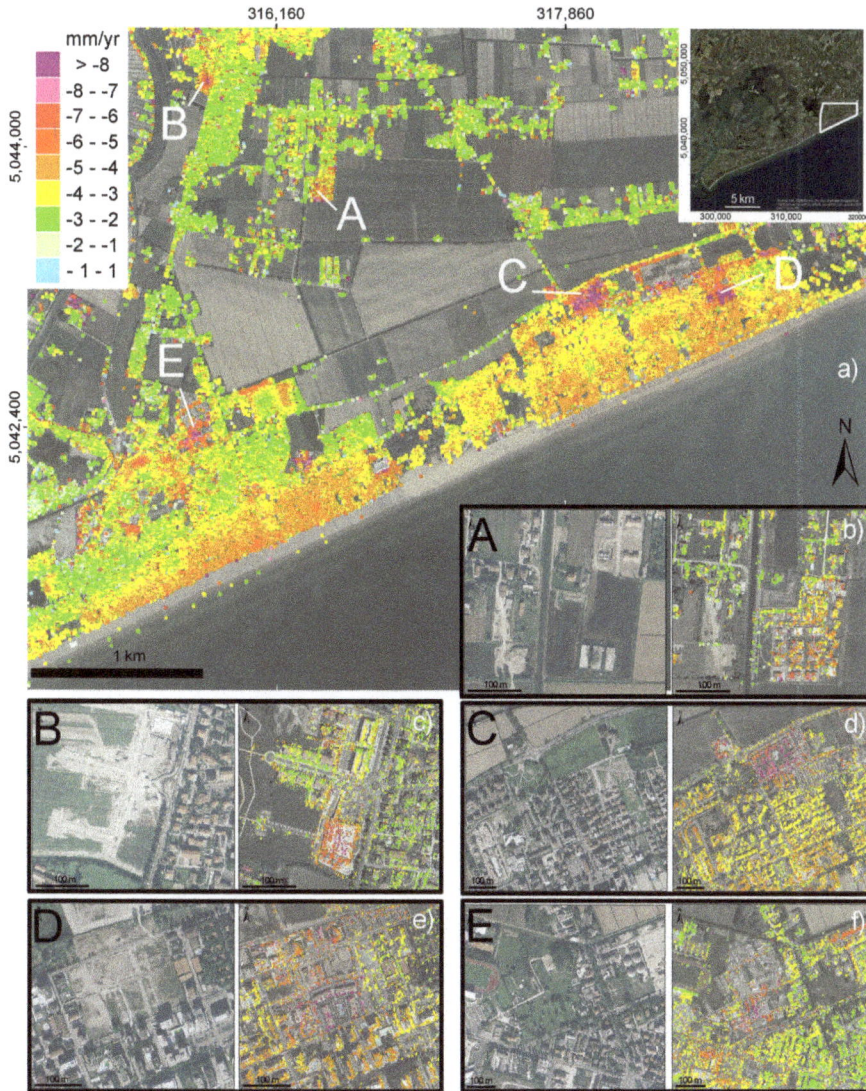

Figure 9. (a) Mean displacement rates from TerraSAR-X interferometry between March 2008 and November 2013 at the Jesolo coastland. A–E labels indicate local displacements induced by new urbanizations. (**b**–**f**) Blow ups of Areas A–E. Left and right background images are aerial photographs taken in 2006 and 2012, respectively. Base map source: http://www.pcn.minambiente.it/mattm/en/view-service-wms/.

5. Conclusive Remarks

Over the last century, eustasy and land subsidence have produced a RSLR on the Northern Adriatic coastlands ranging from centimeters to meters. RSLR seriously increased the vulnerability and the geological hazard (e.g., river flooding, riverbank stability, seawater intrusion, and coastline regression) of low-lying environments such as deltas, lagoons, wetlands, and farmlands, a large portion of which lies 3–4 m below the mean sea level. The city of Venice and its lagoon represent extraordinary environments and human heritages susceptible to loss in surface elevation relative to the mean sea level. Natural processes and anthropogenic activities drive the ground dynamics at regional and local scale and at in the long and short term. Being land subsidence the main component of the RSLR, it has been monitored since the end of the 19th century. Over the new millennium, SAR-based interferometry has been effectively used to detect land displacements in the Venice coastland, leading to a significant improvement in both the spatial coverage and the knowledge on the land subsidence process. The use of TerraSAR-X allowed further advancements in monitor ground movements of single structures as well as at regional scale with millimetric precision and metric spatial resolution. The information on ground movements, from a few hundreds of leveling benchmarks measured in the last century increased to some millions of scatterers detected by TerraSAR-X images. The very high spatial resolution and the short repeat-time interval of TerraSAR-X acquisitions enable investigating land displacements with an unprecedented level of detail, opening new perspectives to geodynamic research and environmental studies. Civil engineering sectors also benefit from TerraSAR-X for the possibility to monitor large infrastructures.

A large experience has been acquired by the authors over the last decade in the use of PSI and the result interpretation in the Venice area. The comparison between PSI outcomes provided by TerraSAR-X and other X-, C-, and L-band SAR satellites over the same area has allowed us to point out the main strength factors of this satellite:

- The short repeat-time interval of 11 days and the impressive regularity of image acquisition allow obtaining sub-millimeter accuracy just with 30 scenes, i.e., about one year only.
- The 3 m spatial resolution allows capturing the high variability of the ground movements rates, revealing details on land subsidence never been obtained with other measuring techniques both at large and local scale analysis.
- The regularity of the image acquisition allows using artificial trihedral corner reflectors to overcome the PSI limitation to detect ground movements in wetlands.

The main knowledge advances in understanding and interpretation of land subsidence in the Venice area are in the following:

- The impressive dataset consisting of more 1,900,000 PS depicts a new image of the land subsidence pattern at the Venice coastland and reveals a very high heterogeneity of the ground dynamics both at the local and the regional scales.
- Short-term local man-induced displacements have been revealed affecting the historical center of Venice. In 2008, about 25% of the city experienced some movements due to anthropogenic causes.
- Natural and man-made salt marshes are characterized by a quite wide range of subsidence rates. Displacements range from small uplifts to subsidence rates of more than 20 mm/year and land subsidence is much larger on man-made than natural salt marshes, with a significant negative correlation with the marsh age.
- The subsidence within the lagoon basin, quantified for the first time by a network of TCRs, is 3–4, 1–2 and 2–3 mm/year in the northern, central and southern portions, respectively.
- Significant settlements of a few centimeters per year have been detected at the three inlets where new structures and restoration works were carried out.
- Newly built-up areas induced local land subsidence with sinking rates up to three times higher than those characterizing the older urban areas.

In the Venice Lagoon, where only a few millimeters of further sinking would drastically enhance the environment deterioration and raise concern for the historical heritage, PSI on TerraSAR-X provides accurate data to study natural and anthropogenic land subsidence, as well as to monitor the settlement of structures and infrastructures, e.g., the mobile barrier works at the lagoon inlets. In addition, the integration of geotechnical, geological, morphological and environmental data with TerraSAR-X

Remote Sens. **2018**, *10*, 1191

interferometric products will allow vulnerability and risk analyses for management plans aimed at a sustainable coastland use.

Author Contributions: L.T., C.D.L. and P.T. calibrated and validated the interferometric solutions by CGPS, processed and analyzed the outcomes, interpreted data and wrote the paper. T.S. performed the SAR processing by GAMMA Software.

Funding: This research received no external funding.

Acknowledgments: This work has been developed in the framework of the Flagship Project RITMARE—The Italian Research for the Sea—coordinated by the Italian National Research Council and funded by the Italian Ministry of Education, University and Research within the National Research Program 2011–2013. Data courtesy: (1) Project LAN0242, "Monitoring land subsidence in Venice"; (2) Project COA0612 © DLR, "Assessing vertical movements of natural tidal landforms and anthropogenic structures at the Venice Lagoon inlets"; and (3) Project COA1800©DLR, "Ground surface dynamics in the Venice Lagoon: five years of monitoring of natural tidal landforms and anthropogenic structures by TerraSAR-X". CGPS Time series, Nevada Geodetic Laboratory (NGL). This work is a contribution to the International Geoscience Programme (IGCP) Project No. 663: "Impact, Mechanism, Monitoring of Land Subsidence in Coastal cities (IM2LSC)".

Conflicts of Interest: The authors declare no conflict of interest.

References

1. Trincardi, F.; Barbanti, A.; Bastianini, M.; Benetazzo, A.; Cavaleri, L.; Chiggiato, J.; Papa, A.; Pomaro, A.; Sclavo, M.; Tosi, L.; et al. The 1966 flooding of Venice: What time taught us for the future. *Oceanography* **2016**, *29*, 178–186. [CrossRef]

2. Carbognin, L.; Teatini, P.; Tomasin, A.; Tosi, L. Global change and relative sea level rise at Venice: What impact in term of flooding. *Clim. Dyn.* **2010**, *35*, 1055–1063. [CrossRef]

3. Carbognin, L.; Tosi, L. Interaction between Climate Changes, Eustacy and Land Subsidence in the North Adriatic Region, Italy. *Mar. Ecol.* **2002**, *23*, 38–50. [CrossRef]

4. Teatini, P.; Strozzi, T.; Tosi, L.; Wegmüller, U.; Werner, C.; Carbognin, L. Assessing short- and long-time displacements in the Venice coastland by synthetic aperture radar interferometric point target analysis. *J. Geophys. Res. Earth Surf.* **2007**, *112*. [CrossRef]

5. Tosi, L.; Teatini, P.; Strozzi, T.; Carbognin, L.; Brancolini, G.; Rizzetto, F. Ground surface dynamics in the northern Adriatic coastland over the last two decades. *Rendiconti Lincei Sci. Fisiche e Nat.* **2010**, *21*, 115–129. [CrossRef]

6. Liu, G.; Jia, G.; Nie, Y.; Li, T.; Zhang, R.; Yu, B.; Li, Z. Detecting Subsidence in Coastal Areas by Ultrashort-Baseline TCPInSAR on the Time Series of High-Resolution TerraSAR-X Images. *IEEE Trans. Geosci. Remote Sens.* **2014**, *52*, 1911–1923.

7. Yang, M.; Yang, T.; Zhang, L.; Lin, J.; Qin, X.; Liao, M. Spatio-Temporal Characterization of a Reclamation Settlement in the Shanghai Coastal Area with Time Series Analyses of X-, C-, and L-Band SAR Datasets. *Remote Sens.* **2018**, *10*, 329. [CrossRef]

8. Wu, J.; Hu, F. Monitoring Ground Subsidence along the Shanghai Maglev Zone Using TerraSAR-X Images. *IEEE Geosci. Remote Sens. Lett.* **2017**, *14*, 117–121. [CrossRef]

9. Luo, Q.; Zhou, G.; Perissin, D. Monitoring of Subsidence along Jingjin Inter-City Railway with High-Resolution TerraSAR-X MT-InSAR Analysis. *Remote Sens.* **2017**, *9*, 717. [CrossRef]

10. Putri, R.F.; Bayuaji, L.; Sumantyo, J.T.S.; Kuze, H. TerraSAR-X DinSAR for land deformation detection in Jakarta Urban area, Indonesia. *J. Urban Environ. Eng.* **2013**, *7*, 195–205. [CrossRef]

11. Le, T.S.; Chang, C.-P.; Nguyen, X.T.; Yhokha, A. TerraSAR-X Data for High-Precision Land Subsidence Monitoring: A Case Study in the Historical Centre of Hanoi, Vietnam. *Remote Sens.* **2016**, *8*, 338. [CrossRef]

12. Kim, S.W.; Dixon, T.; Amelung, F.; Wdowinski, S. A time-series deformation analysis from TerraSAR-X SAR data over New Orleans, USA. In Proceedings of the 3rd International Asia- Pacific Conference on Synthetic Aperture Radar (APSAR), Seoul, Korea, 26–30 September 2011.

13. Antonova, S.; Sudhaus, H.; Strozzi, T.; Zwieback, S.; Kääb, A.; Heim, B.; Langer, M.; Bornemann, N.; Boike, J. Thaw Subsidence of a Yedoma Landscape in Northern Siberia, Measured In Situ and Estimated from TerraSAR-X Interferometry. *Remote Sens.* **2018**, *10*, 494. [CrossRef]

14. Gambolati, G.; Teatini, P.; Ferronato, M.; Strozzi, T.; Tosi, L.; Putti, M. On the uniformity of anthropogenic Venice uplift. *Terra Nova* **2009**, *21*, 467–473. [CrossRef]

15. Strozzi, T.; Teatini, P.; Tosi, L. TerraSAR-X reveals the impact of the mobile barrier works on the Venice coastal stability. *Remote Sens. Environ.* **2009**, *113*, 2682–2688. [CrossRef]

16. Tosi, L.; Teatini, P.; Bincoletto, L.; Simonini, P.; Strozzi, T. Integrating geotechnical and interferometric SAR measurements for secondary compressibility characterization of coastal soils. *Surv. Geophys.* **2012**, *33*, 907–926. [CrossRef]

17. Strozzi, T.; Teatini, P.; Tosi, L.; Wegmüller, U.; Werner, C. Land subsidence of natural transitional environments by satellite radar interferometry on artificial reflectors. *J. Geophys. Res. Earth Surf.* **2013**, *118*, 1177–1191. [CrossRef]

18. Tosi, L.; Teatini, P.; Strozzi, T. Natural versus anthropogenic subsidence of Venice. *Sci. Rep.* **2013**, *3*, 2710. [CrossRef] [PubMed]

19. Tosi, L.; Strozzi, T.; Da Lio, C.; Teatini, P. Regional and local land subsidence at the Venice coastland by TerraSAR-X PSI. *Proc. IAHS* **2015**, *372*, 199–205. [CrossRef]

20. Da Lio, C.; Teatini, P.; Strozzi, T.; Tosi, L. Understanding land subsidence in salt marshes of the Venice Lagoon from SAR Interferometry and ground-based investigations. *Remote Sens. Environ.* **2018**, *205*, 56–70. [CrossRef]

21. Werner, C.; Wegmüller, U.; Strozzi, T.; Wiesmann, A. Interferometric Point Target Analysis for deformation mapping. In Proceedings of the Geoscience and Remote Sensing Symposium (IGARSS 2003), Toulouse, France, 21–25 July 2003; IEEE International: Piscataway, NJ, USA, 2003; Volume 7, pp. 4362–4364.

22. Wegmüller, U.; Werner, C.; Strozzi, T.; Wiesmann, A. Monitoring mining induced surface deformation. In Proceedings of the IEEE International Geoscience and Remote Sensing Symposium (IGARSS 2004), Anchorage, AK, USA, 20–24 September 2004; Volume 3, pp. 1933–1935.

23. Bock, Y.; Wdowinski, S.; Ferretti, A.; Novali, F.; Fumagalli, A. Recent subsidence of the Venice Lagoon from continuous GPS and interferometric synthetic aperture radar. *Geochem. Geophys. Geosyst.* **2012**, *13*, Q03023. [CrossRef]

24. Lan, H.; Gao, X.; Liu, H.; Yang, Z.; Li, L. Integration of TerraSAR-X and PALSAR PSI for detecting ground deformation. *Int. J. Remote Sens.* **2013**, *34*, 5393–5408. [CrossRef]

25. Ciavola, P.; Armaroli, C.; Chiggiato, J.; Valentini, A.; Deserti, M.; Perini, L.; Luciani, P. Impact of storms along the coastline of Emilia-Romagna: The morphological signature on the Ravenna coastline (Italy). *J. Coast. Res.* **2007**, *540–544*, 540–544. [CrossRef]

26. Teatini, P.; Tosi, L.; Strozzi, T.; Carbognin, L.; Cecconi, G.; Rosselli, R.; Libardo, S. Resolving land subsidence within the Venice Lagoon by persistent scatterer SAR interferometry. *Phys. Chem. Earth* **2012**, *40–41*, 72–79. [CrossRef]

27. Hanssen, R.; Feijt, A. A first quantitative evaluation of atmospheric effects on SAR interferometry. In Proceedings of the Fringe 96' Workshop on ERS SAR Interferometry, Zurich, Switzerland, 30 September–2 October 1996; ESA SP-406. ESA Publications Division: Paris, France, 1997.

28. Ketelaar, V.B.H.G. *Satellite Radar Interferometry: Subsidence Monitoring Techniques*; Springer: Berlin, Germany, 2009.

29. Rizzetto, F.; Tosi, L. Aptitude of modern salt marshes to counteract relative sea-level rise, Venice Lagoon (Italy). *Geology* **2011**, *39*, 755–758. [CrossRef]

30. Jankowski, K.L.; Törnqvist, T.E.; Fernandes, A.M. Vulnerability of Louisiana's coastland wetlands to present-day rates of relative sea-level rise. *Nat. Commun.* **2017**, *8*. [CrossRef] [PubMed]

31. Rizzetto, F.; Tosi, L.; Carbognin, L.; Bonardi, M.; Teatini, P. Geomorphic setting and related hydrogeological implications of the coastal plain south of the Venice Lagoon, Italy. *Int. Assoc. Hydrol. Sci.* **2003**, *278*, 463–470.

32. Tosi, L.; Teatini, P.; Carbognin, L.; Brancolini, G. Using high resolution data to reveal depth-dependent mechanisms that drive land subsidence: The Venice coast, Italy. *Tectonophysics* **2009**, *474*, 271–284. [CrossRef]

33. Biessy, G.; Moreau, F.; Dauteuil, O.; Bour, O. Surface deformation of an intraplate area from GPS time series. *J. Geodyn.* **2011**, *52*, 24–33. [CrossRef]

34. Teatini, P.; Tosi, L.; Strozzi, T.; Carbognin, L.; Wegmüller, U.; Rizzetto, F. Mapping regional land displacements in the Venice coastland by an integrated monitoring system. *Remote Sens. Environ.* **2005**, *98*, 403–413. [CrossRef]

35. Lv, X.; Yazici, B.; Bennett, V.; Zeghal, M.; Abdoun, T. Joint pixels InSAR for health assessment of levees in New Orleans. In Proceedings of the Geo-Congress: Stability and Performance of Slopes and Embankments III, Geotechnical Special Publication No. 231, San Diego, CA, USA, 3–6 March 2013; ASCE: Reston, VA, USA, 2013; pp. 179–288.

36. Fiaschi, S.; Tessitore, S.; Bonì, R.; Di Martire, D.; Achilli, V.; Borgstrom, S.; Ibrahim, A.; Floris, M.; Meisina, C.; Ramondini, M.; et al. From ERS-1/2 to Sentinel-1: Two decades of subsidence monitored through A-DInSAR techniques in the Ravenna area (Italy). *GISci. Remote Sens.* **2016**, *54*, 305–328. [CrossRef]

remote sensing

MDPI

Article

Landslide Monitoring Using Multi-Temporal SAR Interferometry with Advanced Persistent Scatterers Identification Methods and Super High-Spatial Resolution TerraSAR-X Images

Feng Zhao [1], Jordi J. Mallorqui [1,*], Rubén Iglesias [2], Josep A. Gili [3] and Jordi Corominas [3]

[1] CommSensLab, Department of Signal Theory and Communications (TSC), Building D3, Universitat Politècnica de Catalunya (UPC), Jordi Girona 1-3, 08034 Barcelona, Spain; feng.zhao@tsc.upc.edu

[2] Dares Technology, 08860 Castelldefels, Spain; riglesias@dares.tech

[3] Department of Civil and Environmental Engineering, Building D2, Universitat Politècnica de Catalunya (UPC), Jordi Girona 1-3, 08034 Barcelona, Spain; j.gili@upc.edu (J.A.G.); jordi.corominas@upc.edu (J.C.)

* Correspondence: mallorqui@tsc.upc.edu; Tel.: +34-93-401-72-29

Received: 25 April 2018; Accepted: 7 June 2018; Published: 11 June 2018

Abstract: Landslides are one of the most common and dangerous threats in the world that generate considerable damage and economic losses. An efficient landslide monitoring tool is the Differential Synthetic Aperture Radar Interferometry (DInSAR) or Persistent Scatter Interferometry (PSI). However, landslides are usually located in mountainous areas and the area of interest can be partially or even heavily vegetated. The inherent temporal decorrelation that dramatically reduces the number of Persistent Scatters (PSs) of the scene limits in practice the application of this technique. Thus, it is crucial to be able to detect as much PSs as possible that can be usually embedded in decorrelated areas. High resolution imagery combined with efficient pixel selection methods can make possible the application of DInSAR techniques in landslide monitoring. In this paper, different strategies to identify PS Candidates (PSCs) have been employed together with 32 super high-spatial resolution (SHR) TerraSAR-X (TSX) images, staring-spotlight mode, to monitor the Canillo landslide (Andorra). The results show that advanced PSI strategies (i.e., the temporal sub-look coherence (TSC) and temporal phase coherence (TPC) methods) are able to obtain much more valid PSs than the classical amplitude dispersion (D_A) method. In addition, the TPC method presents the best performance among all three full-resolution strategies employed. The SHR TSX data allows for obtaining much higher densities of PSs compared with a lower-spatial resolution SAR data set (Sentinel-1A in this study). Thanks to the huge amount of valid PSs obtained by the TPC method with SHR TSX images, the complexity of the structure of the Canillo landslide has been highlighted and three different slide units have been identified. The results of this study indicate that the TPC approach together with SHR SAR images can be a powerful tool to characterize displacement rates and extension of complex landslides in challenging areas.

Keywords: DInSAR; landslide monitoring; PSI; super high-spatial resolution TerraSAR-X images; pixel selection; measurement pixels' density

---

1. Introduction

Every year, with the onset of rains and snow melting, landslides represent one of the major natural threats to human life and infrastructures in natural and urbanized environments. In this context, different surveying techniques, such as inclinometers, extensometers, piezometers, jointmeters, photogrammetry, LiDAR or Global Positioning Satellite System, are typically employed to address landslide monitoring problems [1–8]. Nonetheless, these conventional techniques present several limitations. They are labor intensive, expensive and usually require skillful users for data interpretation. Moreover, they typically provide poor spatial sampling and coverage, which hinder the characterization of complex landslides. Finally, some of these techniques require the direct installation of devices over the landslide surface, which could be a complex task, sometimes impossible to fulfill, in hard-to-reach locations. During the last decade, Synthetic Aperture Radar (SAR) Differential Interferometry (DInSAR) techniques based on space-borne SAR sensors have matured to a widely used geodetic tool for the accurate monitoring of complex displacement phenomena with millimetric accuracy [9–13]. Concretely, the new generation of X-band SAR sensors, like the German TerraSAR-X and TanDEM-X satellites or the Italian constellation Cosmo-Skymed, have led to a scientific breakthrough presenting a lower revisiting time (up to few days) and an improved spatial resolution (even below the meter), compared with their predecessors ERS-1/2, ENVISAT-ASAR and RADARSAT-1 or the recently Sentinel-1, which worked at the C-band.

Despite all these clear advantages, DInSAR solutions present some limitations, especially for the X-band, over vegetated scenarios in mountainous environments, where landslides typically occur. The DInSAR technique takes advantage of a time-series of SAR images but not all pixels of the image are useful for interferometric processing. Only those pixels with enough phase quality along the whole observing period, i.e., the Persistent Scatterers (PSs), can be used as measurement points (MPs) to derive ground displacement. These PSs, which usually correspond to man-made structures (like buildings, bridges or roads), rocky areas and bare surfaces with no vegetation, are usually scarce in mountainous areas [14,15]. In addition, severe limitations arise from temporal decorrelation over vegetated areas, snow episodes typical in mountainous regions, layover and shadowing effects caused by SAR geometrical distortions, the presence of tropospheric atmospheric artifacts or when rapid displacements are faced, making the processing in such areas difficult and challenging at the same time. Finally, it must be taken into account that SAR sensors are only sensitive to the satellite-to-target component of displacement, i.e., line of sight (LOS) direction, which may notably differ from the real one. The measured displacement will be in fact a projection of the real one [9,12]. Many DInSAR, also known as Persistent Scatters Interferometry (PSI), techniques and algorithms, which share similar principles, have been developed. They have been tested in the last twenty years using many different sensors, either orbital, airborne or ground-based, and over many different scenarios, making this technique a powerful and reliable tool for monitoring any kind of ground motion episodes [14–21].

Large landslides constitute a very specific and challenging scenario for DInSAR. As they are located in mountainous areas and the displacement is usually down-slope, the landslides have to be mostly oriented east to west in order to be sensitive to the displacement if polar orbital sensors are going to be used [9,10]. Not all landslides are suitable for being monitored with orbital SAR. On the one hand, to avoid problems with phase ambiguity, the displacement rate of the landslide must be small enough, let us say a few decimetres per year (depending on the wavelength and revisiting period of the radar). In other words, the SAR interferometry is suitable for monitoring landslides "Very slow" to "Extremely slow" according to the standard landslide classifications [22,23]. In addition, foreshortening and layover can jeopardize the performance of the DInSAR processing so the selection of the proper acquisition geometry is also crucial. In order to reduce geometric distortion and, at the same time, maximize the projection of the landslide displacement to the LOS, it is advisable to observe, if possible, the landslide from behind, as it has been done in this paper. However, each case can be different from the other and so it would require a detailed analysis considering the landslide particularities and the surrounding topography [9,10,12,24]. Atmospheric artifacts, caused by both tropospheric stratification and turbulent component, can contaminate the interferometric phase and, as they can be strongly correlated with the topography, they can also be difficult to remove [25–29]. Finally, a landslide can present a quite complex behaviour with different sliding units moving at different velocity rates. A good density of PS is required in order to be able to delimit and characterize the behaviour of the different local displacements, so it would be necessary to use a PSI strategy able to select as much pixels as possible at full resolution in areas where most of the pixels will be severely decorrelated [9,10]. It is evident that the chances of detecting small and isolated PSs within decorrelated areas will arise as the resolution of the images employed increases [11,30,31].

With super high-resolution (SHR) data, the classical Gaussian scattering model used to model speckle is not always fulfilled since it is possible to find resolution cells with few scatterers [24,32]. This approach is known as partially developed speckle [33,34]. In the situation of having an isolated scatterer within the resolution cell, the value is given by the deterministic impulse response of the SAR system, i.e., by a bidimensional sinc response [24,35]. These types of scatterers typically correspond to man-made structures, outcrops, exposed rocks, etc. These objects can be exploited as opportunistic high-quality points for displacement monitoring applications. Of course, in high-resolution SAR images, it is more probable to have this situation in natural environments [11,30]. Taking into account the previous considerations, landslide monitoring will be greatly benefited by the usage of SHR data.

In this paper, 32 Staring Spotlight TerraSAR-X images (acquired from July 2014 to November 2016, with a resolution of 0.23 m in azimuth and 0.59 m in range) and three full-resolution PSI approaches (i.e., the classical amplitude dispersion [14], the temporal sub-look coherence (TSC) [36,37] and the temporal phase coherence (TPC) [38] methods) are employed to monitor a complex landslide located in El Forn de Canillo (Andorran Pyrenees). Although the advantages of the Staring Spotlight TerraSAR-X SAR data have been demonstrated by different applications such as absolute height estimation [39] and measuring rates of archaeological looting [40], the examples in terms of PSI landslide monitoring are still rare. To our knowledge, the work presented in this paper is the first attempt to study the possible benefits of SHR SAR images for landslide monitoring, especially regarding the aspects of pixel density and capability to detect PSs within decorrelated areas. At the same time, the above-mentioned three PS strategies have also been tested to determine the one most suited for this kind of scenario.

The paper is organized as follows. The landslide's geological setting and employed dataset are firstly presented in Section 2. Section 3 introduces the procedures of PSI, where the different strategies are described. Section 4 presents the landslide monitoring results with TerraSAR-X images, which are analyzed and compared with GPS measurements to evaluate their reliability. After that, in Section 5 the advantages of SHR SAR images are highlighted by the comparison of the results with those achieved with lower resolution sensors, Sentinel-1 in this case. Finally, Section 6 presents the conclusions.

2. Study Area and Dataset

2.1. Canillo Landslide

The area selected in this paper corresponds to one of the biggest and ancient landslides of the Andorran Pyrenees. It is located at El Forn de Canillo (42.5610°N, 1.6018°E) in the Principality of Andorra, which is a mountainous country between Spain and France in the Central Pyrenees, as Figure 1a shows. It is a complex structure with deposits composed of overlapped colluvial layers generated by different landslide episodes. It was firstly described by Corominas and Alonso in 1984 [41] and has been the subject of several studies where its morphology, failure mechanisms and evolution has been deeply analyzed. The hillslope of El Forn de Canillo is composed by a sequence of slides and earth-flows with a complex structure, which affects an estimated mass at around 3×10^8 m^3. In this context, different ancient sliding units were identified in 1994 by Santacana [42] (see Figure 1b). The first one corresponds to a slide originated in the area of Pla del Géspit-Costa de les Gerqueres, located in the southeast of the landslide, which reaches the foot of the hillside. A second event was originated under El Pic de Maians, reaching the height of 1540 m, and which overlaps with the previous sliding unit, closing in the Valira river valley. Finally, a third rockslide with a lower extension originated on the hillside known as La Roca del Forn, in the northeast side of the hillslope, was identified. Recent local instabilities have been identified in different locations within the landslide mass [43]. The landslide of El Forn de Canillo was originated as the result of the hillside destabilization, due to a decompression phenomenon after the removal of the Valira Glacier during the Pleistocene, after the Maximum Ice Extent. The Valira River has been progressively eroding the base of the whole mass without reaching the bedrock, and thus originating the landslide [42].

In front of some evidence of displacement (geomorphological signs of instability and some cracking in the road pavement and in a hydroelectric channel that crosses the Forn de Canillo), the authorities promoted several actions in the year 2000 for the management of their geo-hazard threats leading to the monitoring of El Forn de Canillo. Between the years 2007 and 2009, a network of geotechnical devices, including inclinometers, rod extensometers and piezometers, were installed over the landslide surface to characterize and understand the dynamics of the sliding mass. A total of 10 boreholes, reaching typically a depth between 40 and 60 m, were drilled and equipped with this instrumentation [43,44]. The readings recorded have provided evidence that, in addition to a residual movement of some millimeters per year in the main body of the slide, the most active part of the landslide corresponds to the secondary landslide of Cal Borró-Cal Ponet. This area registered a velocity up to roughly 2 cm/month between May and June 2009 when intense sudden rain events and snow melting occurred [44].

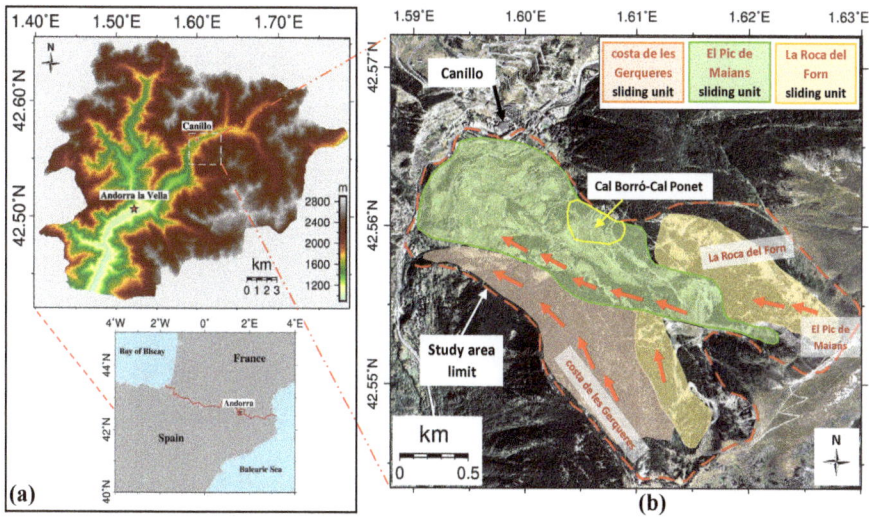

Figure 1. (**a**) location and topography of the Canillo landslide; (**b**) aerial view of the study area (Google Earth, 11 October 2017). The town of Canillo is located on the north border of the landslide. The red arrows indicate the moving directions of the ancient landslide units (modified from Santacana, 1994 [42]).

2.2. SAR Dataset

In this study, the slides' motion is monitored with 32 Staring Spotlight TerraSAR-X (TSX) Single Look Complex (SLC) SAR images. This imaging mode is the classical spotlight mode and it is able to enhance the azimuth resolution, compared with the stripmap mode, by steering the antenna in azimuth to a rotation center within the imaged scene [45]. The coverage of the SAR images is around 6.5 km in length and 3 km in width, which has been plotted in Figure 2a (yellow rectangle). The SAR image main parameters are presented in Table 1.

An amplitude image of the SAR dataset is presented in Figure 2b. As it can be seen, the SAR images' geometric distortion effects (i.e., foreshortening, shadow and layover) are not serious within the study area limit. The extended brighter areas of the image are those affected by the foreshortening and layover, due to the steepest topography. Dark areas are those affected by shadowing. This is favoured by a certain parallelism between the topography of the slope and the LOS from the satellite, thanks to its descending flight direction. The landslide is partially vegetated. Only a few strong scatterers (man-made structures, like buildings and roads, or bare rocks) are sparsely distributed within the study area limit, as is also visible in Figure 1b, thus making it challenging to monitor this landslide with conventional PSI techniques.

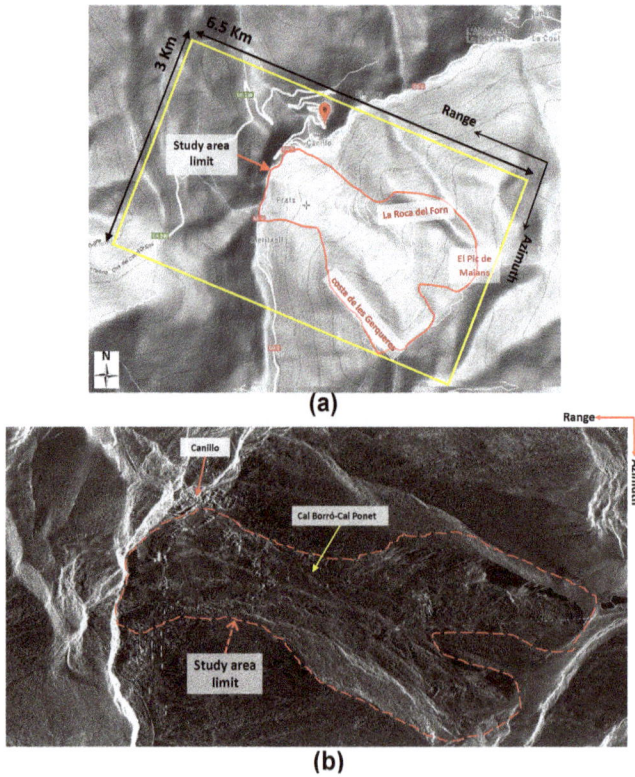

(a)

(b)

Figure 2. (a) coverage of the TerraSAR-X dataset (i.e., the yellow rectangle) displayed on a topographic map of the area (map from https://elevationmap.net); (b) amplitude of an SAR image in radar coordinates (azimuth, slant-range) acquired by the TerraSAR-X sensor in Staring Spotlight mode, and the red line illustrates the boundary of the study area limit.

Table 1. Main parameters of the employed Staring Spotlight TerraSAR-X images. Heading and LOS angles defined clockwise with respect to the north.

Parameter	Value
Acquisition Period	22 July 2014–15 November 2016
Heading Angle	189.8 (degree)
LOS Angle	279.8 (degree)
Incidence Angle	39 (degree)
Azimuth Resolution	0.23 (m)
Slant Range Resolution	0.59 (m)
Wavelength	3.1 (cm)
Revisit Cycle	11 (day)

2.3. GPS Validation Data

The Canillo landslide is monitored with the Global Positioning System (GNSS/GPS) since December 2012. Although several continuous monitoring GPS techniques exist [8], the small rate of displacements justified a discontinuous approach, with yearly field campaigns [7]. A network of 78 GPS points were established at Canillo, covering most of the landslide and the surrounding area as Figure 3 shows. Six points (blue filled triangles in Figure 3) serve as base points to check the stability of the local datum. Once per year, in October, a two day campaign is carried out covering all the control points, spread along the landslide. The GPS method has been the Real Time Kinematic (RTK), with two geodetic-level receivers (Topcon Hiper-Pro, double frequency, double constellation, (Topcon Positioning Systems Inc., Tokyo, Japan)). The final results are the point coordinates in the ETRS89 reference system (Longitude, Latitude and elevation for instance). The estimated accuracy of the resulting coordinate increments is around 1 cm in planimetry and 2 cm in elevation [7].

Three GPS campaigns fit within the study period: October 2014, October 2015 and October 2016. The six base points (E1, E2, E3, E4, E6 and G44 in Figure 3), which are on the assumed stable substrate outside the unstable area, and a total of 72 control points spread over the landslide deposits have been measured. The base points were measured in order to rule out systematic or instrumental errors and thus validate the measures carried out. The control points have been distributed throughout the landslide with the aim of providing a comprehensive overview of its behavior.

The results of the displacement observed at the reference points (points E and G44 in Figure 3), outside the landslide, are within the range of the error and therefore can be considered stable, as expected. Among the 72 GPS control points within the study area limit, 37 are selected for PSI results' validation. The correspondence between GPS points and the PSs has been made with proximity criteria but also discarding any change of geomorphological sub-unit. The difference between GPS and PSI in terms of precision, spatial resolution and temporal resolution is noticeable, but the measured displacement of these selected GPS control points can be used to examine the reliability of the PSI derived ground displacement, as it will be done in Section 4.2.

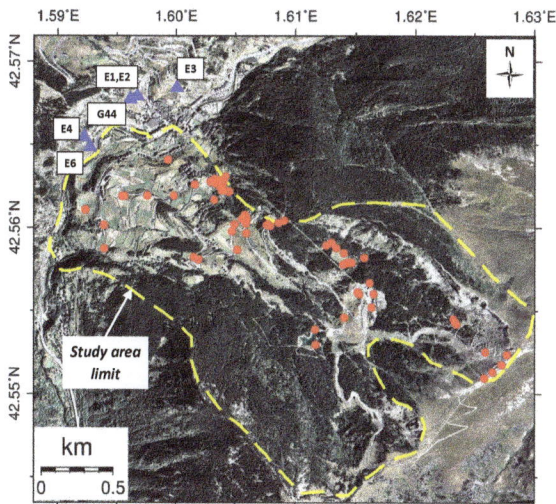

Figure 3. The locations of the GPS measurement points. The filled-in blue triangles and red circles indicate the GPS base points and control points, respectively.

3. Methodology

In this section, the different PSI strategies that will be compared in this paper are introduced. Most of the processing steps are identical for all of them, so the description will be focused on the different PS identification methods used that characterize each strategy.

3.1. Differential SAR Interferometry (DInSAR) Processing

In the conventional strip-map mode, SAR images' azimuth resolution is around half of the azimuth antenna length, which cannot be reduced arbitrarily to improve the resolution without the risk of causing range ambiguities. To overcome this limitation and achieve a higher resolution, the spotlight mode extends the illuminating time of each scatterer by sweeping the azimuth beam backward during imaging [46]. This brings a systematic Doppler centroid drift in the azimuth direction of the focused SAR images.

Prior to the DInSAR processing of the data, the particularities of Staring Spotlight acquisition mode have to be considered during the classical interferometric processing. When performing the image co-registration and common band filtering (if required), all base-banding steps have to consider the azimuth variation of the Doppler spectrum, which is different to the one of the stripmap mode and would require a deramping of the images involved. The details of how to deal with this issue can be found in [37,46]. The other steps of InSAR processing are identical to those of the stripmap case. The spotlight DInSAR processing module, able to work with sliding and staring data, has been implemented in the SUBSOFT-GUI, which is the UPC's DInSAR processing chain based on the Coherent Pixel Technique (CPT) [17,20].

In this study, in order to limit the influence of geometrical and temporal decorrelation on interferograms, we set the interferograms' temporal and spatial baseline thresholds as 365 days and 230 m, respectively. These values have allowed a good interconnection of the images and they act as upper-limits to avoid having interferograms with too long temporal or spatial baselines. The interferograms have been selected using a Delauney triangulation over the SLCs' distribution considering its acquisition time and spatial baselines with respect a master image, as shown in Figure 4. With these restrictions and with the help of an external DEM of the area with 5 m resolution provided by the Government of Andorra, a total of 80 differential interferograms have been generated from the 32 TSX images.

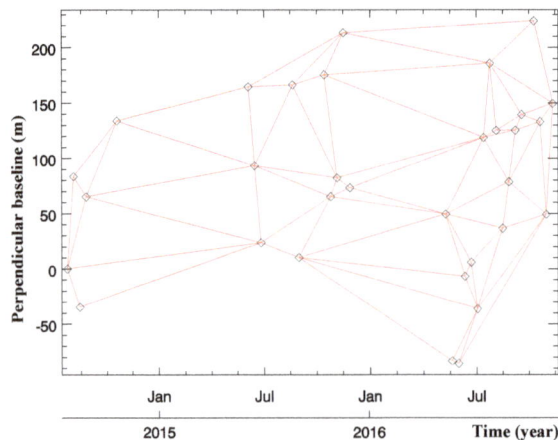

Figure 4. The spatial and temporal baseline distributions of the TerraSAR-X data generated interferograms over the study area. The black diamonds and red lines denote the SAR images and interferograms, respectively.

One of the characteristics of X-band data is that it decorrelates very fast in vegetated areas, but, at the same time, the coherent pixels are able to preserve their phase quality very well over time. In other words, if they are coherent, they keep the coherence well. The main advantage of working with high resolution data is the capability to detect small coherent features embedded in uncorrelated areas. In order to illustrate this, Figure 5 shows two coherence maps obtained from two different interferograms using a multi-look of 5 × 3 (azimuth × range). The resolution of the multi-looked interferogram is 1.15 × 1.77 m. One with a temporal baseline of 11 days and the other with 10 months. The coherence maps look very similar for both cases demonstrating the previous statement.

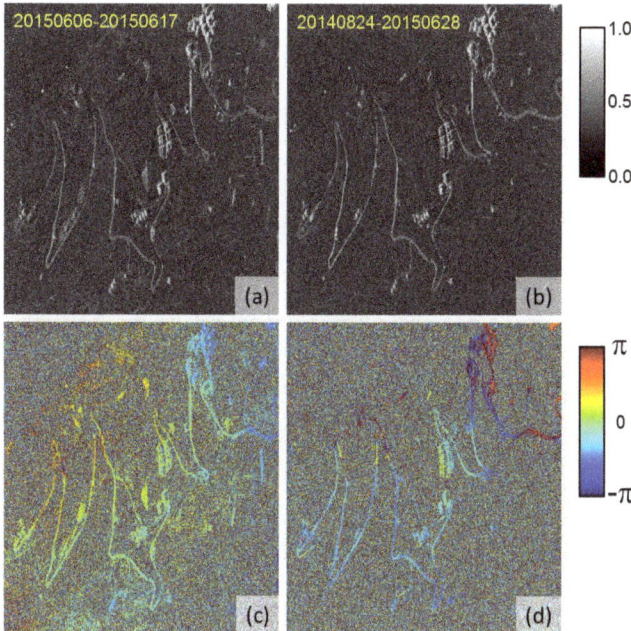

Figure 5. Coherence (**a**,**b**) and differential phase (**c**,**d**) of two interferograms with temporal baselines of 11 days (**a**,**c**) and 10 months (**b**,**d**) over the study area. Despite most of the pixels decorrelating very fast, the coherent ones are able preserve their phase quality very well along time.

3.2. Persistent Scatterers Identification

Together with the classical full-resolution pixel selection method (i.e., the amplitude dispersion (D_A) method), another two techniques (the temporal sublook coherence (TSC) and the temporal phase coherence (TPC) methods) have been used to identify pixels with high phase quality, known as PS Candidates (PSCs). As the D_A approach [14] is very well known by the PSI community, we will only introduce briefly the TSC and TPC approaches, which are two pixel selection methods developed by the authors.

3.2.1. PS Candidates Selection by Temporal Sublook Coherence (TSC)

Different from the D_A method, which selects persistent PSs by exploring pixels' amplitude stability, the TSC method intends to identify those pixels that behave like point scatterers in the spectral domain along time [36]. Any target that presents a correlated spectrum in range, azimuth and elevation along time would be identified as PS. In practice, targets usually present a nonuniform azimuth scattering pattern, worsened in the Staring Spotlight case due to the length of the synthetic aperture, and the assumption of correlated spectrum can only be applied in range. This method presents some advantages. For instance, with this approach, the radiometric calibration of the images is not necessary since amplitude plays no role in the detection and, thus, point-like scatterers that change its amplitude along time can be perfectly selected. An example of the latter case will be highly directive targets whose reflectivity has a strong dependence on the incidence angle. In addition, it was demonstrated in [36] that it is more reliable with reduced sets of images than D_A.

Before TSC estimation, two range sublooks (SL) of each SAR image have to be generated. Focused SAR images are usually tapered with a linear window (Hamming, Hanning, Kaiser, etc.) to reduce the impact of the sidelobes. In order to ensure that the two sublooks in which the spectrum will be divided present a symmetrical shape, the original spectrum has to be unweighted to flatten it. Once the range spectrum has been flattened, two sublooks are generated (each one corresponding to one half of the original spectrum) and base banded to the same central frequency to avoid any undesired linear phase term during the later spectral correlation. To reduce once again the sidelobes, each sublook is tapered with a linear window. Finally, the inverse Fourier transform is applied to get both SLs in the spatial domain. A detailed explanation of the whole process is perfectly detailed in [36]. Once the sublooks of all SAR images are obtained, the TSC of any arbitrary pixel (i, j) can be calculated with Equation (1)

$$\left|\widehat{\gamma}_{tmp}(i,j)\right| = \frac{\left|\sum\limits_{n=1}^{N_{im}} S_1(i,j,n) \cdot S_2^*(i,j,n)\right|}{\sum\limits_{n=1}^{N_{im}} |S_1(i,j,n)|^2 \cdot \sum\limits_{n=1}^{N_{im}} |S_2(i,j,n)|^2}, \tag{1}$$

where S_1 and S_2 are the pixel (i, j) corresponding complex values of the first and second sublook for the acquisition image n, and N_{im} refers to the total number of images. The sketch of the TSC estimation for a generic pixel can be represented by Figure 6.

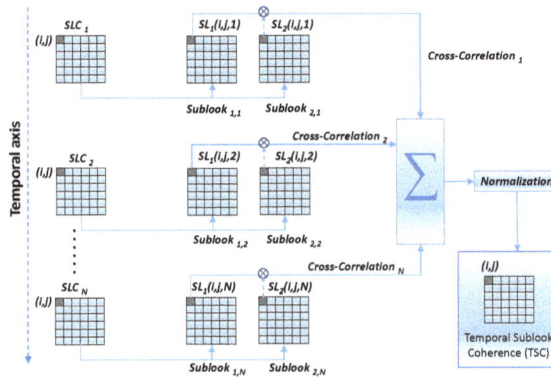

Figure 6. Sketch of the TSC estimation for a generic pixel. From left to right, the Single Look Complex (SLC) images of the dataset, the two sublooks generated from each image, coherence calculation and final TSC [37].

The temporal sublook coherence (TSC) can be regarded as the classical coherence and, similarly, pixels can be selected based on the application of a threshold. High values of TSC would be associated with point-like scatterers. Similarly to the case of classical coherence, relations between the true TSC and the expected one can be established as a function of the number of images employed, as well as the true TSC and the pixel phase standard deviation [36,37]. These relations help to perform the pixel selection based on a phase standard deviation threshold, allowing for using a criterion independent on the number of images. From the phase standard deviation, the corresponding TPC threshold can be calculated. The selected pixels can then be treated as PSs and processed by the DInSAR algorithm to derive the displacement maps and time-series.

3.2.2. PS Candidates Selection by Temporal Phase Coherence (TPC)

After removing the topographic term using an external DEM, the phase of a differential interferogram can be expressed as Equation (2)

$$\psi = \psi_{def} + \psi_{atm} + \psi_{orb} + \psi_{\zeta_{DEM}} + \psi_{noise}, \tag{2}$$

where ψ_{def}, ψ_{atm} and ψ_{orb} denote the phase terms introduced by displacement along the LOS direction, atmospheric artifacts (atmospheric phase screen, APS) and SAR satellite orbit indeterminations. $\psi_{\zeta_{DEM}}$ is the residual phase due to the DEM error, and ψ_{noise} is the noise phase term. This latter term can be assumed to present a random behaviour in the neighbourhood of a given pixel while the other can be assumed to be deterministic. Thus, the noise phase term can be used as a metric of pixel's phase quality. The temporal phase coherence (TPC) can be used to evaluate the quality of a pixel from the behaviour of this phase noise along the stack of interferograms. TPC can be estimated based on ψ_{noise} from all generated interferograms, as Equation (3) shows

$$\gamma_{TPC} = \frac{1}{M} \cdot \left| \sum_{i=1}^{M} e^{j \cdot \psi_{noise,i}} \right|, \tag{3}$$

where M is the number of interferograms and $\psi_{noise,i}$ is the noise phase term of the ith interferogram.

To obtain for each interferogram the noise phase term of a pixel, it is necessary to estimate the deterministic terms. In order to do that, the neighbouring pixels will be used assuming, in theory, a spatial low-pass behaviour of all deterministic terms in the vicinity of the pixel whose TPC is being estimated, a.k.a the central pixel. The phase of the neighbouring pixels is estimated by averaging their complex values, but excluding the central pixel, and then calculating the argument of this complex number. With this approach, similarly to the classical multi-looking in interferometry, the pixels' amplitude is used to give more significance to those pixels with higher amplitude in front of those with lower values that, in principle, can be expected to be noisier and less reliable. The first three terms of Equation (2) can be assumed to be spatially low-pass. Indeed, APS, orbital residues and the phase offset of the interferogram perfectly fulfill this condition while, for the deformation, it would be an acceptable approximation. Then, subtracting the neighbouring phase from the central phase gives Equation (4)

$$\psi^{central} - \psi^{neigh} \equiv \psi^{dif} = \psi_{\zeta_{DEM}}^{dif} + \psi_{noise}^{dif}, \tag{4}$$

where $\psi_{\zeta_{DEM}}^{dif} = \psi_{\zeta_{DEM}}^{central} - \psi_{\zeta_{DEM}}^{neigh}$ and $\psi_{noise}^{dif} = \psi_{noise}^{central} - \psi_{noise}^{neigh}$. Thus, the terms have been grouped in deterministic along the interferometric stack, $\psi_{\zeta_{DEM}}^{dif}$, and random, ψ_{noise}^{dif}. As (4) shows, the estimation of the noise phase of the central pixel, i.e., $\psi_{noise}^{central}$, would be affected by the deterministic terms. The averaging would reduce the noise term of the neighbouring pixels, ψ_{noise}^{neigh}. Thus, we can assume than $\psi_{noise}^{central} \approx \psi_{noise}^{dif}$. Thus, by subtracting the deterministic term $\psi_{\zeta_{DEM}}^{dif}$ from ψ^{dif}, the noise phase of the central pixel can be estimated. In the practical implementation, all phase operations are obviously done in the complex domain.

The phases due to DEM errors ($\varepsilon_{DEM}^{central}$ and $\varepsilon_{DEM}^{neigh}$) of the central and neighboring pixels can be rewritten as Equations (5) and (6), respectively:

$$\psi_{\zeta_{DEM}}^{central} = \frac{4\pi}{\lambda} \cdot \frac{B_n}{R_0 \cdot sin(\vartheta_0)} \cdot \varepsilon_{DEM}^{central}, \tag{5}$$

$$\psi_{\zeta_{DEM}}^{neigh} = \frac{4\pi}{\lambda} \cdot \frac{B_n}{R_0 \cdot sin(\vartheta_0)} \cdot \varepsilon_{DEM}^{neigh}, \tag{6}$$

where λ, B_n, R_0 and ϑ_0 are the wavelength, the perpendicular baseline, the absolute range distance in the LOS direction between the sensor and the target and the incidence angle, respectively. Then, we can derive $\psi_{\zeta_{DEM}}^{dif}$ as (7)

$$\psi_{\zeta_{DEM}}^{dif} = \frac{4\pi}{\lambda} \cdot \frac{B_n}{R_0 \cdot sin(\vartheta_0)} \cdot \triangle\varepsilon_{DEM}, \tag{7}$$

where $\triangle\varepsilon_{DEM} = \varepsilon_{DEM}^{central} - \varepsilon_{DEM}^{neigh}$ is the difference of DEM errors between the central and the averaged error of the neighboring pixels. We use Equation (8) to estimate each pixel's $\triangle\varepsilon_{DEM}$ and then the $\psi_{\zeta_{DEM}}^{dif}$ is calculated by Equation (7):

$$\arg\max_{\triangle\varepsilon_{DEM}} \left\{ \gamma_{TPC} = \frac{1}{M} \cdot | \sum_{i=1}^{M} e^{j \cdot \psi_i^{dif} - j \cdot \psi_{\zeta_{DEM,i}}^{dif}} | \right\}. \tag{8}$$

Until now, $\psi_{\zeta_{DEM}}^{dif}$ has been estimated and then $\psi_{noise}^{central}$ can be derived by Equation (4) under the assumption that $\psi_{noise}^{central} \approx \psi_{noise}^{dif}$. All pixels' noise phase terms of all the interferograms can be estimated by this way and then the TPC can be calculated by Equation (3).

TPC provides a temporal coherence of each pixel and fixing a threshold can perform the identification of PSCs. As in the case of classical coherence or the TSC, a relationship between TPC and the phase standard can be established in order to select a threshold independent on the number of images and interferograms. The derivation of these relations has been discussed in detail in [38].

3.3. Linear and Nonlinear (Time-Series) Displacement Estimation

The linear and nonlinear displacement terms and the DEM error can be estimated by using UPC's ground motion detection software SUBSOFT-GUI (UPC, Barcelona, Spain). SUBSOFT-GUI is a user-friendly software package for PSI processing. It allows for performing all required steps, starting from the image co-registration, differential interferograms generation and filtering, pixel selection and deformation time-series extraction. The software uses a Graphical User Interface (GUI) and most of the steps have been automatized, which facilitates the processing of any dataset. The detailed procedures of the linear and nonlinear blocks in SUBSOFT-GUI can be found by referring to [17,20]. Three independent processes, based on the same set of differential interferograms but with three different PS selection strategies (D_A, TSC and TPC approaches), have been carried out to compare the performance of each pixel selection technique under similar conditions. For each strategy, the measured parameter can be related with a phase standard deviation as shown in Figure 7.

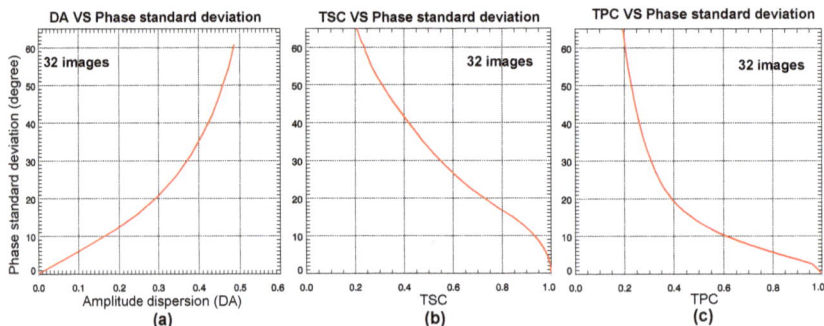

Figure 7. Standard deviation of the interferometric phase as a function of (**a**) D_A, (**b**) TSC and (**c**) TPC for the 32 images set.

The comparison of the different strategies is always a difficult task as there are many parameters that can be adjusted. In this case, the key point that makes the difference is the capability of the different strategies to select PSs. The larger the number, the better performance of the PSI processing, as it allows a better connection of the different areas and reduces the chances of having isolated clusters of PSs. It is also true that the three processes could have been optimized with a fine-tuning of the processing parameters, but, in practice, it is expected that the possible small variations on the final results would not be enough to modify the conclusions.

3.4. Atmospheric Artefacts

InSAR observations are usually plagued by propagation delays, which are also known as atmosphere phase screen (APS). As the atmosphere properties (temperature, pressure, and relative humidity that set the refractive index) between radar platform and the ground targets vary spatially and temporally, the phase delays vary from one day to another. For microwaves, it is well known that propagation delays have two major sources: tropospheric terms and ionosphere effects. At X-band, ionosphere is almost invisible and so the only significant source is the troposphere [26,47]. The atmospheric propagation delay in interferograms can be categorized into vertical stratification and turbulence mixing [26]. While the latter can be compensated, thanks to its random behaviour in time and correlated behaviour in space, with a set of temporal and spatial filters during data processing [14,18,20], the former can be much more difficult. Stratification is prone to occurring in areas with steep topography and the APS appears to be strongly correlated with the elevation. If not properly compensated, APS can be misinterpreted as topography or displacement. Different strategies can be used to characterize and compensate the stratified APS, for instance with models following a linear or quadratic phase-elevation relationship [25,27–29].

The time of the pass of the satellite for the TSX data acquisitions was early in the morning, around 6:03 a.m. UTC (8:03 a.m. in local summer time and 7:03 a.m. in local winter time). At this time of the day, the atmosphere is very stable, compared with the strong fluctuations that can be observed during the day, and stratified APS has not been observed in the dataset.

4. Results and Discussion

4.1. Line-of-Sight (LOS) Monitoring Results

The LOS displacement rate maps derived by the three methods (i.e., the D_A, TSC and TPC) are shown in Figure 8a–c, respectively. To make a fair comparison, the pixel selection thresholds for all the three methods were established based on a phase standard deviation of 15°. Using the plots shown in Figure 7, the corresponding thresholds for each strategy can be selected. Similar displacement trends have been detected by all of them, and the maximum displacement velocity reaches up to −3.5 cm/year (the minus sign means movement away from the satellite, i.e., downslope motion due to the landslide orientation). Within the landslide limits, there are mainly three large displacement subareas (indicated by the red rectangles in Figure 8a–c), located at the El Pic de Maians (subarea A), costa de les Gerqueres (subarea B) and Cal Borró-Cal Ponet (subarea C), respectively. These three subareas' locations and displacement patterns are coincident with the monitoring results obtained with another dataset in 2011 [37]. The dataset consisted on Sliding-spotlight TerraSAR and GB-SAR images, and data from inclinometers deployed in the landslide, all acquired from October 2010 until October 2011. Previous results have confirmed that the location and evolution of the landslide body have not changed significantly during the recent years. This fact is in good agreement with the geological expectations.

Among the three pixel selection methods, D_A and TSC select pixels that behave as point scatterers while TPC can work on both point and distributed scatterers (DSs). Since there are many DS pixels (e.g., the road) in the study area, TPC obtains a much higher density of measurement pixels (MP) than D_A and TSC approaches.

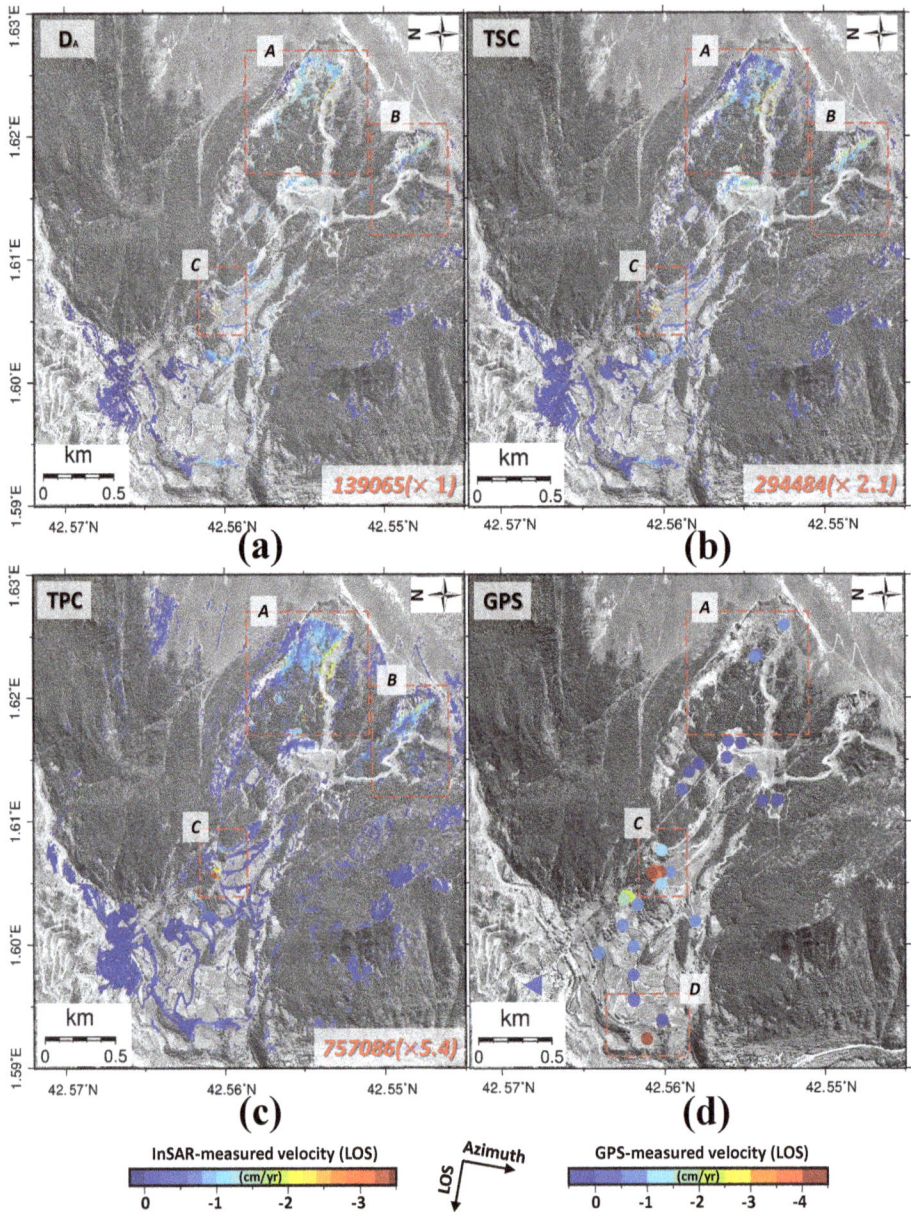

Figure 8. LOS displacement velocity maps derived by (**a**) D_A, (**b**) TSC, (**c**) TPC and (**d**) GPS approaches, respectively. The filled blue triangle in (**d**), i.e., E1, indicates the location of the GPS base point. GPS displacements have been projected to LOS. The red rectangles highlight the areas zoomed in Figure 9. The red numbers at the right bottom corner of (**a**–**c**) represent the amount of valid pixels obtained by each method.

Notice in Figure 8 how well the TPC method has identified those pixels along the downhill road, while the other two have just selected a reduced set of them. At the same time, the TSC method obtains more PSs than D_A. This can be explained by the fact that the D_A method is very sensitive to the amplitude changes that highly directive scatterers produce when the local incidence angle changes from image to image. Specifically, the number of PSs obtained by TPC method is 757,086, the counterparts of TSC and D_A methods are 139,065 and 294,484, respectively. The improvement of the TPC and TSC methods on D_A is around ×5.4 and ×2.1, respectively. The TPC method thus has the best performance in terms of PSs' density.

To better analyse the details of the landslide, the three subareas' monitoring results have been enlarged and plotted in Figure 9. From column A (results of the subarea A), we can find that the displacement velocities obtained by D_A (−1.3 cm/yr) are greater then those of TSC and TPC (−0.6 cm/yr) at the locations highlighted by the red ellipses. Similar differences can be observed between the TPC derived results and the other two methods within the subarea C (along the downhill road). These displacement velocities' differences are mainly caused by the sparsity of selected pixels that reduces the number of connections of D_A (Figure 9a,c) or TSC (Figure 9f) during the linear displacement estimation. Different areas interconnected by low-quality links can lead to small offsets in the velocity results. The sparser the local connections, the more easily the estimated displacement can be affected by nearby lower quality pixels and APS. Therefore, the high estimated displacement velocities in Figure 9a,f are mostly due to the low densities of PSs within these local areas.

As Figure 9g–i shows, thanks to the super high resolution (SHR) of the images and TPC's good performance on pixel selection, the displacement details of the different landslide units are well detected. For instance, more pixels have been selected along the narrow paths (around 1 m in width), as highlighted by red ellipses in Figure 9i. Benefiting from this high density of PSs, the displacement boundaries (illustrated by the yellow dashed lines in Figure 9i) can be clearly determined by the TPC approach in subarea C. These boundaries can hardly be seen from the results of the other two methods, as shown in Figure 9c,f.

Besides the displacement results, PSI techniques can also obtain the DEM error of the selected pixels with respect to the reference DEM used. The inclusion of the retrieved DEM error on the geocoding of the final results largely improves the geolocation quality of the displacement maps. Figure 10 shows some interesting examples that illustrate the capabilities of SHR TSX data to retrieve the vertical distribution of scatterers in manmade structures. The examples shown have been obtained from the TPC processing. Figure 10a shows a communications tower located in Canillo. The vertical distribution of scatterers perfectly follows the tower's structure as the picture validates. It is also interesting, looking at the GoogleEarth image, to compare the distribution of scatterers with the shadow of the tower projected over ground. Figure 10b and c show a couple of chairlifts from the Grandvalira ski station. Once again, the vertical distribution of scatterers perfectly follows the metallic structure, as the pictures and projected shadows demonstrate. Finally, Figure 10d shows a couple of high voltage towers. The good performance of the vertical location of the scatterers, thanks to the inclusion of the calculated DEM error on the geocoding process, can also be used as proof of the reliability of the displacement velocity maps obtained. Both velocity and DEM error have been calculated simultaneously when adjusting the linear model to the interferometric data [17,20].

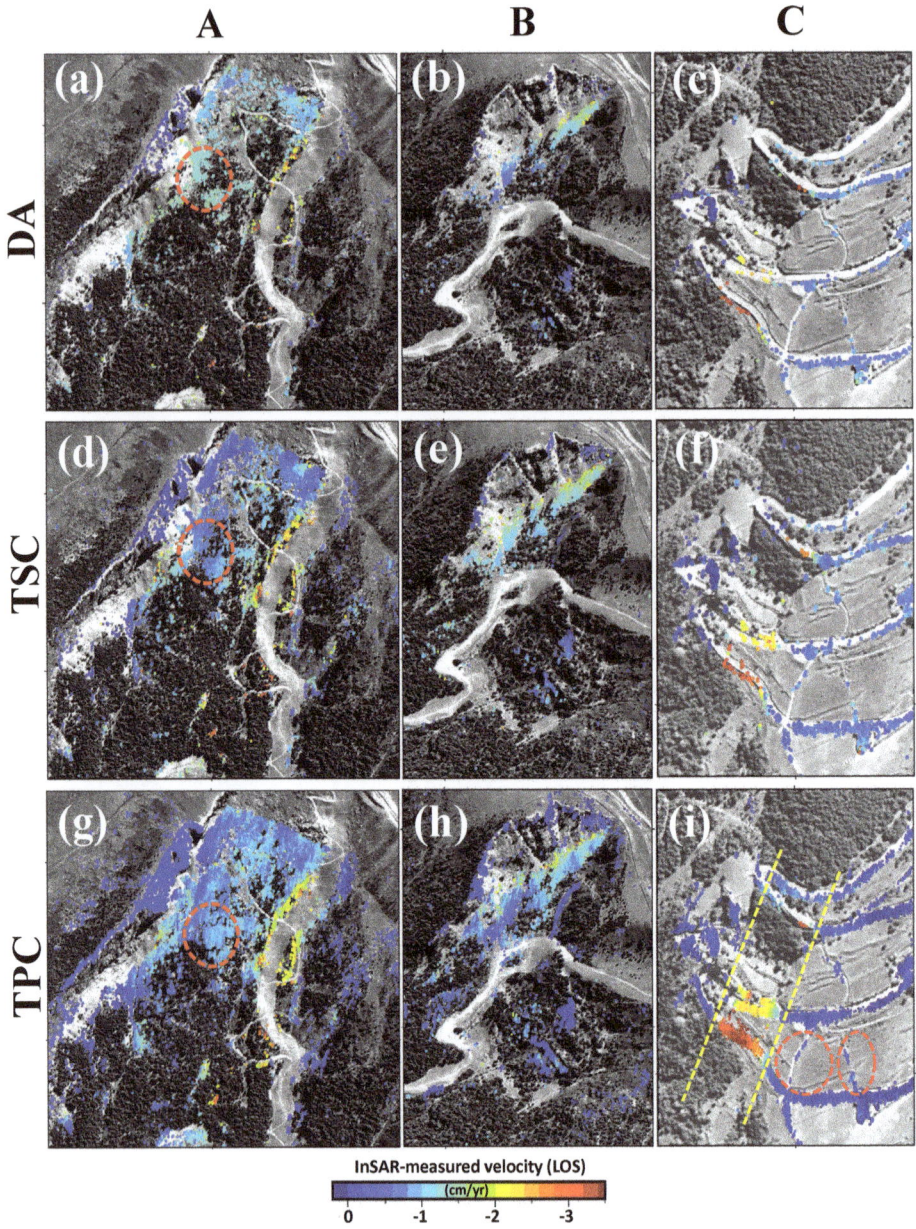

Figure 9. The close-up of the three subareas limited by red rectangles in Figure 8a–c. (**a**–**c**) are the results of the D_A method, (**d**–**f**) obtained by the TSC method and (**g**–**i**) obtained by the TPC method. Red ellipses highlight areas commented in Section 4.1. Yellow dashed lines highlight the edges of the slide.

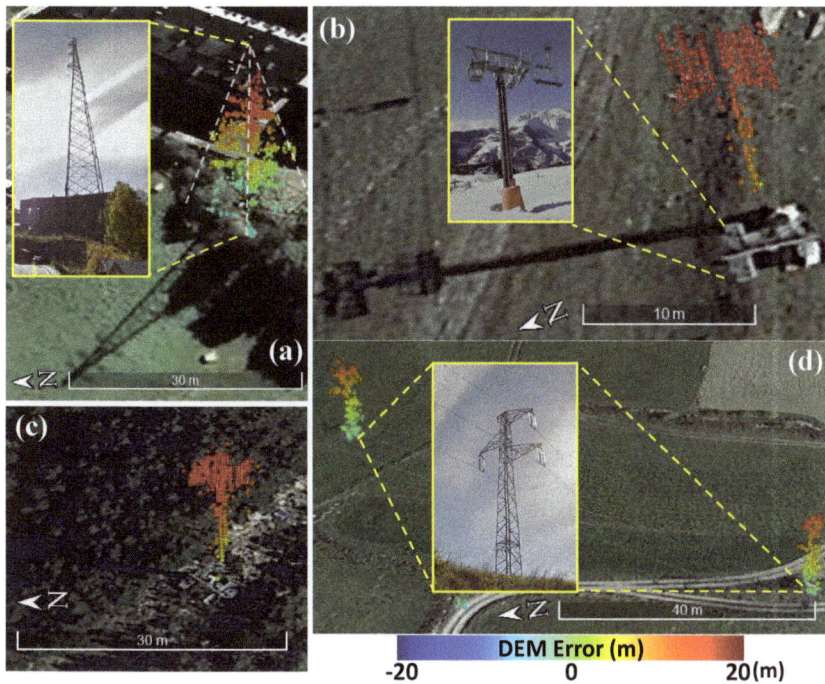

Figure 10. SHR TerraSAR-X data derived DEM errors at the locations of some manmade structures in the study area by the TPC method. (**a**) communications tower, (**b**,**c**) chairlifts towers and (**d**) high voltage towers. PSs have been geocoded over a GoogleEarth image using the retrieved DEM error.

4.2. Comparison with GPS Measurements

The displacement velocities of the 37 GPS control points introduced in Section 2.3 have been projected to the LOS direction [48,49] to compare them with the DInSAR results, as shown in Figure 8d. In subarea A of Figure 8d, a small displacement with a velocity around −1 cm/yr has been detected. In the subarea C, significant movement with velocity around −4 cm/yr has been monitored by the GPS. In the subareas A and C, the GPS and PSI measured displacement velocities are consistent with each other. Unfortunately, no GPS points were available in the subarea B for comparison. On the contrary, large displacements have been recorded by the GPS within the subarea D (highlighted by the red rectangle in Figure 8d), where there are no counterpart PSI pixels in its near vicinity. However, the further neighboring PSI pixels present LOS velocities about −1.5 cm/yr, providing evidence of the agreement of the GPS and PSI results also in this subarea.

To summarize the comparison, a scatter plot with the GPS and PSI derived displacements is shown in Figure 11. In this plot, the PSI displacements are estimated by averaging those of the neighbouring pixels of the related GPS measurement point (less than 50 m apart). In addition, they have been determined from the displacement time-series taking the overall two year displacement from October 2014 to October 2016, as the GPS date campaigns. As Figure 11 reveals, the GPS and PSI displacements follow the same trends and present a correlation coefficient of $R^2 = 0.90$. For GPS measurement points with noticeable displacement (highlighted by the red ellipse in Figure 11), their surrounding PSI pixels show large displacements as well. Meanwhile, for those stable GPS measurement points (limited by the blue rectangle), with displacements between −2 to 2 cm, their corresponding PSI displacements are also within this range.

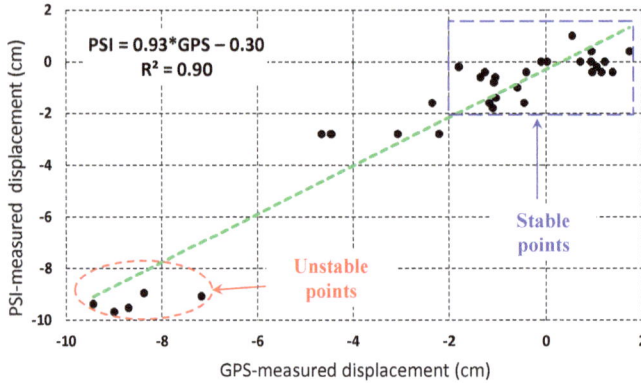

Figure 11. Comparison of PSI and GPS derived displacements (October 2014 to October 2016).

4.3. Down-Slope (DSL) Direction Displacement Monitoring Result

The ground motion derived by DInSAR is along the LOS direction, but it is usually projected to the down-slope (DSL) direction to better interpret the landslide displacement. The detailed LOS to DSL direction projection method can be found by referring to [12,24]. As it is out the scope of this paper, we do not describe it here. We projected the TPC method's ground displacement velocities to the DSL direction, and the result is shown by Figure 12. It has to be noted that, when doing the projection, only those PSs with projection factors smaller than 3 have been preserved to avoid artificially amplifying displacement values and noise when the slope is gentle. Thanks to the relative orientation of the landslide with respect the satellite path, most of the projection factors within this study area are small. Thus, the majority of PSs have been preserved, and the displacement patterns along the LOS and DSL directions are similar (e.g., the neighboring area of P1). Except for a small set of pixels nearby point P4 in Figure 12, the displacement velocities of the previous three displacement subareas (in Figure 8c) have not been heavily amplified via the projection.

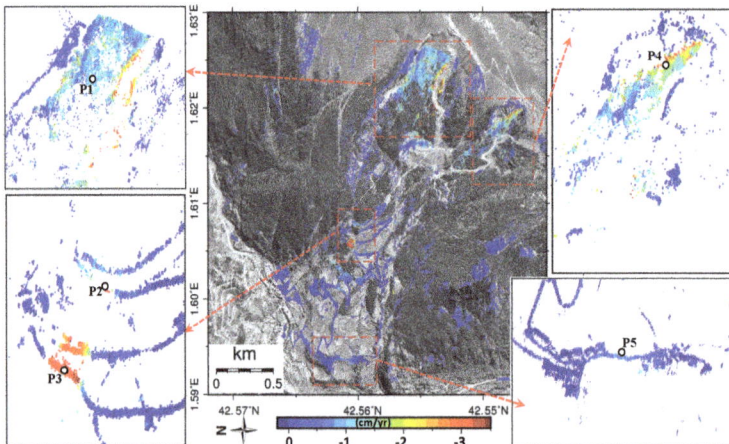

Figure 12. Down-slope displacement velocity map derived by the TPC method. Estimated displacement velocities within subareas A, B, C and D in Figure 8 have been enlarged for a better visualization with a white background. The locations of points P1–P5 in the subareas, which are further analyzed in the text, have also been indicated.

Besides the subareas A, B and C in Figure 8, in Figure 12, we have highlighted another subarea, which is located at the foot of the hill. In this subarea, noticeable displacement has been identified at the location of P5, which may be caused by the extrusion of the landslide main body moving towards the downhill direction.

4.4. PSI Time-Series

To investigate the temporal evolution of the Canillo landslide, the DSL time-series displacement results obtained by the TPC method at two different PSs (P2 and P3 in Figure 12) have been plotted in Figure 13. The displacements observed for both PSs are exhibiting considerable nonlinear components, presenting some acceleration and deceleration periods within each year. From the two PSs' 2016 displacement time-series (Figure 13b,d), we can find that the stable periods start at the beginning of July and end at the middle of August. These periods are coincident with the trend of Canillo averaged monthly precipitation, where the lowest precipitation is in July with an average of 79 mm, as Figure 13e shows. This indicates that the movements of the landslide have some seasonal patterns, which are correlated with the amount of precipitation.

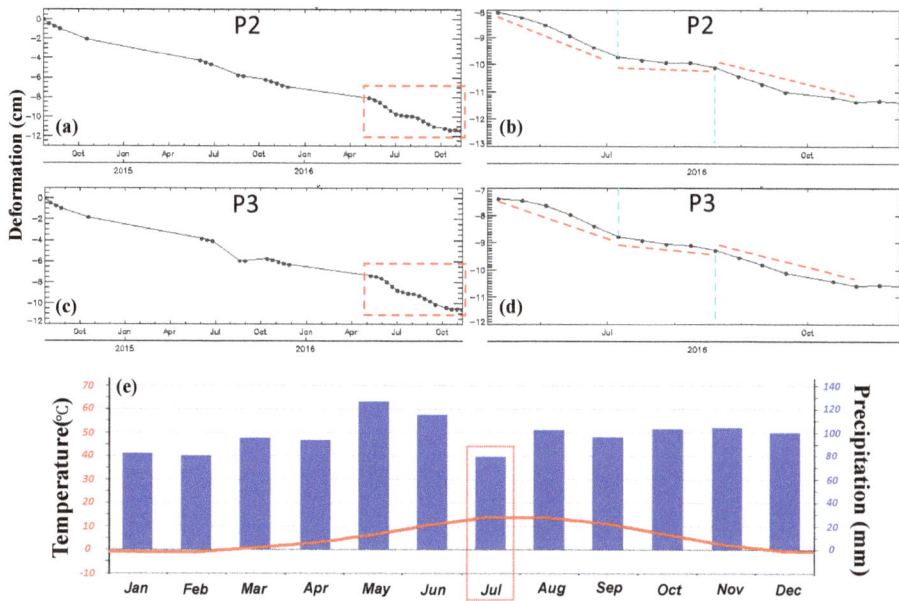

Figure 13. TPC method derived down-slope time-series displacement of P2 and P3, Figure 12. (**a,c**) cover the period 22 July 2014–15 November 2016 whereas (**b,d**) are a close-up of the dashed red rectangles inside (**a,c**), covering the period May 2016–November 2016 approximately. The red lines indicate the different deformation trends while the vertical blue ones the location of trend changes; (**e**) is the averaged monthly temperature (red line) and precipitation (blue bars) of Canillo (CLIMATE-DATA.ORG, https://en.climate-data.org/location/13728/); July has been highlighted with a red rectangle.

5. Comparison with Low-Resolution Data

Sentinel-1A data of the study area have been processed with D_A and TPC methods to highlight the advantages of the SHR data in regional-scale landslide monitoring. TSC has not been included as it provides similar results than TPC. Sentinel-1A images have resolutions of 14 and 2.5 m in azimuth and range directions, respectively. Fourteen Sentinel-1A SAR images acquired from the 11 May 2016 to 19 November 2016 have been employed to generate 33 interferograms. In the pixel selection step, the same phase standard deviation threshold (15°) as with TSX data has been used. The displacement velocity maps obtained using the two PSI strategies, D_A and TPC, are shown in Figure 14.

Similarly to the case of TSX data, TPC is able to obtain much more PSs than D_A (×4.0), and the displacement trends derived are similar to those of TSX but less detailed. For both methods, their PSs' densities have decreased dramatically compared with the TSX data case. Specifically, for D_A and TPC methods, the numbers of PSs are ×146 and ×197 less w.r.t. that of the TSX case. This significant reduction of the PSs' density is mainly due to two reasons that are closely related. In addition to the logical reduction due to the coarse resolution of Sentinel-1A data, there is also the fact that many small PSs surrounded by decorrelated pixels that were detected with SHR data are now mixed all together due to the worse resolution and, consequently, not detected.

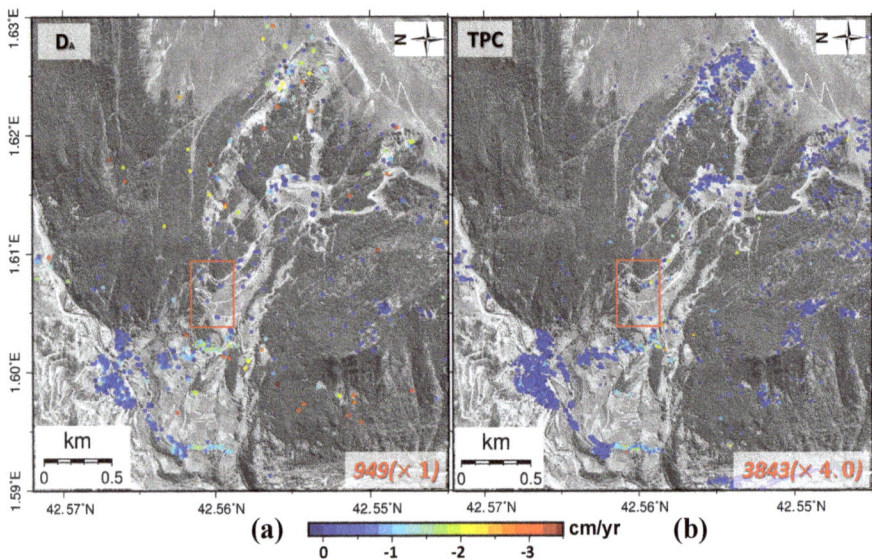

Figure 14. The LOS ground displacement velocity maps derived by (a) D_A and (b) TPC methods with Sentinel-1A SAR images.

The Sentinel-1A data monitoring results of the Cal Borró-Cal Ponet section (subarea C in Figure 8 and where the strongest displacement has been detected) have been highlighted with a red rectangle in Figure 14. In this subsection, the displacement clearly detected with TSX data does not appear in the Sentinel-1A results with none of the pixel selection methods. A detailed view of Cal Borró is shown in Figure 15. Similarly, Figure 14 shows no noticeable displacement in any of the other two subareas (subareas A and B in Figure 8c). However, the small displacement at the base of the landslide is detected with both PSI strategies and agrees with the results of SHR data. Moreover, the sparse distribution of PSs, which can be poorly interconnected, allows the appearance of some outliers, pixels whose velocities are clearly erroneous, scattered along the image. The presence of outliers is more noticeable on the D_A results in the form of isolated red points, those with the highest velocities.

To conclude, for regional-scale landslide monitoring, the TSX SHR SAR images have the advantage of obtaining more detailed monitoring results with better reliability compared with those of lower resolution sensors.

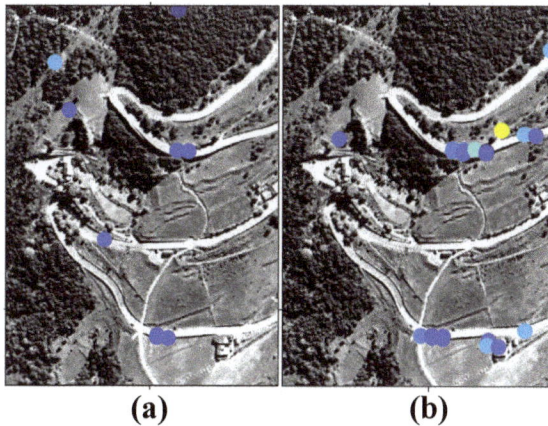

Figure 15. The LOS ground displacement velocity maps, Sentinel-1A SAR images. Enlargement of the red rectangles inside Figure 14. (**a**) D_A method; (**b**) TPC method. The color scale for the displacements is the same as that in Figure 14.

6. Conclusions

In this paper, the ability of super high-spatial resolution (SHR) SAR images together with advanced PS selection strategies for regional-scale landslide monitoring in a challenging area has been studied. Thirty-two SHR TerraSAR-X (TSX) images (July 2014 to October 2016), with resolutions of 0.23 and 0.59 m in azimuth and range directions, have been employed to monitor the Canillo landslide (Andorra) by using PSI techniques with three different pixel selection methods.

This study has demonstrated that improving the number of high-quality pixels for its later PSI processing results of crucial importance in landslide monitoring in natural environments. Under the application point of view, to the authors' knowledge, it is one of the first times when such a high density of PS has been obtained in mountainous areas. SHR SAR data jointly with advanced full-resolution PSI strategies allow the achievement of a more robust network of PS (improving the linear estimation without propagation errors and the reliable estimation of APS) and thus favors the reliable estimation of displacement maps in a major number of points inside a landslide. This is a general conclusion that does not depend on the landslide. A different issue is if the particularities of a given landslide (orientation, type of vegetation coverage, local topography, snow episodes, etc.) made it unsuitable for PSI monitoring. Similarly, well-established interferometric techniques for DEM generation fail on forested areas. It is clear that the particular characteristics of the scenario may limit the application of the technique.

The landslide's overall displacement patterns observed by the three methods in El Forn de Canillo are similar. Three main subareas with noticeable displacement have been detected, which are similar to those obtained in previous PSI monitoring results. This indicates that the evolution of the landslide main body did not change significantly during recent years. The PSI measured displacement rates have been compared with GPS measurements of the same period, and they are both in good agreement. It is worth highlighting the higher information/resolution of the PSI techniques in comparison with the GPS low point density, as it can be appreciated in Figure 8. Although already highlighted in the literature, in the Canillo Landslide, the PSI capability for detecting incipient movements in zones not previously surveyed by the geological engineering specialists has been verified (as the subarea costa de les Gerqueres, red rectangle B in Figure 8).The displacement time-series of two significant pixels are characterized by considerable nonlinear components, exhibiting some acceleration and stabilization periods within each year. These periods can be correlated with the averaged monthly precipitation

Remote Sens. **2018**, *10*, 921

amounts, revealing the important influence of rain/snow melting episodes on the development of this landslide.

SHR SAR data initially designed for improving monitoring capabilities over man-made structures, such as buildings, bridges, railways or highways, have also demonstrated an outstanding performance over natural reflectors, such as outcrops or exposed rocks with the proper PSs selection strategy. Indeed, this improvement in terms of density allows a better characterization and delineation of complex landslides. Among the three full-resolution PSC selection strategies, the advanced ones (i.e., the TSC and TPC) are able to obtain much more valid PSs than the classical D_A method. The TPC method presents the best performance. Thanks to these huge amount of PSs, the displacement details of the regional-scale landslides can be characterized with better precision when combining the TPC method with SHR TSX data. Compared with the lower-spatial resolution SAR data (Sentinel-1A in this study), SHR data can better characterize the landslide, particularly if the different subareas are small.

The results of this work show that the density of valid PSs can be greatly enhanced by using the TPC method together with SHR SAR images. Thus, they can together be used as a powerful tool for detailed landslide monitoring in difficult areas.

Author Contributions: F.Z., J.J.M. and R.I. developed the methodologies and designed the experiments; F.Z. performed the experiments; J.J.M., J.A.G. and J.C. analyzed and validated the results.

Acknowledgments: This research work has been supported by the China Scholarship Council (Grant 201606420041), by the Spanish Ministry of Economy, Industry and Competitiveness (MINECO), the State Research Agency (AEI) and the European Funds for Regional Development (EFRD) under project TEC2017-85244-C2-1-P and by the National Natural Science Foundation of China (Grant 51574221). CommSensLab is Unidad de Excelencia Maria de Maeztu MDM-2016-0600 financed by the Agencia Estatal de Investigación, Spain. TerraSAR-X data were provided by the German Aerospace Center (DLR) in the scope of the project GEO2468. Sentinel-1A data were provided by the European Space Agency (ESA). Some figures were prepared using the public domain GMT software (Wessel and Smith, 1998).

Conflicts of Interest: The authors declare no conflict of interest.

References

1. Dunnicliff, J.; Green, G.E. *Geotechnical Instrumentation for Monitoring Field Performance*; John Wiley & Sons: Hoboken, NJ, USA, 1993.
2. Pinyol, N.M.; Alonso, E.E.; Corominas, J.; Moya, J. Canelles landslide: Modelling rapid drawdown and fast potential sliding. *Landslides* **2012**, *9*, 33–51. [CrossRef]
3. Ramesh, M.V. Design, development, and deployment of a wireless sensor network for detection of landslides. *Ad Hoc Netw.* **2014**, *13*, 2–18. [CrossRef]
4. Uhlemann, S.; Smith, A.; Chambers, J.; Dixon, N.; Dijkstra, T.; Haslam, E.; Meldrum, P.; Merritt, A.; Gunn, D.; Mackay, J. Assessment of ground-based monitoring techniques applied to landslide investigations. *Geomorphology* **2016**, *253*, 438–451. [CrossRef]
5. Zhang, Y.; Tang, H.; Li, C.; Lu, G.; Cai, Y.; Zhang, J.; Tan, F. Design and Testing of a Flexible Inclinometer Probe for Model Tests of Landslide Deep Displacement Measurement. *Sensors* **2018**, *18*, 224. [CrossRef] [PubMed]
6. Calcaterra, S.; Cesi, C.; Di Maio, C.; Gambino, P.; Merli, K.; Vallario, M.; Vassallo, R. Surface displacements of two landslides evaluated by GPS and inclinometer systems: A case study in Southern Apennines, Italy. *Nat. Hazards* **2012**, *61*, 257–266. [CrossRef]
7. Gili, J.A.; Corominas, J.; Rius, J. Using Global Positioning System techniques in landslide monitoring. *Eng. Geol.* **2000**, *55*, 167–192. [CrossRef]
8. Malet, J.P.; Maquaire, O.; Calais, E. The use of Global Positioning System techniques for the continuous monitoring of landslides: Application to the Super-Sauze earthflow (Alpes-de-Haute-Provence, France). *Geomorphology* **2002**, *43*, 33–54. [CrossRef]
9. Colesanti, C.; Wasowski, J. Investigating landslides with space-borne Synthetic Aperture Radar (SAR) interferometry. *Eng. Geol.* **2006**, *88*, 173–199. [CrossRef]
10. Wasowski, J.; Bovenga, F. Investigating landslides and unstable slopes with satellite Multi Temporal Interferometry: Current issues and future perspectives. *Eng. Geol.* **2014**, *174*, 103–138. [CrossRef]

11. Bovenga, F.; Wasowski, J.; Nitti, D.; Nutricato, R.; Chiaradia, M. Using COSMO/SkyMed X-band and ENVISAT C-band SAR interferometry for landslides analysis. *Remote Sens. Environ.* **2012**, *119*, 272–285. [CrossRef]

12. Hu, X.; Wang, T.; Pierson, T.C.; Lu, Z.; Kim, J.; Cecere, T.H. Detecting seasonal landslide movement within the Cascade landslide complex (Washington) using time-series SAR imagery. *Remote Sens. Environ.* **2016**, *187*, 49–61. [CrossRef]

13. Confuorto, P.; Di Martire, D.; Centolanza, G.; Iglesias, R.; Mallorqui, J.J.; Novellino, A.; Plank, S.; Ramondini, M.; Thuro, K.; Calcaterra, D. Post-failure evolution analysis of a rainfall-triggered landslide by multi-temporal interferometry SAR approaches integrated with geotechnical analysis. *Remote Sens. Environ.* **2017**, *188*, 51–72. [CrossRef]

14. Ferretti, A.; Prati, C.; Rocca, F. Permanent scatterers in SAR interferometry. *IEEE Trans. Geosci. Remote Sens.* **2001**, *39*, 8–20. [CrossRef]

15. Ferretti, A.; Fumagalli, A.; Novali, F.; Prati, C.; Rocca, F.; Rucci, A. A new algorithm for processing interferometric data-stacks: SqueeSAR. *IEEE Trans. Geosci. Remote Sens.* **2011**, *49*, 3460–3470. [CrossRef]

16. Berardino, P.; Fornaro, G.; Lanari, R.; Sansosti, E. A new algorithm for surface deformation monitoring based on small baseline differential SAR interferograms. *IEEE Trans. Geosci. Remote Sens.* **2002**, *40*, 2375–2383. [CrossRef]

17. Mora, O.; Mallorqui, J.J.; Broquetas, A. Linear and nonlinear terrain deformation maps from a reduced set of interferometric SAR images. *IEEE Trans. Geosci. Remote Sens.* **2003**, *41*, 2243–2253. [CrossRef]

18. Lanari, R.; Mora, O.; Manunta, M.; Mallorquí, J.J.; Berardino, P.; Sansosti, E. A small-baseline approach for investigating deformations on full-resolution differential SAR interferograms. *IEEE Tran. Geosci. Remote Sens.* **2004**, *42*, 1377–1386. [CrossRef]

19. Hooper, A.; Zebker, H.; Segall, P.; Kampes, B. A new method for measuring deformation on volcanoes and other natural terrains using InSAR persistent scatterers. *Geophys. Res. Lett.* **2004**, *31*, doi:10.1029/2004GL021737.

20. Blanco-Sanchez, P.; Mallorquí, J.J.; Duque, S.; Monells, D. The coherent pixels technique (CPT): An advanced DInSAR technique for nonlinear deformation monitoring. *Pure Appl. Geophys.* **2008**, *165*, 1167–1193. [CrossRef]

21. Iglesias, R.; Monells, D.; Fabregas, X.; Mallorqui, J.J.; Aguasca, A.; Lopez-Martinez, C. Phase quality optimization in polarimetric differential SAR interferometry. *IEEE Trans. Geosci. Remote Sens.* **2014**, *52*, 2875–2888. [CrossRef]

22. Cruden, D.M.; Varnes, D.J. Landslide types and processes. In *Landslides: Investigation and Mitigation*; Turner, A., Schuster, R., Eds.; Transportation Research Board, US National Research Council: Washington, DC, USA, 1996; Volume 247, Chapter 3; pp. 36–75.

23. Hungr, O.; Leroueil, S.; Picarelli, L. The Varnes classification of landslide types, an update. *Landslides* **2014**, *11*, 167–194. [CrossRef]

24. Iglesias, R. High-Resolution Space-Borne and Ground-Based SAR Persistent Scatterer Interferometry for Landslide Monitoring. Ph.D. Thesis, Universitat Politècnica de Catalunya, Barcelona, Spain, 2015.

25. Beauducel, F.; Briole, P.; Froger, J.L. Volcano-wide fringes in ERS synthetic aperture radar interferograms of Etna (1992–1998): Deformation or tropospheric effect? *J. Geophys. Res. Solid Earth* **2000**, *105*, 16391–16402. [CrossRef]

26. Hanssen, R.F. *Radar Interferometry: Data Interpretation and Error Analysis*; Vol. 2, Springer Science & Business Media: Dordrecht, Tthe Netherlands, 2001.

27. Elliott, J.; Biggs, J.; Parsons, B.; Wright, T. InSAR slip rate determination on the Altyn Tagh Fault, northern Tibet, in the presence of topographically correlated atmospheric delays. *Geophys. Res. Lett.* **2008**, *35*, doi:10.1029/2008GL033659.

28. Iglesias, R.; Fabregas, X.; Aguasca, A.; Mallorqui, J.J.; López-Martínez, C.; Gili, J.A.; Corominas, J. Atmospheric phase screen compensation in ground-based SAR with a multiple-regression model over mountainous regions. *IEEE Trans. Geosci. Remote Sens.* **2014**, *52*, 2436–2449. [CrossRef]

29. Hu, Z.; Mallorquí, J.J.; Centolanza, G.; Duro, J. Insar atmospheric delays compensation: Case study in tenerife island. In Proceedings of the 2017 IEEE International Geoscience and Remote Sensing Symposium (IGARSS), Fort Worth, TX, USA, 23–28 July 2017; pp. 3167–3170.

30. Bamler, R.; Eineder, M.; Adam, N.; Zhu, X.; Gernhardt, S. Interferometric potential of high resolution spaceborne SAR. *Photogramm.-Fernerkund.-Geoinf.* **2009**, *2009*, 407–419. [CrossRef]

31. Prati, C.; Ferretti, A.; Perissin, D. Recent advances on surface ground deformation measurement by means of repeated space-borne SAR observations. *J. Geodyn.* **2010**, *49*, 161–170. [CrossRef]

32. Lee, J.S.; Jurkevich, L.; Dewaele, P.; Wambacq, P.; Oosterlinck, A. Speckle filtering of synthetic aperture radar images: A review. *Remote Sens. Rev.* **1994**, *8*, 313–340. [CrossRef]

33. Daba, J.S.; Jreije, P. Advanced stochastic models for partially developed speckle. *World Acad. Sci. Eng. Technol.* **2008**, *41*, 566–570.

34. Lopes, A.; Nezry, E.; Touzi, R.; Laur, H. Structure detection and statistical adaptive speckle filtering in SAR images. *Int. J. Remote Sens.* **1993**, *14*, 1735–1758. [CrossRef]

35. Curlander, J.C.; McDonough, R.N. *Synthetic Aperture Radar*; John Wiley & Sons: New York, NY, USA, 1991; Volume 396.

36. Iglesias, R.; Mallorqui, J.J.; López-Dekker, P. DInSAR pixel selection based on sublook spectral correlation along time. *IEEE Trans. Geosci. Remote Sens.* **2014**, *52*, 3788–3799. [CrossRef]

37. Iglesias, R.; Mallorqui, J.J.; Monells, D.; López-Martínez, C.; Fabregas, X.; Aguasca, A.; Gili, J.A.; Corominas, J. PSI deformation map retrieval by means of temporal sublook coherence on reduced sets of SAR images. *Remote Sens.* **2015**, *7*, 530–563. [CrossRef]

38. Zhao, F.; Mallorqui, J.J. A temporal phase coherence estimation algorithm and its application on DInSAR pixel selection. *IEEE Trans. Geosci. Remote Sens.* **2018**, Undergoing Review.

39. Duque, S.; Breit, H.; Balss, U.; Parizzi, A. Absolute height estimation using a single TerraSAR-X staring spotlight acquisition. *IEEE Geosci. Remote Sens. Lett.* **2015**, *12*, 1735–1739. [CrossRef]

40. Tapete, D.; Cigna, F.; Donoghue, D.N. 'Looting marks' in space-borne SAR imagery: Measuring rates of archaeological looting in Apamea (Syria) with TerraSAR-X Staring Spotlight. *Remote Sens. Environ.* **2016**, *178*, 42–58. [CrossRef]

41. Corominas, J.; Alonso, E. Inestabilidad de Laderas en el Pirineo Catalán. Tipología y causas. In Proceedings of the Inestabilidad de laderas en el Pirineo, Barcelona, Spain, 16–17 January 1984; pp. 1–53.

42. Santacana, N. Estudi dels Grans Esllavissaments d'Andorra: Els Casos del Forn i del Vessant d'Encampadana. Master's Thesis, Department of Dynamic Geology, Geophysics and Paleontology, Faculty of Geology, University of Barcelona, Barcelona, Spain, 1994.

43. Corominas, J.; Iglesias, R.; Aguasca, A.; Mallorquí, J.J.; Fàbregas, X.; Planas, X.; Gili, J.A. Comparing satellite based and ground based radar interferometry and field observations at the Canillo landslide (Pyrenees). In *Engineering Geology for Society and Territory-Volume 2*; Springer: Cham, Switzerland, 2015; pp. 333–337.

44. Torrebadella, J.; Villaró, I.; Altimir, J.; Amigó, J.; Vilaplana, J.; Corominas, J.; Planas, X. El deslizamiento del Forn de Canillo en Andorra. Un ejemplo de gestión del riesgo geológico en zonas habitadas en grandes deslizamientos. In Proceedings of the VII Simposio Nacional Sobre Taludes y Laderas Inestables, Barcelona, Spain, 27–30 October 2009; pp. 403–414.

45. Mittermayer, J.; Wollstadt, S.; Prats-Iraola, P.; Scheiber, R. The TerraSAR-X staring spotlight mode concept. *IEEE Trans. Geosci. Remote Sens.* **2014**, *52*, 3695–3706. [CrossRef]

46. Eineder, M.; Adam, N.; Bamler, R.; Yague-Martinez, N.; Breit, H. Spaceborne spotlight SAR interferometry with TerraSAR-X. *IEEE Trans. Geosci. Remote Sens.* **2009**, *47*, 1524–1535. [CrossRef]

47. Davis, J.; Herring, T.; Shapiro, I.; Rogers, A.; Elgered, G. Geodesy by radio interferometry: Effects of atmospheric modeling errors on estimates of baseline length. *Radio Sci.* **1985**, *20*, 1593–1607. [CrossRef]

48. Cascini, L.; Fornaro, G.; Peduto, D. Advanced low-and full-resolution DInSAR map generation for slow-moving landslide analysis at different scales. *Eng. Geol.* **2010**, *112*, 29–42. [CrossRef]

49. Monserrat, O.; Moya, J.; Luzi, G.; Crosetto, M.; Gili, J.; Corominas, J. Non-interferometric GB-SAR measurement: Application to the Vallcebre landslide (eastern Pyrenees, Spain). *Nat. Hazards Earth Syst. Sci.* **2013**, *13*, 1873. [CrossRef]

remote sensing

MDPI

Review

A Review of Ten-Year Advances of Multi-Baseline SAR Interferometry Using TerraSAR-X Data

Xiao Xiang Zhu [1,2,*], Yuanyuan Wang [2], Sina Montazeri [1] and Nan Ge [1]

[1]   Remote Sensing Technology Institute, German Aerospace Center, 82234 Weßling, Germany;
     Sina.Montazeri@dlr.de (S.M.); Nan.Ge@dlr.de (N.G.)
[2]   Signal Processing in Earth Observation, Technical University of Munich, 80333 Munich, Germany;
     wang@bv.tum.de
*   Correspondence: xiaoxiang.zhu@dlr.de; Tel.: +49-8153-28-3531

Received: 30 June 2018; Accepted: 25 August 2018; Published: 30 August 2018

Abstract: Since its launch in 2007, TerraSAR-X has continuously provided spaceborne synthetic aperture radar (SAR) images of our planet with unprecedented spatial resolution, geodetic, and geometric accuracy. This has brought life to the once inscrutable SAR images, which deterred many researchers. Thanks to merits like higher spatial resolution and more precise orbit control, we are now able to indicate individual buildings, even individual floors, to pinpoint targets within centimeter accuracy. As a result, multi-baseline SAR interferometric (InSAR) techniques are flourishing, from point target-based algorithms, to coherent stacking techniques, to absolute positioning of the former techniques. This article reviews the recent advances of multi-baseline InSAR techniques using TerraSAR-X images. Particular focus was put on our own development of persistent scatterer interferometry, SAR tomography, robust estimation in distributed scatterer interferometry and absolute positioning using geodetic InSAR. Furthermore, by introducing the applications associated with these techniques, such as 3D reconstruction and deformation monitoring, this article is also intended to give guidance to wider audiences who would like to resort to SAR data and related techniques for their applications.

Keywords: multi-baseline; multi-pass; PS; DS; geodetic; TomoSAR; D-TomoSAR; PSI; robust estimation; covariance matrix; InSAR; SAR; review

1. Introduction

1.1. Overview of Multi-Baseline InSAR

Since its launch in 2007, TerraSAR-X has continuously revealed synthetic aperture radar (SAR) images of unprecedented high resolution from space. This has brought life to the once obscure and sometimes inscrutable SAR images that deterred many researchers. Figure 1 shows a comparison of the medium resolution ERS image and a high resolution TerraSAR-X spotlight image of the same area in Las Vegas. Individual buildings are for the first time interpretable by the naked eye from spaceborne SAR images, because the 1-m resolution in spotlight mode is well beyond the inherent scale of the 3-m floor height typical of urban infrastructure. This marks the start of an era of urban infrastructure monitoring using spaceborne SAR images. Currently, the staring spotlight mode provides images with a resolution up to 25 cm, from which the mapping of individual window edges is even possible. This breakthrough in spatial resolution, together with the precise orbit determination with sub-centimeter accuracy [1,2], positions TerraSAR-X images as a perfect dataset for long-term repeated monitoring of large areas with precision and high resolution.

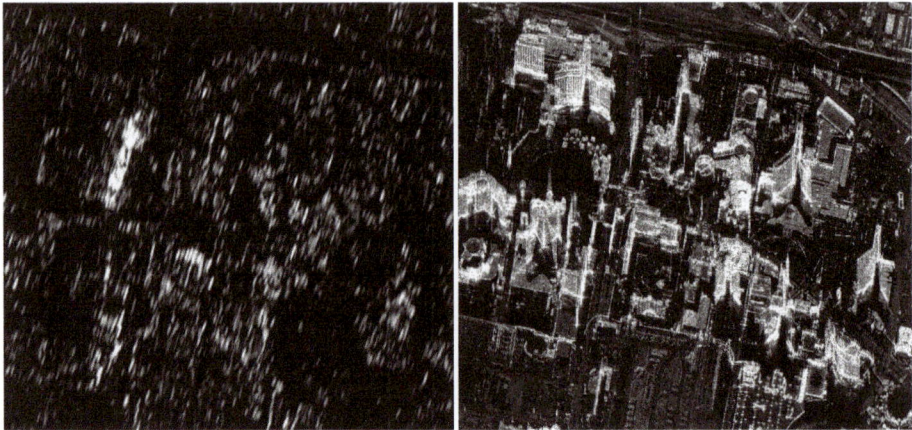

Figure 1. Comparison of medium (ERS) and high (TerraSAR-X) resolution SAR images of downtown Las Vegas [3].

Among the many promising InSAR techniques that prospered in the past decade, multi-baseline, especially multi-pass, InSAR techniques are undoubtedly one of the jewels in the crown. They build up invaluable data cubes of long-term image time series. For example, the TerraSAR-X revisit time of 11 days allows monthly deformation signals of the Earth's surface, such as ground subsidence, to be retrieved using techniques like persistent scatterer interferometry (PSI). For monitoring dense urban areas, SAR tomography (TomoSAR) and its differential form, D-TomoSAR inversion, are the most competent techniques because of their capability of layover separation. They generate point clouds with density comparable to that of a LiDAR. Both PSI and TomoSAR produce highly accurate parameter estimates, because they work on highly coherent point targets. Therefore, they are often the workhorses for deformation monitoring and 3D reconstruction, especially in urban areas. To complement these techniques, distributed scatterer (DS)-based techniques, such as SqueeSAR [4], robust InSAR optimization (RIO) [5] and coherence tomography enable dense monitoring of deformation in areas of low interferometric coherence, such as volcanic areas. Among them, some algorithms, such as RIO, address the statistical robustness of estimators to ensure the reliability of the accuracy of the estimates in operational processing over large areas. Despite the many advantages of multi-baseline InSAR, they are still relative measures, as the estimates are often relative to a local reference point whose 3D position is unknown. Such differential operation is often performed in multi-baseline InSAR in order to mitigate some common phase errors, such as atmospheric delay. It was only until recently, that geodetic InSAR [6] bridged the gap between multi-baseline InSAR techniques and absolute positioning using SAR imaging geodesy [7] to produce absolute 3D (and higher dimensional) InSAR point clouds. It is an important piece of the components of the ecosystem of Earth observation using SAR data. Multi-baseline InSAR techniques that were once only a relative measure can now be employed as geodetic techniques to provide centimeter-level absolute positioning and millimeter-level relative deformation monitoring.

1.2. Principle of Multi-Baseline InSAR

InSAR is the technique of using SAR as an interferometer. Multi-baseline InSAR techniques exploit the interferometric phase (i.e., the phase difference) of multiple complex-valued SAR images. These images are acquired at different satellite positions, time, or frequency, and hence, they create spatial, temporal baselines, or Δk-radar when forming interferograms. For TerraSAR-X images, such multi-baseline configuration is usually acquired in a repeat-pass manner (hence "multi-pass"), except if the twin satellite TanDEM-X was employed. Figure 2 shows the multi-baseline InSAR configuration in an urban scenario at a fixed azimuth position. The TerraSAR-X satellite flies perpendicular into the screen/paper. The term *r* indicates the line of sight (LOS), i.e., the slant range direction, of the sensor; *s* is the elevation direction that is perpendicular to the range and azimuth. The blue outline on the surface indicates the area illuminated by radar pulses. The elongated ellipse is the range-elevation tube within which all the objects are imaged into a single pixel in the focused SAR image. The cross-section of the tube naturally depends on the range and azimuth resolution of the sensor. The extent of the tube Δ*s* is much larger than the dimension of the cross-section because of the large distance between the sensor and the object, as well as the small angular diversity among different acquisitions. Therefore, it is common that several objects, such as a building roof, tree and ground, are layovered in a single pixel in a TerraSAR-X image.

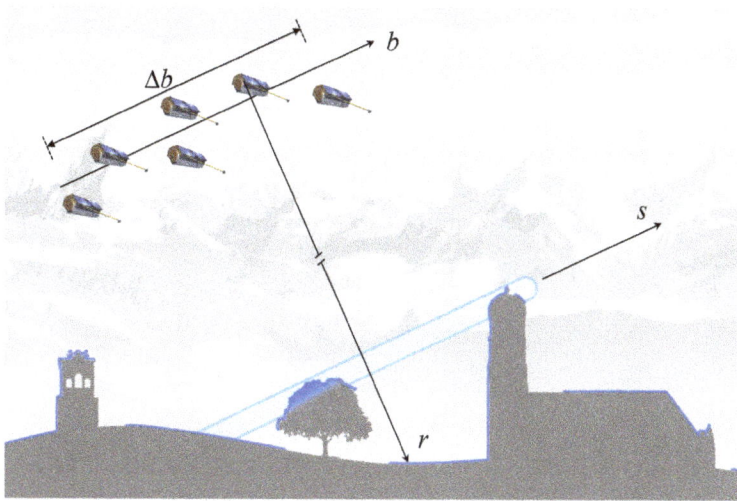

Figure 2. Schematic drawing of the principle of multi-baseline InSAR at a fixed azimuth position (modified after [8]). The TerraSAR-X satellite flies away from the reader into the screen/paper. The line-of-sight, i.e., the range direction, of the sensor is indicated by *r*. The range timing is always delayed after propagation through the atmosphere. The term *s* is the elevation direction that is perpendicular to the range. The blue outline on the surface indicates the area illuminated by radar pulses. The elongated ellipse is the range-elevation resolution cell with in which all the objects are imaged into a single pixel in the final SAR image. It is very common that several objects, such as a building roof, tree and ground, are layovered in a single pixel in a TerraSAR-X image.

If one considers a single phase center in the range-azimuth-elevation tube without layover (i.e., single scatterer model), the absolute interferometric phase of the *n*-th measurement in a multi-baseline InSAR stack is [9]:

$$\phi_n = -\frac{4\pi}{\lambda}\frac{b_n s}{R} + \phi_{defo} + \phi_{atmo} + \phi_{error},\tag{1}$$

where λ is the wavelength of the SAR electromagnetic wave, b_n is the baseline of the *n*-th image, *s* is the elevation of the single scatterer and R is the nominal range which is the distance of the SAR sensor to a zero-elevation point. The deformation phase ϕ_{defo} is often modeled as a function $d(t_n)$ (e.g., linear or periodic) of the acquisition time t_n. The interferometric phase is always delayed due to atmospheric propagation. In multi-baseline InSAR, such atmospheric phase delay ϕ_{atmo} is mitigated by subtracting a nearby reference point. This renders multi-baseline InSAR a relative measure, unless the absolute position of the reference point is known a priori.

Based on Equation (1), the forward system model of multi-baseline InSAR measurement can be expressed as Equation (2), where g_n is the pixel value at the n-th image, and $\gamma(s)$ is the reflectivity profile along the elevation direction. Since a far-field antenna acts like a Fourier transform to the signal in the resolution cell, each measurement is actually the Fourier transform at a specific frequency that is linearly proportional to the perpendicular baseline b_n to the master satellite position. This is also known as the system model for TomoSAR [10–13].

$$g_n = \int_{\Delta s} \gamma(s) \exp\left(-j\frac{4\pi b_n}{\lambda R} s\right) ds. \tag{2}$$

In the case of differential TomoSAR (D-TomoSAR), Equation (2) is extended into higher dimensions [14–16].

Equation (2) can be written in a more compact matrix form as:

$$g = R\gamma, \tag{3}$$

where R and γ are the discretized Fourier matrix and the reflectivity profile along the direction s, respectively. Estimating γ is essentially a spectral estimation problem. In the case of PSI- or DS-based interferometry that assumes a single phase center in the resolution cell, it is basically a spectral estimation of a single frequency. Equation (3) will degenerate to either Equation (4) for the deterministic PS mode or Equation (5) for the stochastic DS model.

$$g = r(s_1)\gamma_1, \tag{4}$$

$$E\left(gg^H\right) = r(s_1) C_{gg} r(s_1)^H, \tag{5}$$

where s_1 is the elevation of the single phase center, $r(s_1)$ is the column in R associated with s_1, γ_1 is the complex-valued brightness of the PS and C_{gg} is the covariance matrix of the DS.

1.3. The Structure of This Paper

The rest of this article introduces the recent development of the aforementioned techniques, each in its respective section. In Section 2, we introduce the development of PS-based methods following their improvements in estimates accuracy that in turn refers to the reconstructed point density, as well as the reduction of the required number of images for a reliable estimation. Section 3 focuses on the development of robust InSAR techniques based on DS. Section 4 focuses on the evolution of TerraSAR-X absolute positioning from a single target to many targets and eventually to the fusion with multi-baseline InSAR.

2. Advances in Point Scatterer-Based Methods

This section focuses on the advances of PS-based methods, i.e., PSI and TomoSAR/D-TomoSAR in urban areas. Their development mainly focuses on the improvement of estimation accuracy, which in turn increases the density of the retrieved point cloud or reduces the number of interferograms required for a reliable estimation.

Both PSI and TomoSAR utilize a single-master configuration to extract time-coherent scatterers from SAR images. The major difference between the two methods is the number of scatterers that are assumed within a resolution cell, which requires different spectral estimators to be employed in the parameter retrieval. However, over the past two decades, PSI has made substantial development, so that it usually refers to a full processing chain including interferogram formation and reference network construction. Therefore, PSI is often employed as a preprocessing step for TomoSAR. Several variations of PSI that differ in algorithmic details have been introduced in recent years. For a full review of these techniques the reader is referred to [17]. Although the specifics of existing PSI algorithms are different, the following workflow is widely acknowledged:

Step 1   Differential interferogram formation: From a stack of $N + 1$ co-registered SAR images, a master acquisition is selected. Subsequently N interferograms are computed, while their topographic phase components are removed using a reference digital elevation model (DEM).

Step 2   Reference network construction: Scatterers presumed to be the most phase-stable ones are selected. The detection can be carried out using various methods, such as thresholding on the amplitude dispersion index (ADI) [18] or on the signal-to-clutter ratio (SCR) [19]. These PS candidates are connected to form a reference network. Through the PS double-difference phase measurements, i.e., difference in time and space, differential topography and differential motion parameters are estimated on arcs.

Step 3   Atmospheric phase estimation: The differential topography estimates are integrated with respect to an arbitrarily chosen reference point so that the topographic phase components are removed from the interferometric phases. The remaining phase contributions include deformation, atmosphere, and noise. Then a low-pass filtering in the spatial domain and a high-pass filtering in the temporal domain extracts the atmospheric component, which is interpolated over the entire scene and subtracted from the differential interferograms.

Step 4   PS densification: Additional PS are computed from the corrected differential interferograms. These PS are connected to the nearest point(s) in the reference network and their modeled parameters are estimated.

Step 5   PS geocoding: The DEM height of each PS is added to its differential height estimate. The radar timing of each PS and its updated height are geocoded using satellite orbit and a reference ellipsoid to represent the PS coordinates in a common geodetic coordinate system.

The processing steps for TomoSAR are similar to those of PSI, except the fourth step is replaced with higher-order spectral estimators that can be enumerated as followings.

- The full reflectivity profile is reconstructed using higher-order spectral estimation techniques.
- The scatterers' positions and motion parameters are determined by detecting maxima on the reflectivity profile.

2.1. Overview of Advances

For each of the steps delineated above, numerous improvements have been suggested in the literature. For example, in the reference network step, [20,21] consider the geometry of the connections among arcs to construct a redundant reference network, while dense differential PS pairs were used in [22] to form the network. In terms of network inversion, to robustly retrieve the topography and deformation estimates of the PS in the reference network, a ℓ_1 norm outlier rejection scheme was proposed after the LAMBDA estimation [23]. In [24,25], numerical weather data were used to simulate and mitigate tropospheric delay. For a detailed comparison of widely used PSI techniques, the interested reader is referred to [21].

The development of TomoSAR has been mainly focused on the improvement of the spectral estimator and the scatterer detector. Studies have been conducted to improve the maximum likelihood estimator (MLE) by restricting the support of the signal (i.e., nonlinear least square) [3,13], by ℓ_2 norm regularization (i.e., the Tikhonov method) [15], and by ℓ_1 norm regularization (i.e., compressive sensing-based method) [26,27]. The SL1MMER algorithm proposed in [27] was also recently extended to the M-SL1MMER [28], which exploits group sparsity in the urban environment. M-SL1MMER achieves a comparable result with far fewer images than SL1MMER and other algorithms. Several studies have also addressed the efficiency and robustness of the detection of scatterers. For example, [29] describes the optimal detection of multiple scatterers, and [27,30–32] address scatterer detection in the super-resolution regime where the distance among scatterers is less than the elevation resolution.

In general, TomoSAR is so far the most competent multi-baseline InSAR method for urban area monitoring. However, the relatively high computational cost limits it for extensive uses like PSI, especially for the CS-based TomoSAR algorithms. Therefore, combining PSI and TomoSAR has also been proposed to improve the computational efficiency of TomoSAR processing [22,33,34]. Only recently, an efficient sparse recovery algorithm was proposed, which made city-scale 3D/4D reconstruction directly using SL1MMER operational [35].

2.2. Very High Resolution PSI

PSI is undoubtedly the workhorse for deformation monitoring of large areas, owing to its computational efficiency and reliability in the accuracy of the deformation estimates. As mentioned earlier, estimating the unknown elevation and deformation parameters in PSI is a spectral estimation of a single frequency. The spectral estimator is essentially a periodogram that can be expressed as follows.

$$\hat{\theta} = \arg\max_{\theta} \left\{ \left| \frac{1}{N} \sum_{n=1}^{N} g_n \exp\left(-j\phi_n\left(\theta\right)\right) \right| \right\} \approx \arg\max_{\theta} \left\{ \left| \frac{1}{N} \sum_{n=1}^{N} \frac{g_n}{|g_n|} \exp\left(-j\phi_n\left(\theta\right)\right) \right| \right\}, \quad (6)$$

where θ denotes the parameters, including the elevation s and the deformation parameters, and $\phi_n\left(\theta\right)$ is the modeled phase of the PS in the n-th image (i.e., Equation (1)). Often, the amplitude of g is dropped in the estimation [18], since it barely changes the estimates for PS of high signal-to-noise ratio (SNR).

Employing very high resolution (VHR) PSI, it is now possible to detect very localized deformation patterns even on different parts of a single building [36]. Apart from its deformation monitoring capability, VHR PSI leads to detailed 3D reconstruction of urban areas owing to the high density PSI point clouds. It can typically produce 40,000 to 100,000 PS per square kilometer using TerraSAR-X high resolution spotlight images [37,38]. The 3D reconstruction capability has even been strengthened by the geometrical fusion of PSI point clouds obtained from different viewing geometries, i.e., along-heading and cross-heading orbits [39]. Especially in the case of cross-heading orbits, that is, the combination of point clouds from ascending and descending orbits, point cloud fusion provides a shadow-free point cloud of the observed area. It also allows a decomposition of the raw LOS PSI deformation measurements into 3D displacement vectors in geodetic coordinate system [36,40,41].

2.3. Differential TomoSAR

Unlike PSI, D-TomoSAR retrieves the full reflectivity profile γ, and detects prominent peaks from it. Therefore, D-TomoSAR is inherently a more competent method for urban area monitoring than PSI. The MLE (under complex Gaussian noise) of γ can be expressed as follows.

$$\hat{\gamma}_{MLE} = \arg\min_{\gamma} \frac{1}{2} \|\mathbf{g} - \mathbf{R}\gamma\|_2^2. \quad (7)$$

During the last decade, we have developed a suite of algorithms named Tomo-GENESIS [42] to address both the methodological and practical aspects of D-TomoSAR. For example, the Tomo-GENESIS suite includes both conventional linear estimators [15] and the compressive sensing (CS)-based estimator that works in the superresolving regime [27,30,31], as well as a computationally efficient processing pipeline [22], the fusion of TomoSAR point clouds from multiple aspects [43] and 3D object reconstruction from TomoSAR point clouds [44–46].

2.3.1. Conventional (Non-Superresolving) D-TomoSAR

For spaceborne data, the number of acquisitions is usually far less than the discretization of γ. Therefore, Equation (3) is often under-determined. A popular method before the invention of CS-based

TomoSAR techniques to regularize the equation system was to employ the ℓ_2 norm regularization that is also known as Tikhonov regularization. The regularized estimator is shown as follows.

$$\hat{\gamma}_{\ell_2} = \arg\min_{\gamma} \frac{1}{2} \|g - R\gamma\|_2^2 + \lambda_{\ell_2} \|\gamma\|_2^2,\qquad(8)$$

where λ_{ℓ_2} is the regularization parameter. We have implemented the estimator using singular value decomposition with Wiener filtering on the system matrix **R**. Therefore this algorithm is also known as SVD-Wiener in the community [15].

This type of estimator is also a maximum a posteriori (MAP) estimator. It is the optimal Bayesian estimator that minimizes posterior expected loss. Experiments showed promising performance on TerraSAR-X image stacks [13]. However, in the classical Nyquist–Shannon sampling regime, the resolution of the reconstructed reflectivity profile is limited by the so-called Rayleigh resolution (see Equation (9)) [15] that is governed by the spread of the baseline Δb.

$$\rho_s = \frac{\lambda R}{2\Delta b}.\qquad(9)$$

2.3.2. Super-Resolving D-TomoSAR

For dense urban areas, closely spaced objects often coexist in a range-azimuth-elevation resolution cell. These objects cannot be resolved by conventional tomographic inversion algorithms. This is where CS-based super-resolving tomographic inversion comes to play, as it can achieve super-resolution in the estimate of γ, if it is sparse. The CS-based TomoSAR estimator can be generally expressed in a similar form as Equation (8), except that the ℓ_2 regularization term is replaced by the signal sparsity term, i.e., the ℓ_0 norm. Because of the nonconvexity of the ℓ_0 norm, it is often relaxed by the ℓ_1 norm in optimization, such as the SL1MMER "scale-down by ℓ_1 norm minimization, model selection, and estimation reconstruction" algorithm proposed in [27]. The $\ell_2 + \ell_1$ norm estimator can be expressed as follows.

$$\hat{\gamma}_{\ell_0} = \arg\min_{\gamma} \frac{1}{2} \|g - R\gamma\|_2^2 + \lambda_K \|\gamma\|_1,\qquad(10)$$

where λ_K is a regularization parameter (K being the sparsity, i.e., the number of discrete scatterers). In practice, the minimization of the ℓ_0 norm is often relaxed by the ℓ_1 norm for better convexity in the optimization.

Because of their super-resolving ability and the robustness of the ℓ_1 norm minimization, CS-based D-TomoSAR algorithms are the state of the art in term of the accuracy of the parameter estimate and the performance of scatterer detection. This in turn increases the density of the reliable points. Figure 3 is a comparison of the point cloud retrieved by PSI and SL1MMER of the same building (Bellagio Hotel, Las Vegas). SL1MMER retrieves many more points than PSI. Yet, CS-based algorithms are less computationally efficient than the conventional TomoSAR. To cope with large area processing, we enriched Tomo-GENESIS with an approach [22,47] that integrates PSI, conventional TomoSAR, and super-resolving TomoSAR. Recently, we have developed a fast and accurate ℓ_1-regularized least square solver with application to D-TomoSAR [35]. This new solver offers a speedup of one or two orders of magnitude than typical second order cone programming. With above-mentioned advances, we are able to reconstruct a high-quality TomoSAR point cloud of an entire city with density comparable to that of LiDAR. For a better overview of the capability of the aforementioned methods, Table 1 summarizes the typical density of the point cloud reconstructed by PSI and D-TomoSAR using a TerraSAR-X high resolution sliding spotlight image stack.

Table 1. Comparison of the typical density of the point cloud reconstructed by PSI and D-TomoSAR using TerraSAR-X high resolution spotlight image stack.

	Density (thousand/km^2)
PSI [38]	40–100
D-TomoSAR (non-superresolving) [15]	150–250
D-TomoSAR (SL1MMER) [48]	500–1500

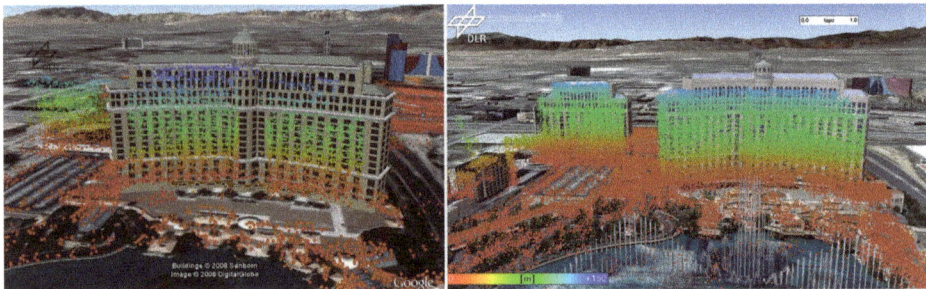

Figure 3. Comparison of the density of the 3D point cloud retrieved by PSI (**left**) and TomoSAR (**right**) of Bellagio Hotel, Las Vegas [8].

2.3.3. Staring Spotlight TomoSAR

In spotlight modes, the radar beam is steered back and forth toward a common reference target in order to increase its illumination time t_{AP} (see Figure 4). The beam sweep rate controls the balance between the scene spatial extent and the azimuth resolution. In the TerraSAR-X sliding spotlight mode, the radar beam is swept at a moderate rate with a squint angle range up to ± 0.75 degrees [49]. While in its staring spotlight mode, the beam sweep rate is set to equal the frequency modulation (FM) rate of the reference target. In other words, the radar beam is configured to exactly follow the target over time and the squint angle range can be up to ca. ± 2.2 degrees. As a result, the azimuth resolution is maximized. Nevertheless, the improved azimuth resolution comes at the cost of reduced scene extent: the time span of a focused image Δt_{image} is considerably shorter. Needless to say, the slant range resolution stays unchanged for both modes, as long as the same range bandwidth is employed during imaging.

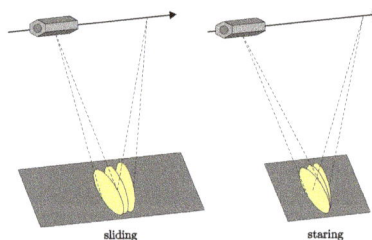

Figure 4. TerraSAR-X sliding (**left**) and staring (**right**) spotlight imaging geometry [48].

The transition from sliding to staring spotlight requires several adaptations in SAR focusing and InSAR processing. In the staring spotlight mode, the satellite can no longer be assumed to be standing still during chirp transmission and reception, or to follow a linear trajectory. In addition, variations of tropospheric and ionospheric delay within the large squint angle span also need to be corrected. Another major challenge is to estimate Doppler centroid frequency as a function of focused image time.

The TerraSAR-X multimode SAR processor [50,51] and the integrated wide area processor [24,25] were revised accordingly.

For PSI and TomoSAR in urban areas, the improved azimuth resolution has at least two advantages. PSs in the same resolution cell in the sliding spotlight mode may be resolved in different resolution cells in the staring spotlight mode. This leads to an increase in the density of the resulted 4D point cloud. Furthermore, the clutter in each resolution cell may be significantly suppressed, thanks to the increased azimuth resolution. Consequently, the SCR of PSs increases, which in turn leads to a better lower bound on the variance of elevation estimates [52].

In order to demonstrate these improvements, we processed two interferometric stacks of the City of Las Vegas in the sliding and staring spotlight modes using the SL1MMER algorithm. Each stack consists of 12 scenes acquired from October 2014 to February 2015. In each mode, 11 coregistered complex interferograms were used for the TomoSAR reconstruction of two regions of interests (ROIs).

One of the ROIs is a relatively flat region that was selected mainly for the assessment of relative vertical accuracy. The mean intensity maps in both modes are shown in Figure 5a. In the staring spotlight mode, point-like targets, such as the six bright points aligned at each side of the central field, are more focused. Even for DS, clutter appears to be more suppressed and thus the boundaries between areas of different smoothness are easier to recognize. This indicates an increased SCR. As a result, the reconstructed TomoSAR point cloud from staring spotlight images has a significant increase in the number of points compared to that of the sliding spotlight images. Indeed, the total number of scatterers in the staring spotlight is approximately 5.5 times as high, and the scatterer density is up to circa 13.5 million points per km^2 in this small region. In addition, the better SCR also improves the relative accuracy of height estimates. In fact, the relative accuracy of height estimates using staring spotlight images is approximately 1.7 times as high [48].

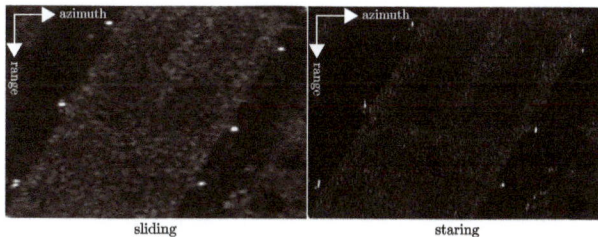

(a) ROI #1 that contains a flat area.

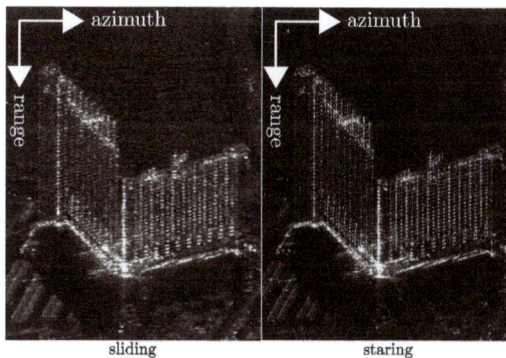

(b) ROI #2 that contains two high-rise buildings.

Figure 5. Mean intensity map of two ROIs in the sliding (a) and staring (b) spotlight modes [48].

Another ROI contains two high-rise buildings (Hilton Grand Vacations on the Las Vegas Strip), which were chosen as a demonstration of layover separation. The mean intensity maps are shown in Figure 5b. Similarly, point-like targets stand out more prominently from clutter in the staring spotlight mode and the regularities on the building facades are more clearly visible. The TomoSAR point clouds of single and double scatterers are shown in Figure 6. For this ROI, a substantial increase in the number of (single and double) scatterers was also observed. The scatterer density in the staring spotlight mode is approximately 5.1 times as high, see Table 2. The number of double scatterers in the staring spotlight mode almost rivals the number of single scatterers in the sliding spotlight mode.

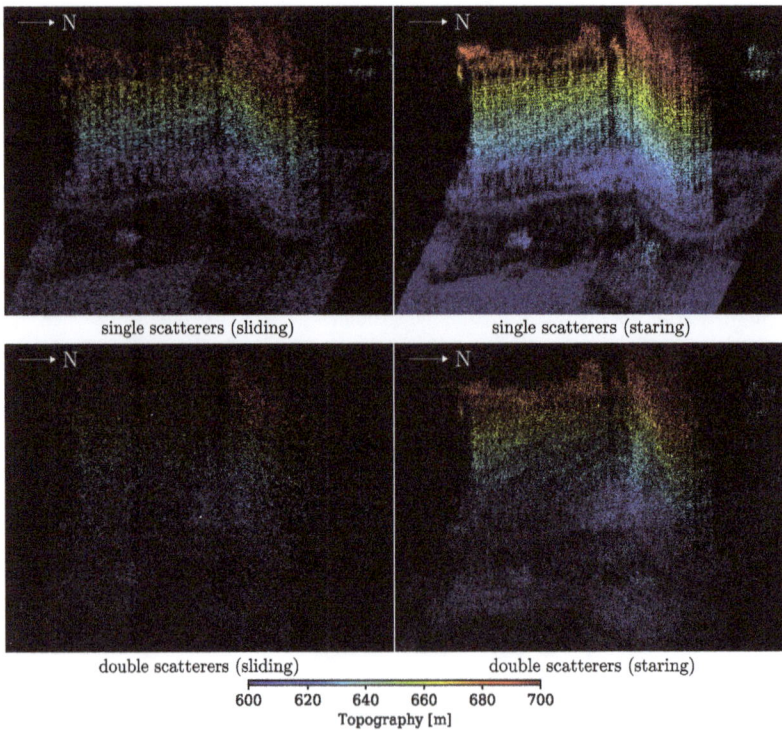

Figure 6. Updated topography (m) of the region in Figure 5b with 12 TerraSAR-X images in the sliding (left column) and staring (right column) spotlight modes, respectively. The upper and lower rows show single and double scatterers, respectively [48].

Table 2. Statistics of the point clouds in Figure 6.

	Sliding	Staring	Ratio [1]
No. of single scatterers	148, 646	740, 656	4.98
No. of double scatterers	21, 576	124, 546	5.77
Total no. of scatterers	170, 222	865, 202	5.08
Single-to-double-scatterer ratio	6.89	5.95	1.16
Scatterer density (million/km^2)	1.56	7.91	5.08

[1] The ratio was calculated by dividing the larger by the smaller value.

2.3.4. Point Cloud Fusion

Both PSI and D-TomoSAR deliver 4D point clouds relative to their reference points. They need to be co-registered when considering the results from multiple SAR image stacks. Although general point cloud fusion is a classic topic in the computer vision field, there is very little literature addressing InSAR point cloud fusion, especially for point clouds from image stacks of cross-heading orbits. This is because the fusion of two point clouds requires the identification of common points in the two point clouds. There is theoretically no common point from such two point clouds due to the cross-heading geometry.

The first attempt to fuse cross-heading TerraSAR-X point clouds in an urban area was presented in [36]. This algorithm employs RANSAC to robustly match the ground points of two cross-heading TerraSAR-X PSI point clouds. The point correspondences are found by searching closely spaced point pairs on the ground surface. Therefore, this algorithm does not address the exact point correspondence. To find the exact point correspondence, Wang and Zhu detected the end positions of L-shaped facades in the two TomoSAR point clouds where the two point clouds converge [43]. In [6,53], the authors located dozens of pairs of street lampposts in two point clouds as point correspondences, additionally taking into account the diameter of the lampposts.

The fusion of along-heading (either both ascending or both descending) InSAR point clouds is less challenging. Classical point cloud co-registration methods such as iterative closest point (ICP) can be directly applied. Gernhardt et al. have demonstrated the direct application of ICP on multiple InSAR point clouds of a volcano [39].

2.3.5. 3D Motion Decomposition

A natural step after the fusion of multiple D-TomoSAR point clouds from different aspects is the decomposition of the 1D LOS displacement vector into its original 3D motion components. A 3D deformation vector in a geographic coordinate system is highly beneficial to improving the interpretation of the deformation pattern. A 3D motion decomposition algorithm was proposed and validated on four TomoSAR point clouds in [54]. The method relies on either geometrical [39,43] or geodetic fusion [6,55] of multi-aspect TomoSAR point clouds as input. It estimates the 3D motion components of the queried point target by inclusion of observations from all different viewing geometries and robust inversion with ℓ_1 norm minimization in a local neighborhood. The method allows for highly detailed and shadow-free 3D deformation monitoring, as has been demonstrated in [54]. An example of seasonal motion decomposition on a small test site in Berlin is demonstrated in Figure 7, where it shows the vertical (up), and horizontal (east-west) linear deformation of a railway bridge.

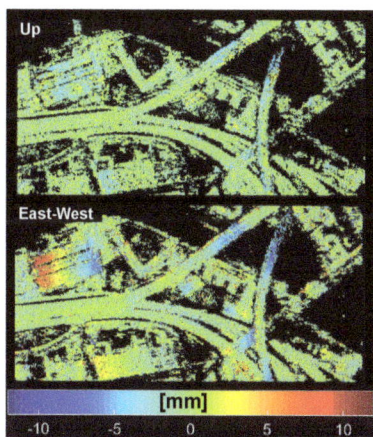

Figure 7. Decomposed seasonal deformation of a railway bridge located in the northeast of Berlin, Germany [54].

2.3.6. Object Reconstruction

Due the development described above, the quality of TomoSAR point cloud, including point density and relative accuracy, has become sufficient for the reconstruction of 3D models of individual objects. We have developed a suite of algorithms that have proved effective for tasks ranging from reconstructing vertical facade [44,45] (see Figure 8), to the detection and reconstruct of a LOD1 model of individual buildings [46,56].

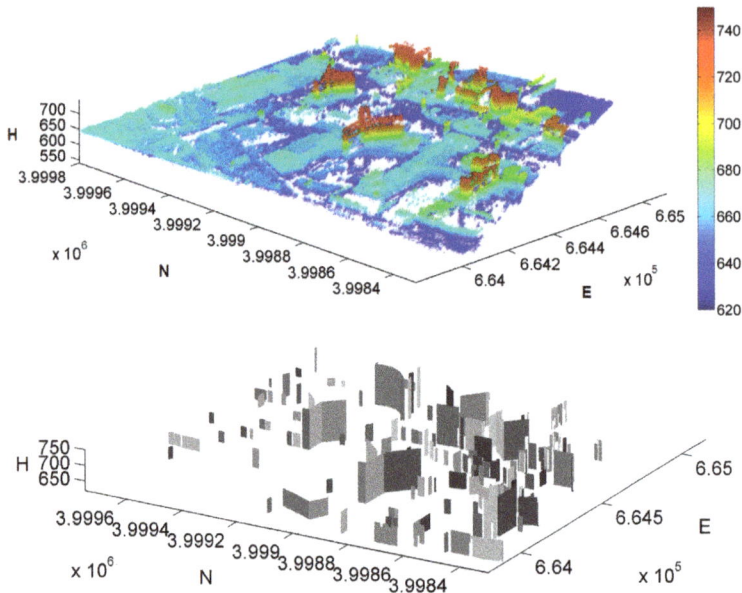

Figure 8. A TomoSAR point cloud of Las Vegas (upper), and the reconstructed facades (lower) [45]. The color of the point cloud represent its height above ellipsoid.

2.4. Object-Based InSAR Algorithms

The reconstruction of such high quality dense point clouds, as in aforementioned examples, are only possible with a stack of fairly high number of images. In practice, we are often faced with a limited number of images. In such situations, a proper algorithm should exploit information from neighboring pixels in order to reduce the number of images needed for a reliable reconstruction, such as adaptive filtering and nonlocal filtering that have been extensively described in previous literature, such as [4,57,58] and [59–61], respectively. However, this section goes beyond these pixel cluster-based methods. It focuses on the recent development of object-based algorithms that explicitly exploit geometric and semantic information to support parameter retrieval in multi-baseline InSAR. To this end, this section introduces the M-SL1MMER algorithm [28], which exploits the freely available building footprint from OpenStreetMap (OSM), and RoMIO (Robust Multi-pass InSAR via Object-based low rank decomposition) [62,63], which exploits the smoothness prior and low rank property of the InSAR data stack of individual objects.

2.4.1. M-SL1MMER

Multiple-snapshot SL1MMER (M-SL1MMER) is an extension of the original SL1MMER algorithm for joint tomographic reconstruction of resolution cells containing scatterers that share, up to quantization errors, the same height (hereafter referred to as "iso-height resolution cells") [28].

Similar approaches based on multiple snapshots or polarimetric channels can be found, for example, in [64–66].

In M-SL1MMER, the iso-height resolution cells are detected by projecting the freely available OSM building footprint [67] to the SAR image, and shifting it toward the near range direction until it reaches the top of the building. Each shifted position of the footprint represents a cluster of iso-height resolution cells. Let a specific iso-height cluster contain M resolution cells; the InSAR measurements $\mathbf{g}_m \in \mathbb{C}^N$ (N being the number of interferograms) of the m-th resolution cell can be approximated with the linear model (see Equation (3)) $\mathbf{g}_m \approx \mathbf{R}_m \gamma_m$ for all $m = 1, \ldots, M$. In addition, we assume without loss of generality that $\mathbf{R}_1 \approx \cdots \approx \mathbf{R}_M$ and rewrite the M linear models in the more compact form $\mathbf{G} \approx \mathbf{R}\boldsymbol{\Gamma}$, where the m-th columns of $\mathbf{G}$ and $\boldsymbol{\Gamma}$ equal $\mathbf{g}_m$ and γ_m, respectively. A key element of M-SL1MMER involves solving the following $\ell_{2,1}$ regularization problem:

$$\hat{\boldsymbol{\Gamma}} = \arg\min_{\boldsymbol{\Gamma}} \frac{1}{2}\|\mathbf{R}\boldsymbol{\Gamma} - \mathbf{G}\|_F^2 + \lambda_{2,1}\|\boldsymbol{\Gamma}\|_{2,1}, \tag{11}$$

where $\|\cdot\|_F$ and $\|\boldsymbol{\Gamma}\|_{2,1} = \sum_{i=1}^{L}\|\gamma^i\|_2$ denote the Frobenius and $\ell_{2,1}$ norms, respectively, and γ^i is the i-th row of $\boldsymbol{\Gamma}$. The $\ell_{2,1}$ norm is known to promote the entries of $\boldsymbol{\Gamma}$ to be jointly sparse among columns. In other words, nonzero rows can be expected in $\hat{\boldsymbol{\Gamma}}$ or its submatrices. Solving the minimization problem in (11) is followed by model selection and amplitude debiasing independently for each resolution cell, as in the SL1MMER algorithm (see Section 2.3).

As a practical demonstration, we reconstructed the elevation of two high-rise buildings using 6 TanDEM-X bistatic sliding spotlight interferograms. The elevation estimates of the upper and lower layers are depicted in Figure 9. In the case of layover, the higher and lower scatterers can be found in the upper and lower layers, respectively. The smooth color transition on the reconstructed building facades already indicates its high quality. Roof-facade and facade-ground interactions are clearly visible in the near and far range, respectively. This can also be observed in the elevation difference map under layover (see Figure 10). The color change from deep blue (near range) to cyan (far range) corresponds to increasing elevation distance between building roof and facade.

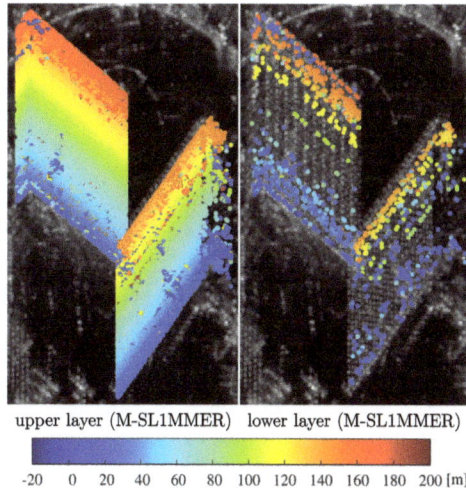

upper layer (M-SL1MMER)   lower layer (M-SL1MMER)

-20   0   20   40   60   80   100   120   140   160   180   200 [m]

Figure 9. Elevation estimates of two test buildings with M-SL1MMER using 6 TanDEM-X bistatic sliding spotlight interferograms. In the case of layover, the higher and lower scatterers appear in the upper and lower layers, respectively [28].

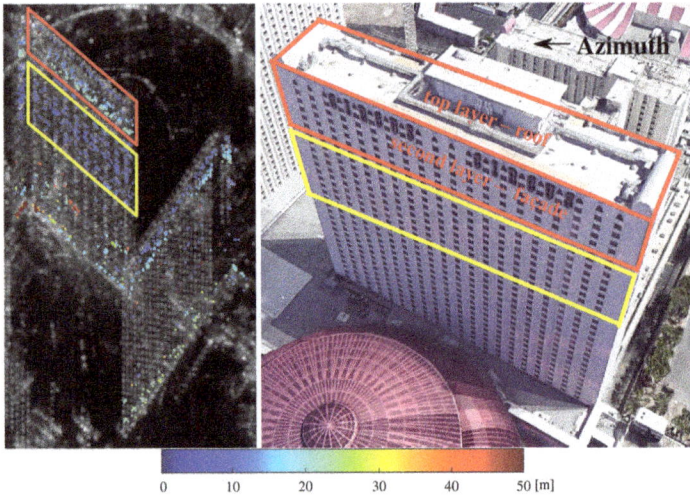

Figure 10. Difference of elevation estimates of higher and lower scatterers in Figure 9 subject to layover. The red and yellow rectangles mark areas where roof-facade and facade-facade interactions are expected, respectively [28].

2.4.2. RoMIO

As a complement to M-SL1MMER, RoMIO does not necessarily require explicit information of the footprints of the objects in the image. It is a more general framework that exploits the low rank property of InSAR phase tensor, because the low rankness of a tensor describes its information entropy, which requires looser signal support than the explicit iso-height line required in M-SL1MMER. RoMIO filters the InSAR data tensor by robustly minimizing its rank. Therefore, it can be regarded as a filtering step in prior to multi-baseline InSAR algorithms. The core RoMIO estimator can be expressed as follows.

$$\{\hat{\mathcal{X}},\hat{\mathcal{E}}\} = \arg\min_{\mathcal{X},\mathcal{E}} \operatorname{rank}(\mathcal{X}) + \lambda_{rank}\|\mathcal{E}\|_0, \quad s.t. \ \mathcal{X}+\mathcal{E}=\mathcal{G}, \tag{12}$$

where $\mathcal{G}$ is the observed InSAR phase tensor, $\mathcal{X}$ and $\mathcal{E}$ model the tensor of the true signal, and the sparse outliers, respectively, $\hat{\mathcal{X}}, \hat{\mathcal{E}}$ are the recovered outlier-free phase tensor and the estimated outlier tensor, respectively, rank($\mathcal{X}$) refers to the multilinear rank of $\mathcal{X}$, and λ_{rank} is the regularization parameter. In practice, the multilinear rank and the ℓ_0 norm are relaxed by the tensor nuclear norm $\|\mathcal{X}\|_*$ and ℓ_1 norm, respectively.

RoMIO reaches filtering performance comparable to state-of-the-art filtering algorithms, i.e., nonlocal means filtering [59,61]. However, it outperforms nonlocal means filtering by a factor of two in terms of the interferometric phase variance when the interferogram is corrupted by 50% outliers [63]. The merit of this extreme robustness in turn improves parameter estimation in multi-baseline InSAR algorithms. In typical settings of the TerraSAR-X high-resolution spotlight image stack, i.e., 10–20 images, SNR of 0–5 dB, a combination RoMIO and PSI outperforms the original PSI by a factor of 10 to 30 in the accuracy of the linear deformation estimates [63].

While optimizing the deformation parameters using multi-baseline InSAR algorithms, e.g., PSI, RoMIO can also make use of the explicit support of objects, such as a given segmentation mask of the SAR image. RoMIO includes a spatial regularization term, e.g., smoothness, of the 2D matrices of the parameters in the estimator [62]. A general form of such regularized estimators can be expressed as follows.

$$\{\hat{\mathbf{S}}, \hat{\mathbf{P}}\} = \arg\min_{\mathbf{S},\mathbf{P}} \frac{1}{2} \|\mathcal{W} \odot (\mathcal{G} - \overline{\mathcal{G}}(\mathbf{S},\mathbf{P}))\|_F^2 + \lambda_{TV} f(\mathbf{S},\mathbf{P}), \tag{13}$$

where $\mathbf{S}$ and $\mathbf{P}$ are the matrices of the elevation and deformation parameters. Similar to other MAP estimators, e.g., Equation (8), the first term on the right-hand side of the estimator is a data fidelity term that calculates a weighted log likelihood between the observed InSAR phase tensor $\mathcal{G}$ and the modeled tensor $\overline{\mathcal{G}}$, where $\mathcal{W}$ denotes an optional weighting tensor, and $\odot$ denotes the element-wise product between two tensors. An example of the weighting tensor can be a tensor comprised of coherence matrices of each interferogram. Pixels of higher coherence are given higher weights. The function $f(\mathbf{S}, \mathbf{P})$ denotes the regularization term that represents the spatial prior of $\mathbf{S}$ and $\mathbf{P}$. The regularization parameter $\lambda_T V$ controls the balance between these two terms. In [62], we made use of the popular total variation as a smoothness prior.

3. Advances in Robust Estimation

3.1. Overview of Advances

Robust estimation in multi-baseline InSAR was sporadically mentioned in previous literature. Some examples include using an adaptive window to improve the covariance matrix estimation [4,57,68], improving the PSI reference network by ℓ_1 norm minimization [23,24], and robust detection of multiple scatterers in TomoSAR [31,33]. However, it was not systematically addressed until [5]. Wang and Zhu pointed out that, due to the existence of non-Gaussian samples and unmodeled phase, e.g., the atmospheric phase, robust estimation in multi-baseline InSAR lies on the following two fundamental problems:

- covariance matrix estimation for DS, due to the existence of non-Gaussian and nonstationary samples
- phase history parameters estimation for both DS and PS, due to observations with large unmodeled phase

The impact of non-robust covariance estimation and the existence of nonstationary phase on parameter estimation in multi-baseline InSAR has been confirmed in several recent works, such as [69–72], and [58,73], respectively. The following sections will elaborate on these two points. The development of robust estimation is greatly associated with DS-based InSAR. Please refer to [74] for a recent review of DS-based InSAR techniques.

3.2. Robust Covariance Matrix Estimation

The estimation of the covariance matrix of a pixel is usually carried out by the sample covariance matrix. Its estimator is shown in Equation (14), where $\mathbf{g}$ is the multivariate observation, and $\mathbf{G}$ is the matrix consisting of M spatial samples, that is $\mathbf{G} = [\mathbf{g}_1, \mathbf{g}_2, ..., \mathbf{g}_M]$. Equation (14) is also the MLE if the samples are complex circular Gaussian (CCG) distributed. Unfortunately, this equation does not always hold in real data. This is why a robust estimator is necessary. A robust covariance estimator should consider the following two scenarios (and the mixture of both):

- the selected samples are non-Gaussian (possibly heavily tailed distribution)
- the expected interferometric phase of the samples is nonstationary, e.g., very strong underlying topographic phase

$$\hat{\mathbf{C}}_{gg} = \frac{1}{M} \sum_{m=1}^{M} \mathbf{g}_m \mathbf{g}_m^H = \frac{1}{M} \mathbf{G} \mathbf{G}^H \tag{14}$$

The following content will summarize the robust covariance estimators, focusing on the points above.

3.2.1. Non-Gaussian Samples

For the first scenario, [5] proposed that the sample covariance matrix can be made more robust by an M-estimator, which is essentially an iterative reweighted sample covariance matrix [75,76]:

$$\hat{\mathbf{C}}_{k+1} = \frac{1}{M} \sum_{m=1}^{M} w \left(\mathbf{g}_m^H \hat{\mathbf{C}}_k^{-1} \mathbf{g}_m \right) \mathbf{g}_m \mathbf{g}_m^H, \tag{15}$$

where m and k are the sample index and the iteration index, respectively, and $w(x)$ is a weighting function of the negative log-likelihood of the sample $\mathbf{g}_m$ to the CCG probability density function (PDF). The weighting function down-weights highly deviated samples whose log-likelihood is small. Equation (15) is solved iteratively. The authors of [5] also proposed an approximation to drop the iterative process, which is the sign covariance matrix (SCM) [77,78]. Extending it to complex number, it is:

$$\hat{\mathbf{C}}_{SCM} = \frac{1}{M} \sum_{m=1}^{M} \|\mathbf{g}_m\|^{-2} \mathbf{g}_m \mathbf{g}_m^H \tag{16}$$

SCM is an engineering solution for fast processing under the general M-estimator's framework. The weighting function is replaced by the inverse of the ℓ_2 norm of the sample. Therefore, only the direction (or sign) of each multivariate sample is considered.

3.2.2. Non-Gaussian Samples with Nonstationary Interferometric Phase

It is often the case that the interferometric phase of the selected samples are not stationary, due to varying topography and motion or other factors. Usually, this type of deterministic phase is estimated and mitigated in prior to covariance estimation. For example, [58] proposed a multi-resolution defringe algorithm to mitigate such nonstationary phase.

Nevertheless, poor estimates significantly affect the covariance matrix estimation. Therefore, [5] proposed a new covariance estimator rank M-estimator (RME) for complex multivariate. The RME is derived by replacing the multivariate $\mathbf{g}$ with its rank $\mathbf{r}$ in Equation (15):

$$\hat{\mathbf{C}}_{RME,k+1} = \frac{1}{M} \sum_{m=1}^{M} w \left(x_m \left(\hat{\mathbf{C}}_{RME,k} \right) \right) \hat{\mathbf{r}}_m . \hat{\mathbf{r}}_m^H \tag{17}$$

The complex rank vector $\mathbf{r}$, analogous to its real number version [78], is defined as follows:

$$\hat{\mathbf{r}}_m = \frac{1}{J} \sum_{j=1}^{J} \frac{\mathbf{g}_m \odot \mathbf{g}_j^*}{\left\| \mathbf{g}_m \odot \mathbf{g}_j^* \right\|}, \tag{18}$$

where $\mathbf{g}_j$ is a direct neighborhood sample of $\mathbf{g}_m$, and $\odot$ denotes the Hadamard product. The multiplication of the complex conjugate of a direct neighbor mitigates the nonstationary interferometric phase of $\mathbf{g}_m$. Due to the multiplication, the RME is a fourth-order descriptor of the sample statistics. An element-wise square root on $|\hat{\mathbf{C}}_{RME}|$ should be performed in order to obtain the second-order momentum. It was proven that the element-wise square root of $|\hat{\mathbf{C}}_{RME}|$ approaches $|\hat{\mathbf{C}}_{\mathbf{gg}}|$ asymptotically under CCG distribution when calculating the rank using one neighborhood sample [5].

3.2.3. Comparison

We compared the sample covariance matrix, M-Estimator, and the RME under three different scenarios: (1) multivariate CCG, (2) a heavily tailed multivariate distribution (complex t-distribution), and (3) nonstationary multivariate complex t-distribution. For each scenario, 1000 ten-acquisition vectors were simulated according to the distribution and a predefined coherence matrix that has a exponential decay of the coherence w.r.t. the temporal baseline. In the last scenario, linear phase fringes with ten different fringe frequencies randomly picked within $[0\ \pi/10]$ were added to the phases of the ten acquisitions.

The results are shown in Figure 11, where each row corresponds to the three scenarios, respectively. The top left subplot can be regarded as the ground truth, because MLE is the optimal estimator under CCG, and is asymptotically unbiased. All three estimators can preserve the correct shape of the covariance matrix under CCG. The MLE fails in the second scenario, where the samples are contaminated by outliers. The coherence is usually overestimated because of the large amplitude of the outliers. In the last scenario, both MLE and M-estimator are not capable of dealing with nonstationary phases. Heavy underestimation occurs because of the summation of the complex numbers with non-constant phases. The estimates of M-estimator are extremely low due to more summation operations caused by the iterative process. Last but not least, RME is invariant to such nonstationary phase, and hence maintains good performance in all conditions.

A quantitative experiment shows that the robust estimator is extremely effective for samples with low coherence. At true coherence of 0.2, M-estimator outperforms the Gaussian MLE by a factor of 1.1 to 2.3, and a factor of 1 to 10, in terms of the accuracy and the bias, respectively, under a wide range of outlier percentages [5].

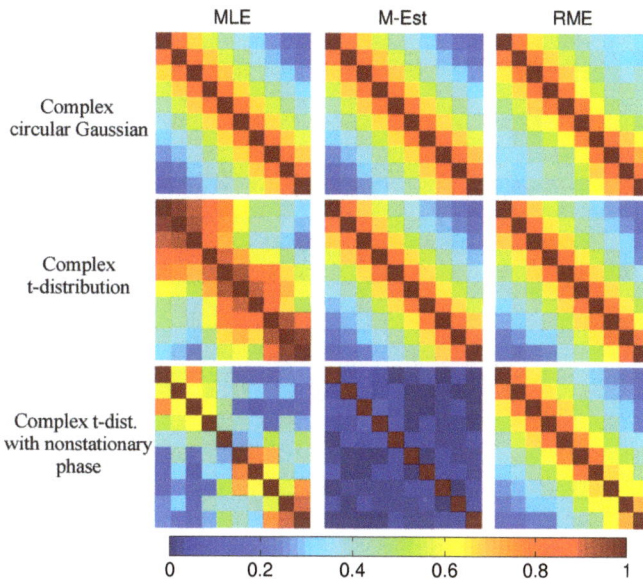

Figure 11. Comparison of three covariance matrix estimators under three different observation cases: first row: complex circular Gaussian, second row: complex t-distribution with one degree of freedom, and third row: nonstationary complex t-distribution with one degree of freedom. First column: MLE (under Gaussian), second column: M-estimator with t-distribution weighting, and third column: rank M-estimator with t-distribution weighting.

3.3. Robust Phase History Parameters Retrieval

A robust covariance matrix estimate alone is not sufficient for a robust estimation of the phase history parameters, i.e., elevation, and motion parameters, because a multi-pass InSAR observation $\mathbf{g} \in \mathbb{C}^N$ may contain an unmodeled phase, e.g., uncompensated atmospheric phase, unmodeled motion phase, etc. The following content provides examples of robust estimators for the retrieval of the phase history parameters of both PS and DS.

3.3.1. Robust PS Estimator

The general form of the MLE of PS phase history parameters can be expressed as follows:

$$\hat{\theta}_{MLE} = \arg\min_{\theta} \|\mathbf{g} - \bar{\mathbf{g}}(\theta)\|_2^2, \tag{19}$$

where $\bar{\mathbf{g}}(\theta)$ is the modeled PS signal. Equation (19) is shown to be equivalent to Equation (6) in [79]. Similar to the robust covariance estimator, it can be robustified by an M-estimator:

$$\hat{\theta}_{M-est} = \arg\min_{\theta} \sum_{i=1}^{N} \rho\left(\text{Re}\left[\varepsilon_i(\theta)\right]/\sigma_R\right) + \rho\left(\text{Im}\left[\varepsilon_i(\theta)\right]/\sigma_I\right), \tag{20}$$

where the residual $\varepsilon_i(\theta)$ equals $g_i - \bar{g}_i(\theta)$, $\text{Re}[\cdot]$, $\text{Im}[\cdot]$ are the real and imaginary parts of a complex number, and σ_R and σ_I are the standard deviations of the real and imaginary parts of the residual, respectively. The function $\rho(x)$ is the so-called robust loss function that can be derived from the PDF of the contaminated distribution of $\mathbf{g}$, if it is known. However, it is usually unknown in practice. We shall use stable empirical functions instead, e.g., the Tukey biweight function.

3.3.2. Robust DS Estimator

According to [80], the MLE of DS phase history parameters can be expressed as follows:

$$\hat{\theta}_{MLE} = \arg\min_{\theta} \left\{ \mathbf{g}^H \mathbf{\Phi}(\theta) |\mathbf{C}|^{-1} \mathbf{\Phi}(\theta)^H \mathbf{g} \right\} \tag{21}$$

where $\mathbf{\Phi}$ is a diagonal matrix of the modeled interferometric phase. If stationarity is assumed for a DS and its neighborhood, one can treat a cluster of DSs as a single PS by averaging them, as proposed in SqueeSAR [4]. Then, the robustified DS estimator is identical to Equation (20).

However, if the objective is a full inversion of individual single-look DS observation (without averaging) without the strict assumption of phase stationarity, the robustified estimator is shown in [5] to be in the following form:

$$\hat{\theta}_{M-est} = \arg\min_{\theta} \left\{ \varepsilon^H(\theta) \mathbf{W}(\bar{\varepsilon}) \varepsilon(\theta) \right\}, \tag{22}$$

where the residuals $\varepsilon(\theta)$ is shown in Equation (23). It is whitened by a robust covariance matrix estimate, e.g., $\hat{\mathbf{C}}_{RME}$. The matrix $\mathbf{W} \in \mathbb{R}^{N \times N}$ is a diagonal robust weighting matrix computed from the mean residual $\bar{\varepsilon}$. Because of possible outliers in the residual, $\bar{\varepsilon}$ should also be robustly estimated, for example by a robust weighted averaging of $\varepsilon(\theta)$ of the selected samples.

$$\varepsilon(\theta) = |\hat{\mathbf{C}}_{RME}|^{-0.5} \mathbf{\Phi}(\theta)^H \mathbf{g} \tag{23}$$

To summarize, Equation (22) is a joint estimation of the phase parameters of individual single-look DS observations in a neighborhood. It is solved iteratively. Its computation should begin with initial estimates of each sample in the neighborhood (assumed to be the same), which jointly determine the initial weighting matrix. The same weighting matrix is used to retrieve the parameters of each single-look DS in the neighborhood, and is updated on the basis of all the estimates upon finishing one iteration.

To demonstrate the robustness of the estimator, Figure 12 shows the linear deformation rate of the volcano Stromboli, Italy, estimated by the robust DS phase history parameter retrieval method. Parameter estimation in active volcanic areas is challenging due to strong decorrelation, and the varying deformation model. In the experiment, only 16 interferograms acquired in 2008 were used. We can see that scatterers over 50% of the surface area were retrieved, although most of them did not undergo any significant deformation. The crater shows an uplift of 10 cm/year, and the southern slope undergoes a subsidence of up to 20 cm/year. This may suggest certain displacement of the magma underneath the volcano.

Figure 12. The linear deformation rate of the volcano Stromboli, in Italy, estimated by the robust DS phase history parameter retrieval method. In total, 16 interferograms acquired in 2008 were used. The crater shows an uplift of 10 cm/year, and the southern slope undergoes a subsidence of up to 20 cm/year. This may suggest certain displacement of the magma underneath the volcano. Courtesy: the tropospheric correction was done by Cong et al.

4. Advances in Absolute Positioning

A unique feature of TerraSAR-X is its precise orbit determination and high precision range measurements, which allows for an unprecedented 2D localization accuracy of image pixels below one meter. In recent years, this level of accuracy has been further improved by thorough consideration of the most prominent error factors affecting range and azimuth measurements of SAR, a method termed SAR imaging geodesy [7,81]. SAR imaging geodesy is seen as a great leap in SAR technology, because it extends the applications of SAR to the geodetic positioning domain rather than the imaging domain. Two of the numerous application examples of SAR imaging geodesy are geodetic stereo SAR [82], a method that retrieves the precise 3D absolute position of a target by combining its 2D radar timings from different orbit tracks, and a framework called geodetic InSAR [6], in which multi-baseline InSAR and stereo SAR are combined to achieve absolute 4D InSAR point clouds. A brief introduction to the two methods is given below, and the most recent advances of these techniques and their new applications are described.

The SAR imaging geodesy method aims at attaining 2D absolute pixel localization [7]. A single pixel in a focused complex SAR image is localized, in across-track, by range τ_{rg} and, in along-track, by azimuth t_{az} times. For a point target inside the mentioned pixel, the following equations read:

$$\tau_{rg} = \frac{2R}{c} + \delta\tau_{SD} + \delta\tau_O + \delta\tau_F + \delta\tau_I + \delta\tau_T + \delta\tau_G$$
$$t_{az} = t + \delta t_{SD} + \delta t_O + \delta t_F + \delta t_G, \tag{24}$$

where R is the geometric distance from the sensor to the center of the pixel in meters and c is the speed of light in vacuum; the other terms are all expressed in seconds. The raw acquisition time is denoted by t and the timing error terms subscripted by SD, O, F, I, T and G represent delays caused by satellite dynamics, orbit inaccuracies, feature localization error, ionospheric delay, tropospheric delay, and geodynamic effects, respectively. The magnitude of individual errors range from a couple of centimeters for the ionospheric effect, if the satellite operates in X-band, followed by decimeter regimes for satellite dynamic effects and geodynamic effects for both components, to up to a few meters for the tropospheric effect, depending on the weather conditions and the average incidence angle of the acquired SAR images. Some of the mentioned errors and their effects on SAR measurements are shown in Figure 13. The curved propagation path shown in Figure 13 is highly exaggerated for visualization purposes only. In order to remove the mentioned timing errors, the imaging geodesy method exploits the highly precise orbit data of TerraSAR-X and Tandem-X [1,2,83], utilizes a highly sophisticated SAR processor to avoid unnecessary approximations [84], precisely extracts targets with sub-pixel sensitivity [85,86], and corrects the path delay and geodynamic errors by global numerical weather data [81,87] and state-of-the-art geodetic models [88].

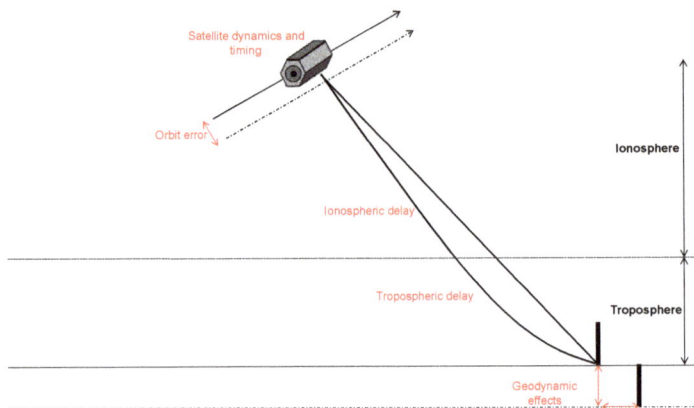

Figure 13. The errors affecting range and azimuth timings of SAR measurements, colorized in red. Orbit errors cause the satellite trajectory to deviate from the true track, while satellite dynamics and atmospheric disturbances cause delays in the timings, which lead to incorrect annotation of τ_{rg} and t_{az}. Geodynamic effects change the position of a target on the ground, which again hampers the accuracy of the timings. Please note that the atmospheric effect shown in the figure is highly exaggerated for visualization purposes only. The main cause for atmospheric delay is the decrease of the speed of light.

By combining the τ_{rg} and t_{az} of the same target visible in SAR images acquired from two or more different viewing geometries, the stereo SAR method determines the 3D position of the target (see Figure 14). The 2D radar timing coordinates of a particular target in the SAR image $x_T = (t_{az}, \tau_{rg})$ are linked to their corresponding 3D coordinates on the surface of the Earth $\mathbf{X}_T = (X, Y, Z)$ by the range-Doppler equation system [85]:

$$|\mathbf{X}_S - \mathbf{X}_T| - c \cdot \tau_{rg} = 0$$
$$\frac{\dot{\mathbf{X}}_S (\mathbf{X}_S - \mathbf{X}_T)}{|\dot{\mathbf{X}}_S||\mathbf{X}_S - \mathbf{X}_T|} = 0 \tag{25}$$

with $\mathbf{X}_S$ and $\dot{\mathbf{X}}_S$ denoting the position and the velocity vector of the satellite relative to t_{az}, and τ_{rg} being the calibrated two-way traveled time from the satellite to the target. The variable t_{az} is implicitly included in the second equation relating the state-vector of the satellite to the time of the acquisition using a polynomial model [82]. The estimation of the coordinates is performed by least squares adjustment plus stochastic modeling of timing observations using the variance component estimation (VCE) [82]. The relative accuracy of the estimated coordinates depends on the SCR of the target, the precision of the external atmospheric and geodynamic corrections, the degree of difference in the combined viewing geometries, and the number of SAR acquisitions.

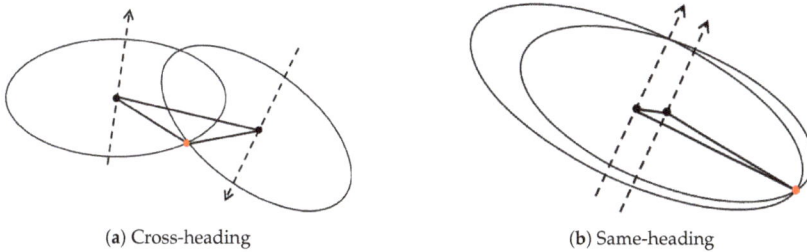

(a) Cross-heading   (b) Same-heading

Figure 14. Localization of a point target (red dot) from (**a**) cross-heading and (**b**) same-heading satellite tracks. The satellites are shown by black dots; their trajectories are presented by dashed lines and the baselines are depicted by solid lines between the satellite positions. The black circles are defined by the range-Doppler equations and their intersection leads to the 3D position of the target [89].

4.1. Overview of Advances

SAR imaging geodesy was first named in 2011 by Eineder et al., since the method incorporates correction principles used in geodesy with SAR [7]. Schubert et al. reported on the correction of atmospheric delays by local height dependent models in [90]. Gisinger studied the effect of utilizing different mapping functions for converting the zenith atmospheric delays into the radar line of sight in [91]. These methods used the local GNSS zenith path delays for atmospheric corrections. Cong et al. introduced atmospheric correction through the 3D integration of weather data obtained from the European Center for Medium-Range Weather Forecasts (ECMWF) and using global TEC maps [81]. Apart from atmospheric errors, calibration of internal electronic delays of the SAR sensor was investigated in [7] and the precision of azimuth timing was improved by calibrating the sensor's internal clock rate [92]. The most prominent geodynamic effects, such as solid earth tides, pole tides, and continental drifts, were included in further studies [7,81,86,93]. In order to improve the localization precision into sub-centimeter regimes, Balss et al. further modeled geodynamic effects with smaller magnitudes, such as atmospheric pressure loading, ocean tidal loading, ocean pole tides, and atmospheric tidal loading [94]. In all the studies, the geodynamic effects were considered by the state-of-the-art models of the IERS 2010 convention [88]. The already precise orbit of TerraSAR-X [83] has been further improved by modeling the non-gravitational forces and also solar radiation pressure modeling [1]. The world-wide reproducibility of high precision measurements was demonstrated in [95] and an operational processor called the SAR Geodesy Processor (SGP) was introduced in [87]. Relative to applications, the high precision ranging measurement of TerraSAR-X has been exploited for maritime purposes [96,97]. In terms of achievable accuracy, SAR imaging geodesy is capable of localizing corner reflectors with 1.16 cm and 1.85 cm range and azimuth standard deviations, respectively [98].

The first results on 3D localization of CRs by means of stereo SAR was reported in [99]. Although 3D positioning using multi-aspect TerraSAR-X images had been previously demonstrated in [100–102], the results in [99] were unique in the sense that the stereo processing was carried out on thoroughly calibrated range and azimuth timings. Gisinger et al. demonstrated the applicability of the geodetic stereo SAR method not only on CRs but on opportunistic non-ideal scatterers such as PS in an urban area [82]. The manually extracted scatterers could be localized with 3D precision better than 10 cm [82], which paved the way for new geodetic applications such as secular ground movement estimation using natural PS [103,104], high precision mapping of road networks (DriveMark) [105], and highly precise automatic SAR Ground Control Point (GCP) generation [89,106–109]. In terms of achievable accuracy, geodetic stereo SAR is able to localize corner reflectors with 3D precision better than 4 cm and an absolute accuracy of 2–3 cm when compared to independently surveyed reference positions [82].

4.2. Geodetic InSAR

The geodetic InSAR approach integrates the capabilities of multi-baseline InSAR with SAR imaging geodesy and stereo SAR techniques. The goal of the framework is to tackle the shortcomings of both methods: the relative estimates of all InSAR approaches and the small number of points that can be absolutely localized by geodetic stereo SAR. Therefore, it tends to achieve absolute positioning of a large number of scatterers by exploiting the advantages of both techniques. In the following, the workflow of the geodetic InSAR technique is described and some example applications are demonstrated.

4.2.1. SAR GCP Generation

The first major part of the procedure is concerned with extraction of GCPs from multi-aspect SAR images. This includes [89]:

Step 1　Detection and matching of identical PS from SAR images acquired from different orbits. In the reference geodetic SAR tomography technique this task was performed manually [6]. At the current state of the framework, the identification of common PS can be carried out using the PSI multi-track fusion algorithm [39] for same-heading tracks and utilizing high resolution optical data [106] or external geospatial road network data [109] for cross-heading tracks. A combination of all the mentioned methods for automatic detection of large number of GCPs was used in [89].

Step 2　Precise timing extraction of PS from stacks of non-coregistered SLC images. This is done by PTA [85,86].

Step 3　PS visibility check and initial outlier removal. The time series of phase noise approximated by SCR of each PS [19] is analyzed and the outliers are robustly removed by the adjusted box plot method [110].

Step 4　Correction of PS timings in the stack of images using imaging geodesy.

Step 5　Absolute 3D positioning of each PS by the stereo SAR method [82]. The posterior quality measures of the observations and the estimates are also reported in this step.

4.2.2. Absolute Localization of InSAR Point Clouds

The main objective of the geodetic InSAR framework is to resolve the DEM error of the reference point with respect to which the topography and deformation parameters are estimated [20,21,40]. The geodetic InSAR approach can overcome this problem, to some extent, in two ways dependent on the number of available GCPs. If only a small number of GCPs are available, the best candidate will be chosen as the reference point during PSI/TomoSAR processing and at the final stage the geocoded coordinates of all points in the point clouds are shifted toward the absolute coordinates of this point [6]. If a large number of GCPs are available, for instance using the GCP generation approaches in [89,107], the DEM error of the reference point is approximated as a post-processing step. Therefore, the difference in ellipsoidal heights of GCPs and their corresponding geocoded PS heights are calculated and a height offset is robustly estimated. The height offset is added to the geocoded PS heights and an updated geocoding is carried out which results in absolute coordinates of the InSAR point cloud [111].

4.2.3. Applications

To conclude this subsection, a few examples and applications of the geodetic InSAR framework are demonstrated below.

Figure 15 shows the city of Oulu in Finland overlaid by 2049 GCPs obtained from four stacks of TerraSAR-X high resolution spotlight images.

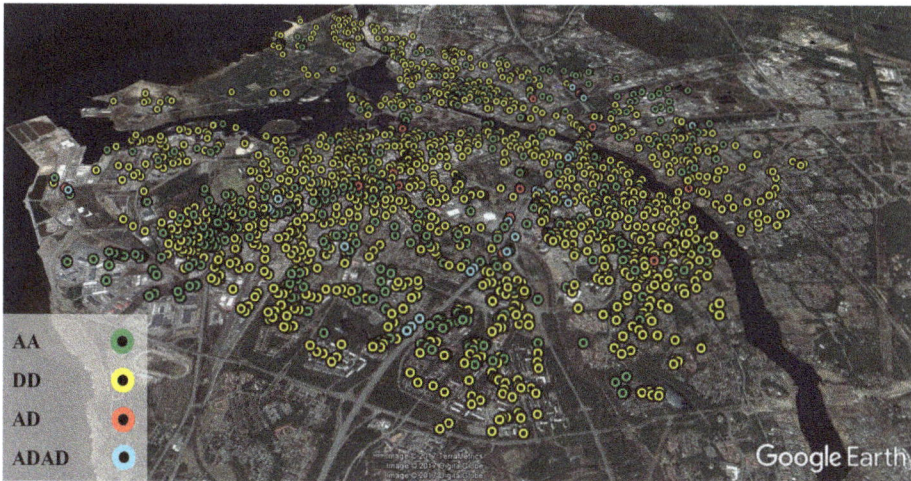

Figure 15. Total number of 2049 GCPs in Oulu, color-coded based on the geometry configuration used for their positioning (AA: ascending-ascending, DD: descending-descending, AD: ascending-descending and ADAD: quad geometry) [89]. The underlying optical image is taken from Google Earth.

The GCPs are color-coded based on the underlying geometry configuration used for their localization, where AA, DD, and AD stand for ascending-ascending, descending-descending and ascending-descending orbits, respectively; ADAD means that scatterers were localized from all the four viewing geometries. It is observed that the entire central area of Oulu is covered with the generated GCPs. The candidates from the same-heading geometries stem from built areas, while the ones from cross-heading orbits include the bases of lamp poles, street lights, and traffic lights. The statistics of the generated GCPs are reported in Table 3, which demonstrate the extremely high potential of TerraSAR-X for precise 3D positioning.

Comparison with a reference LiDAR point cloud shows that we can achieve a horizontal absolute accuracy of 20 cm using just a single GCP to correct the geocoding of an InSAR point cloud [6,55]. Therefore, employing over one thousand GCPs, as shown previously, can achieve extremely high absolute accuracy, presumably in the order of centimeter. In order to demonstrate this, a close comparison of two cross-heading InSAR point clouds before and after height correction is shown in Figure 16, where the red and green points represent the PS of descending and ascending tracks, respectively. It can be seen that after the calibration of the height of the reference point using the GCPs, the endpoint of building facades correctly match.

Table 3. Averaged statistics based on the stereo SAR least squares estimated 3D coordinate standard deviations in Oulu. The letters A and D stand for ascending and descending geometries, respectively. The sample mean and standard deviation are denoted by μ and σ and $S_{[ENH]}$ represents the local coordinates standard deviations within a 95% confidence level [89].

Geometry	Number of Scatterers	μ_{s_E} (cm)	μ_{s_N} (cm)	μ_{s_H} (cm)	σ_{s_E} (cm)	σ_{s_N} (cm)	σ_{s_H} (cm)
AA	565	17.73	5.04	15.87	11.98	2.63	11.09
DD	1417	15.08	3.80	16.71	10.38	2.10	11.30
AD	24	2.26	2.50	1.75	0.99	1.11	0.83
ADAD	43	1.17	1.40	1.12	0.42	0.55	0.37

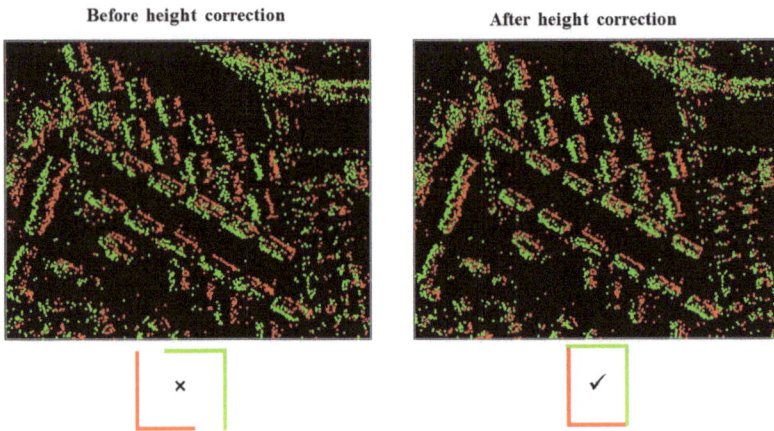

Before height correction   After height correction

Figure 16. Demonstration of absolute localization of PSI point clouds obtained from an ascending and a descending orbit track of Oulu. The endpoints of buildings visible from each geometry match correctly with the endpoints from the opposing geometry.

To give an impression of the fused TomoSAR point cloud of a large area, Figure 17 shows a result obtained by fusing four TomoSAR point clouds of Berlin obtained from two pairs of cross-heading high resolution TerraSAR-X spotlight images that are fused by selecting an identical GCP as the reference point of all point clouds. The point cloud has in total 63 million scatterers in an area of 50 km^2. Such shadow-free highly detailed TomoSAR point clouds can be further utilized to reconstruct dynamic 3D and 4D city models [44–46,112].

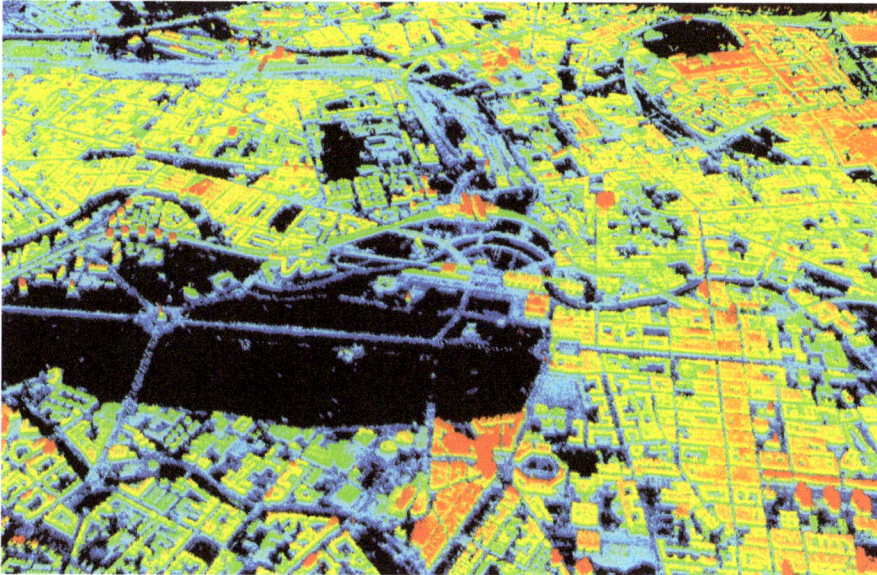

Figure 17. 3D view of central Berlin after geodetic registration of four TomoSAR point clouds obtained from a pair of cross-heading high resolution TerraSAR-X spotlight data. The height is color-coded and ranges between 70 m and 110 m [6].

5. Conclusions and Outlook

This paper provides a review of the multi-baseline InSAR techniques in the scope of TerraSAR-X data. It covers the evolution of multi-baseline InSAR techniques, particularly with respect to improving the relative estimation accuracy, introducing robustness to the estimators, and achieving accurate absolute positioning of scatterers, which includes bridging the absolutely located scatterers with the relative measures obtained from multi-baseline techniques. Particular focus was placed on our own development work, specifically SL1MMER, M-SL1MMER, Tomo-GENESIS (TomoSAR), RIO (robust estimation), RoMIO (object-based InSAR), and geodetic InSAR (absolute positioning).

Looking into the future, the next generation spaceborne SAR missions, including high resolution wide swath (HRWS) and Tandem-L, will simultaneously possess high resolution and global coverage, which would enable novel applications such as monitoring global changes. Retrieving geo-parameters from these data will require not only new technological approaches to manage large amounts of data, but also new analysis methods. In the following, we would like to point out some promising future directions:

- Big data management technologies: So far, besides big missions, such as global TanDEM-X DEM generation, scientists are dealing with SAR data in the order of up to terabytes. However, this is about to change. Already today, petabytes of Sentinel-1 data are openly accessible to the public. Yet, only very limited groups are capable of national-scale InSAR data processing, to say nothing about global. To be prepared for the future, novel big geo-data management technologies are of high relevance.

- Fast and accurate parameter inversion algorithms: The development of inversion algorithms should keep up with the pace of data growth. For example, as a pre-study of Tandem-L, sequential interferometric phase estimators are proposed instead of full covariance matrix inversion to tackle the challenge of big InSAR data [72]. Fast solvers are demanded for many advanced parameter inversion models that often involve non-convex, nonlinear, and complex-valued optimization problem, such as CS-based tomographic inversion, or low rank complex tensor decomposition. Besides aforementioned model-based inversion methods, recently, data-driven machine learning/deep learning methods have boosted the baseline performances in many remote sensing problems [113], mostly in classification and detection tasks, yet its potential in InSAR processing or more generally in geoparamater estimation is not yet exploited at all. This deserves more attention of the community.

- Complicated motion: Up to now, only limited motion models, such as linear, seasonal or a combination of several basic models, are considered for deformation estimations of InSAR. There are also studies using model order selection to detect different types of motion either being embedded in the estimation [114] or considered as a post-processing [115]. However, the actual motion can be far more complex than any model can describe. The weekly repeat cycle and long-term monitoring capability of future sensors will enable retrieving much more complex motion models, and even allow performing classification of different types of motions and detecting anomalies. This calls for more sophisticated algorithms.

- Data assimilation: At present, the interferometric stack is usually a static cube of interferograms. As Sentinel-1 provides global coverage every six days, new stacking and multi-pass InSAR concepts should be able to include new images without excessive computational burden. This requires development of the data assimilation strategy, as well as novel inversion algorithms that only require the new measurements and the previous estimates for updating the parameters of interest.

- Multi-sensor data fusion: In the Copernicus era, it is standard that more than one data source, such as SAR and optical, is available at any test site. Intelligent use of the complementary peculiarities of the ever-increasing number of diverse remote sensing sensors and other geo-data sources has become the natural choice for many applications [116]. Some preliminary work in the community demonstrated that introducing the geometric prior or semantic prior to InSAR or TomoSAR reconstruction could significantly reduced the number of required SAR data while retaining the estimation accuracy [28,63]. This is definitely a promising future direction.

Author Contributions: X.X.Z. conceived of and designed most of the algorithms mentioned in this paper, e.g., Tomo-GENESIS, SL1MMER, M-SL1MMER and geodetic InSAR, as well as contributed to the structure design and material selection of the paper, and thoughts of future research directions and partly the paper writing. Y.W. performed partly the experiments described in Sections 2 and 3, partly contributed to the paper writing, and contributed to organizing the contents of the paper. S.M. contributed partly to the writing of Section 2, carried out experiments described in Section 4, and wrote this section in its entirety. He also proofread the manuscript. N.G. contributed "Staring Spotlight TomoSAR" in Section 2.3 and "M-SL1MMER" in Section 2.4.

Funding: This research was funded by the European Research Council (ERC) under the European Union's Horizon 2020 research and innovation program with the grant number ERC-2016-StG-714087 (Acronym: So2Sat, project website: www.so2sat.eu), and the Helmholtz Association under the framework of the Young Investigators Group "Signal Processing in Earth Observation (SiPEO)" with the grant number VH-NG-1018 (project website: www.sipeo.bgu.tum.de). **Acknowledgments:** The authors gratefully acknowledge the Gauss Centre for Supercomputing e.V. (www.gauss-centre.eu) for funding this project by providing computing time on the GCS Supercomputer SuperMUC at Leibniz Supercomputing Centre (www.lrz.de).

Conflicts of Interest: The authors declare no conflict of interest. The founding sponsors had no role in the design of the study; in the collection, analyses or interpretation of data; in the writing of the manuscript; nor in the decision to publish the results.

Abbreviations

The following abbreviations are used in this manuscript:

ADI	Amplitude dispersion index
CCG	Complex circular Gaussian
CS	Compressive sensing
DEM	Digital elevation model

DS	Distributed scatterer
D-TomoSAR	Differential TomoSAR
ECMWF	European Center for Medium-Range Weather Forecasts
GCP	Ground control point
HRWS	High resolution wide swath
ICP	Iterative closest point
IERS	International Earth Rotation and Reference Systems Service
InSAR	SAR interferometry
LOS	Line of sight
MAP	Maximum a posteriori
MLE	Maximum likelihood estimator
M-SL1MMER	Multiple-snapshot SL1MMER
OSM	OpenStreetMap
PDF	Probability density function
PSI	Persistent scatter interferometry
PS	Point/Persistent Scatterer
RIO	Robust InSAR optimization
RME	Rank M-estimator
ROI	Region of interest
ROMIO	Robust multi-pass InSAR via object-based low rank decomposition
SAR	Synthetic aperture radar
SCR	Signal-to-clutter ratio
SL1MMER	Scale-down by L1 norm minimization, model selection, and estimation reconstruction
SNR	Signal-to-noise ratio
TEC	Total electron content
TomoSAR	SAR tomography
VHR	Very high resolution

References

1. Hackel, S.; Montenbruck, O.; Steigenberger, P.; Balss, U.; Gisinger, C.; Eineder, M. Model improvements and validation of TerraSAR-X precise orbit determination. *J. Geodesy* **2017**, *91*, 547–562. [CrossRef]
2. Hackel, S.; Gisinger, C.; Balss, U.; Wermuth, M.; Montenbruck, O. Long-Term Validation of TerraSAR-X Orbit Solutions with Laser and Radar Measurements. *Remote Sens.* **2018**, *10*, 762. [CrossRef]
3. Zhu, X.X. Very High Resolution Tomographic SAR Inversion for Urban Infrastructure Monitoring—A Sparse and Nonlinear Tour. Ph.D. Thesis, Technische Universität München, München, Germany, 2011.
4. Ferretti, A.; Fumagalli, A.; Novali, F.; Prati, C.; Rocca, F.; Rucci, A. A New Algorithm for Processing Interferometric Data-Stacks: SqueeSAR. *IEEE Trans. Geosci. Remote Sens.* **2011**, *49*, 3460–3470. [CrossRef]
5. Wang, Y.; Zhu, X.X. Robust Estimators for Multipass SAR Interferometry. *IEEE Trans. Geosci. Remote Sens.* **2016**, *54*, 968–980. [CrossRef]
6. Zhu, X.X.; Montazeri, S.; Gisinger, C.; Hanssen, R.F.; Bamler, R. Geodetic SAR Tomography. *IEEE Trans. Geosci. Remote Sens.* **2016**, *54*, 18–35. [CrossRef]
7. Eineder, M.; Minet, C.; Steigenberger, P.; Cong, X.Y.; Fritz, T. Imaging Geodesy—Toward Centimeter-Level Ranging Accuracy with TerraSAR-X. *IEEE Trans. Geosci. Remote Sens.* **2011**, *49*, 661–671. [CrossRef]
8. Zhu, X.X.; Bamler, R. Superresolving SAR Tomography for Multidimensional Imaging of Urban Areas: Compressive sensing-based TomoSAR inversion. *IEEE Signal Process. Mag.* **2014**, *31*, 51–58. [CrossRef]
9. Bamler, R.; Hartl, P. Synthetic aperture radar interferometry. *Inverse Probl.* **1998**, *14*, R1. [CrossRef]
10. Reigber, A.; Moreira, A. First demonstration of airborne SAR tomography using multibaseline L-band data. *IEEE Trans. Geosci. Remote Sens.* **2000**, *38*, 2142–2152. [CrossRef]
11. Lombardini, F.; Montanari, M.; Gini, F. Reflectivity estimation for multibaseline interferometric radar imaging of layover extended sources. *IEEE Trans. Signal Process.* **2003**, *51*, 1508–1519. [CrossRef]
12. Fornaro, G.; Serafino, F.; Soldovieri, F. Three-dimensional focusing with multipass SAR data. *IEEE Trans. Geosci. Remote Sens.* **2003**, *41*, 507–517. [CrossRef]

13. Zhu, X. Spectral Estimation for Synthetic Aperture Radar Tomography. Master's Thesis, Lehrstuhl für Methodik der Fernerkundung, TU München, Munich, Germany, 2008.

14. Fornaro, G.; Reale, D.; Serafino, F. Four-dimensional SAR imaging for height estimation and monitoring of single and double scatterers. *IEEE Trans. Geosci. Remote Sens.* **2009**, *47*, 224–237. [CrossRef]

15. Zhu, X.X.; Bamler, R. Very high resolution spaceborne SAR tomography in urban environment. *IEEE Trans. Geosci. Remote Sens.* **2010**, *48*, 4296–4308. [CrossRef]

16. Zhu, X.X.; Bamler, R. Let's do the time warp: Multicomponent nonlinear motion estimation in differential SAR tomography. *IEEE Geosci. Remote Sens. Lett.* **2011**, *8*, 735–739. [CrossRef]

17. Crosetto, M.; Monserrat, O.; Cuevas-González, M.; Devanthéry, N.; Crippa, B. Persistent Scatterer Interferometry: A review. *ISPRS J. Photogramm. Remote Sens.* **2016**, *115*, 78–89. [CrossRef]

18. Ferretti, A.; Prati, C.; Rocca, F. Permanent scatterers in SAR interferometry. *IEEE Trans. Geosci. Remote Sens.* **2001**, *39*, 8–20. [CrossRef]

19. Adam, N.; Kampes, B.M.; Eineder, M. The development of a scientific persistent scatterer system: Modifications for mixed ERS/ENVISAT time series. In Proceedings of the Envisat and ERS Symposium, Salzburg, Austria, 6–10 September 2004; pp. 1–9.

20. Kampes, B.M. *Radar Interferometry: Persistent Scatterer Technique*; Remote Sensing and Digital Image Processing; Springer: Dordrecht, The Netherlands, 2006.

21. Van Leijen, F. Persistent Scatterer Interferometry Based on Geodetic Estimation Theory. Ph.D. Thesis, Delft University of Technology, Delft, The Netherlands, 2014.

22. Wang, Y.; Zhu, X.X.; Bamler, R. An Efficient Tomographic Inversion Approach for Urban Mapping Using Meter Resolution SAR Image Stacks. *IEEE Geosci. Remote Sens. Lett.* **2014**, *11*, 1250–1254. [CrossRef]

23. Rodriguez Gonzalez, F.; Bhutani, A.; Adam, N. L1 network inversion for robust outlier rejection in persistent Scatterer Interferometry. In Proceedings of the 2011 IEEE International Geoscience and Remote Sensing Symposium, Vancouver, BC, Canada, 24–29 July 2011; pp. 75–78. [CrossRef]

24. Adam, N.; Gonzalez, F.R.; Parizzi, A.; Brcic, R. Wide area Persistent Scatterer Interferometry: Current developments, algorithms and examples. In Proceedings of the 2013 IEEE International Geoscience and Remote Sensing Symposium, Melbourne, VIC, Australia, 21–26 July 2013; pp. 1857–1860. [CrossRef]

25. Rodriguez Gonzalez, F.; Adam, N.; Parizzi, A.; Brcic, R. The Integrated Wide Area Processor (IWAP): A Processor For Wide Area Persistent Scatterer Interferometry. In Proceedings of the ESA Living Planet Symposium 2013, Edinburgh, UK, 9–13 September 2013; pp. 1–4.

26. Zhu, X.; Bamler, R. Very high Resolution SAR tomography via Compressive Sensing. In Proceedings of the Fringe 2009 Workshop, Frascati, Italy, 30 November–4 December 2009; pp. 1–7.

27. Zhu, X.; Bamler, R. Tomographic SAR Inversion by L1-Norm Regularization—The Compressive Sensing Approach. *IEEE Trans. Geosci. Remote Sens.* **2010**, *48*, 3839–3846. [CrossRef]

28. Zhu, X.X.; Ge, N.; Shahzad, M. Joint sparsity in SAR tomography for urban mapping. *IEEE J. Sel. Top. Signal Process.* **2015**, *9*, 1498–1509. [CrossRef]

29. Pauciullo, A.; Reale, D.; De Maio, A.; Fornaro, G. Detection of Double Scatterers in SAR Tomography. *IEEE Trans. Geosci. Remote Sens.* **2012**, *50*, 3567–3586. [CrossRef]

30. Zhu, X.X.; Bamler, R. Demonstration of Super-Resolution for Tomographic SAR Imaging in Urban Environment. *IEEE Trans. Geosci. Remote Sens.* **2012**, *50*, 3150–3157. [CrossRef]

31. Zhu, X.; Bamler, R. Super-Resolution Power and Robustness of Compressive Sensing for Spectral Estimation With Application to Spaceborne Tomographic SAR. *IEEE Trans. Geosci. Remote Sens.* **2012**, *50*, 247–258. [CrossRef]

32. Budillon, A.; Johnsy, A.C.; Schirinzi, G. A Fast Support Detector for Superresolution Localization of Multiple Scatterers in SAR Tomography. *IEEE J. Sel. Top. Appl. Earth Obs. Remote Sens.* **2017**, *10*, 1–12. [CrossRef]

33. Ma, P.; Lin, H. Robust Detection of Single and Double Persistent Scatterers in Urban Built Environments. *IEEE Trans. Geosci. Remote Sens.* **2016**, *54*, 2124–2139. [CrossRef]

34. Siddique, M.A.; Wegmüller, U.; Hajnsek, I.; Frey, O. Single-Look SAR Tomography as an Add-On to PSI for Improved Deformation Analysis in Urban Areas. *IEEE Trans. Geosci. Remote Sens.* **2016**, *54*, 6119–6137. [CrossRef]

35. Shi, Y.; Zhu, X.X.; Yin, W.; Bamler, R. A fast and accurate basis pursuit denoising algorithm with application to super-resolving tomographic SAR. *IEEE Trans. Geosci. Remote Sens.* **2018**, in press. [CrossRef]

36. Gernhardt, S.; Bamler, R. Deformation monitoring of single buildings using meter-resolution SAR data in PSI. *ISPRS J. Photogramm. Remote Sens.* **2012**, *73*, 68–79. [CrossRef]

37. Gernhardt, S.; Adam, N.; Hinz, S.; Bamler, R. Appearance of Persistent Scatterers for Different TerraSAR-X Acquisition Modes. In Proceedings of the ISPRS Workshop, Hannover, Germany, 2–5 June 2009; Volume XXXVIII-1-4-7/W5, pp. 1–5.

38. Gernhardt, S.; Adam, N.; Eineder, M.; Bamler, R. Potential of very high resolution SAR for persistent scatterer interferometry in urban areas. *Ann. GIS* **2010**, *16*, 103–111. [CrossRef]

39. Gernhardt, S.; Cong, X.Y.; Eineder, M.; Hinz, S.; Bamler, R. Geometrical Fusion of Multitrack PS Point Clouds. *IEEE Geosci. Remote Sens. Lett.* **2012**, *9*, 38–42. [CrossRef]

40. Gernhardt, S. High Precision 3D Localization and Motion Analysis of Persistent Scatterers using Meter-Resolution Radar Satellite Data. Ph.D. Thesis, Technische Universität München, München, Germany, 2012.

41. Kalia, A.C. Value-added Products in the Framework of the German Ground Motion Service. In Proceedings of the FRINGE 2017: 10th International Workshop on Advances in the Science and Applications of SAR Interferometry and Sentinel-1 InSAR, Helsinki, Finland, 5–9 June 2017.

42. Zhu, X.X.; Wang, Y.; Gernhardt, S.; Bamler, R. Tomo-GENESIS: DLR's tomographic SAR processing system. In Proceedings of the Joint Urban Remote Sensing (JURSE), Sao Paulo, Brazil, 2013; pp. 159–162. [CrossRef]

43. Wang, Y.; Zhu, X.X. Automatic Feature-Based Geometric Fusion of Multiview TomoSAR Point Clouds in Urban Area. *IEEE J. Sel. Top. Appl. Earth Obs. Remote Sens.* **2015**, *8*, 953–965. [CrossRef]

44. Zhu, X.X.; Shahzad, M. Facade Reconstruction Using Multiview Spaceborne TomoSAR Point Clouds. *IEEE Trans. Geosci. Remote Sens.* **2014**, *52*, 3541–3552. [CrossRef]

45. Shahzad, M.; Zhu, X. Robust Reconstruction of Building Facades for Large Areas Using Spaceborne TomoSAR Point Clouds. *IEEE Trans. Geosci. Remote Sens.* **2015**, *53*, 752–769. [CrossRef]

46. Shahzad, M.; Zhu, X.X. Automatic Detection and Reconstruction of 2-D/3-D Building Shapes from Spaceborne TomoSAR Point Clouds. *IEEE Trans. Geosci. Remote Sens.* **2016**, *54*, 1292–1310. [CrossRef]

47. Wang, Y.; Zhu, X.X.; Bamler, R.; Gernhardt, S. Towards TerraSAR-X Street View: Creating City Point Cloud from Multi-Aspect Data Stacks. In Proceedings of the 2013 Joint Urban Remote Sensing Event (JURSE), Sao Paulo, Brazil, 21–23 April 2013; pp. 198–201. [CrossRef]

48. Ge, N.; Rodriguez Gonzalez, F.; Wang, Y.; Zhu, X.X. Spaceborne Staring Spotlight SAR Tomography— A First Demonstration with TerraSAR-X. *IEEE J. Sel. Top. Appl. Earth Obs. Remote Sens.* **2018**, in press.

49. Eineder, M.; Adam, N.; Bamler, R.; Yague-Martinez, N.; Breit, H. Spaceborne spotlight SAR interferometry with TerraSAR-X. *IEEE Trans. Geosci. Remote Sens.* **2009**, *47*, 1524–1535. [CrossRef]

50. Breit, H.; Fritz, T.; Balss, U.; Lachaise, M.; Niedermeier, A.; Vonavka, M. TerraSAR-X SAR processing and products. *IEEE Trans. Geosci. Remote Sens.* **2010**, *48*, 727–740. [CrossRef]

51. Duque, S.; Breit, H.; Balss, U.; Parizzi, A. Absolute Height Estimation Using a Single TerraSAR-X Staring Spotlight Acquisition. *IEEE Geosci. Remote Sens. Lett.* **2015**, *12*, 1735–1739. [CrossRef]

52. Bamler, R.; Eineder, M.; Adam, N.; Zhu, X.; Gernhardt, S. Interferometric Potential of High Resolution Spaceborne SAR. *Photogramm. Fernerkund. Geoinf.* **2009**, *2009*, 407–419. [CrossRef]

53. Zhu, X.; Montazeri, S.; Gisinger, C.; Hanssen, R.; Bamler, R. Geodetic TomoSAR—Fusion of SAR Imaging Geodesy and TomoSAR for 3D Absolute Scatterer Positioning. In Proceedings of the 2014 IEEE International Geoscience and Remote Sensing Symposium (IGARSS), Quebec City, QC, Canada, 13–18 July 2014.

54. Montazeri, S.; Zhu, X.X.; Eineder, M.; Bamler, R. Three-Dimensional Deformation Monitoring of Urban Infrastructure by Tomographic SAR Using Multitrack TerraSAR-X Data Stacks. *IEEE Trans. Geosci. Remote Sens.* **2016**, *54*, 6868–6878. [CrossRef]

55. Montazeri, S. The Fusion of SAR Tomography and Stereo-SAR for 3D Absolute Scatterer Positioning. Master's Thesis, Delft University of Technology, Delft, The Netherlands, 2014.

56. Sun, Y.; Shahzad, M.; Zhu, X. First Prismatic Building Model Reconstruction from TomoSAR Point Clouds. *ISPRS Int. Arch. Photogramm. Remote Sens. Spat. Inf. Sci.* **2016**, *XLI-B3*, 381–386. [CrossRef]

57. Parizzi, A.; Brcic, R. Adaptive InSAR Stack Multilooking Exploiting Amplitude Statistics: A Comparison Between Different Techniques and Practical Results. *IEEE Geosci. Remote Sens. Lett.* **2011**, *8*, 441–445. [CrossRef]

58. Wang, Y.; Zhu, X.; Bamler, R. Retrieval of Phase History Parameters from Distributed Scatterers in Urban Areas Using Very High Resolution SAR Data. *ISPRS J. Photogramm. Remote Sens.* **2012**, *73*, 89–99. [CrossRef]

59. Deledalle, C.A.; Denis, L.; Tupin, F. NL-InSar: Nonlocal interferogram estimation. *IEEE Trans. Geosci. Remote Sens.* **2011**, *49*, 1441–1452. [CrossRef]

60. Zhu, X.X.; Bamler, R.; Lachaise, M.; Adam, F.; Shi, Y.; Eineder, M. Improving TanDEM-X DEMs by Non-local InSAR Filtering. In Proceedings of the 10th European Conference on Synthetic Aperture Radar, Berlin, Germany, 2–6 June 2014; pp. 1–4.

61. Baier, G.; Rossi, C.; Lachaise, M.; Zhu, X.X.; Bamler, R. Nonlocal InSAR filtering for high resolution DEM generation from TanDEM-X interferograms. *IEEE Trans. Geosci. Remote Sens.* **2018**, in press. [CrossRef]

62. Kang, J.; Wang, Y.; Körner, M.; Zhu, X. Robust Object-based Multi-pass InSAR Deformation Reconstruction. *IEEE Trans. Geosci. Remote Sens.* **2017**, *55*, 4239–4251. [CrossRef]

63. Kang, J.; Wang, Y.; Schmitt, M.; Zhu, X.X. Object-based Multipass InSAR via Robust Low Rank Tensor Decomposition. *IEEE Trans. Geosci. Remote Sens.* **2018**, *56*, 3062–3077. [CrossRef]

64. Aguilera, E.; Nannini, M.; Reigber, A. Multisignal compressed sensing for polarimetric SAR tomography. *IEEE Geosci. Remote Sens. Lett.* **2012**, *9*, 871–875. [CrossRef]

65. Schmitt, M.; Stilla, U. Compressive sensing based layover separation in airborne single-pass multi-baseline InSAR data. *IEEE Geosci. Remote Sens. Lett.* **2013**, *10*, 313–317. [CrossRef]

66. Fornaro, G.; Verde, S.; Reale, D.; Pauciullo, A. CAESAR: An approach based on covariance matrix decomposition to improve multibaseline–multitemporal interferometric SAR processing. *IEEE Trans. Geosci. Remote Sens.* **2015**, *53*, 2050–2065. [CrossRef]

67. OpenStreetMap Contributors. Planet Dump. 2017. Available online: https://www.openstreetmap.org (accessed on 16 April 2018).

68. Schmitt, M.; Schonberger, J.L.; Stilla, U. Adaptive Covariance Matrix Estimation for Multi-Baseline InSAR Data Stacks. *IEEE Trans. Geosci. Remote Sens.* **2014**, *52*, 1–11. [CrossRef]

69. Samiei-Esfahany, S.; Martins, J.E.; van Leijen, F.; Hanssen, R.F. Phase Estimation for Distributed Scatterers in InSAR Stacks Using Integer Least Squares Estimation. *IEEE Trans. Geosci. Remote Sens.* **2016**, *54*, 5671–5687. [CrossRef]

70. Cao, N.; Lee, H.; Jung, H.C. A Phase-Decomposition-Based PSInSAR Processing Method. *IEEE Trans. Geosci. Remote Sens.* **2016**, *54*, 1074–1090. [CrossRef]

71. Even, M. A Study on Spatio-Temporal Filtering in the Spirit of SqueeSAR. In Proceeding of the 11th European Conference on Synthetic Aperture Radar (EUSAR), Hamburg, Germany, 6–9 June 2016; pp. 462–465.

72. Ansari, H.; De Zan, F.; Bamler, R. Sequential Estimator: Toward Efficient InSAR Time Series Analysis. *IEEE Trans. Geosci. Remote Sens.* **2017**, *55*, 5637–5652. [CrossRef]

73. Verde, S.; Reale, D.; Pauciullo, A.; Fornaro, G. Improved Small Baseline processing by means of CAESAR eigen-interferograms decomposition. *ISPRS J. Photogramm. Remote Sens.* **2018**, *139*, 1–13. [CrossRef]

74. Even, M.; Schulz, K. InSAR Deformation Analysis with Distributed Scatterers: A Review Complemented by New Advances. *Remote Sens.* **2018**, *10*, 744. [CrossRef]

75. Ollila, E.; Koivunen, V. Influence functions for array covariance matrix estimators. In Proceedings of the 2003 IEEE Workshop on Statistical Signal Processing, St. Louis, MO, USA, 28 September–1 October 2003; pp. 462–465.

76. Zoubir, A.M.; Koivunen, V.; Chakhchoukh, Y.; Muma, M. Robust Estimation in Signal Processing: A Tutorial-Style Treatment of Fundamental Concepts. *IEEE Signal Process. Mag.* **2012**, *29*, 61–80. [CrossRef]

77. Visuri, S.; Koivunen, V.; Oja, H. Sign and rank covariance matrices. *J. Stat. Plan. Inference* **2000**, *91*, 557–575. [CrossRef]

78. Croux, C.; Ollila, E.; Oja, H. Sign and rank covariance matrices: Statistical properties and application to principal components analysis. In *Statistical Data Analysis Based on the L1-Norm and Related Methods*; Springer: Basel, Switzerland, 2002; pp. 257–269.

79. Rife, D.; Boorstyn, R.R. Single tone parameter estimation from discrete-time observations. *IEEE Trans. Inf. Theory* **1974**, *20*, 591–598. [CrossRef]

80. De Zan, F. Optimizing SAR Interferometry for Decorrelating Scatterers. Ph.D. Thesis, Politecnico di Milano, Milan, Italy, 2008.

81. Cong, X.Y.; Balss, U.; Eineder, M.; Fritz, T. Imaging Geodesy—Centimeter-Level Ranging Accuracy with TerraSAR-X: An Update. *IEEE Geosci. Remote Sens. Lett.* **2012**, *9*, 948–952. [CrossRef]

82. Gisinger, C.; Balss, U.; Pail, R.; Zhu, X.X.; Montazeri, S.; Gernhardt, S.; Eineder, M. Precise Three-Dimensional Stereo Localization of Corner Reflectors and Persistent Scatterers with TerraSAR-X. *IEEE Trans. Geosci. Remote Sens.* **2015**, *53*, 1782–1802. [CrossRef]

83. Yoon, Y.; Eineder, M.; Yague-Martinez, N.; Montenbruck, O. TerraSAR-X Precise Trajectory Estimation and Quality Assessment. *IEEE Trans. Geosci. Remote Sens.* **2009**, *47*, 1859–1868. [CrossRef]

84. Breit, H.; Börner, E.; Mittermayer, J.; Holzner, J.; Eineder, M. The TerraSAR-X Multi-Mode SAR Processor—Algorithms and Design. In Proceedings of the 5th European Conference on Synthetic Aperture Radar, Ulm, Germany, 25–27 May 2004.

85. Cumming, I.G.; Wong, F.H.C. *Digital Processing of Synthetic Aperture Radar Data: Algorithms and Implementation*; Artech House Remote Sensing Library: Boston, MA, USA, 2005.

86. Balss, U.; Cong, X.Y.; Brcic, R.; Rexer, M.; Minet, C.; Breit, H.; Eineder, M.; Fritz, T. High precision measurement on the absolute localization accuracy of TerraSAR-X. In Proceedings of the 2012 IEEE International Geoscience and Remote Sensing Symposium (IGARSS), Munich, Germany, 22–27 July 2012; pp. 1625–1628. [CrossRef]

87. Eineder, M.; Balss, U.; Suchandt, S.; Gisinger, C.; Cong, X.; Runge, H. A definition of next-generation SAR products for geodetic applications. In Proceedings of the 2015 IEEE International Geoscience and Remote Sensing Symposium (IGARSS), Milan, Italy, 26–31 July 2015; pp. 1638–1641. [CrossRef]

88. Petit, G.; Luzum, B. *IERS Conventions*; IERS Technical Note No. 36; Verlag des Bundesamts für Kartographie und Geodäsie: Frankfurt am Main, Germany, 2010.

89. Montazeri, S.; Gisinger, C.; Eineder, M.; Zhu, X.X. Automatic Detection and Positioning of Ground Control Points Using TerraSAR-X Multiaspect Acquisitions. *IEEE Trans. Geosci. Remote Sens.* **2018**, *56*, 2613–2632. [CrossRef]

90. Schubert, A.; Jehle, M.; Small, D.; Meier, E. Influence of Atmospheric Path Delay on the Absolute Geolocation Accuracy of TerraSAR-X High-Resolution Products. *IEEE Trans. Geosci. Remote Sens.* **2010**, *48*, 751–758. [CrossRef]

91. Gisinger, C. Atmospheric Corrections for TerraSAR-X Derived from GNSS Observations. Master's Thesis, Technische Universität München, München, Germany, 2012.

92. Balss, U.; Breit, H.; Fritz, T.; Steinbrecher, U.; Gisinger, C.; Eineder, M. Analysis of internal timings and clock rates of TerraSAR-X. In Proceedings of the 2014 IEEE International Geoscience and Remote Sensing Symposium (IGARSS), Quebec City, QC, Canada, 13–18 July 2014; pp. 2671–2674, doi:10.1109/IGARSS.2014.6947024. [CrossRef]

93. Schubert, A.; Jehle, M.; Small, D.; Meier, E. Mitigation of atmospheric perturbations and solid Earth movements in a TerraSAR-X time-series. *J. Geodesy* **2012**, *86*, 257–270. [CrossRef]

94. Balss, U.; Gisinger, C.; Cong, X.; Eineder, M.; Fritz, T.; Breit, H.; Brcic, R. GNSS-Based Signal Path Delay and Geodynamic Corrections for Centimeter Level Pixel Localization with TerraSAR-X. In Proceedings of the 5th TerraSAR-X Science Team Meeting, Oberpfaffenhofen, Germany, 10–14 June 2013; pp. 1–4.

95. Balss, U.; Gisinger, C.; Cong, X.Y.; Brcic, R.; Hackel, S.; Eineder, M. Precise measurements on the absolute localization accuracy of TerraSAR-X on the base of far-distributed test sites. In Proceeding of the EUSAR 2014: 10th European Conference on Synthetic Aperture Radar, Berlin, Germany, 2–4 June 2014.

96. Eineder, M.; Balss, U.; Biarge, S.D. Water level measurement by controlled radar reflection and TerraSAR-X imaging geodesy. In Proceedings of the 2014 IEEE International Geoscience and Remote Sensing Symposium (IGARSS), Quebec City, QC, Canada, 13–18 July 2014; pp. 5141–5143. [CrossRef]

97. Duque, S.; Balss, U.; Cong, X.Y.; Yague-Martinez, N.; Fritz, T. Absolute Ranging for Maritime Applications using TerraSAR-X and TanDEM-X Data. In Proceedings of the EUSAR 2014: 10th European Conference on Synthetic Aperture Radar, Berlin, Germany, 2–4 June 2014; pp. 1–4.

98. Eineder, M.; Gisinger, C.; Balss, U.; Cong, X.Y.; Montazeri, S.; Hackel, S.; Rodriguez Gonzalez, F.; Runge, H. SAR Imaging Geodesy—Recent Results for TerraSAR-X and for Sentinel-1. In Proceedings of the FRINGE 2017: 10th International Workshop on Advances in the Science and Applications of SAR Interferometry and Sentinel-1 InSAR, Helsinki, Finland, 5–9 June 2017.

99. Balss, U.; Gisinger, C.; Cong, X.; Eineder, M.; Brcic, R. Precise 2-D and 3-D Ground Target Localization with TerraSAR-X. *ISPRS Int. Arch. Photogramm. Remote Sens. Spat. Inf. Sci.* **2013**, *XL-1/W*, 23–28. [CrossRef]

100. Eldhuset, K.; Weydahl, D.J. Geolocation and Stereo Height Estimation Using TerraSAR-X Spotlight Image Data. *IEEE Trans. Geosci. Remote Sens.* **2011**, *49*, 3574–3581. [CrossRef]

101. Raggam, H.; Perko, R.; Gutjahr, K.; Kiefl, N.; Koppe, W.; Hennig, S. Accuracy Assessment of 3D Point Retrieval from TerraSAR-X Data Sets. In Proceedings of the 8th European Conference on Synthetic Aperture Radar, Aachen, Germany, 7–10 June 2010; pp. 1–4.

102. Koppe, W.; Wenzel, R.; Hennig, S.; Janoth, J.; Hummel, P.; Raggam, H. Quality assessment of TerraSAR-X derived ground control points. In Proceedings of the 2012 IEEE International Geoscience and Remote Sensing Symposium (IGARSS), Munich, Germany, 22–27 July 2012; pp. 3580–3583. [CrossRef]

103. Gisinger, C.; Gernhardt, S.; Auer, S.; Balss, U.; Hackel, S.; Pail, R.; Eineder, M. Absolute 4-D positioning of persistent scatterers with TerraSAR-X by applying geodetic stereo SAR. In Proceedings of the 2015 IEEE International Geoscience and Remote Sensing Symposium (IGARSS), Milan, Italy, 26–31 July 2015; pp. 2991–2994. [CrossRef]

104. Gisinger, C.; Montazeri, S.; Balss, U.; Cong, X.Y.; Hackel, S.; Zhu, X.X.; Pail, R.; Eineder, M. Applying geodetic SAR with TerraSAR-X and TanDEM-X. In Proceedings of the TerraSAR-X/TanDEM-X Science Team Meeting, Oberpfaffenhofen, Germany, 17–20 October 2016.

105. Runge, H.; Balss, U.; Suchandt, S.; Klarner, R.; Cong, X. DriveMark – Generation of High Resolution Road Maps with Radar Satellites. In Proceedings of the 11th ITS European Congress, Glasgow, Scotland, 6–9 June 2016; pp. 1–6.

106. Montazeri, S.; Zhu, X.X.; Balss, U.; Gisinger, C.; Wang, Y.; Eineder, M.; Bamler, R. SAR ground control point identification with the aid of high resolution optical data. In Proceedings of the 2016 IEEE International Geoscience and Remote Sensing Symposium (IGARSS), Beijing, China, 10–15 July 2016; pp. 3205–3208. [CrossRef]

107. Balss, U.; Runge, H.; Suchandt, S.; Cong, X.Y. Automated extraction of 3-D Ground Control Points from SAR images—An upcoming novel data product. In Proceedings of the 2016 IEEE International Geoscience and Remote Sensing Symposium (IGARSS), Beijing, China, 10–15 July 2016; pp. 5023–5026. [CrossRef]

108. Nitti, D.O.; Morea, A.; Nutricato, R.; Chiaradia, M.T.; La Mantia, C.; Agrimano, L.; Samarelli, S. Automatic GCP extraction with high resolution COSMO-SkyMed products. In Proceedings of the SPIE Remote Sensing 2016, Edinburgh, UK, 26–29 September 2016; Volume 10003. [CrossRef]

109. Montazeri, S.; Gisinger, C.; Zhu, X.X.; Eineder, M.; Bamler, R. Automatic positioning of SAR ground control points from multi-aspect TerraSAR-X acquisitions. In Proceedings of the 2017 IEEE International Geoscience and Remote Sensing Symposium (IGARSS), Fort Worth, TX, USA, 23–28 July 2017; pp. 961–964. [CrossRef]

110. Hubert, M.; Vandervieren, E. An adjusted boxplot for skewed distributions. *Comput. Stat. Data Anal.* **2008**, *52*, 5186–5201. [CrossRef]

111. Montazeri, S.; Zhu, X.X.; Gisinger, C.; Rodriguez Gonzalez, F.; Eineder, M.; Bamler, R. Towards Absolute Positioning of InSAR Point Clouds. In Proceedings of the FRINGE 2017: 10th International Workshop on Advances in the Science and Applications of SAR Interferometry and Sentinel-1 InSAR, Helsinki, Finland, 5–9 June 2017.

112. Wang, Y.; Zhu, X.X.; Zeisl, B.; Pollefeys, M. Fusing Meter-Resolution 4-D InSAR Point Clouds and Optical Images for Semantic Urban Infrastructure Monitoring. *IEEE Trans. Geosci. Remote Sens.* **2017**, *55*, 14–26. [CrossRef]

113. Zhu, X.X.; Tuia, D.; Mou, L.; Xia, G.S.; Zhang, L.; Xu, F.; Fraundorfer, F. Deep Learning in Remote Sensing: A Comprehensive Review and List of Resources. *IEEE Geosci. Remote Sens. Mag.* **2017**, *5*, 8–36. [CrossRef]

114. Ge, N.; Zhu, X. Bistatic-like Differential SAR Tomography—A Preliminary Framework for Tandem-L. *IEEE Trans. Geosci. Remote Sens.* **2018**, submitted.

115. Chang, L.; Hanssen, R.F. A Probabilistic Approach for InSAR Time-Series Postprocessing. *IEEE Trans. Geosci. Remote Sens.* **2016**, *54*, 421–430. [CrossRef]

116. Schmitt, M.; Zhu, X.X. Data Fusion and Remote Sensing: An ever-growing relationship. *IEEE Geosci. Remote Sens. Mag.* **2016**, *4*, 6–23. [CrossRef]

remote sensing

MDPI

Article

Remote Sensing Intertidal Flats with TerraSAR-X. A SAR Perspective of the Structural Elements of a Tidal Basin for Monitoring the Wadden Sea

Winny Adolph [1,*], Hubert Farke [1], Susanne Lehner [2] and Manfred Ehlers [3]

[1] Wadden Sea National Park Authority of Lower Saxony (NLPV), Virchowstr.1, 26382 Wilhelmshaven, Germany; hubert.farke@t-online.de

[2] Susanne Lehner, German Aerospace Center (DLR), Remote Sensing Technology Institute (IMF), Oberpfaffenhofen, 82234 Weßling, Germany; Susanne.Lehner@dlr.de

[3] Institute for Geoinformatics and Remote Sensing (IGF), University of Osnabrück, Wachsbleiche 27, 49090 Osnabrück, Germany; manfred.ehlers@uos.de

* Correspondence: mail@winny-adolph.de; Tel.: +49-4421-911-155

Received: 1 May 2018; Accepted: 5 July 2018; Published: 7 July 2018

Abstract: Spatial distribution and dynamics of intertidal habitats are integral elements of the Wadden Sea ecosystem, essential for the preservation of ecosystem functions and interlocked with geomorphological processes. Protection and monitoring of the Wadden Sea is mandatory and remote sensing is required to survey the extensive, often inaccessible tidal area. Mainly airborne techniques are carried out for decades. High-resolution satellite-borne sensors now enable new possibilities with satellite synthetic aperture radar (SAR) offering high availability of acquisitions during low water time due to independence from daylight and cloud cover. More than 100 TerraSAR-X images from 2009 to 2016 were used to examine the reproduction of intertidal habitats and macrostructures from the flats south of the island of Norderney and comparative areas in the Lower Saxony Wadden Sea. As a non-specific, generic approach to distinguish various and variable surface types continuously influenced by tidal dynamics, visual analysis was chosen which was supported by extensive in situ data. This technically unsophisticated access enabled us to identify mussel beds, fields of shell-detritus, gully structures, mud fields, and bedforms, the latter detected in the upper flats of every East Frisian island. Based on the high frequency of TerraSAR-X recordings for the Norderney area, a bedform shift was observed in a time-series from 2009 to 2015. For the same period, the development of a mud field with an adjoining depression was traced. Beside seasonal variations of the mud field, the formation of a mussel bed settling in the depression was imaged over the years. This study exemplifies the relevance of TerraSAR-X imagery for Wadden Sea remote sensing. Further development of classification methods for current SAR data together with open access availability should contribute to large-scale surveys of intertidal surface structures of geomorphic or biogenic origin and improve monitoring and long-term ecological research in the Wadden Sea and related tidal areas.

Keywords: synthetic aperture radar; TerraSAR-X; habitat mapping; monitoring; remote sensing; Wadden Sea; mussel beds; intertidal bedforms; tidal gullies

1. Introduction

Tidal flat areas off shallow coasts can be found worldwide. The world's largest coherent intertidal area, the Wadden Sea, is stretching for over 500 km along the North Sea coast of The Netherlands, Germany, and Denmark with a width of up to 20 km. The system of barrier islands, intertidal flats and sandbanks, channels, gullies, and salt marshes forms the transition between the mainland and the open North Sea. The Wadden Sea is one of the last large-scale and near-natural ecosystems in Central Europe, whose ecological functions are supraregional and of far-reaching importance, e.g., as an indispensable stepping stone for birds migrating on the East Atlantic Flyway. In addition, it also plays an important role in coastal protection. The Wadden Sea is protected by a high national as well as international conservation status and it is listed as UNESCO world heritage site. Regular monitoring of the area is mandatory but complex and expensive because of the large area of rough terrain which is accessible only in tight timeframes due to the changing tides. Therefore, remote sensing techniques are required and aerial mapping and photography (e.g., mussel beds, seagrass meadows, salt marsh vegetation) as well as airborne lidar (light detection and ranging) provide the high resolution needed to determine surface structures of the tidal flats or salt marsh vegetation types. Today they are applied in operational monitoring programs of the Wadden Sea. Aerial photographs have been used in the Lower Saxony Wadden Sea since the 1990s for the monitoring of mussel beds [1–3] and for biotope mapping of the salt marshes [4–6]. Aerial photography and lidar are also used by the responsible authorities for coastal protection (NLWKN, unpublished data). A possible support of the seagrass mappings by aerial photographs was examined by Ref. [7].

Advances in synthetic aperture radar (SAR) technology have enabled a high level of spatial resolution also for satellite-borne sensors implemented by a new class of high-resolution SAR satellites. Since the launch of TerraSAR-X in 2007, followed by TanDEM-X (both X-band), the COSMO-SkyMed satellite constellation (X-band), and Radarsat-2 (C-band), these satellites provide SAR data with resolutions in the scale of meters [8]. The radar satellite Sentinel-1 with a slightly lower resolution (5 m in stripmap mode) has been available since 2014 with open data policy.

With the technical improvements not only offered by the SAR sensors, but also with increasing spatial and/or spectral resolution of electro-optical systems (e.g., Landsat-8, RapidEye, SPOT-4, World-View, or currently, Sentinel-2), the use of satellite data for the protection and management of coastal areas such as the Wadden Sea is becoming increasingly realistic. Against this background, a number of German research projects such as OFEW (2005–2007), DeMarine-1-Environment TP4 (2008–2011), DeMarine-2 SAMOWatt (2012–2015), and WIMO (2011–2015) was conducted to apply high-resolution satellite data for the requirements of monitoring and long-term ecological research in the Wadden Sea; for overviews, see [9–13]. Various authors have demonstrated the value of state-of-the-art satellite data for the exploration of tidal areas [10,11,14–20]. The further development of image classification methods designed for the tidal area has gained pace, for example, regarding the exploitation of polarimetric information from SAR data, whose potential has already been documented by, e.g., Refs. [21–23].

The aim of the present study is to determine the potential of the high-resolution intensity images acquired by TerraSAR-X to identify the distribution and development of the main geomorphological structures and habitats in a whole tidal basin and their dynamics which are of utmost importance for monitoring and long term ecological research but also for the management of the area. In order to recognize as many different surface structures as possible, this study focuses on visual image interpretation, which takes into account backscatter intensity and contrast as well as shapes, patterns, and textures of surface features reflected by the SAR data, but also their configuration or surrounding context. This is of particular importance because the Wadden Sea, characterized by a flat topography, a very dynamic variability, the variety of gradients, transitional zones, and surface structures under the influence of constantly changing water levels and weather conditions, poses great challenges to classification of intertidal surfaces. In this context, visual analysis should provide technically unsophisticated access to as much of the information contained in the SAR data as possible.

Previous knowledge and experience play an important role in the visual interpretation process, with recognition and interpretation running through an iterative process, where both steps heavily rely on one another [24]. That is, context information, such as environmental conditions (acquisition time related to tidal cycle, water level, weather conditions) and processes, field experience, and in situ data, which is difficult to quantify automatically, are essential components flowing into the analysis.

Therefore, in this study, extended field observations partly synchronous to the satellite acquisition are carried out throughout the period of the investigations to validate the image interpretation results. An initial basis of the terrain knowledge was laid during the comprehensive mapping of the main study area as part of the DeMarine-1 project.

TerraSAR-X spotlights and high resolution spotlights proved most suitable to investigate typical intertidal surface structures and habitats such as mussel beds, shell-detritus, gully systems, mud fields, and bedforms which are clearly reproduced and can be drawn from the intensity images by visual analysis. Regarding intertidal bedforms, visual image analysis raised the assumption of bedform movement, therefore the positions of bedform structures in the upper flats of Norderney were further analyzed using the extensive time series of satellite images available for this study. For this purpose, the water-covered bedform troughs were extracted from the TerraSAR-X images using an automated method developed by Ref. [19].

The studies presented here were part of the German research project WIMO (Scientific monitoring concepts for the German Bight) with subproject TP 1.4 (Application of high resolution SAR-data (TerraSar-X) for monitoring of eulittoral surface structures and habitats). In addition, data from the DeMarine-1 and DeMarine-2 projects with the subprojects TP4 (Integration of Optical and SAR Earth Observation Data and in situ Data into Wadden Sea Monitoring) and SAMOWatt (Satellite data for Monitoring in the Wadden Sea) have been included in the investigations.

2. Materials and Methods

2.1. Study Site in the Tidal Flats of Norderney (German Wadden Sea)

The study was carried out in the East Frisian Wadden Sea, which forms the western part of the German North Sea coast between the river Ems and the Weser estuary. Towards the open North Sea, the Wadden Sea is bordered by a chain of barrier islands (Figure 1a). The tidal flats between the island of Norderney and the mainland coast were selected as the main study area. For comparative purposes, surface structures from other parts of the East Frisian Wadden Sea are also included (Figure 1b).

The back-barrier tidal basin of Norderney covers the geomorphic structures and habitats which are frequent and characteristic for Wadden Sea flats: mussel beds, fields of shell detritus, seagrass beds, low lying areas collecting residual waters, a drainage system of channels and gullies and the tidal flats varying in sediment composition from the more sheltered muddy regions near the mainland coast and the watershed to the more exposed sandflats close to the Norderney inlet which connects the tidal basin with the open sea. The different sediment types on the tidal flats and in the subtidal significantly influence the environmental conditions for the organisms living in or on the bottom of the Wadden Sea thus forming habitats with typical species communities. With a mean tidal range of 2.4 m $\pm$ 0.7 m [25], the back-barrier tidal basin of Norderney is classified as upper mesotidal according to Ref. [26].

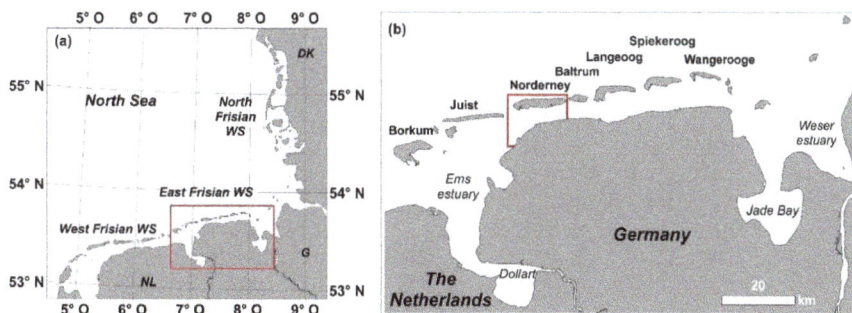

Figure 1. The study area in the German Wadden Sea: (**a**) the Trilateral Wadden Sea in the German Bight; (**b**) the main investigation area located at Norderney in the East Frisian Wadden Sea.

2.2. TerraSAR-X Data Base

SAR data were acquired by the high-frequency (9.6 GHz) X-band sensor of TerraSAR-X with a wavelength of 3.1 cm, operating at 514 km altitude. The data were collected in Spotlight (SL) and High Resolution Spotlight (HRS) mode, which provide ground range resolutions of 1.5–3.5 m [27], few images were taken in stripmap mode with a resolution of 3 m. Since the SAR data should be combined with extensive in situ data and to perform spatio-temporal analyses, Geocoded Ellipsoid Corrected images (GEC) were chosen which can be easily imported into geographic information systems (GIS). To allow acquisition times close to the time of low tide on the one hand, and to obtain a sufficient amount of data on the other hand, SAR data had to be collected at varying orbits and incidence angles. This enabled us to acquire extensive and detailed time series, as well as recordings before and after events such as storm and storm tide or ice drift. From the resulting set of more than 100 TerraSAR-X images available for the years 2009 to 2016, the SAR data documented in this study are listed in Table 1. These images were acquired within the time period 1.5 h before and after low tide and apart from the stripmap image recorded in HH-polarization, the data were taken vertically co-polarized (VV).

Table 1. TerraSAR-X acquisitions used in this study. Image mode: HRS = High Resolution Spotlight, SL = Spotlight, SM = Stripmap, Inc. = Incidence angle, Orbit direction: A = ascending, D = descending, Δ tLT = Acquisition time related to low tide (*positive values*: acquisition at rising tide), Gauge level related to normal height null (NHN), WS, WD = wind speed, wind direction.

Site	Date	Image Mode	Rel. Orbit	Inc. [°]	Orbit Dir.	Δ tLT [min]	Gauge [cm < NHN]	WS [m/s]	WD [°]
Norderney	21/07/2009	HRS	131	20.8	A	63	111 [2]	3.9 [7]	60
Juist/Borkum	05/04/2011	SM	63	37.4	D	9	136 [1]	10.9 [6]	210
Spiekeroog	17/05/2011	SL	40	37.0	A	14	142 [3]	7.6 [8]	270
Norderney	02/06/2011	HRS	116	45.1	A	11	145 [2]	5.4 [7]	360
Norderney	04/06/2011	SL	139	23.3	D	0	160 [2]	5.5 [7]	60
Norderney	16/07/2011	HRS	116	45.1	A	−18	152 [2]	3.2 [7]	160
Norderney	19/07/2011	SL	154	46.6	D	−82	106 [2]	5.5 [7]	190
Norderney	14/10/2011	SL	139	23.6	D	15	174 [2]	3.2 [7]	130
Norderney	10/01/2012	HRS	139	23.5	D	43	116 [2]	3.6 [7]	270
Wangerooge	19/05/2012	SL	116	47.9	A	50	144 [5]	4.8 [8]	30
Baltrum	07/06/2012	SL	63	35.3	D	−52	144 [2]	3.1 [7]	150
Wangerooge	15/10/2012	HRS	40	38.1	A	−2	142 [4]	6.2 [8]	160
Norderney	30/11/2012	SL	63	36.4	D	21	129 [2]	5.4 [7]	10
Norderney	09/06/2013	HRS	131	21.1	A	−23	144 [2]	6.9 [7]	360
Norderney	28/02/2014	HRS	131	21.1	A	63	67 [2]	3.4 [7]	60
Norderney	14/06/2014	HRS	63	36.1	D	46	132 [2]	9.9 [7]	350
Norderney	11/08/2014	HRS	116	45.1	A	6	111 [2]	8.5 [7]	220
Norderney	07/12/2014	HRS	63	36.1	D	56	102 [2]	7.6 [7]	190
Norderney	19/04/2015	HRS	78	54.3	D	40	166 [2]	2.9 [7]	260
Langeoog	21/06/2016	HRS	78	54.2	D	26	105 [2]	2.5 [7]	310

Water level data (source: Federal Waterways and Shipping Administration WSV, provided by Federal Institute for Hydrology BfG) are from the gauges: [1] Borkum Fischerbalje, [2] Norderney Riffgat, [3] Spiekeroog, [4] Wangerooge West, and [5] Wangerooge East. Wind speed and wind direction (source: German Weather Service DWD) are from the weather stations: [6] Borkum, [7] Norderney, and [8] Spiekeroog.

2.3. Image Analysis

The TerraSAR-X data were calibrated to "sigma naught" (σ0), the radar reflectivity per unit area in ground range using ERDAS Imagine (version 2013–2016), to correct for geometry of acquisition cf. [28]. For image interpretation and analysis the intensity images were directly imported into the geographic information system (GIS) of ESRI ArcGIS 10.1 where the data was repetitively verified with geospatial in situ data or compared to monitoring results.

According to initial tests, statistical analysis of backscatter differences such as height, mean value, amplitude, or variance seemed not sufficient for the clear demarcation of most intertidal surfaces. Therefore, in this study the images are analyzed via visual interpretation integrating i.a. the patterns of internal structures or textures characterizing the surface structures reflected by TerraSAR-X data as well as contextual data including extensive in situ data or weather and gauge level data (cf. Introduction).

2.3.1. Visual Image Analysis

The radar backscatter recorded by the SAR sensor can be considered as a measure of the surface roughness, with smoother surfaces rendered dark in the resulting image and rougher surfaces appearing brighter. Characteristic surface properties of the various structures and habitats in the tidal area therefore lead to corresponding patterns and textures in the radar image. A major difference is seen between water-covered areas and exposed areas such as sediment surfaces and biogenic structures. Although the water surface appears highly variable due to currents, wind, and waves—sometimes in interaction with surface active agents such as biofilms—it can be clearly distinguished from the emerged tidal flats, especially if the edges are markedly distinct. Even from gradual transitions, which are also common in tidal areas, visual references to the surface morphology can be obtained. On the flats, residual water trapped in hollow surface structures helps to detect or identify geomorphic surface characteristics from TerraSAR-X images, such as depressed areas, bedforms, or draining systems. Residual water also contributes to the identification of typical large-scale structures and habitats with specific roughness properties such as mussel beds, fields of shell detritus or mud fields. Associated puddles caught in the humpy sediment surface of a mud field or pools within mussel beds are characteristic features reflected by specific patterns of backscatter in the SAR image.

2.3.2. Digital Image Analysis

Visual image analysis raised the assumption of bedform movement in the upper flats of the island of Norderney, therefore a spatio-temporal analysis of bedform positions was performed using the extensive time series of satellite images collected during this study. To extract relevant markers from the SAR data, bedform positions were determined by detection of the water-covered troughs according to the method proposed by Ref. [19] which is based on textural analysis combined with an unsupervised classification: For comparison with the complete set of TerraSAR-X data, the images were re-sampled at their highest common resolution, a pixel size of 1.25 m. Speckle reduction was performed by edge-preserving Frost and Median filtering and followed by a textural analysis calculating Gray Level Co-occurrence Matrix (GLCM) statistical parameters (variance, homogeneity, and mean) according to Ref. [29]. From the resulting feature images, the water-covered troughs were derived by means of unsupervised ISODATA classification. Image processing was carried out using ERDAS imagine (2013–2015) and ENVI 4.7 software. The classification output was vectorized and imported into ESRI ArcGIS 10.1 for further analysis. The correct assignment of classes was verified by regularly collected in situ data combined with visual interpretation of the SAR images.

2.4. Ground Truth, Monitoring and Environmental Data

Visual image interpretation was performed in conjunction with extensive ground truth data. The background of the terrain knowledge comes from a survey carried out as part of the DeMarine-1 project in 2008/2009 [10,30]. In this context, the tidal areas of Norderney were surveyed according to a comprehensive protocol and photographically documented in a 300 × 300 m grid of stations. In 2014, sections of the grid were revisited for comparison with the 2008/2009 situation. During the current research on the WIMO project, all of the structures described below have been extensively validated by GPS measurements and photo documentation, in part simultaneously with SAR acquisition (cf. Figure 2). Garmin's GPSmap 62s was used for the GPS measurements, the photos were taken with cameras with GPS functionality. Additionally, the bedforms in the upper flats of Norderney were validated by high-precision height measurements recorded by Real Time Kinematic Differential GPS (RTK-DGPS) with a Leica Differential-GPS SR530 and AT 502 antenna type, see [19,20]. Furthermore, data from the annual mussel monitoring program of the National Park authority in Lower Saxony (NLPV) were used which are obtained by interpretation of aerial photographs. These data are available as shapefiles and indicate the location and areal extent of the intertidal mussel beds of Lower Saxony [31]. Environmental background information included water level data from the gauges at Borkum Fischerbalje, Norderney Riffgat, Spiekeroog, Wangerooge West and Wangerooge East (source: Federal Waterways and Shipping Administration WSV, provided by Federal Institute for Hydrology BfG), as well as wind speed and direction data from the weather stations on Borkum, Norderney and Spiekeroog (source: German Weather Service DWD).

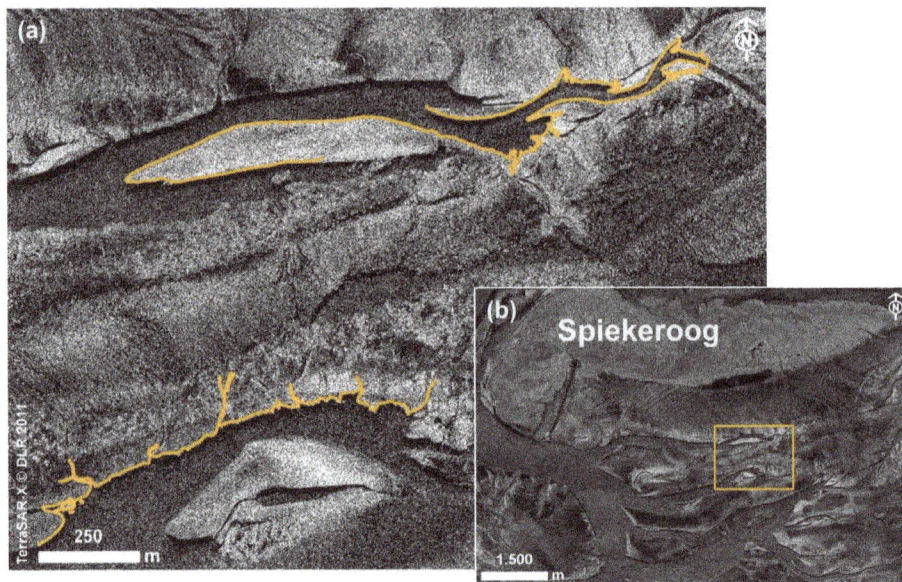

Figure 2. In situ verification of land-water-lines: (**a**) GPS measurement of channel edges synchronous to satellite overflight at 14 min. after low tide (yellow line); (**b**) location of study area (rectangle) in the tidal flats of Spiekeroog (SL of 17/05/2011, ascending orbit).

3. Results

Many characteristic habitats and large-scale surface structures of the tidal flats are clearly reproduced by the TerraSAR-X data. They can be visually identified and analyzed from high-resolution (HRS), spotlight (SL), and, depending on the size of the structure, even in stripmap (SM) images. Figure 3 gives an overview of the main study area, the tidal flats south of the island of Norderney, reproduced by TerraSAR-X. The added in situ photographs illustrate some of the macrostructures imaged by the SAR sensor. Some of them, such as mussel beds or fields of shell detritus, are usually displayed very clearly due to their outstanding surface roughness and specific textures. Edges also, in particular the steeper slopes of channels and gullies or the steeply sloping edges of high sandflats, are clearly shown depending on their orientation relative to the sensor.

Other intertidal structures, however, are specifically reproduced due to the contrasting water and sediment surfaces. Most obvious, water level lines delineate the sub-littoral from the tidal area at low tide or flooded areas from exposed flats in the course of the tides. But also residual water caught in troughs and depressions helps to recognize the relief of tidal flat surfaces indicating structures such as intertidal bedforms, depressed areas, or mud fields.

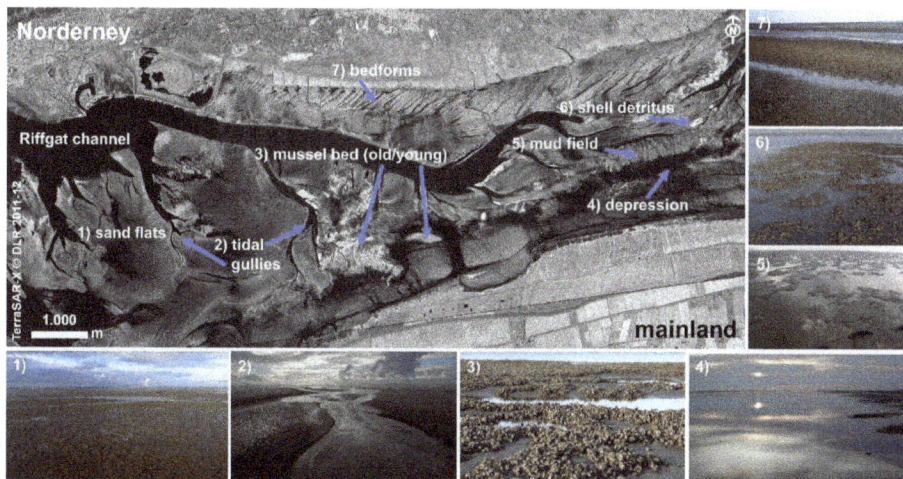

Figure 3. Overview of characteristic habitats and large-scale surface structures of the tidal flats south of Norderney, imaged by TerraSAR-X (02/06/2011 and 30/11/2012, large picture) and the corresponding in situ photographs (small pictures).

3.1. Tidal Channels and Gullies

Twice a day, in the course of the tides, the tidal flats are flooded and drained through the system of tideways, such as channels and gullies. Depending on their position in this system these tideways are exposed to high flow velocities, which especially in sandy environments, causes regular shifts of the edges and leads to highly dynamic channel courses. These tideways can be identified from TerraSAR-X imagery (Figure 4) and over time, also the shifting of their courses or their positional stability. Furthermore, characteristic shapes formed by the branches may provide information about the surrounding sediment.

Channels and gullies are mapped in the TerraSAR-X data depending on their width, the shape of the edges and the surface of the water they contain. In case of water-filled channels, the waterline will mark the edge, whereas for smaller and dry-fallen gullies especially the steep edges eaten into the sediment will be reproduced. The intertidal area shown in Figure 4a exposed to the direct influence of the inlet between the islands of Norderney and Juist open to the North Sea exemplifies morphological development in dynamic tidal areas. TerraSAR-X data from 2009–2012 enables one to observe the shifting of the channel section during that period. Over the entire time, the channel has been relocated by a maximum of over 100 m locally (Figure 4c). The branching arms, by contrast, have remained largely stable. Part of the channel, north of the first branch in the upper part of the image section, is stabilized by an adjoining mussel bed (indicated by internal structures and high backscatter).

By means of visual analysis, the channels are clearly visible in the TerraSAR-X data (Figure 4b,c). However, Figure 4b also illustrates how automatic channel detection may be difficult due to the varying representation of the water surface and to internal patterns e.g., depending on the presence of surface-active agents, weather, or flow conditions at the time of acquisition.

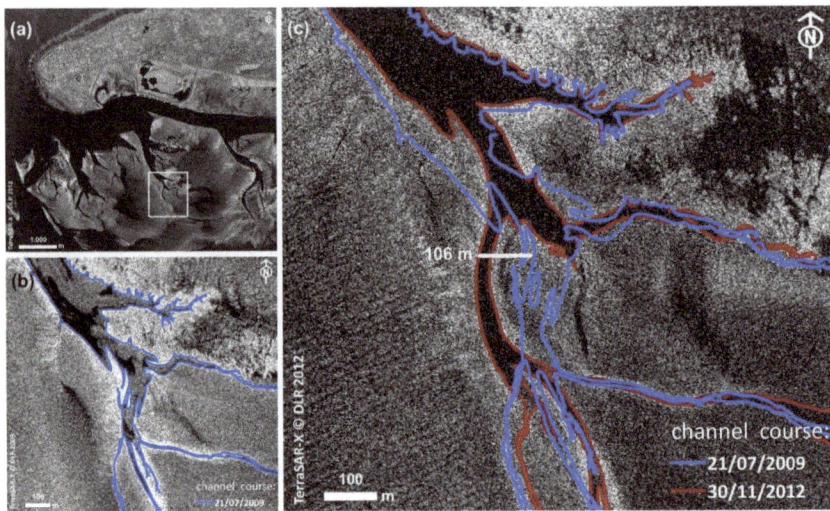

Figure 4. Relocation of tidal channels imaged by TerraSAR-X: (**a**) location of study area (rectangle) in the tidal flats of Norderney (SL of 30/11/2012); (**b**) channel course in 2009 (HRS of 21/07/2009, 111 cm < NHN, ascending orbit); (**c**) shifted channel course in 2012 (red lines) compared to course of 2009 (blue lines) (SL of 30/11/2012, 129 cm < NHN, descending orbit).

3.2. Intertidal Bedforms

3.2.1. Intertidal Bedforms in the Upper Island Flats of the East Frisian Islands

In large areas of the upper back-barrier tidal flats of the East Frisian islands, the sediment surface forms a pattern of periodic crests and troughs thus creating bedform fields of considerable size. The troughs are covered with water throughout the whole time of emergence, therefore the bedforms are clearly reproduced by TerraSAR-X imagery and they can be detected in the whole set of images (acquired from 2009 to 2016) and in the upper island flat of each of the East Frisian islands. In Figure 5, an overview of the bedform fields of the East Frisian islands is given, it shows the bedforms directly adjoining the southern island's shores are generally oriented in a north-easterly direction, but especially in the lower flats also cross-profiles can appear. The dimensions and the exact orientations may vary from island to island.

In the study area at Norderney the bedform positions and their dynamics were examined in detail. The photograph (Figure 6a) gives an impression of their appearance in the field. The bedforms imaged by TerraSAR-X and the vectorized classification result for the water-covered troughs are given in Figure 6b,c.

The results of the survey are described in detail by Ref. [19], who demonstrate that visual trough detection as well as results from unsupervised ISODATA classification of textural parameters from TerraSAR-X data are in good accordance with the in situ measurements of the bedforms. Spatio-temporal GIS-analysis of trough positions extracted from a time-series of TerraSAR-X images then revealed a shifting of the bedforms in an easterly direction during the study period from 2009–2015. This general bedform shift is demonstrated for the years 2012–2015 in Figure 7. The western trough edges are highlighted because in situ measurements as well as TerraSAR-X data reproducing variable states of water-cover indicate an asymmetry of the bedforms leading to steeper western trough edges and smoother eastern edges. Therefore the waterlines of the western edges proved to be a better indicator for the trough positions even with a slightly varying amount of residual water on the exposed flats due to environmental conditions or tidal state.

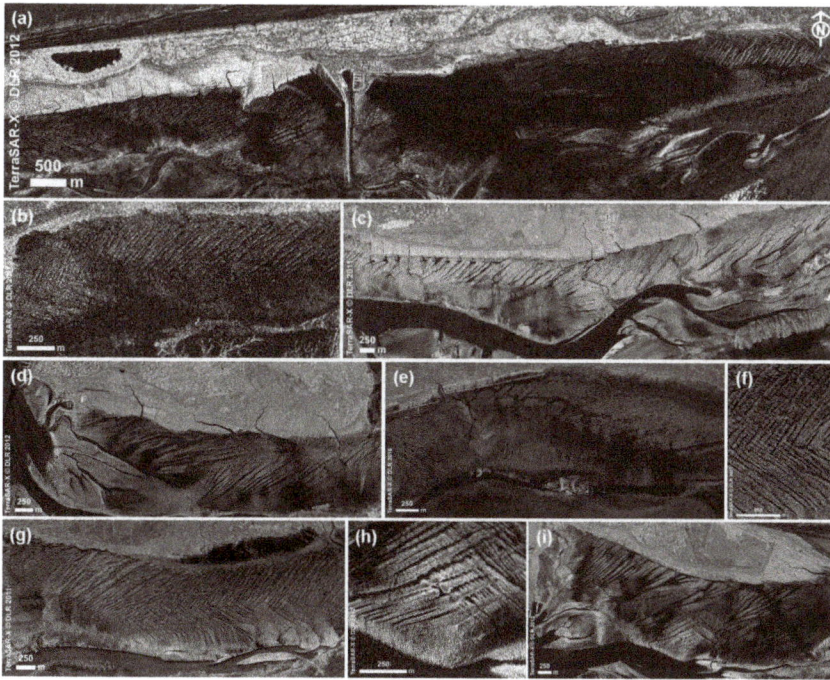

Figure 5. Bedforms in the upper flats of the East Frisian islands imaged by TerraSAR-X: (**a**) Juist (SM of 05/04/2011, desc.); (**b**) Borkum (SM of 05/04/2011, desc.); (**c**) Norderney (HRS of 02/06/2011, asc.); (**d**) Baltrum (SL of 07/06/2012, desc.); (**e**) Langeoog (HRS of 21/06/2016, desc.); (**f**) cross-patterns, detail of (**g**); (**g**) Spiekeroog (SL of 17/05/2011, asc.); (**h**) cross-patterns, detail of (**i**);(**i**) Wangerooge (SL of 15/10/2012, asc.). HRS = High resolution Spotlight, SL = Spotlight, SM = Stripmap acquisition mode of TerraSAR-X, asc. = ascending orbit, desc. = descending orbit.

Figure 6. Intertidal bedforms at Norderney: (**a**) photography of intertidal bedforms in the test area (26/03/2014); (**b**) image section: test area in the flats of Norderney (HRS of 02/06/2011, asc.); (**c**) Trough extraction result (white lines) from the same TerraSAR-X data.

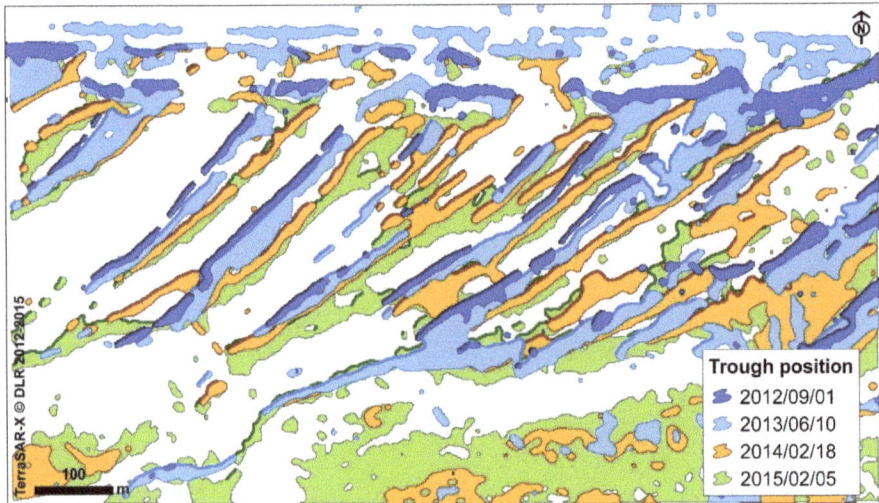

Figure 7. Trough positions extracted from TerraSAR-X images of 2012–2015. Western trough edges are highlighted (reprinted by permission from Springer Nature Terms and Conditions for RightsLink Permissions Springer Customer Service Centre GmbH: Springer Nature, Geo-Marine Letters: Monitoring spatiotemporal trends in intertidal bedforms of the German Wadden Sea in 2009–2015 with TerraSAR-X, including links with sediments and benthic macrofauna, Adolph et al. 2016).

The high frequency of TerraSAR-X data acquisition also enabled us to study the bedform positions in the course of the year and in connection with the effects of storm events. Adolph et al. [19] showed that the trough positions extracted from TerraSAR-X data generally remained stable from late winter to late summer and a shift to the east regularly occurred during winter. The change from the summer to the winter situation in 2013 provides a good insight into the shifting forces. In that year, the troughs kept their positions in every TerraSAR-X acquisition from February to August. However, the TerraSAR-X data from mid-December show a clear bedform shift which is most likely the effect of two very heavy gales in late October and early December with maximum wind speeds exceeding 130 and 120 km/h, respectively [19].

3.2.2. Temporary Surface Structures

Observing the tidal areas by means of TerraSAR-X data, different types of linear structures of the sediment surface were identified. So far, no further investigation of any of these structures has been carried out but similar characteristics were found in TerraSAR-X images of tidal areas throughout the East Frisian Wadden Sea, and also in transition from the tidal to the subtidal areas. In this way, TerraSAR-X imagery opens up new insights into large-scale tidal flat morphology and provides an opportunity to examine its genesis, development, and the significance for the tidal areas.

Southeast of the island of Wangerooge, as an example, in the near-shore area close to the mainland, linear surface structures were detected on a TerraSAR-X image from 2012 and verified in situ. A field of common cockle (*Cerastoderma edule*) apparently stabilized the linear structures and made them both more durable and more conspicuous in the TerraSAR-X data. Additionally, the elevated ridges of the sediment and cockle surface were covered by green algae, which contributed to the clear picture (Figure 8). In situ observations in 2016 have shown that in the meantime, the cockle field had been occupied by blue mussels (*Mytilus edulis*) and turned into a patchy mussel bed.

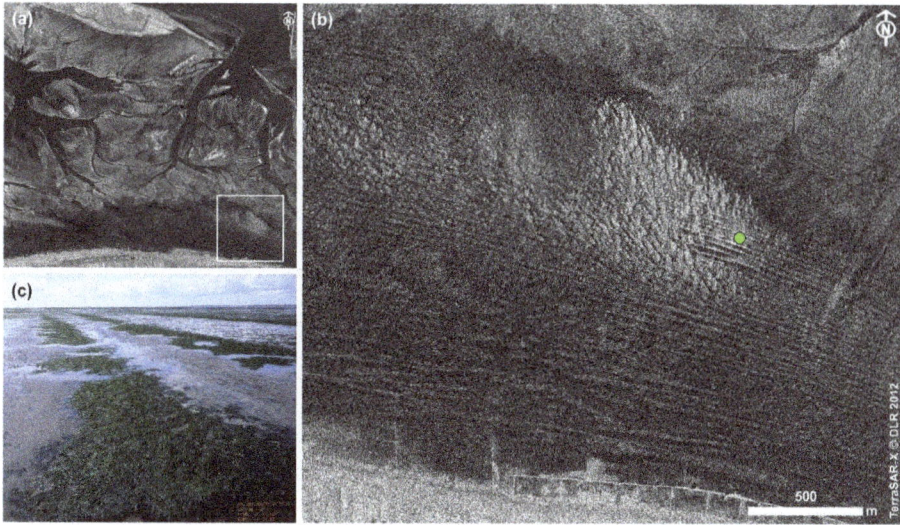

Figure 8. Temporary linear surface structures in the tidal area south of Wangerooge: (**a**) Tidal flats between the island of Wangerooge and the mainland coast, rectangle marks image Section b (SL of 19/05/2012, asc.); (**b**) linear surface structures within and in the surroundings of a cockle field (*C. edule*), point marks position of photographer; (**c**) photography of surface structures (02/06/2012).

3.3. Mud Field

The large mud field close to the watershed of the tidal flats beneath the island of Norderney extends over ca. 1.8 km along the Riffgat channel (see Figure 9) with a width of 300–400 m. On the wavy to humpy sediment surface, water puddles formed between the muddy humps resulting in a characteristic pattern (see Figure 9e), leading to a relatively high backscatter in the TerraSAR-X data. In addition, the mud field is traversed by a dense network of highly branched gully structures which drain the water from the adjacent depression in the south of the mud field to the channel in the north. These properties lead to a specific reproduction of the mud field in the TerraSAR-X images characterized by a high backscatter and the recognizable texture of the many gully structures (see Figure 9b). The contrast with the Riffgat channel and the water covering the area of the depression also facilitate to determine the contours of this mud field. In situ the southern edge of the mud field is clearly marked by the finely branched gullies originating from the water-covered depression. Here, the surface of the muddy deposits stands out from the more solid, smoother sediment surface of the depressed area (see Figure 9c,d). Therefore, GPS measurements of the mud field's edge carried out in summer 2011 (27/07/2011) show very good agreement with the contours reproduced by the TerraSAR-X acquisition recorded within a short time frame (16/07/2011). In Figure 9b, the yellow line represents the GPS measurement.

Figure 9. Large mud field in the tidal area of Norderney: (**a**) location of the mud field close to the watershed, marked by oval line (HRS of 02/06/2011, asc.); (**b**) GPS measurement of mud field's edge taken on 27/07/2011 (yellow line) compared to TerraSAR-X HRS of 16/07/2011, asc.; (**c**) photography along the mud field's edge (27/07/2011); (**d**) gully delta at the mud field's edge (27/07/2011); (**e**) humpy mud field surface with water puddles (27/07/2011).

Seasonal Aspects

In situ studies show variations in the surface form of the mud field. Extent and height of the muddy humps vary as well as their shape, which can be smooth and wavy or in contrast have steep erosive edges. These variations may occur locally, e.g., the silt surface always tends to be smoother on the edge of the mud field towards the depression. Overall, however, the field surveys showed that the mud field surface was more pronounced during the calmer season of the year (usually the summer) than after the stormy time of winter. When GPS-measuring the mud field's edge in summer 2011 (27/07/2011), the mud deposits clearly stood out from the depressed area covered with water. Thus, the boundary of the mud field was obvious and also well defined by the waterline (see Figure 9b,c). In the following January (17/01/2012), after two storms had passed through in the first days of the month (03–06/01/2012), a much more gradual transition was observed from the depression to the mud field. While the gully deltas where still in place, the smooth, slightly wavy surface of the silt accumulation began to emerge only gradually from the lower area, just beyond the ends of the gully deltas.

In fact, regarding the mud field over several years (2011–2015) in the TerraSAR-X data, seasonal changes are observed. During summer, the mud field surface is displayed in full width with high backscatter, in winter, however, the area of high backscattering retreats towards the Riffgat channel. The internal gully structures, on the other hand, remain visible throughout the year, across the entire width from channel to depression. This can be seen in Figure 10a–d showing the reproduction of the mud field in TerraSAR-X acquisitions from summer 2013, the following winter (02/2014) and the next summer (06/14) and winter (12/2014).

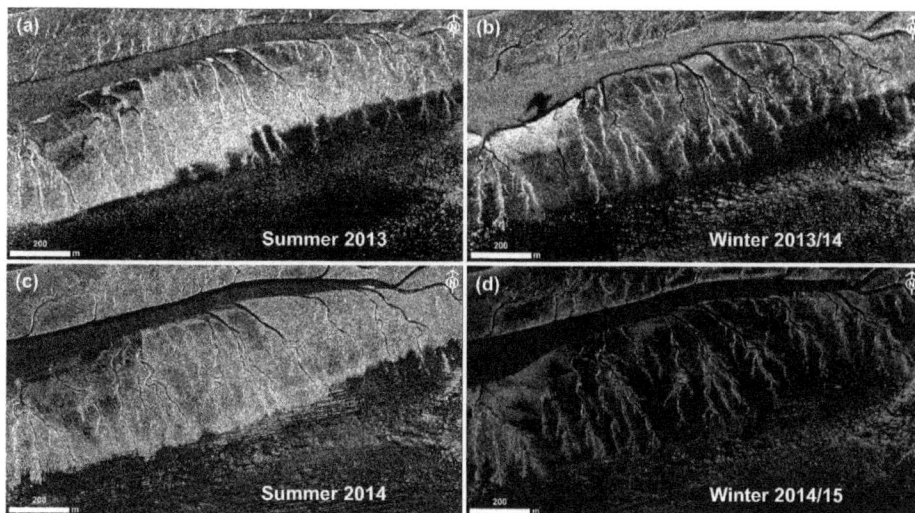

Figure 10. Seasonal aspects of the large mud field close to the watershed of the tidal area of Norderney reproduced by TerraSAR-X HRS acquisitions, VV polarised, ©DLR: (**a**) 09/06/2013, orbit 131, asc., 144 cm < NHN; (**b**) 28/02/2014, orbit 131, 67 cm < NHN; (**c**) 14/06/2014, orbit 63, desc., 132 cm < NHN; (**d**) 07/12/2014, orbit 63, 102 cm < NHN.

The formation of similar mud fields between tidal channels or expanded gully deltas and low-lying, often water-covered flat areas (depressions) with a network of gullies connecting both across the mud field, can also be seen in SAR images covering the tidal areas of other East Frisian islands.

3.4. Mussel Beds

Intertidal settlements of blue mussels (*M. edulis*) associated with Pacific oysters (*Crassostrea gigas*) form solid structures sticking out above the sediment surface. These biogenic structures are characterized by a high surface roughness caused by the mussels and by the larger Pacific oysters often growing upright. They are reflected with high backscatter in the SAR images and the varying forms of appearance in which mussel beds occur in situ are also reproduced by the TerraSAR-X data: Young beds that have settled during an actual spat fall are relatively homogeneously occupied by mussels or by homogeneously distributed smaller patches. Over the years, a typical structure of mature beds develops, in which more or less elevated areas covered by mussels form an irregular pattern with open interspaces. This is reflected in the TerraSAR-X data accordingly, with young beds showing homogenous backscatter, while old mussel beds have characteristic internal structures (Figure 11a). In most cases the mussel beds reflected by TerraSAR-X are in good agreement with field observations or with the monitoring results currently obtained from aerial photographs. This is exemplified in Figure 11b, where the yellow line represents the monitoring result from the year of the TerraSAR-X acquisition.

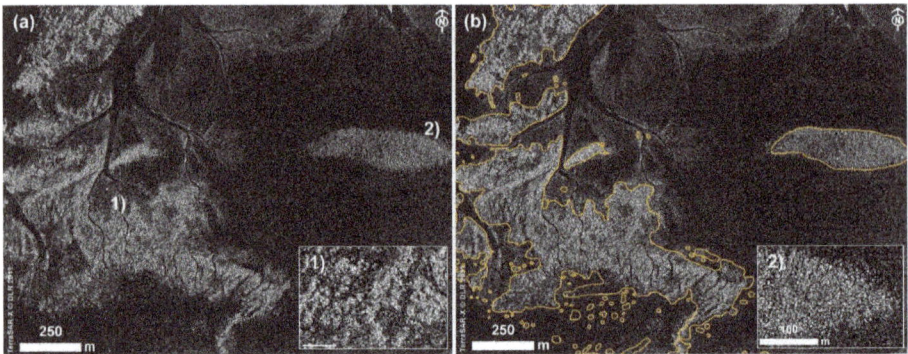

Figure 11. Mussel beds in the central area of the tidal flats south of Norderney imaged by TerraSAR-X (SL of 19/07/2011, desc.): (**a**) established old mussel bed (1) and young mussel bed (2); (**b**) yellow line represents monitoring result from aerial photography interpretation (2011).

3.5. Tidal Flat Dynamics Imaged by TerraSAR-X

The tidal area close to the watershed of the Norderney basin between the eastern Riffgat channel and the mainland coast may serve as an example to demonstrate both the stability and the variability of tidal areas and their reproduction in the TerraSAR-X data. A time series of TerraSAR-X images shows the developments taking place in this area from 2009–2015 (Figure 12). The branches of the Riffgat channel, at the top of the picture, do not change their courses during this period. Likewise, the large mud field (Figure 12, Region 1) remains as such, only the shape of the southern edge, constituting the boundary to the adjacent depression, changes slightly. The gully structures within the mud field remain essentially the same, even if displacements occur in the course of the smaller branches. Since the TerraSAR-X data were recorded in April to July, the mud field is shown in the aspect of the calmer season in each of the four SAR acquisitions. Compared to 2009, the area increased slightly in 2011, 2014, and 2015.

Most obvious in the SAR data, however, is the development of mussel beds in the low lying area south of the mud field (Figure 12, Region 2-4): in the field surveys of 2008/2009, this area proved to be a depression with open sediment surface, often water-covered, and in wide areas densely populated by common cockles (*C. edule*) and the polychaete worm sand mason (*Lanice conchilega*). Sand masons build tubes protruding up to a few centimeters above the sediment surface which leads to an increased roughness, particularly when they break through the surface of shallow water covering the flats (Figure 12a, Region 4, see also photography in Figure 12a). Just like the shell detritus of cockles (cf. chap. 3.2), sand masons can serve as a substrate for the settlement of blue mussels. In 2011, the first mussel settlement in this location was reproduced in the TerraSAR-X image (Figure 12b, Region 4), and in the data from 2014, the mussel bed with its internal structures is already well recognizable as such (Figure 12c, Region 4, photography in Figure 12c). The typical pattern of an established mussel bed can be seen here in 2015 (Figure 12b–d). Southwest of the mud field however, a mussel bed with open structures developed from 2009 to 2011, which in the following years recedes and confines to a few central bed structures in 2014/2015 (Figure 12, Region 3).

In effect, the area of the extensive depression clearly discernible in 2008–2011, has narrowed until 2015. It has been taken up, in particular, by scattered mussel settlements but also by accumulations of muddy sediment, partly forming temporary linear structures (Figure 12c,d, Region 2) which are visible at the mud field's edge in 2014 and throughout the area of the formerly water covered depression. These may be due to the unusually turbulent summer season of that year [32,33].

South of the mussel bed in Region 3, higher backscatter is visible especially in the 2014 acquisitions (Figure 12c). From the field surveys it is known that, in this area, fields of seagrass patches occur.

Seagrass itself was not detected in the TerraSAR-X data according to this study, as it lies flat on the sediment at low tide and is characterized mainly by its spectral features. In some cases, though, the seagrass vegetation leads to the formation of elevated surface structures, which are reflected in the SAR data.

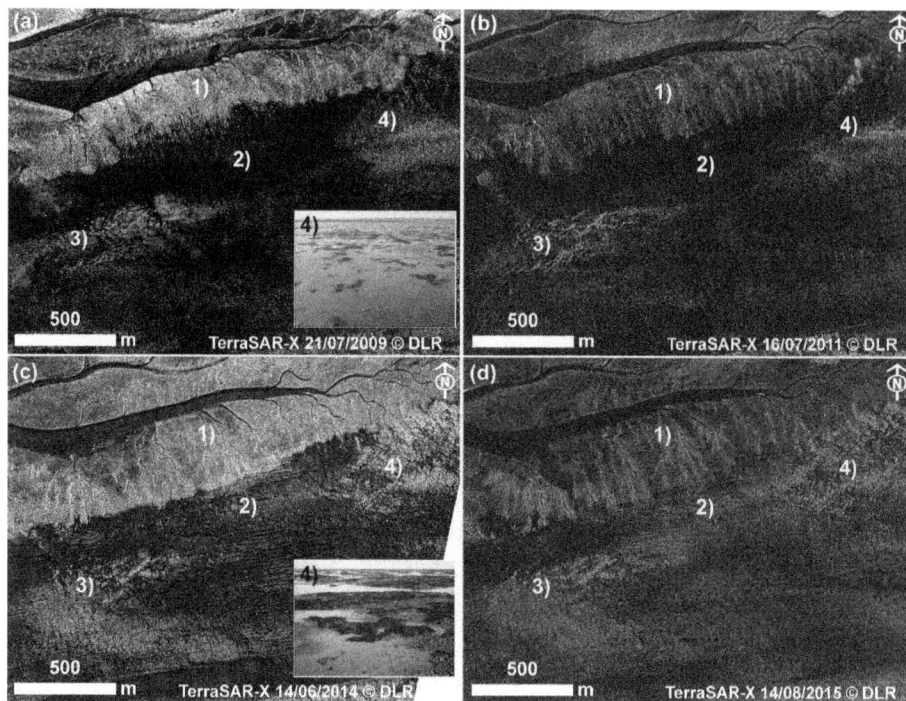

Figure 12. Time series 2009–2015 of tidal area imaged by TerraSAR-X, HRS: (**a**) 21/07/2009, asc., 111 cm < NHN and photograph of 14/07/2008; (**b**) 16/07/2011, asc., 152 cm < NHN; (**c**) 14/06/2014, desc., 132 cm < NHN and photograph of 17/10/2014; (**d**) 19/04/2015, desc., 166 cm < NHN. (1) mud flat; (2) depression; (3) area of patchy mussel bed; (4) area of solid mussel bed.

In summary, certain habitats and structures such as the mud field, mussel bed, or the water-covered depression are clearly recognizable in the TerraSAR-X data due to typical characteristics and patterns. Intermediate states of developments or vague surface structures, on the other hand, can only be identified through field observations or context knowledge. This applies, for example, to the extensive fields of sand mason, which can be recognized at the appropriate level of residual water due to the disturbance of the smooth water surface, to scattered young mussel settlements and oyster scree scattered by winter storms, or to the surface structures sometimes generated by seagrasses.

For monitoring, often it is sufficient to carry out a correct identification of a structure in situ once, to determine its characteristics and boundaries. Further development can then be monitored via TerraSAR-X data.

4. Discussion

The results of the present study show the great potential of satellite SAR data to contribute to the monitoring of the tidal Wadden Sea area. Visual image interpretation of TerraSAR-X data combined with extensive in situ data enable the detection and observation of various large-scale surface structures and characteristic habitats. This is to be emphasized as the smooth and dynamic relief of the Wadden

Sea, influenced by variable water levels and weather conditions, places great demands on classification methods in general.

4.1. Geometry of Acquisition

In general, using different geometries of acquisition, different angles of incidence, and ascending and descending orbit directions, we found that the reproduction of surface structures indicated or amplified by the contrast of sediment and water surfaces is relatively insensitive to geometry of acquisition when making use of visual image interpretation. The same holds for habitats with an extensive three-dimensional surface roughness, such as mussel beds, mud fields, and fields of shell detritus which can be visually identified by their specific patterns and textures under the differing geometries we used.

However, we found some variations in the characteristics of the TerraSAR-X images are due to varying incidence angles of the geometry of acquisition. In near range, that is at small incidence angles <24° (relative orbits 131, 139), the images show sharp contrasts and widespread high backscatter. Therefore, strongly scattering structures are not well demarcated from each other: Mussel beds, humpy mud fields with a dense network of gullies, sediment surfaces roughened by sandworm (*Arenicola marina*) heaps, and steep sandy slopes (depending on exposition in relation to sensor, orbit direction) are displayed similarly brightly which makes the differentiation of these surfaces more difficult. Furthermore, in mussel beds, internal structures are less recognizable. However, when surrounded by smooth surfaces, e.g., smooth water cover, these scatterers stand out sharply. Any roughness of the water surface, on the other hand, is also highlighted and eddies and currents can clearly be seen when biofilms or other surface-active agents are present. As backscatter values of the flooded areas can be quite high, they often exceed those of smooth intertidal surfaces.

With incidence angles of 30–40° (rel. orbits 40, 63), the water surface becomes more uniform and scatters less, the images are less sharp in contrast and more differentiated in the backscatter values. Mussel beds and other structures with high backscatter are better distinguished from each other and from rougher surroundings.

Increasing incidence angles of 40–47° (rel. orbits 116, 154), amplify further differentiation of backscatter intensities. Mussel beds, for example, stand out more clearly from their surroundings, from other rougher surfaces, or from steep edges with high backscatter, which is also due to the fact that the internal structures are better recognizable. Fine linear structures of the sediment surfaces are clearly visible.

All of this is reinforced with incidence angles above 50° (rel. orbits 25, 78). Mussel beds are clearly recognizable. However, gradual transitions are now displayed very fluently and demarcations are therefore less obvious. Under good environmental conditions, i.e., with well drained flats, fine surface structures are clearly visible (e.g., linear structures).

In summary, for most intertidal surface types acquisitions at incidence angles between 30–47° therefore are most suitable. For specific questions smaller or higher incidences can be useful.

4.2. Environmental Influences—Water Cover

An essential aspect in the interpretation of SAR images from tidal areas is the varying presence of water. Apart from the tidal cycle, the water regime is influenced by external conditions affecting tidal water level and, to a lesser extent, also residual water remaining on the flats. Therefore, knowledge of weather and environmental conditions at the time of recording or in the time before may be essential for image analysis. Time of exposure, wind speed and direction, as well as spring/neap tides affect the water coverage in the area but they may also influence the roughness of water and sediment surfaces or sediment moisture. Hence, the same area may appear partly different in SAR acquisitions taken at different times. In general, the flats are better drained after low tide compared to the time of falling tide, even at the same gauge level. Such effects should be taken into account as well as knowledge about general processes and phenomena occurring in tidal areas.

As an example, the appearance of tidal channels, creeks, and gullies reproduced in SAR images is heavily dependent on water level as soon as the water reaches the tideway's edges or goes beyond. Therefore, changes in water level due to weather conditions or even wind-drift may have an effect especially on the gently rising slip-off slopes in contrast to the steep edges of the eroding banks, whose positions will not be markedly affected even for larger variations of the water levels. Thus, to observe and compare the courses of channels and gullies, they should on the one hand be imaged close to maximum drained stage, and on the other hand the steep eroding banks should be used as markers. The same applies when determining migration rates for bedforms in the upper island flats, whose slopes have been found to be slightly asymmetrical [19]. Conversely, in the case of multi-temporal acquisitions, the magnitude of changes in the water level lines would indicate the slope inclination.

On the whole, the SAR data used in this study not only image the channel network and drainage system in tidal areas but they also provide a valuable source of insight into surface morphology of tidal flats mapped due to accumulation of residual water on the exposed flats. Such are the distribution of depressed areas indicated by frequent water coverage or troughs marking bedforms of the sediment surface.

4.3. Visual Analysis and Classification

Overall, the visual approach proved generic enough to provide an overview of most elements structuring the main research area at Norderney by taking into account not only statistical parameters such as backscatter intensity and contrast, but also shapes, sizes, patterns, and textures of surface features reflected by the SAR data, as well as their spatial distribution and surroundings. Especially patterns, texture, and context information proved to have a great significance for the image interpretation. The importance of contextual information—site and time specific—in SAR image interpretation is also emphasized by Ref. [34] who studied the effects of environmental factors and natural processes on radar backscattering in the Korean tidal areas.

Although successful application of TerraSAR-X images is shown, the results also indicate problem zones and variations with the risk of misinterpretation. Areas with no clearly distinctive features or with broad transition zones between habitats may demonstrate the limits of exact demarcation. Mussel beds can exemplify both clearly identifiable areas and problem zones, which will be discussed in the following.

Mussel beds are particularly well recognized by their specific internal structures of beds and interspaces, which also allow a certain understanding of their maturity and compactness. Still, due to variability in the appearance of habitats and structures, misinterpretation can occur, e.g., with shell detritus. Mostly, fields of shell detritus are easy to distinguish from mussel beds, because they lack the characteristic internal structures. However, for young mussel beds or very densely covered areas of mussel beds, internal structures may be similar. In these cases, supplementary in situ data is necessary for correct interpretation of the SAR data. Likewise, steeply sloping edges of high sand flats exposed to the sensor could be mistaken for dense beds of mussels or shell detritus, although these are often recognizable from their location, or from comparison with acquisitions of a different recording geometry. In case of doubt, the structure should be clarified on site. Surfaces that have been identified and verified can then be tracked over time in the SAR data with little effort. Or they can also be identified in other places with this acquired knowledge.

The sole visual analysis of the TerraSAR-X images may also reach its limits when it comes to determining the exact demarcation of surfaces which directly merge into each other with flowing transitions e.g., where mussel beds are directly surrounded by humpy mud flats or fields of shell detritus that extend far beyond the mussel bed and represent their own habitats. In such cases, again, field observations are needed. Preferably, additional distinguishing characteristics are to be found to design a specific classification method, for example, by exploiting the polarized information of the SAR data. Various authors have shown the additional potential of multi-polarization SAR imagery for the detection of bivalve beds, using fully polarimetric e.g., [35,36] or dual-copolarized SAR data [37,38].

Wang et al. [22] discriminated bivalve beds from the surrounding bare sediments through polarimetric decomposition based on dual-copolarized SAR data. Further research is needed to investigate to what extent polarimetric information can be used for the detection of other surface types in the tidal area. Geng et al. [21] identified different surface cover types (i.e., seawater, mud flats, and aquaculture algae farms) through polarimetric decomposition, and Ref [39], and recently Ref [23] pointed to the potential of fully polarimetric interpretation of SAR imagery for classification purposes in tidal areas.

Gade et al. [23] found evidence of mapping characteristics of seagrass beds in SAR data from the Schleswig-Holstein Wadden Sea, whereas in the present study, no general detection of seagrass is proven. In some cases, areas of which seagrass vegetation is known from the field surveys, were characterized by diffusely elevated backscatter values, which may be due to elevated structures of the sediment surface induced by the seagrass cover. Comparative areas vegetated by seagrass, however, could be completely inconspicuous in the SAR data. However, the seagrass stocks in the Lower Saxony Wadden Sea are smaller and of significantly lower density than those in the Schleswig-Holstein Wadden Sea. For these reasons, recognition of seagrass was not pursued in this study. Still, seagrass is a parameter required for Wadden Sea monitoring. For test areas in the Schleswig-Holstein Wadden Sea, Ref. [11] showed that seagrass meadows can be classified based on optical satellite data with a high degree of detail. At present, electro-optical sensors seem to be essential for the detection of seagrass—respectively of vegetated areas—but merging with SAR data could also include surface roughness information. If it is proven that seagrass meadows produce characteristic surface structures reflected by the radar return, this could facilitate their differentiation from green algae or diatoms.

Sediment distribution on tidal flats is another information that would be important for monitoring, but could not be directly obtained from the SAR data by visual interpretation. In some cases indirect detection methods are conceivable, e.g., for mud fields, which are characterized by a humpy surface with puddles and dense gully structures. Also channel network features i.e., the meandering patterns, density and complexity of creeks and gullies and their branches provide, among others, information about the surrounding sediment. The authors of Refs. [40,41], who extracted the geometric information of tidal channels from aerial photography as well as Ref. [42], using electro-optical satellite data (KOMPSAT-2), found lower tidal channel density in areas of higher sand percentage, while complex and dendritic channel patterns were found in mud flat areas. Regarding movement of sediment, Ref. [43] applied the waterline technique to satellite SAR to form a Digital Elevation Model (DEM) of the intertidal zone of Morecambe Bay, U.K. for measurement of long-term morphological change in tidal flat areas. Automated waterline extraction from SAR imagery is used by Ref. [44] for determination of changes in coastal outlines.

The present study shows that visual interpretation of SAR imagery has its own value in support of monitoring and questions of ecological or morphological research in tidal areas. It provides technically unsophisticated access to remote sensing information about characteristic surface structures which can be used by nature protection managers or researches of various disciplines. As a first analysis approach, visual interpretation can also indicate the potential of the satellite SAR data for further investigation and thus may provide pointers for the development of automatable classification methods with regard to monitoring requirements. Currently, Sentinel-1 is providing an increasing base of SAR data available with open and free access which can be screened for monitoring and research for the Wadden Sea.

4.4. Contribution of Satellite SAR for Future Monitoring of Tidal Flats

The regular recording of position, area, and status of characteristic spatial structures in defined time intervals is an indispensable condition for monitoring tidal flat areas such as the Wadden Sea. Monitoring this area has to integrate differing requirements which cannot be provided by a single sensor system. Therefore, a spatially and temporally differentiated monitoring concept combining the benefits of different sensor classes has to be developed. This study has shown that particularly habitats and geomorphic structures characterized by their surface roughness combined with specific textures and patterns are clearly recognizable in TerraSAR-X acquisitions. Other surface structures are virtually marked by residual water, which is an outstanding advantage of this sensor technology. Because of the high temporal availability, SAR data are also predestined to cover periods between sumptuous in situ campaigns, expensive recordings such as lidar scans, or electro-optical acquisitions dependent on daylight and weather. Thus, continuity in the tracking of dynamic structures can be ensured or new events can be discovered in a timely manner by the SAR sensors.

Another approach to meet the monitoring requirements is to directly fuse data from different sensor systems to leverage their respective benefits concerning areal coverage, spatial and temporal resolution, sensitivity, and geometric accuracy while also taking into account financial aspects. In this regard, the advances in satellite technology and the open data policy for imagery from an increasing number of sensors, such as recently the sentinel satellites from the ESA Copernicus program, has already promoted the development of image classification methods. Against the background of the different sensor properties, the combination of different SAR sensors as well as SAR and optical sensors has been examined to refine the differentiation between scatterers or to obtain high-resolution multispectral images e.g., [45–48]. For tidal areas, e.g., Ref. [49] used information from both space-borne microwave (SAR) and optical/shortwave infrared remote sensing to determine sediment grain-size of tidal flats in the Westerschelde, and Ref. [50] investigated the use of multi-frequency SAR data for sediment classification and for the detection of bivalve beds. Results from the DeMarine projects [10,11,14] and the WIMO project [15,17,20] have also shown that a combination of high spatial resolution SAR data and specific spectral resolution (specific wavelengths) benefits the classification of intertidal habitats. New algorithms and procedures in the area of neural network deep learning [51,52] may also bring advances in information extraction from satellite data.

So, to develop operational methods that harness satellite-based remote sensing data for Wadden Sea monitoring and meet the requirements of monitoring obligations from national and international legislation, the advantages of various sensor classes and methods of information extraction will have to be combined. Regarding the exponential growth of technology and methods of information extraction, interdisciplinary research as well as collaboration of nature protection managers with experts in electro-optical and SAR remote sensing will be absolutely beneficial.

5. Conclusions

- High-resolution SAR data as recorded by TerraSAR-X enables identification of essential geomorphic surface structures and habitats of the Wadden Sea ecosystem and their dynamics.
- Independence of SAR sensors from daylight and weather and a high repetition rate (11 days for TerraSAR-X) offer high temporal availability of data and allow to record long-term developments, short-term (e.g., seasonal) developments, and also event effects (e.g., storms, human intervention).
- Even in the spotlight modes providing highest spatial resolution, the footprint of one acquisition covers about the area of a tidal basin. This allows one to determine the status, size, and distribution of the intertidal macrostructures and habitats of a whole sub-unit of the Wadden Sea ecosystem.
- Visual interpretation of TerraSAR-X data combined with context information such as ground truth, monitoring results, or data on environmental conditions, both integrated in a GIS, proved to be a technically unsophisticated access to the information contained in the SAR data. As a first analysis approach, it can also provide basics for the further development of automatable classification methods.
- High-resolution SAR sensors can contribute relevant data for remote sensing the Wadden Sea. For future Wadden Sea monitoring or long-term ecological research, the combination or fusion of appropriate sensor data (e.g., SAR, multi-spectral data) is promising to significantly expand the interpretation options of advanced satellite-borne remote sensing techniques and to develop automated classification methods.
- In this study, the integration of diverse spatial data (such as large-scale remote sensing data and local sampling data) in a GIS has emerged as an essential component assisting the visual analysis. Beyond that, in a broader context, GIS allow to merge classification results and thus to compose

a multifarious overall picture (respectively data base) of the Wadden Sea ecosystem which can support the inter-disciplinary analysis of complex relationships and processes.

- The overview of the geomorphic and biogenic structural elements and habitats of the Wadden Sea ecosystem, their spatial arrangement and dynamics, seen from the perspective of satellite remote sensing using both optical and SAR sensors should be used to contribute to a holistic approach to monitor and further explore the eco-morphological evolution of the tidal system of the Wadden Sea and related tidal systems worldwide.

Author Contributions: H.F. and S.L. conceived and designed the basic project, managed the funding acquisition (WIMO) and provided resources. W.A. refined the project, conducted the research and investigation, performed the data collection and analysis and synthesized the data in the GIS. W.A. wrote the paper, which was commented on by H.F. and reviewed by S.L. and M.E. Validation of the data in the field were carried out by W.A. and H.F. M.E. was involved as external mentor and contributed to the financing.

Funding: This research was carried out as part of the project "Scientific Monitoring Concepts for the German Bight" (WIMO) funded by the Lower Saxony Ministry for the Environment, Energy and Climate Change and the Lower Saxony Ministry for Research and Culture (grant numbers VWZN2564., VWZN2869, VWZN2881). The APC was funded by DLR-MF.

Acknowledgments: The Ministries are highly acknowledged for the funding. We thank DLR for providing the TerraSAR-X data via project COA1075. We are grateful to Ursula Marschalk und Achim Roth (DLR) for their support in acquiring data, which had to be in sync with field campaigns, and also to the colleagues from the DLR Forschungsstelle Maritime Sicherheit, Bremen for the workplace, help and fruitful discussions. We also thank the Institute for Advanced Study (HWK) in Delmenhorst, especially Doris Meyerdierks and Verena Backer for the WIMO overall project management and the function of HWK as platform for discussions and workshops. In the WIMO subproject "Remote sensing", the cooperation with the colleagues from the University of Osnabrück (IGF), especially Richard Jung, the University of Hannover (IPI) with Alena Schmidt and the Senckenberg Institute in Wilhelmhaven with Ruggero Capperucci and Alexander Bartholomä gave insights into the possibilities of the different sensory systems and their pro and cons. We thank our colleagues for the good teamwork especially in the joint field campaigns and the many fruitful discussions. Equally we thank our colleagues from the projects DeMarine-1 and DeMarine-2, Martin Gade, Kerstin Stelzer, Gabriele Müller, Kai Eskildsen, and Jörn Kohlus and also the colleagues from the DLR, Andrey Pleskachevsky, Stephan Brusch and Wolfgang Rosenthal for close cooperation over many years. Thanks also to the reviewers for their advice which helped to improve the text. Personal thanks to Gerald Millat und Gregor Scheiffarth from the National Park Administration Lower Saxon Wadden Sea, whose support of our work was of particular importance and who were always willing to discuss or help.

Conflicts of Interest: The authors declare no conflict of interest. The founding sponsors had no role in the design of the study; in the collection, analyses, or interpretation of data; in the writing of the manuscript, and in the decision to publish the results.

References

1. Millat, G. *Entwicklung Eines Methodisch-Inhaltlichen Konzeptes zum Einsatz von Fernerkundungsdaten für ein Umweltmonitoring im Niedersächsischen Wattenmeer*; Schriftenreihe der Nationalparkverwaltung Niedersächsisches Wattenmeer: Wilhelmshaven, Germany, 1996; Volume 1, pp. 1–125.

2. Herlyn, M.; Millat, G. *Wissenschaftliche Begleituntersuchungen zur Aufbauphase des Miesmuschelmanagements im Nationalpark "Niedersächsisches Wattenmeer"*; Abschlussbericht der Niedersächsischen Wattenmeerstiftung; Wilhelmshaven, Germany, 2004, (unpublished data).

3. Herlyn, M. Quantitative assessment of intertidal blue mussel (*Mytilus edulis* L.) stocks: Combined methods of remote sensing, field investigation and sampling. *J. Sea Res.* **2005**, *53*, 243–253. [CrossRef]

4. Ringot, J.L. Erstellen eines Interpretationsschlüssels und Kartierung der Biotoptypen terrestrischer Bereiche des Nationalparks Niedersächsisches Wattenmeer auf der Basis des CIR-Bildfluges vom 21.08.1991. 1992/1993, (unpublished data).

5. Esselink, P.; Petersen, J.; Arens, S.; Bakker, J.P.; Bunje, J.; Dijkema, K.S.; Hecker, N.; Hellwig, U.; Jensen, A.-V.; Kers, A.S.; et al. Salt Marshes. Thematic Report No. 8. In *Quality Status Report 2009—Wadden Sea Ecosystem 25*; Marencic, H., Vlas, J., Eds.; Common Wadden Sea Secretariat, Trilateral Monitoring and Assessment Group: Wilhelmshaven, Germany, 2009.

6. Petersen, J.; Dassau, O.; Dauck, H.-P.; Janinhoff, N. Applied vegetation mapping of large-scale areas based on high resolution aerial photographs—A combined method of remote sensing, GIS and near comprehensive field verification. *Wadden Sea Ecosyst.* **2010**, *26*, 75–79.

7. Kolbe, K. Erfassung der Seegrasbestände im niedersächsischen Wattenmeer über visuelle Luftbildinterpretation—2008. *Küstengewässer und Ästuare* **2011**, *4*, 1–35.

8. Moreira, A.; Prats-Iraola, P.; Younis, M.; Krieger, G.; Hajnsek, I.; Papathanassiou, K.P. A tutorial on synthetic aperture radar. *IEEE Geosci. Remote Sens. Mag.* **2013**, *1*, 6–43. [CrossRef]

9. Stelzer, K.; Brockmann, C. Operationalisierung von Fernerkundungsmethoden fürs Wattenmeermonitoring (OFEW). Abschlussbericht, 2007. Available online: http://docplayer.org/7506004-Operationalisierung-von-fernerkundungsmethoden-fuer-das-wattenmeermonitoring-zusammenfassung.html (accessed on 27 April 2018).

10. Stelzer, K.; Geißler, J.; Gade, M.; Eskildsen, K.; Kohlus, J.; Farke, H.; Reimers, H.-C. *DeMarine Umwelt: Operationalisierung Mariner GMES-Dienste in Deutschland. Integration optischer und SAR Erdbeobachtungsdaten für das Wattenmeermonitoring*; Jahresbericht 2009–2010; Bundesamt für Seeschifffahrt und Hydrographie: Hamburg, Germany, 2010; pp. 37–55.

11. Müller, G.; Stelzer, K.; Smollich, S.; Gade, M.; Adolph, W.; Melchionna, S.; Kemme, L.; Geißler, J.; Millat, G.; Reimers, H.-C.; et al. Remotely sensing the German Wadden Sea—A new approach to address national and international environmental legislation. *Environ. Monit. Assess.* **2016**, *188*, 595. [CrossRef] [PubMed]

12. Winter, C.; Backer, V.; Adolph, W.; Bartholomä, A.; Becker, M.; Behr, D.; Callies, C.; Capperucci, R.; Ehlers, M.; Farke, H.; et al. *WIMO—Wissenschaftliche Monitoringkonzepte für die Deutsche Bucht*; Abschlussbericht; 2016; pp. 1–159. Available online: http://dx.doi.org/10.2314/gbv:860303926 (accessed on 05 July 2018). [CrossRef]

13. Winter, C. Monitoring concepts for an evaluation of marine environmental states in the German Bight. *Geo-Mar. Lett.* **2017**, *37*, 75–78. [CrossRef]

14. Gade, M.; Alpers, W.; Melsheimer, C.; Tanck, G. Classification of sediments on exposed tidal flats in the German Bight using multi-frequency radar data. *Remote Sens. Environ.* **2008**, *112*, 1603–1613. [CrossRef]

15. Jung, R.; Adolph, W.; Ehlers, M.; Farke, H. A multi-sensor approach for detecting the different land covers of tidal flats in the German Wadden Sea—A case study at Norderney. *Remote Sens. Environ.* **2015**, *170*, 188–202. [CrossRef]

16. Gade, M. A polarimetric radar view at exposed intertidal flats. In Proceedings of the 2016 IEEE International Geoscience and Remote Sensing Symposium (IGARSS), Beijing, China, 10–15 July 2016.

17. Jung, R. A Multi-Sensor Approach for Land Cover Classification and Monitoring of Tidal Flats in the German Wadden Sea. Ph.D. Dissertation, University of Osnabrueck, Osnabrueck, Germany, December 2015.

18. Wang, W.; Yang, X.; Liu, G.; Zhou, H.; Ma, W.; Yu, Y.; Li, Z. Random Forest Classification of Sediments on Exposed Intertidal Flats Using ALOS-2 Quad-Polarimetric SAR Data. *Int. Arch. Photogramm. Remote Sens. Spat. Inf. Sci.* **2016**, *8*, 1191–1194. [CrossRef]

19. Adolph, W.; Schückel, U.; Son, C.S.; Jung, R.; Bartholomä, A.; Ehlers, M.; Kröncke, I.; Lehner, S.; Farke, H. Monitoring spatiotemporal trends in intertidal bedforms of the German Wadden Sea in 2009–2015 with TerraSAR-X, including links with sediments and benthic macrofauna. *Geo-Mar. Lett.* **2017**, *37*, 79–91. [CrossRef]

20. Adolph, W.; Jung, R.; Schmidt, A.; Ehlers, M.; Heipke, C.; Bartholomä, A.; Farke, H. Integration of TerraSAR-X, RapidEye and airborne lidar for remote sensing of intertidal bedforms on the upper flats of Norderney (German Wadden Sea). *Geo-Mar. Lett.* **2017**, *37*, 193–205. [CrossRef]

21. Geng, X.-M.; Li, X.-M.; Velotto, D.; Chen, K.-S. Study of the polarimetric characteristics of mud flats in an intertidal zone using C- and X-band spaceborne SAR data. *Remote Sens. Environ.* **2016**, *176*, 56–68. [CrossRef]

22. Wang, W.; Gade, M. A new SAR classification scheme for sediments on intertidal flats based on multi-frequency polarimetric SAR imagery. *Int. Arch. Photogramm. Remote Sens. Spat. Inf. Sci.* **2017**, *XLII-3/W2*, 223–228. [CrossRef]

23. Gade, M.; Wang, W.; Kemme, L. On the imaging of exposed intertidal flats by single- and dual-co-polarization Synthetic Aperture Radar. *Remote Sens. Environ.* **2018**, *205*, 315–328. [CrossRef]

24. Albertz, J. *Einführung in die Fernerkundung. Grundlagen der Interpretation von Luft- und Satellitenbildern*, 4th ed.; Wissenschaftliche Buchgesellschaft: Darmstadt, Germany, 2009; ISBN 978-3-534-23150-8.

25. Eitner, V.; Kaiser, R.; Niemeyer, H.D. Nearshore sediment transport processes due to moderate hydrodynamic conditions. In *Geology of Siliciclastic Shelf Seas*; de Batist, M., Jacobs, P., Eds.; Geological Society: London, UK, 1996; pp. 267–288.

26. Hayes, M.O. Barrier islands morphology as a function of tidal and wave regime. In *Barrier Islands*; Leatherman, S.P., Ed.; Academic Press: New York, NY, USA, 1979; pp. 1–28.

27. Fritz, T.; Eineder, M. TerraSAR-X Ground Segment Basic Product Specification Document. Available online: http://sss.terrasar-x.dlr.de/docs/TX-GS-DD-3302.pdf (accessed on 27 April 2018).

28. Airbus Defence & Space. Radiometric Calibration of TerraSAR-X Data. Beta Naught and Sigma Naught Coefficient Calculation. TSXX-ITD-TN-0049-radiometric_calculations_I3.00.doc. Available online: https://dep1doc.gfz-potsdam.de/attachments/download/365/r465_9_tsx-x-itd-tn-0049-radiometric_calculations_i3.00.pdf (accessed on 27 April 2018).

29. Haralick, R.M.; Shanmugam, K.; Dinstein, I. Textural Features for Image Classification. *IEEE Trans. Syst. Man. Cybern.* **1973**, *3*, 610–621. [CrossRef]

30. Farke, H. *DeMarine-Umwelt: TP 4—Integration Optischer und SAR Beobachtungsdaten für das Wattenmeermonitoring*; Schlussbericht; 2011. Available online: http://dx.doi.org/10.2314/gbv:722405367 (accessed on 05 July 2018). [CrossRef]

31. NLPV. Monitoring Data: Aerial Mapping for Annual Mussel Monitoring. Available online: http://www.nationalpark-wattenmeer.de/nds/service/publikationen/1130_muschelwildbänke-von-borkum-bis-cuxhaven-gis-daten (accessed on 26 April 2018).

32. Deutscher Wetterdienst (DWD). Available online: www.dwd.de/DE/presse/pressemitteilungen/DE/2014/20140730_Deutschlandwetter_Juli_2014.html (accessed on 28 April 2018).

33. UnwetterZentrale. Available online: http://www.unwetterzentrale.de/uwz/955.html (accessed on 28 April 2018).

34. Lee, H.; Chae, H.; Cho, S.-J. Radar Backscattering of Intertidal Mudflats Observed by Radarsat-1 SAR Images and Ground-Based Scatterometer Experiments. *IEEE Trans. Geosci. Remote Sens.* **2011**, *49*, 1701–1711. [CrossRef]

35. Choe, B.-H.; Kim, D.-J.; Hwang, J.-H.; Oh, Y.; Moon, W.M. Detection of oyster habitat in tidal flats using multi-frequency polarimetric SAR data. *Estuar. Coast. Shelf Sci.* **2012**, *97*, 28–37. [CrossRef]

36. Cheng, T.-Y.; Yamaguchi, Y.; Chen, K.-S.; Lee, J.-S.; Cui, Y. Sandbank and Oyster Farm Monitoring with Multi-Temporal Polarimetric SAR Data Using Four-Component Scattering Power Decomposition. *IEICE Trans. Commun.* **2013**, *96*, 2573–2579. [CrossRef]

37. Gade, M.; Melchionna, S.; Kemme, L. Analyses of multi-year synthetic aperture radar imagery of dry-fallen intertidal flats. *Int. Arch. Photogramm. Remote Sens. Spat. Inf. Sci.* **2015**, *XL-7/W3*, 941–947. [CrossRef]

38. Gade, M.; Melchionna, S. Joint use of multiple Synthetic Aperture Radar imagery for the detection of bivalve beds and morphological changes on intertidal flats. *Estuar. Coast. Shelf Sci.* **2016**, *171*, 1–10. [CrossRef]

39. Park, S.-E.; Moon, W.M.; Kim, D.-J. Estimation of Surface Roughness Parameter in Intertidal Mudflat Using Airborne Polarimetric SAR Data. *IEEE Trans. Geosci. Remote Sens.* **2009**, *47*, 1022–1031. [CrossRef]

40. Ryu, J.-H.; Eom, J.A.; Choi, J.-K. Application of airborne remote sensing to the surface sediment classification in a tidal flat. In Proceedings of the IGARSS 2010: 2010 IEEE International Geoscience and Remote Sensing Symposium, Honolulu, HI, USA, 25–30 July 2010; pp. 942–945. [CrossRef]

41. Eom, J.A.; Choi, J.-K.; Ryu, J.-H.; Woo, H.J.; Won, J.-S.; Jang, S. Tidal channel distribution in relation to surface sedimentary facies based on remotely sensed data. *Geosci. J.* **2012**, *16*, 127–137. [CrossRef]

42. Choi, J.-K.; Eom, J.A.; Ryu, J.-H. Spatial relationships between surface sedimentary facies distribution and topography using remotely sensed data: Example from the Ganghwa tidal flat, Korea. *Mar. Geol.* **2011**, *280*, 205–211. [CrossRef]

43. Mason, D.C.; Scott, T.R.; Dance, S.L. Remote sensing of intertidal morphological change in Morecambe Bay, U.K., between 1991 and 2007. *Estuar. Coast. Shelf Sci.* **2010**, *87*, 487–496. [CrossRef]

44. Wiehle, S.; Lehner, S.; Pleskachevsky, A. Waterline detection and monitoring in the German Wadden Sea using high resolution satellite-based Radar measurements. *Int. Arch. Photogramm. Remote Sens. Spat. Inf. Sci.* **2015**, *XL-7/W3*, 1029–1033. [CrossRef]

45. Klonus, S.; Rosso, P.; Ehlers, M. Image Fusion of High Resolution TerraSAR-X and Multispectral Electro-Optical Data for Improved Spatial Resolution. In *Remote Sensing—New Challenges of High Resolution*; Jürgens, C., Ed.; EARsel Joint Workshop: Bochum, Germany, 2008; pp. 249–264. ISBN 978-3-925143-79-3.

46. Klonus, S.; Ehlers, M. Additional Benefit of Image Fusion Method from Combined High Resolution TerraSAR-X and Multispectral SPOT Data for Classification. In Proceedings of the 29th Annual EARSeL Symposium, Chania, Kreta, 15–18 June 2009.

47. Rosso, P.H.; Michel, U.; Civco, D.L.; Ehlers, M.; Klonus, S. Interpretability of TerraSAR-X fused data. In Proceedings of the SPIE Europe Remote Sensing, Berlin, Germany, 31 August 2009. [CrossRef]

48. Metz, A.; Schmitt, A.; Esch, T.; Reinartz, P.; Klonus, S.; Ehlers, M. Synergetic use of TerraSAR-X and Radarsat-2 time series data for identification and characterization of grassland types—A case study in Southern Bavaria, Germany. In Proceedings of the 2012 IEEE International Geoscience and Remote Sensing Symposium (IGARSS), Munich, Germany, 22–27 July 2012; pp. 3560–3563. [CrossRef]

49. Van der Wal, D.; Herman, P.M.J. Regression-based synergy of optical, shortwave infrared and microwave remote sensing for monitoring the grain-size of intertidal sediments. *Remote Sens. Environ.* **2007**, *111*, 89–106. [CrossRef]

50. Gade, M.; Melchionna, S.; Stelzer, K.; Kohlus, J. Multi-frequency SAR data help improving the monitoring of intertidal flats on the German North Sea coast. *Estuar. Coast. Shelf Sci.* **2014**, *140*, 32–42. [CrossRef]

51. Luus, F.P.S.; Salmon, B.P.; van den Bergh, F.; Maharaj, B.T.J. Multiview Deep Learning for Land-Use Classification. *IEEE Geosci. Remote Sens. Lett.* **2015**, *12*, 2448–2452. [CrossRef]

52. Cheng, G.; Yang, C.; Yao, X.; Guo, L.; Han, J. When Deep Learning Meets Metric Learning: Remote Sensing Image Scene Classification via Learning Discriminative CNNs. *IEEE Trans. Geosci. Remote Sens.* **2018**, *99*, 2811–2821. [CrossRef]

remote sensing

MDPI

Article

Remote Sensing of Organic Films on the Water Surface Using Dual Co-Polarized Ship-Based X-/C-/S-Band Radar and TerraSAR-X

Stanislav A. Ermakov [1],*, Irina A. Sergievskaya [1], José C.B. da Silva [2], Ivan A. Kapustin [1], Olga V. Shomina [1], Alexander V. Kupaev [1] and Alexander A. Molkov [1]

[1]   Institute of Applied Physics, Russian Academy of Sciences, 46 Uljanova St., 603950 Nizhny Novgorod, Russia; onw2009@mail.ru (I.A.S.); kapustin-i@yandex.ru (I.A.K.); seamka@yandex.ru (O.V.S.); sant3@mail.ru (A.V.K.); wave3d@mail.ru (A.A.M.);

[2]   Department of Geoscience, Environment & Spatial Planning, University of Porto, Rua do Campo Alegre, 687, 4169-007 Porto, Portugal; jdasilva@fc.up.pt

*   Correspondence: stas.ermakov8@gmail.com; Tel.: +7-831-416-4935

Received: 30 April 2018; Accepted: 5 July 2018; Published: 10 July 2018

Abstract: Microwave radar is a well-established tool for all-weather monitoring of film slicks which appear in radar imagery of the surface of water bodies as areas of reduced backscatter due to suppression of short wind waves. Information about slicks obtained with single-band/one-polarized radar seems to be insufficient for film characterization; hence, new capabilities of multi-polarization radars for monitoring of film slicks have been actively discussed in the literature. In this paper the results of new experiments on remote sensing of film slicks using dual co-polarized radars— a satellite TerraSAR-X and a ship-based X-/C-/S-band radar—are presented. Radar backscattering is assumed to contain Bragg and non-Bragg components (BC and NBC, respectively). BC is due to backscattering from resonant cm-scale wind waves, while NBC is supposed to be associated with wave breaking. Each of the components can be eliminated from the total radar backscatter measured at two co-polarizations, and contrasts of Bragg and non-Bragg components in slicks can be analyzed separately. New data on a damping ratio (contrast) characterizing reduction of radar returns in slicks are obtained for the two components of radar backscatter in various radar bands. The contrast values for Bragg and non-Bragg components are comparable to each other and demonstrate similar dependence on radar wave number; BC and NBC contrasts grow monotonically for the cases of upwind and downwind observations and weakly decrease with wave number for the cross-wind direction. Reduction of BC in slicks can be explained by enhanced viscous damping of cm-scale Bragg waves due to an elastic film. Physical mechanisms of NBC reduction in slicks are discussed. It is hypothesized that strong breaking (e.g., white-capping) weakly contributes to the NBC contrast because of "cleaning" of the water surface due to turbulent surfactant mixing associated with wave crest overturning. An effective mechanism of NBC reduction due to film can be associated with modification of micro-breaking wave features, such as parasitic ripples, bulge, and toe, in slicks.

Keywords: remote sensing; film slicks on the sea surface; dual co-polarized microwave radar; surface wind waves; wave breaking

---

1. Introduction

Pollution of the sea surface is an imminent threat for the ecological state of open ocean, coastal zones, and inland waters. Remote sensing of marine films, both biogenic pollutions and oil spills, is aimed to identify the films and to quantify their characteristics, and is a very important and urgent problem actively discussed in the literature (see, [1–7], and references therein). This problem, however, is still far from a comprehensive solution. Microwave radar, as a day-and-night/all-weather tool, is particularly interesting for remote sensing of the sea surface. Slicks, associated with surfactants and/or mineral oil films can be easily detected, basically at low-to-moderate wind speeds, using both side-looking real aperture radar (RAR) onboard ships, aircrafts and marine platforms, and satellite synthetic aperture radar (SAR) (see, e.g., [4–7]). Film slicks appear in radar imagery of the sea surface as areas of reduced radar backscatter, and the latter is essentially a result of enhanced suppression of short wind waves.

One of the difficulties to resolve the problem is that mechanisms of suppression of wind waves by films are still not completely understood, particularly for crude oil films. Theoretical analysis has shown that the wave damping coefficient depends on physical characteristics of surface films, such as the surface tension and film elasticity (or viscoelasticity) for monomolecular surfactant films (see e.g., [8–10]), and on the surface and interfacial tensions and viscoelasticities, as well as the volume viscosity for oil layers of finite thickness ("thick oil films") [11,12]. The physical characteristics for monomolecular organic films were experimentally investigated in detail in [13–15], while properties of "thick" oil films have been studied insufficiently (see, [12,16]).

Another difficulty is how to correctly describe the action of films on short wind waves, keeping in mind that physical mechanisms of generation of wind waves, their nonlinear interactions and dissipation, and, hence, formation of the wind wave spectrum are poorly known and parameterized. As a result, existing models of the wind wave spectrum [17–20] are mostly empirical and cannot be considered as very reliable.

The third difficulty is that the very mechanisms of microwave scattering at the sea surface are still under investigation and existing models of radar returns cannot explain experiments properly. Conventional models, a Bragg scattering model and its extended version—a two-scale model (see, e.g., [21,22])—are unable, in general, to explain some important characteristics of microwave radar returns. In particular, it has been demonstrated in [23] when analyzing TerraSAR-X and Envisat ASAR VV-polarized images that reduction of radar return (radar contrast) in film slicks at incidence angles from 20° to 40°, and at low-to-moderate wind, is relatively weak (about 2–3 times) and almost independent both on incidence angle and film elasticity. A conventional composite radar model, taking into account Bragg (resonance) and specular (Kirchhoff) scattering mechanisms [21,22] has underestimated the radar contrast at incidence angles less than 25–30° and overestimated the contrasts at larger angles [23]. A possible reason of the model drawbacks is the existence of an additional component of radar returns, associated with breaking of wind waves, as it has been hypothesized in [24]. This hypothesis is supported, first of all, by the occurrence of strong "spikes" in radar returns, particularly in HH-polarization, which are not consistent with the Bragg scattering mechanism [25–27]. Secondly, experimentally measured values of a ratio of radar backscatter at VV and HH polarizations are smaller than those predicted by a two-scale model. The third inconsistency is that microwave radar Doppler shifts do not accurately correspond to the phase velocities of linear gravity-capillary Bragg waves. This may indicate that so-called bound or parasitic waves, which are generated by breaking waves, longer than Bragg ones, contribute to the radar Doppler shifts (see, e.g., [28–30]).

In [20] a physical radar model has been developed, following the hypothesis of [24], that the radar return is a sum of a Bragg ("polarized") component and a non-Bragg ("non-polarized") component, the latter appearing due to quasi-specular reflection from some facets on the profile of steep (breaking) waves. The non-Bragg component (NBC) has been described as similar to a Kirchhoff specular scattering model with some empirical coefficients. The latter have been introduced to make the model consistent with experiment, largely limited to C-band radar observations. It has been suggested in [31] to analyze different combinations of VV and HH signals, such as the polarization difference (PD) and polarization ratio (PR), thus eliminating either Bragg or non-Bragg backscatter components from the total radar return. The two components respond differently on non-uniform currents, low wind areas, slicks, etc., and potential capabilities to emphasize that one or the other surface signature in radar imagery can be realized. In particular, to develop more effective approaches to solve the problem of film slick remote sensing some observations with dual- and quad-polarized single-band radars have been carried out (see, e.g., [31–33]). In [32] where Radarsat (RS) and TerraSAR-X (TS) acquisitions of slicks were analyzed any noticeable effect of surface film on the non-Bragg component was not revealed. However, a significant reduction of NBC in films slicks was reported later in [33]. It was obtained in [33], based on analysis of Radarsat-2 observations, that NBC and BC contrasts in film slicks of different origin are comparable to each other.

A limitation of the experiments was the use of single band radars and poor knowledge of slick properties (film elasticity, surface tension, viscosity, etc.), responsible for suppression of wind waves. Accordingly, further, more detailed, quantitative studies of different components of radar return from the areas covered with films are needed for better understanding of mechanisms of slick radar imaging and for elaboration of models and methods of slick detection and identification. In the context of the problem solution it is crucially important (a) to check whether damping of Bragg waves can be described by hydrodynamic theory which predicts the damping as a function of film parameters, and (b) to understand how breaking wave features, i.e., crest overturning and micro-breaking, are modified by film. As it was mentioned above, dependences of wave damping on film characteristics (and on wave number, too) were studied, e.g., in [12–16]. Films with pre-measured surface tension and elasticity can be used as calibrated ones in experiment to study, in particular, reduction of BC due to damping of Bragg waves. As for the modification of steep wave profiles due to films, which might be responsible for the reduction of non-Bragg radar returns, the problem is practically open. Wave tank experiments [34] revealed that the film was destroyed in the area of spilling wave breakers, which resulted in "cleaning" of the water surface in the vicinity of breaking crests of m-scale waves. Accordingly, one can assume that strong wave breaking of m-scale waves weakly contributes to the reduction of the non-Bragg radar component, and modification of micro-breaking of cm-dm-scale waves in slicks can play the most important role in non-Bragg contrasts in slicks. The latter is characterized by the appearance near wave crests of structures with high curvature, so-called bulge and toe [35], and by parasitic capillary ripples [36,37] propagating on the forward slopes of steep waves.

This paper presents new results of studies of film slicks using dual co-polarized radar: the satellite TerraSAR-X and an X-/C-/S-band scatterometer mounted onboard a research vessel. The paper is organized as follows: Section 2 presents the theoretical background, introducing Bragg and non-Bragg scattering components and their extraction from the total normalized radar cross-section (NRCS) at VV and HH polarizations. Section 2 also describes the apparatus and methodology of the experiments on radar probing of slicks formed by "calibrated" organic films. The obtained experimental results are presented in Section 3, and possible physical mechanisms responsible for suppression of Bragg and non-Bragg radar components are discussed in Section 4. Conclusions are given in Section 5.

2. Materials and Methods

2.1. Theoretical Background

Radar observations of the sea surface indicate that microwave backscattering is characterized by the occurrence of short pulses (spikes) which are significantly larger than some mean ("background") level between the spikes. The latter, as it has been hypothesized in [24], can be associated with breaking of wind waves (either micro- or strong breaking), while the lower background level corresponds basically to Bragg scattering. Thus, the total NRCS was supposed to be a sum of Bragg and non-Bragg components:

$$\sigma_{pp}^0 = \sigma_{BC\_pp} + \sigma_{NBC} \qquad (1)$$

where σ_{pp}^0 is the total NRCS, p denotes vertical (V), or horizontal (H) transmit/receive polarizations, σ_{BC_pp} is the Bragg (VV or HH) component of NRCS described by the two-scale model, σ_{NBC} denotes the NBC associated with quasi-specular scattering from surface facets on micro- or strong wave breakers.

The Bragg component, according to a two-scale radar model (see, e.g., [21,22]) can be expressed as:

$$\sigma^0_{BC\_pp} = 16\pi k^4_{em} R_{pp}(\theta) F(\vec{k}_B) \tag{2}$$

where $F(\vec{k}_B)$ is the spectrum of wind waves at a Bragg wave vector $\vec{k}_B = 2k_{em}\vec{n}_s$, k_{em} is the wave number of an incident electromagnetic wave, $\vec{n}_s$ is a projection of the unit wave vector of the incident wave on the sea surface, $\left|\vec{n}_s\right| = \sin\theta$, θ is an incidence angle, $R_{pp}(\theta)$ is a reflection coefficient which depends on polarizations of the incident/reflected electromagnetic waves, and in general on root mean square (r.m.s.) slopes of long surface waves, i.e., those which are longer than the antenna footprint [21]. Non-Bragg scattering from the areas of wave breaking is assumed to be independent of polarization, and σ_{NBC} is characterized in the literature as non-polarized.

It has been suggested in [31] to remove the NBC from the total NRCS when subtracting σ^0_{HH} from σ^0_{VV}. Thus obtained backscatter polarization difference σ_{PD} is:

$$\sigma_{PD} = \sigma^0_{VV} - \sigma^0_{HH} = \sigma^0_{BC\_VV} - \sigma^0_{BC\_HH} = (R_{VV} - R_{HH}) F(\vec{k}_B) \tag{3}$$

The non-Bragg radar backscatter component, σ_{NBC} can be found from Equations (1) and (3) as:

$$\sigma_{NBC} = \sigma^0_{VV} - (\sigma^0_{VV} - \sigma^0_{HH})/(1 - R_{HH}/R_{VV}) \tag{4}$$

Slick contrasts for the total NRCS, Bragg and non-Bragg components are defined as:

$$K_{pp} = \frac{\sigma^0_{pp\_nonslick}}{\sigma^0_{pp\_slick}}, K_{BC} = K_{PD} = \frac{\sigma_{PD\_nonslick}}{\sigma_{PD\_slick}}, K_{NBC} = \frac{\sigma_{NBC\_nonslick}}{\sigma_{NBC\_slick}} \tag{5}$$

Since the polarization difference is supposedly proportional to the Bragg components of NRCS, the contrast K_{PD} is equal to BC contrast K_{BC}. If one assumes that that film does not affect the scattering coefficients R_{VV} and R_{HH}, then the BC contrast is determined by the wind wave spectrum outside/inside slick $K_{PD} = F_{nonslick}(k_B)/F_{slick}(k_B)$. Transformation of the spectrum of short wind waves due to film can be considered in the frame of a local balance model in [2,4,18], or using an improved physical model [20]. Without going into detail it can be said that the BC contrast for both the models depends on the energy sources, sinks, and nonlinear terms in the kinetic equation for the wind wave spectrum. In a particular case of low wind wave input that is realized, e.g., for wind wave components propagating across the wind, the BC contrast can be expressed as a ratio γ_s/γ_{nsl} of surface wave damping coefficients inside the slick (s) and in a surrounding non slick area (nsl). The ratio γ_s/γ_{nsl} can be described using the theory of wave damping in the presence of an elastic film (see, e.g., [8–10]). Then the BC contrast for the cross wind direction is:

$$K_{BC}(crosswind) \approx \gamma_s/\gamma_0 = \frac{1 - X + XY}{1 - 2X + 2X^2} \tag{6}$$

where $X = \frac{Ek^2}{\rho(2\nu)^{1/2}\omega^{3/2}}$ and $Y = \frac{Ek}{4\rho\nu\omega}$. Here ρ denotes the water density, ν the kinematic water viscosity, E the film elasticity, k and ω are the wave frequency and wave number of the wind wave spectrum component, respectively.

2.2. Experiment

Experiments on radar probing of film slicks were carried out in the southern part of the Gorky Water Reservoir (GWR) of the Volga River, Nizhny Novgorod Region, Russia. GWR extends approximately from north to south for about 100 km, and its width is 5–15 km. Artificial film slicks in the experiments of 2014 and 2015 were observed with satellite TerraSAR-X operating at a frequency of 9.65 GHz. Experiments of 2016 were performed using a new three-band radar designed at the

Institute of Applied Physics, Russian Academy of Sciences (IAP RAS). It is a Doppler radar, operating in the X-, C-, and S-bands at frequencies of 10 GHz, 6 GHz, and 3 GHz, respectively, and at vertical transmit/receive (VV) and horizontal transmit/receive (HH) polarizations of electromagnetic waves in each band. The three-band radar operates in a pulse regime radiating 30-ns pulses. The beam width of the radar pattern is about 0.03 rad (X band), 0.05 rad (C-band), and 0.1 rad (S-band). The dynamic range of the radar signal is about 55 dBs in electronic channels, and is enhanced after digital processing. The radar was mounted onboard a research vessel at a height of about 7 m, looking at an incidence angle of 60° and at an azimuth angle of about 40° to the left from the ship's heading. A photograph of a ship based X-/C-/S-band radar is shown in Figure 1.

Figure 1. A ship based X-/C-/S-band radar.

The radar footprints and, accordingly, radar spatial resolutions, were as follows: the slant range footprints varied from about 2.8 m in the S-band to 0.8 m in the X-band, and the azimuth range footprints from 1.4 m in the S-band to 0.4 m in the X-band. OLE slick sizes in the experiment were typically 200–300 m, i.e., least two orders of magnitude of the radar footprints. The signal-to-noise ratio (SNR) in the experiments with the three-band radar ranged from about 15–17 dB in the slick to 23–27 dB in the background, depending on radar bands and polarizations.

The wind velocity/direction were measured with an acoustic anemometer (WindSonic ®Gill Instruments Limited, Lymington, Hampshire, United Kingdom) mounted onboard the research vessel at a height of about 6 m. Oleic acid (OLE), supposedly simulating natural biogenic films, was used to create film slicks on the water surface. Some amounts of the surfactant (about half a liter of OLE dissolved in one liter of ethanol) were poured on the water from a motor boat to an inflatable trimaran. During the experiments the boat was moving along spiral trajectories creating slicks at least 20 min before satellite overpasses or before slick transects made by the research vessel.

Physical characteristics (the surface tension and film elasticity) of artificial surfactant films used in the experiment were studied in the laboratory using a method of parametric waves [15]. The method is based on measuring the wavelength and the damping coefficient of gravity-capillary standing waves parametrically generated at certain frequencies in a small vertically-oscillating container. From these measurements the surface tension coefficient and the dynamic film elasticity were retrieved (the latter is the main parameter which determines the damping of surface waves). The dynamic elasticity for an oleic acid film retrieved at wave frequencies of 10 Hz, 15 Hz, 20 Hz, and 25 Hz, and the surface tension at a wave frequency of 25 Hz are shown in Figure 2 (cf. [15]). The elasticity of OLE films grows with surfactant concentration and tends to a constant value for a saturated monomolecular film, which is characterized by dense packing of surfactant molecules oriented nearly vertically on the water surface. At mean surfactant concentrations larger than that of saturated monomolecular films ("oversaturated" films) the excess of surface active materials is contained within microscopic drops, and the dynamic elasticity and surface tension in this case remain practically constant. OLE films in our experiments were normally "oversaturated" and could be characterized by constant elasticity and surface tension values, which were about 25–40 mN/m and 30 mN/m, respectively. Supporting measurements of wind velocity and current profiles, as well as sampling of natural biogenic films from the water surface, were conducted from aboard the trimaran. The physical characteristics of natural films were studied in the laboratory by the method of parametric waves in order to filter out the cases of a highly-contaminated background water surface. In the reported experiments the elasticity of the background biogenic films did not exceed 3–5 mN/m.

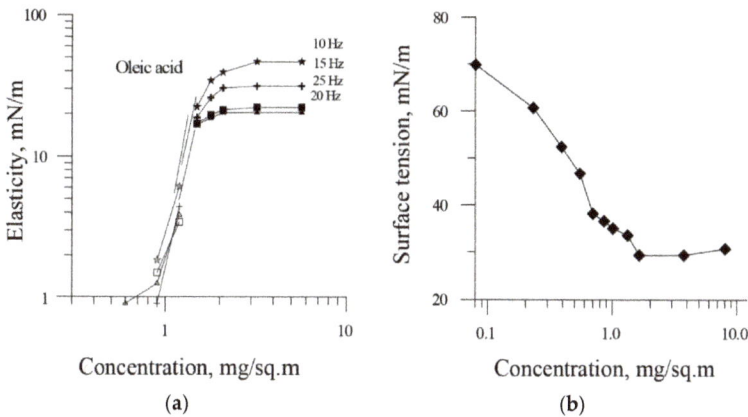

Figure 2. Elasticity at several wave frequencies (**a**) and surface tension at wave frequency of 25 Hz (**b**) as a function of the surfactant concentration for OLE films (cf., [15]).

Experiments with artificial slicks were carried out at moderate winds, and wind speeds velocities were in the range of about 5 m/s to 7 m/s. The radar look azimuth angles kV, i.e., angles between the horizontal projection of the incident electromagnetic wave vector and the wind velocity, varied from about 180° (upwind) to 40° from downwind direction. Characteristics of the experiments are summarized in Table 1.

Table 1. Characteristics of experiments on radar probing of OLE film slicks on the Gorky Water Reservoir.

Date, Sensor	Inc. Angle	Wind Velocity, Dir.	Azimuth Angle (kV)	Bragg Wavenumber k_{Bragg}, rad/cm
31.08.2014, TerraSAR-X	37°	7 m/s, NW	(kV) ≈ 40° (40° from downwind)	2.43
03.08.2015, TerraSAR-X	32.5°	5 m/s, W	(kV) ≈ 180° (upwind)	2.17
22 07.2016, 3-band radar, transect 1	60°	7 m/s, E	(kV) ≈ 80°	1.01, 2.17, 3.63
22 07.2016, 3-band radar, transect 2	60°	7 m/s, E	(kV) ≈ 40°	1.01, 2.17, 3.63
22 07.2016, 3-band radar, transect 3	60°	7 m/s, E	(kV) ≈ 180°	1.01, 2.17, 3.63

3. Results

3.1. Satellite Experiment of 31.08.2014

A fragment of a VV polarization image of an OLE slick in the experiment of 31.08.2014 (TerraSAR-X ascending pass) is shown in Figure 3.

Profiles of NRCS at VV and HH polarizations, BC and NBC contrasts and their ratio for the case of 31.08.2014 are presented in Figure 4. Note, that the contrasts in Figure 4 are current contrasts, which were calculated as ratios of mean background values, chosen well outside a slick to the current BC or NBC values along a slick transect.

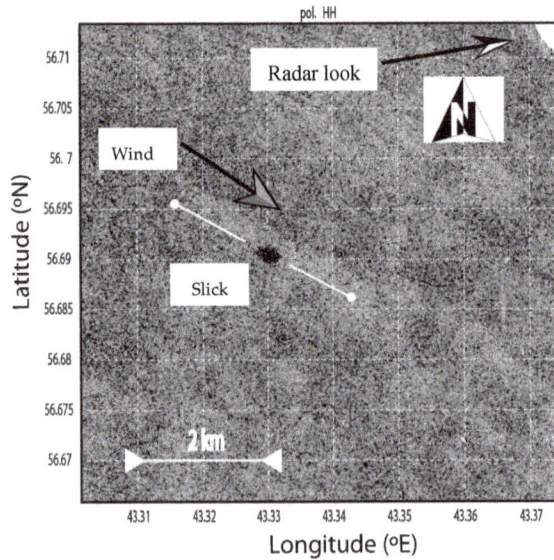

Figure 3. Fragment of the HH image of 31.08.2014 with an OLE slick (a dark spot) in the center. The arrow in the upper right corner denotes the radar look, and the grey arrow in the center is the wind direction. A transect along the slick is shown with the white line segment.

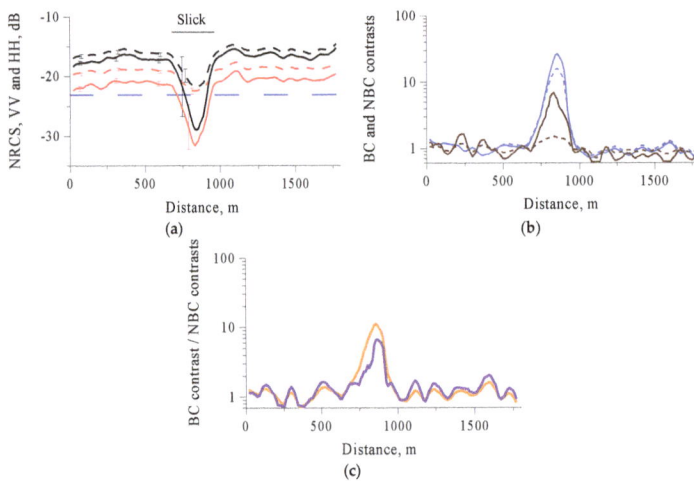

Figure 4. Characteristics of a radar return along the slick transect indicated in Figure 3 for the satellite experiment of 31.08.2014. (**a**) TerraSAR-X NRCS at VV and HH polarizations (black and red lines, respectively) before and after noise floor subtraction (dashed and solid lines, respectively). The dashed blue line is the noise floor. The vertical bars are a 95% confidence limit. (**b**) Current contrasts K_{BC} with and without noise floor subtraction (solid and dashed blue lines, respectively) and K_{NBC} with and without noise floor subtraction (solid and dashed brown lines). (**c**) Contrast ratio K_{BC}/K_{NBC} of radar backscatter with and without noise floor subtraction (violet and orange lines, respectively).

It is seen that the total NRCS drops in the slick by several decibels. The NRCS contrasts, e.g., at VV polarization, are consistent with those obtained in our previous experiments (see, [23]). A drawback of the experiment of 31.08.2014 is that the NRCS values in the slick area are only 1–2 dBs above the noise floor, particularly for HH-polarization. Note that the noise floor is practically the same for both VV and HH polarizations. As a result, the NRCS contrasts after the noise floor subtraction are significantly larger than those without the subtraction and an error of the contrast estimate can be large, so the contrasts in Figure 4 should be considered mostly as rough estimates. Contrasts for the Bragg component, however, are practically unchanged after the noise floor subtraction, while the non-Bragg contrasts are affected by the noise floor. In the considered case the BC contrast is about twice the NBC contrast. The difference between BC and NBC contrasts is even larger if the noise floor is not subtracted.

3.2. Satellite Experiment of 03.08.2015

More reliable contrast estimates were obtained in the experiment of 03.08.2015. A fragment of a VV-polarized image an OLE slick for this case is shown in Figure 5. The slick was elongated due to wind; a slick transect in the image is shown in Figure 5.

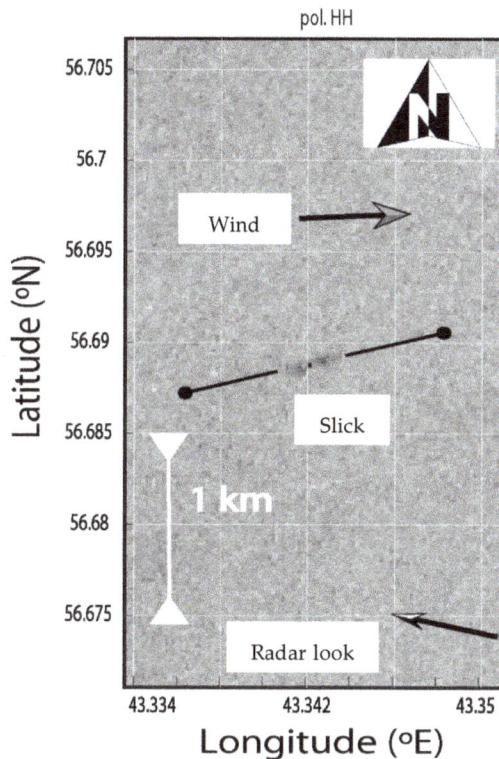

Figure 5. Fragment of a VV polarized image for the experiment of 03.08.2015. The black/white arrow in the bottom right corner denotes radar look, the grey arrow in the upper part denotes the wind direction. Solid and dashed arrows are wind and radar look directions, respectively. A transect along the slick is shown with the black line segment.

An example of NRCS at VV and HH polarizations, as well as Bragg and NBC contrasts and their ratio are shown in Figure 6.

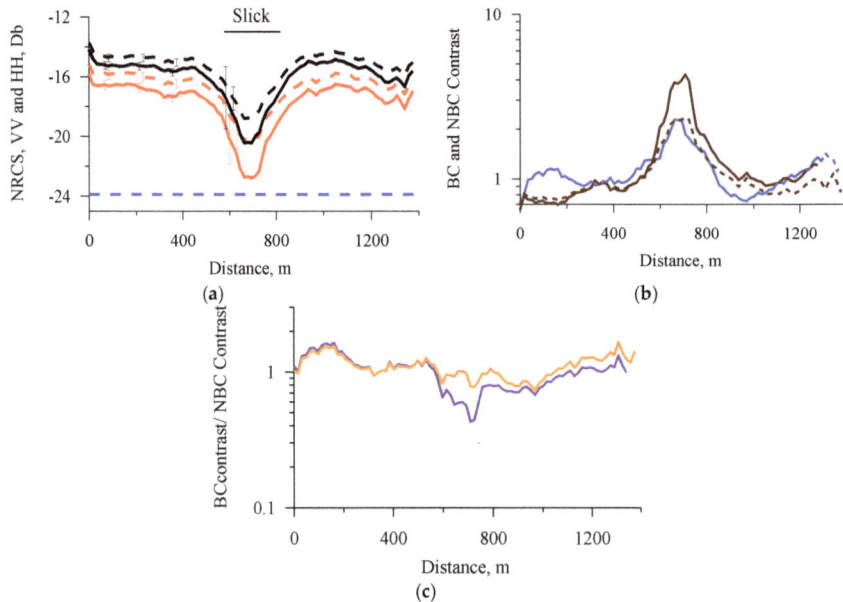

Figure 6. Characteristics of radar returns along the slick transect indicated in Figure 5 for the satellite experiment of 03.08.2015. (**a**) NRSC at VV-pol and HH-pol before noise floor subtraction (black and red dashed lines, respectively), and after noise floor subtraction (black and red solid lines), the blue dashed line is the noise floor level, the vertical bars are a 95% confidence limit. (**b**) BC (blue lines) and NBC (brown lines) are current contrasts without noise floor subtraction (dashed lines) and after noise floor subtraction (solid lines). The blue dashed and solid lines are overlapped. (**c**) Contrast ratio K_{BC}/K_{NBC} (violet: after noise floor subtraction; orange: without noise floor subtraction).

Figure 6 indicates that NRCS both for VV and HH-polarizations for the experiment of 03.08.2015 are well above the noise floor, so that the BC and NBC contrast values are more reliable than for the previous case. It follows from Figure 6c that the NBC contrast, unlike the case of 31.08.2014, is larger than the BC contrast.

3.3. Boat Experiment of 22.07.2016 with a Three-Band Radar

There were three transects through a slick made by a research vessel in order to obtain data at different angles between radar look and wind directions. A scheme illustrating the vessel trajectory and locations of slick transects is shown in Figure 7.

An example of profiles of the radar backscatter at VV and HH polarizations, and of BC and NBC components and their ratios is given in Figure 8 for transect 1, the intensities of radar return in Figure 8 are in arbitrary units, but the same for both VV and HH channels in each radar band.

It is seen that the noise floor of the three-band radar was well below the radar returns in the slick area for both VV and HH polarizations in all bands, so that one can reliably affirm that BC and NBC contrasts are close to each other. In more detail, the NBC contrast in S-band is practically equal to the BC contrast, in C- and X-bands the reduction of NBC is slightly larger than for BC.

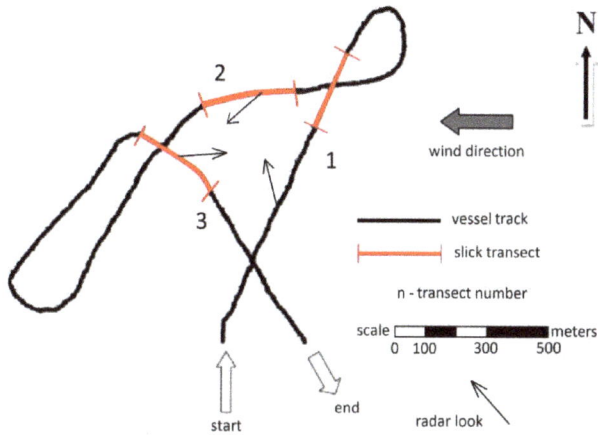

Figure 7. A scheme of a vessel trajectory in the boat experiment of 2016.

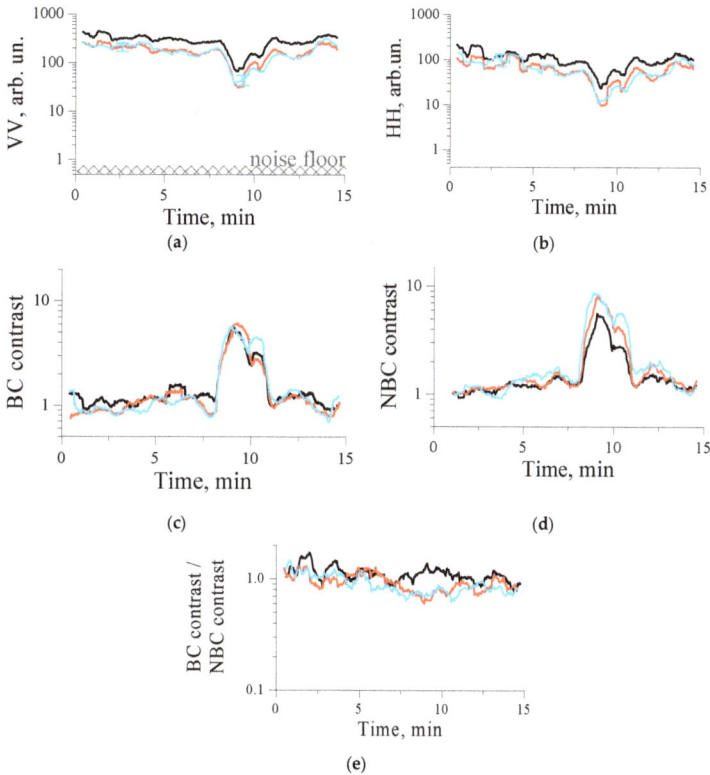

Figure 8. Radar backscatter at (**a**) VV and (**b**) HH polarizations (in arbitrary, but the same units) in the boat slick experiment of 22.07.2016 with a three-band radar. Current contrasts K_{BC} (**c**), contrasts K_{NBC} (**d**) and K_{BC}/K_{NBC} ratio (**e**). Black, blue, and red curves correspond to data for S-, C-, and X-bands, respectively. A noise floor level is depicted in (**a**), and a 95% confidence limit, which is the same for all of the channels, is shown with the vertical bars.

4. Discussion

Let us consider the suppression of different components of radar backscattering estimated from experimental data (see Table 2). Mean contrasts for Bragg and non-Bragg components are presented in Figure 9, and 95% confidence intervals are depicted with the vertical bars. For comparison, some of BC and NBC contrasts obtained with RADARSAT-2 in [33] are shown for similar radar look directions relative to wind direction (kV).

Table 2. Contrasts in slicks for total NRCS at VV and HH polarizations, and for Bragg and NP components in experiments with OLE slicks (see Table 1).

Date, Experiment	Radar Band	k_B, rad/cm	Contrasts				
			K_{VV}	K_{HH}	K_{Bragg}	K_{NBC}	K_{Bragg}/K_{NBC}
22.07.2016 Transect1, (kV) ≈ 80°	S	1.09	4.3	4.1	4.5	4.1	1.1
	C	2.17	5.4	7.6	4.5	8	0.6
	X	3.63	5.7	8.2	4.6	8.5	0.5
22.07.2016 Transect2, (kV) ≈ 40°	S	1.09	3.3	3	3.5	3	1.2
	C	2.17	3.9	3.5	4	3.5	1.1
	X	3.63	5	4.2	5.6	4.1	1.4
22.07.2016 Transect3, (kV) ≈ 180°	S	1.09	4.5	2.3	8.8	2.1	4.2
	C	2.17	8	5.2	9.9	4.9	2
	X	3.63	18	20.5	17.5	20.7	0.8
31.08.2014 TerraSAR-X, (kV) ≈ 40°	X	2.43	7.4	5.8	9	3.8	2.4
03.08.2015 TerraSAR-X, (kV) ≈ 180°	X	2.17	2.9	2.5	2.0	3.5	0.6

Figure 9 clearly indicates that BC contrasts increase with wave number, except for the cross-wind case, and are highest for an upwind look direction. Similar behavior can be noted for contrasts in the directions close to the downwind, although the downwind contrasts are somewhat smaller than for the upwind case. The theoretical contrasts for up/downwind cases are estimated according to [2] at the conditions of experiments (at a wind velocity 7 m/s, E= 20 mN/m, and at two different empirical coefficient values—0.04 and 0.06 in the formula for a wind wave growth rate (see, [2]). The theory [2] has an obvious drawback of being invalid if the wind wave growth rate is close to the damping coefficient. In our case this occurs at wave numbers larger than 1 rad/cm, so at higher wave numbers the theory can be substituted by an empirical model [20]. Without going into detail of the models one can consider that the reduction of BC radar returns in the slick is due to enhanced viscous damping of short wind waves in the presence of film. This conclusion is also supported by an analysis of cross-wind contrasts. The latter, calculated according to Equation (6), are plotted in Figure 9c. A decreasing tendency of cross-wind BC contrasts with wave number is qualitatively consistent with the experiment. One should note that the elasticity of crude oil films according to recent laboratory measurements [16] has been estimated roughly as 25 mN/m, which is compatible with the OLE film elasticity and, thus, the contrast values [33] shown in Figure 9c can be considered as complementary to our data.

The NBC contrast values and their dependence on the wave number and on the azimuth angle are quite similar to the BC contrasts, thus indicating that film significantly influences the processes of wave breaking. One should recall that strong breaking with crest overturning is typical for surface waves of about 1 m in length and larger. This breaking is supposedly weakly affected by film, since the film can be essentially destroyed due to turbulence and air bubbles in "white caps" mixing down surfactants to the subsurface water layers. The effect of "cleaning" of the water surface in the area of strongly breaking wave crests was demonstrated in our laboratory experiments [34]. It was obtained in [34] when measuring the surface tension in the different phases of m-scale surface waves that, in the presence of OLE-film, the surface tension increased in the area of breaking crests. This proved the effect of film destruction by strong wave breakers.

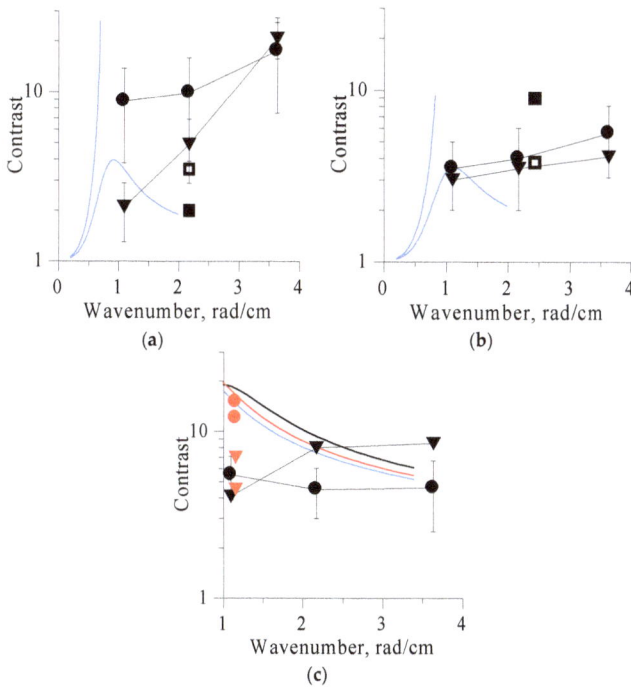

Figure 9. BC and NBC contrasts vs. Bragg wave number for OLE slicks in experiments of 2014–2016 obtained with a three-band radar (●BC, ▼ NBC) and with TerraSAR-X (■ BC, □ NBC). (**a**) Upwind look direction, (**b**) 40° from downwind, the blue curves: theoretical BC contrasts at two different wind wave growth rates; (**c**) cross-wind observations; curves: theoretical BC contrasts at elasticity 20 mN/m (black), 30 mN/m (red), and 40 mN/m (blue), the red symbols: Radarsat data [33] for BC (●) and NBC (▼) contrasts for crude oil/emulsion slicks.

One can assume, however, that NBC contrasts in slicks are basically associated with the suppression of wave micro-breaking. The latter is realized for surface waves shorter than 1 m, and micro-breaking features, such as "toe"/"bulge" structures [35] and "parasitic" capillary ripples (see, e.g., [36,37]) occur in the vicinity of wave crests. The structures can be effectively suppressed by film, even at relatively small mean surfactant concentrations. This is because the surfactant concentration $\Gamma(x, t)$ is modulated by the surface wave orbital velocity $U(x - Ct)$ and achieves maximum values near the wave crests. $\Gamma(x, t)$ in the field of a surface wave travelling in the x-direction at a phase velocity C can be written as (see, e.g., [2,4]):

$$\Gamma(x,t) = \Gamma_0 \frac{C}{C - U(x - Ct)} \tag{7}$$

Since $U(x - Ct)$ is in phase with the surface wave elevation, the Γ values increase at the wave crests, of course, if the film is not destroyed due to strong breaking. Accordingly, the surface tension decreases and the elasticity in general increases at the crest. Significant reduction of the surface tension and enhanced elasticity on wave crests results in the abatement of the source of parasitic capillary ripples, i.e., the Laplace pressure [36], and also leads to "smoothing" of the bulge/toe structures. As a result, the quasi-specular reflection of electromagnetic waves from the micro-breaking structures can be strongly reduced.

To study the effect of suppression of micro-breaking in slicks some wave tank experiments have been carried out. Here, first, qualitative results are presented, and a more detailed analysis will be done elsewhere. Our experiments were conducted in an oval wave tank of IAP RAS, where surface gravity-capillary waves were generated with a mechanical wavemaker. The wave steepness was about 0.1 and larger, which corresponded to the generation of parasitic ripples and the formation of toe/bulge structures (see, e.g., [37]). The wave height was measured with a wire gauge and a fine structure of wave profile was studied using photographs. In order to obtain high-contrast wave profiles photo recording in a dark room was performed. A camera was placed opposite the plexiglas tank window. An optical lens system was used to obtain a laser "knife-shape" beam, which was directed downward onto the water. Typical profiles of short gravity waves at two wavelengths (about 10 cm and 20 cm) are depicted in Figure 10. Parasitic capillary ripples were effectively generated by a 10-cm steep wave and were propagating along the forward wave slope. For longer, 20-cm waves a bulge/toe structure dominated. The structures are characterized by rather large slopes, so that quasi-specular reflection of the incident electromagnetic radiation can occur. Film action on the micro-breaking structures was studied using monomolecular OLE-films at concentrations about 1 mg/m^2. In the presence of film the wave profiles were smoothed and the micro-breaking structures practically disappeared, as illustrated in Figure 10.

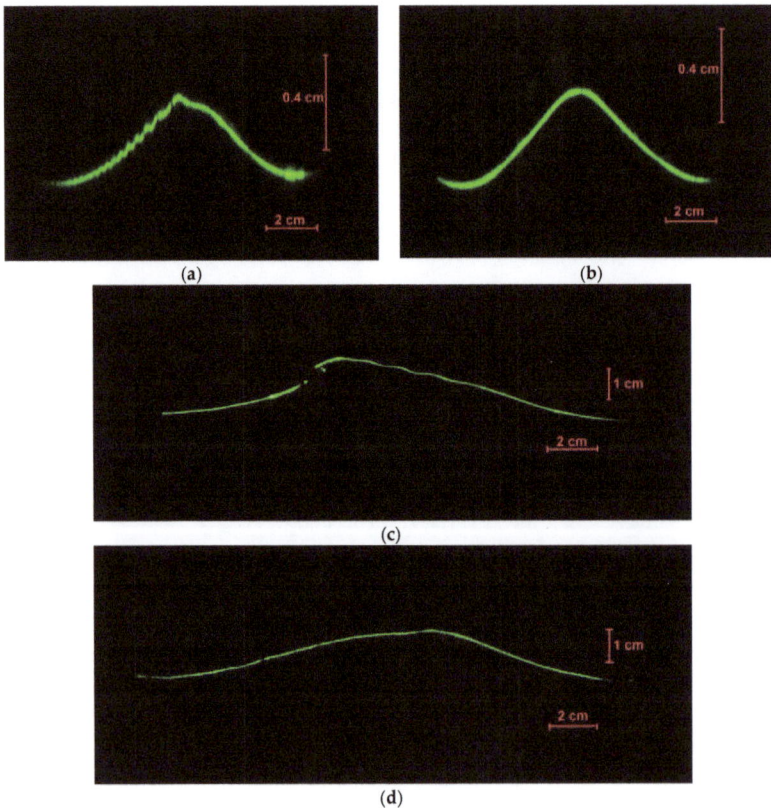

Figure 10. Elevation profiles of steep, short gravity waves of 10 cm (**a,c**) and 20 cm (**b,d**) wavelength (left and right columns, respectively), the upper row—clean water (**a,b**), the bottom row—water contaminated by an oleic acid film (**c,d**). Waves travel from right to left, and vertical and horizontal scales are extended for clarity.

One can, thus, assume that film can influence micro-breakers more effectively than strong breakers. This influence, however, has been insufficiently studied up to now, as well as the impact of wind on micro-breakers. It has been revealed in wave tank experiments [37] that parasitic ripples in the presence of wind were generated at smaller wave crest curvatures than for mechanically-generated waves. As for bulge/toe micro-breakers, no information about the role of wind in their generation has been found in the literature, and further studies of the dynamics of micro-breakers in the presence of film and wind should be carried out.

5. Conclusions

Organic films of oleic acid deployed on the water surface were observed using dual-polarized TerraSAR-X and an X-/C-/S-band dual-polarized microwave radars. Experiments were carried out at moderate wind (wind velocity about 5–7 m/s), at different azimuth angles, and the incidence angles ranged from about 32 to 60 degrees. The Bragg and non-Bragg components of radar returns were obtained from experimentally measured total radar backscatter at two co-polarizations. Suppression of radar returns of film slicks was characterized by radar contrasts, i.e., by ratios of radar return values for NRCS, BC, and NBC, outside and inside slicks in X- to S-radar bands.

It has been concluded that the Bragg scattering component is suppressed in slicks due to enhanced viscous damping of resonant Bragg cm-scale waves. The contrast values of the Bragg component are highest for an upwind look direction and grow with wave number. The contrasts for downwind directions grow with the wave number, too, but are smaller than for the upwind case. For cross-wind observations, the Bragg contrast values are comparable with up- and downwind contrasts, but slowly decrease with wave number.

An important conclusion is that the non-Bragg component is significantly reduced in slicks. Dependences of the NBC contrasts on radar wave number are qualitatively similar to the Bragg ones, and the NBC contrast values are comparable to ones for BC. Assuming that the non-Bragg component is associated with wave breaking one can conclude that the film essentially influences the wave breaking processes. The effect of the reduction of NBC in slicks can hardly be explained by strong wave breaking which is typical for gravity waves longer than 0.5–1 m since the processes of wave crest overturning can destroy the film. Thus, the action of the film on micro-breaking of cm-dm-scale wind waves and the modification of micro-breaking features—parasitic ripples, and toe/bulge structures—can determine the reduction of NBC in slicks. Although this hypothesis is supported by preliminary wave tank experiments, further studies of an impact of the films on wave breaking processes are to be carried out, aimed to better understand the radar returns from the sea surface and to elaborate the radar imaging models of film slicks.

Author Contributions: S.E. conceived, designed, and took part in the experiments, analyzed the results, and wrote the paper; I.S., J.d.S., and O.S. processed and analyzed the data; and I.K., A.K., and A.M. performed the experiments.

Funding: This research regarding Sections 1–3, Section 5, and partly 4 was funded by the Russian Science Foundation (project RSF 18-17-00224). The results of the laboratory optical measurements of micro-breaking wave profiles presented in Section 4 were obtained by I. Kapustin and A. Molkov in the frame of project 17-05-00448 funded by the Russian Foundation of Basic Research.

Acknowledgments: We are grateful to Jorge M. Magalhaes for codes provided to process SAR data, and to G. Leshchov for his help in the experiment, as well as to O. Danilicheva for her help in the preparation of the paper.

Conflicts of Interest: The authors declare no conflict of interest.

References

1. Alpers, W.; Huehnerfuss, H. The damping of ocean waves by surface films: A new look at an old problem. *J. Geophys. Res.* **1989**, *94*, 6251–6266. [CrossRef]

2. Ermakov, S.A.; Panchenko, A.R.; Salashin, S.G. Film Slicks on the Sea Surface and Some Mechanisms of their Formation. *Dyn. Atmos. Oceans* **1992**, *16*, 279–304. [CrossRef]
3. Scott, J.C.; Thomas, N.H. Sea surface slicks–surface chemistry and hydrodynamics in radar remote sensing. In *Wind-Over-Wave Couplings. Perspectives and Prospects*; Sajjadi, S.G., Thomas, N.H., Hunt, J.C.R., Eds.; Clarendon Press: New York, NY, USA; Oxford, UK, 1999; pp. 221–229, ISBN 0-19-850192-7.
4. Da Silva, J.C.; Ermakov, S.A.; Robinson, I.S.; Jeans, D.R.G.; Kijashko, S.V. Role of surface films in ERS SAR signatures of internal waves on the shelf. 1. Short-period internal waves. *J. Geophys. Res.* **1998**, *103*, 8009–8031. [CrossRef]
5. Gade, M.; Alpers, W.; Hühnerfuss, H.; Masuko, H.; Kobayashi, T. Imaging of biogenic and anthropogenic ocean surface films by the multifrequency/multipolarization SIR-C/X-SAR. *J. Geophys. Res.* **1998**, *103*, 18851–18866. [CrossRef]
6. Brekke, C.; Solberg, A.H.S. Oil spill detection by satellite remote sensing. *Remote Sens. Environ.* **2005**, *95*, 1–13. [CrossRef]
7. Minchew, B.; Jones, C.E.; Holt, B. Polarimetric analysis of backscatter from the Deepwater horizon oil spill using L-band synthetic aperture radar. *IEEE Trans. Geosci. Remote Sens.* **2012**, *50*, 3812–3830. [CrossRef]
8. Levich, V.G. *Physicochemical Hydrodynamics*; Prentice-Hall: Englewood Cliffs, NJ, USA, 1962; ISBN 9780136744405.
9. Lucassen-Reynders, E.H.; Lucassen, J. Properties of capillary waves. *Adv. Coll. Int. Sci.* **1970**, *2*, 347–395. [CrossRef]
10. Ermakov, S.A. Resonance damping of gravity-capillary waves on the water surface covered with a surface-active film. *Izv. Atmos. Ocean. Phys.* **2003**, *39*, 624–628.
11. Jenkins, A.D.; Jacobs, S.J. Wave damping by a thin layer of viscous fluid. *Phys. Fluids* **1997**, *9*, 1256–1264. [CrossRef]
12. Ermakov, S.A.; Sergievskaya, I.A.; Gushchin, L.A. Damping of gravity-capillary waves in the presence of oil slicks according to data from laboratory and numerical experiments. *Izv. Atmos. Ocean. Phys.* **2012**, *48*, 565–572. [CrossRef]
13. Huehnerfuss, H.; Lange, P.; Walter, W. Wave damping by monomolecular surface films and their chemical structure. Pt.II. Variation of the hydtophylic part of the film molecules including natural substances. *J. Mar. Res.* **1984**, *42*, 737–759. [CrossRef]
14. Loglio, G.; Noskov, B.; Pandolfini, P.; Miller, P. Static and dynamic surface tension of marine water: Onshore or platform-based measurements by the oscillating bubble tensionmeter. In *Marine Surface Films*; Gade, M., Hühnerfuss, H., Korenowski, G.M., Eds.; Springer: New York, NY, USA, 2006; pp. 93–103, ISBN 3-540-33270-7.
15. Ermakov, S.A.; Kijashko, S.V. Laboratory study of the damping of parametric ripples due to surfactant films. In *Marine Surface Films*; Gade, M., Hühnerfuss, H., Korenowski, G.M., Eds.; Springer: New York, NY, USA, 2006; pp. 113–128, ISBN 3-540-33270-7.
16. Sergievskaya, I.A.; Ermakov, S.A. Damping of gravity–capillary waves on water surface covered with a visco-elastic film of finite thickness. *Izv. Atmos. Ocean. Phys.* **2017**, *53*, 650–658. [CrossRef]
17. Phillips, O.M. Spectral and equilibrium properties of the equilibrium range in the wind-generated gravity waves. *J. Fluid Mech.* **1985**, *156*, 505–531. [CrossRef]
18. Ermakov, S.A.; Zujkova, E.M.; Panchenko, A.R.; Salashin, S.G.; Talipova, T.G.; Titov, V.I. Surface film effect on short wind waves. *Dyn. Atmos. Oceans* **1986**, *10*, 31–50. [CrossRef]
19. Donelan, M.A.; Pierson, W.J., Jr. Radar scattering and equilibrium ranges in wind-generated waves with application to scatterometry. *J. Geophys. Res.* **1987**, *92*, 4971–5029. [CrossRef]
20. Kudryavtsev, V.; Hauser, V.; Caudal, D.; Caudal, G.; Chapron, B. A semiempirical model of the normalized radar cross-section of the sea surface: 1. Background model. *J. Geophys. Res.* **2003**, *108*, 8054. [CrossRef]
21. Valenzuela, G.R. Theories for the interaction of electromagnetic and oceanic waves—A review. *Bound.-Layer Meteorol.* **1978**, *13*, 61–85. [CrossRef]
22. Bass, F.G.; Fuks, M. *Wave Scattering from Statistically Rough Surfaces*; Pergamon: Oxford, UK, 1979; 540p, ISBN 9781483187754.
23. Ermakov, S.; Kapustin, I.; Sergievskaya, I.; Da Silva, J.C.B. Remote sensing of oil films on the water surface using radar. In Proceedings of the SPIE, Remote Sensing of the Ocean, Sea Ice, Coastal Waters and Large Water Regions 2012, Edinburgh, UK, 19 October 2012; p. 85320M. [CrossRef]

24. Phillips, O.M. Radar returns from the sea surface—Bragg scattering and breaking waves. *J. Phys. Oceanogr.* **1988**, *18*, 1065–1074. [CrossRef]

25. Kwoh, D.S.; Lake, B.M. The nature of microwave backscattering from water waves. In *The Ocean Surface*; Toba, Y., Mitsuyasu, H., Eds.; D. Reidel Publishing Company: Dordrecht, The Netherlands, 1985; pp. 249–256.

26. Kwoh, D.S.; Lake, B.M. A deterministic, coherent, and dual-polarized laboratory study of microwave backscattering from water waves, 1. Short gravity waves without wind. *IEEE J. Ocean. Eng.* **1984**, *OE-9*, 291–308. [CrossRef]

27. Jessup, A.T.; Keller, W.C.; Melville, W.K. Measurements of Sea Spikes in Microwave Backscatter at Moderate Incidence. *J. Geophys. Res.* **1990**, *95*, 9679–9688. [CrossRef]

28. Gade, M.; Alpers, W.; Ermakov, S.A.; Huehnerfuss, H.; Lange, P. Wind-wave tank measurements of bound and freely propagating short gravity-capillary waves. *J. Geophys. Res.* **1998**, *103*, 21697–21709. [CrossRef]

29. Ermakov, S.A.; Kapustin, I.A.; Kudryavtsev, V.N.; Sergievskaya, I.A.; Shomina, O.V.; Chapron, B.; Yurovskiy, Y.Y. On the Doppler Frequency Shifts of Radar Signals Backscattered from the Sea Surface. *Radiophys. Quantum Electron.* **2014**, *57*, 239–250. [CrossRef]

30. Ermakov, S.A.; Kapustin, I.A.; Sergievskaya, I.A. On peculiarities of scattering of microwave radar signals by breaking gravity-capillary waves. *Radiophys. Quantum Electron.* **2012**, *55*, 239–250. [CrossRef]

31. Kudryavtsev, V.; Chapron, B.; Myasoedov, A.; Collard, F.; Johannessen, J. On dual co-polarized SAR measurements of the Ocean surface. *IEEE Geosci. Remote Sens. Lett.* **2013**, *10*. [CrossRef]

32. Skrunes, S.; Brekke, C.; Eltoft, T.; Kudryavtsev, V. Comparing near coincident C- and X-band SAR acquisitions of marine oil spills. *IEEE Trans. Geosci. Remote Sens.* **2015**, *53*, 1958–1975. [CrossRef]

33. Hansen, M.W.; Kudryavtsev, V.; Chapron, B.; Brekke, C.; Johannessen, J.A. Wave Breaking in Slicks: Impacts on C-Band Quad-Polarized SAR Measurements. *IEEE J. Sel. Top. Appl. Earth Obs. Remote Sens.* **2016**, *9*, 4929–4940. [CrossRef]

34. Ermakov, S.A.; Kapustin, I.A.; Lazareva, T.N.; Shomina, O.V. Experimental investigation of surfactant film destruction due to breaking gravity waves. Preliminary results. *Sovrem. Probl. Distantsionnogo Zondirovaniya Zemli iz Kosmosa* **2015**, *12*, 72–79.

35. Longuet Higging, M.; Cleaver, R. Crest instability of gravity waves. Part 1. The almost highest wave. *J. Fluid Mech.* **1994**, *258*, 115–129. [CrossRef]

36. Longuet Higgins, M.S. Parasitic capillary waves: A direct calculation. *J. Fluid Mech.* **1995**, *301*, 79–107. [CrossRef]

37. Yermakov, S.A.; Ruvinsky, K.D.; Salashin, S.G. Local correlation of the characteristics of ripples on the crest of capillary gravity waves with their curvature. *Izv. Atmos. Ocean. Phys.* **1988**, *24*, 561–563.

remote sensing

MDPI

Article

Assessing Single-Polarization and Dual-Polarization TerraSAR-X Data for Surface Water Monitoring

Katherine Irwin [1], Alexander Braun [1,*], Georgia Fotopoulos [1], Achim Roth [2] and Birgit Wessel [2]

[1] Department of Geological Sciences and Geological Engineering, Queen's University, Kingston, ON K7L 3N6, Canada; 0kei@queensu.ca (K.I.); gf26@queensu.ca (G.F.)

[2] German Aerospace Center (DLR), D-82234 Wessling, Germany; Achim.Roth@dlr.de (A.R.); Birgit.Wessel@dlr.de (B.W.)

* Correspondence: braun@queensu.ca; Tel.: +1-613-533-6621

Received: 5 May 2018; Accepted: 13 June 2018; Published: 14 June 2018

Abstract: Three synthetic aperture radar (SAR) data classification methodologies were used to assess the ability of single-polarization and dual-polarization TerraSAR-X (TSX) data to classify surface water, including open water, ice, and flooded vegetation. Multi-polarization SAR observations contain more information than single-polarization SAR, but the availability of multi-polarization data is much lower, which limits the temporal monitoring capabilities. The study area is a principally natural landscape centered on a seasonally flooding river, in which four TSX dual-co-polarized images were acquired between the months of April and June 2016. Previous studies have shown that single-polarization SAR is useful for analyzing surface water extent and change using grey-level thresholding. The H-Alpha–Wishart decomposition, adapted to dual-polarization data, and the Kennaugh Element Framework were used to classify areas of water and flooded vegetation. Although grey-level thresholding was able to identify areas of water and non-water, the percentage of seasonal change was limited, indicating an increase in water area from 8% to 10%, which is in disagreement with seasonal trends. The dual-polarization methods show a decrease in water over the season and indicate a decrease in flooded vegetation, which agrees with expected seasonal variations. When comparing the two dual-polarization methods, a clear benefit of the Kennaugh Elements Framework is the ability to classify change in the transition zones of ice to open water, open water to marsh, and flooded vegetation to land, using the differential Kennaugh technique. The H-Alpha–Wishart classifier was not able to classify ice, and misclassified fields and ice as water. Although single-polarization SAR was effective in classifying open water, the findings of this study confirm the advantages of dual-polarization observations, with the Kennaugh Element Framework being the best performing classification framework.

Keywords: synthetic aperture radar; PolSAR; TerraSAR-X; surface water monitoring; flooded vegetation; classification; segmentation

---

1. Introduction

Synthetic aperture radar (SAR) is an active remote sensing technique and can penetrate cloud cover and operate during the day or night. Often, flood events occur during unfavourable weather conditions when optical visibility is low, which allows SAR to be a useful sensor for surface water classification [1]. SAR missions and products vary in polarization, incidence angle, frequency, and resolution, which allows the user to select the most suitable SAR observations to the application at hand, which in this case is the classification of surface water and its varying states. Currently, there exists a trade-off between spatio-temporal coverage and information content with respect to single-polarization and multi-polarization SAR data. While multi-polarization contains more information about the scattering mechanism of the target, the temporal coverage over a single target is limited and prohibits monitoring over regional scales. Many studies addressed the use of single- or multi-polarization SAR individually using diverse classification algorithms, but few addressed the comparison of single- and multi-polarization SAR data for surface water monitoring [1–4]. These comparative studies mainly address single-polarization and quad-polarization processing methods, or address which dual-polarization channels are most effective for surface water monitoring, but lack discussion of the dual-polarization processing methods. In this study, there are three main objectives. The first is to create classified models of surface water using both single- and dual-polarization TerraSAR-X (TSX) data. The second is to compare these models to better understand the extent of the limitations of single-polarization data and to what extent they are aided by dual-polarization data. The third objective is to create a surface water extent time series from the initial snow melt period into spring, to demonstrate the feasibility of near-continuous surface water monitoring from space. The current fleet of ~10 civil SAR missions in orbit and the planned missions up to 2020 provide and will continue to provide an unprecedented amount of observations in the X-, C-, and L-band, mostly in single-polarization mode, which will lead to near-continuous monitoring capabilities. Different radar frequencies (X-, C-, and L-band) interact differently with vegetation, and thus wetlands. The shorter the wavelength (X-band) of the radar, the less penetration into the canopy, while L-band radar penetrates through the canopy to the ground. However, X-band SAR is the preferred sensor for open water mapping as the shorter wavelength leads to increased diffuse scattering when compared with the C-band and L-band radar, which suffer from specular reflections leading to low energy return.

Previous studies have shown that single-polarization SAR data are a viable technology for analyzing surface water extent and spatio-temporal change [4–16]. Single-polarization SAR provides one channel of intensity data in either HH (horizontal linear transmission and horizontal linear reception), HV (horizontal linear transmission and vertical linear reception), VH (vertical linear transmission and horizontal linear reception), or VV (vertical linear transmission and vertical linear reception). One of the most common and effective classification techniques is grey-level thresholding, which can be applied to differentiate areas of water and non-water [4,17]. This has proved successful to delineate open water bodies, but limitations arise with more heterogeneous targets, such as ice covered water bodies and surface water beneath vegetation—which are abundant in Canada, with approximately 14% of the Canadian landscape being covered in wetlands [4]. Vegetation cover leads to misclassification due to shadow and layover effects, and ice can often be misclassified because of the similar backscatter response to rough surface water [4,9]. Wind effects can also lead to misclassification causing water to be mistaken as land, rough vegetation, or ice [4]. Multi-temporal SAR acquisitions often occur during ice-off conditions to avoid misclassification. However, it is important to document the initial stages of snowmelt and spring flooding, because the subsequent hydrological conditions rely on this initial process [4]. The limitations of single-polarization SAR could be aided by the use of dual-polarization SAR, which provides two channels of intensity and phase information (HH/HV, VV/VH, or HH/VV). Having two channels of intensity and phase information allows for discerning among scattering mechanisms, such as surface scattering, double bounce, and volume scattering. Dual-polarization SAR data can be used to distinguish ice and vegetated areas from open water, while those land cover types often lead to misclassification in single-polarization data [4,18,19]. Several studies have identified the significant double-bounce component originating from flooded vegetation and hence, quad-polarization SAR must often be used for classification [20–22]. Quad-polarization data are often not available, while dual-polarization have been demonstrated to be sufficient [18,23–26].

2. Data Description and Methodology

2.1. Study Area and Data Description

The ~6 km × 12 km study area is located south of Lac-Simon in Quebec, Canada (Figure 1). This area was chosen because it is a principally natural landscape with open bodies of water, marshland, and forested areas. Most importantly, it includes the Ruisseau Schryer, which flows from A (Lac-Schryer) to B (Baie-de-l'Ours in Lac-Simon) in Figure 1, and is flooded during the snowmelt each year. This could affect the urban areas located near the river, including the town of Montpellier located on the north side of the stream. Montpellier experienced flooding due to seasonally high water levels and overflowing rivers in April of 2017. These events threatened and damaged multiple homes and a state of emergency was declared in towns near to Montpellier. The orange box in Figure 1 zooms in on the river entering Baie-de-l'Ours, showing a distinct flood plain surrounding both sides of the meandering river.

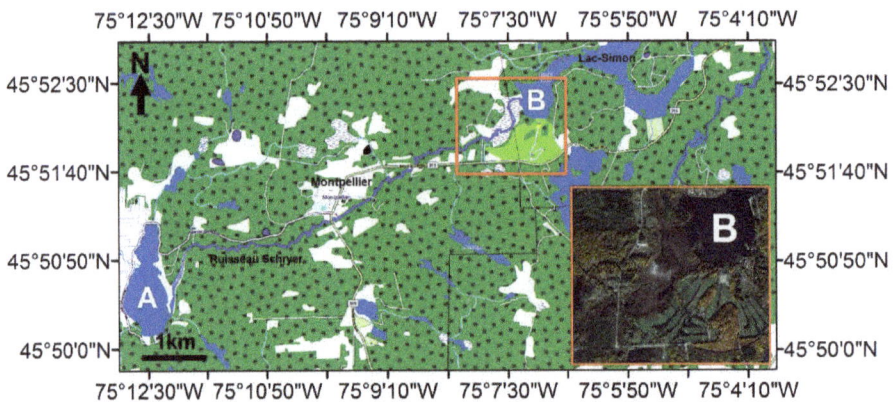

Figure 1. Map of study area showing the town of Montpellier, southern extent of Lac-Simon, and stream (Ruisseau Schryer) flowing from A (Lac-Schryer) to B (Baie-de-l'Ours). Base map is provided by the Quebec Ministry of Energy and Natural Resources (MERN) showing urban areas (white), water (blue), forest (dotted green), low vegetation (tan), golf course (light green), and marshlands (dashed areas). Zoomed in orange box of Google Earth imagery from July 2017 shows a flood plain surrounding the stream entering Lac-Simon at B.

Four TSX dual-co-polarization (HH/VV) strip-map mode scenes were acquired through the spring and summer of 2016. Each scene has an areal extent of 15 km by 50 km and a slant range and azimuth resolution of 1.2 m and 6.6 m, respectively. The incidence angle was chosen to be low enough to be able to penetrate the vegetation types on site and cover the target area. Details of each product are provided in Table 1.

Table 1. Specifications of the four dual-polarization TerraSAR-X strip-map beam mode scenes used herein. VV—vertical linear transmission and vertical linear reception; HH—horizontal linear transmission and horizontal linear reception.

ID	Date (2016)	Mode	Polarization	Product	Look Direction	Path	Incidence Angle (°)
1	2 April	stripmap	HH/VV	SSC	Right	Descending	39
2	24 April	stripmap	HH/VV	SSC	Right	Descending	39
3	5 May	stripmap	HH/VV	SSC	Right	Descending	39
4	18 June	stripmap	HH/VV	SSC	Right	Descending	39

Three eight-band multispectral Landsat 8 images from 13 April, 29 April, and 16 June 2016, with a resolution of 30 m were used to aid in visual comparison and validation of the TSX data (Figure 2). It is important to note the change in ice cover between the 13 April scene and the 29 April scenes. By the 16 June scene, the vegetation canopy is fully developed, shown in green. In the 29 April scene, cloud cover can be noticed, which is a clear limitation of optical imagery. In situ data of the region does not exist, which means that the classification results have to be validated based on the satellite image only. Hydrometric data for Lac Simon and Lac Schryer exist only until 2006, which is before the SAR scenes were acquired. This compromises the evaluation of the absolute performance of the classification methods, however, the relative comparison reveals performance differences that can be validated with existing knowledge of the land cover types and the satellite imagery in the area.

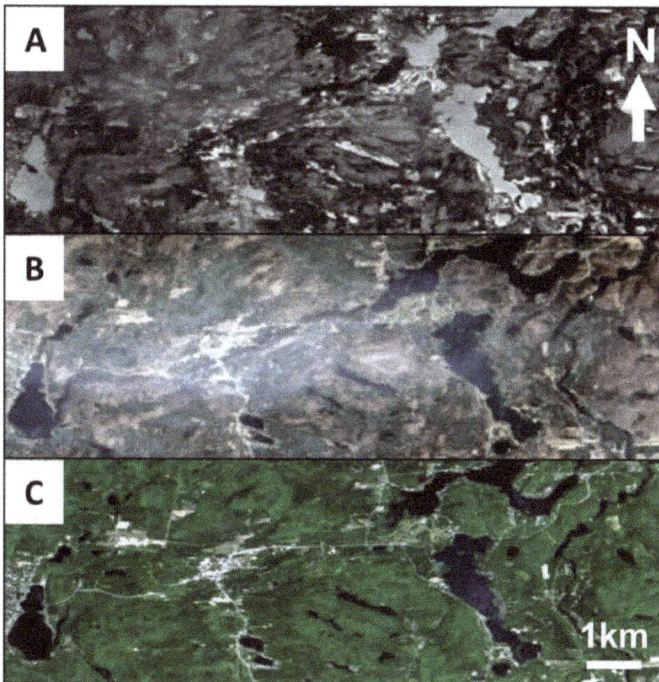

Figure 2. Time series of Landsat 8 true colour optical images (red-green-blue (RGB): 4-3-2) from the following scenes: (**A**) 13 April 2016; (**B**) 29 April 2016; and (**C**) 16 June 2016.

2.2. Classification Methods

Each TSX scene was processed using three different methods: grey-level thresholding applied to single-polarization data (HH), the Kennaugh Element Framework [27] applied to dual-polarization data (HH, VV), and H-Alpha-Dual-Polarization decomposition (HH, VV). The processing workflow for each method can be seen in Figure 3.

The HH band of the TSX data was used to simulate single-polarization data. HH polarization tends to be used over HV or VV, because the difference in backscatter response between land and water are greatest for HH polarization [4,17,28–31]. Grey-level thresholding was used to classify each SAR scene because it is a simple and effective way to map surface water [7,32]. Because the histogram of the intensity data is bimodal, a value in between the two modes can be chosen, in which everything below this threshold is classified as water and everything above is classified as non-water. In this study, the threshold value used was the minimum between the two modes. This method was applied to all four TSX images, producing four models each with two classes (water and other). The processing work flow for this method is shown in Figure 3. The processing software used herein included SNAP, Matlab code developed in house, and the Kennaugh Element Framework by Schmitt et al. [27]. The Shuttle Radar Topography Mission (SRTM) finished Digital Elevation Model (DEM) was used for the terrain correction and standard speckle filters were applied.

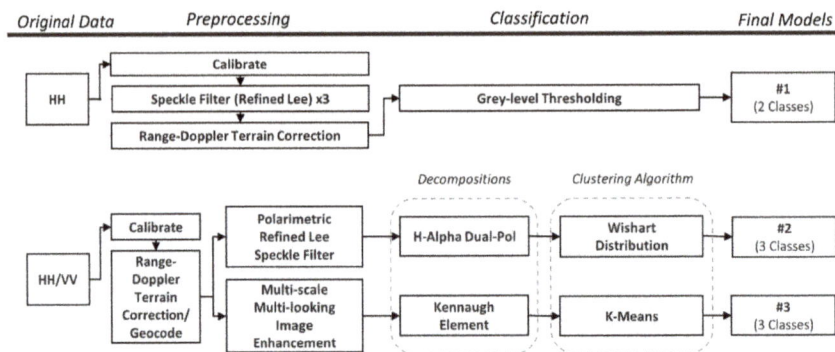

Figure 3. Processing workflow for single- and dual-polarization data to create three final models for each TerraSAR-X (TSX) scene.

Studies have shown successful results using dual-co-polarization HH/VV data for flooded vegetation mapping [18,20,33]. Several common decomposition types exist to break down the data into polarimetric parameters, but there are few that are adapted for dual-polarization data. In this study, two dual-polarization decompositions will be used to classify water and flooded vegetation. The first is the H-Alpha decomposition, which was developed in 2007 and modified for dual-polarization data [34]. The second method is the Kennaugh Element Framework developed by Schmitt et al. [27]. A few studies have researched the use of this technique for mapping wetlands and have proved it successful [18,26,27].

The H-Alpha decomposition for dual-polarization data uses an eigenvector analysis of the coherency matrix [T2], which separates the parameters into scattering processes (the eigenvectors) and their relative magnitudes (the eigenvalues) [34]. There are two parameters outputted from the H-Alpha decomposition, entropy (H) and alpha (α). Entropy is calculated from the eigenvalue information and represents the degree of randomness in the scattering. Alpha (α) is calculated from the eigenvectors and represents a rotation that can indicate the type of scattering mechanism [35]. Figure 4 shows the entropy and alpha parameters for the scene of 2 April 2016. Water is shown with orange colors in Figure 4A (entropy), while ice is shown in blue colors. The alpha parameter shows water along the river, but does not outline the ice-covered regions in the scene. In combination, entropy and alpha parameters are able to distinguish water- and ice-covered areas. The quad-polarization decomposition of H-Alpha has shown to work well in natural environments, because this method is an incoherent decomposition that can characterize distributed targets [1]. However, the modified dual-polarized H-Alpha decomposition is less well-studied. The unsupervised Wishart distribution clustering algorithm is applied to the H-Alpha decomposition, as the coherency matrix can be modelled by this algorithm. The Wishart distribution is robust and can be applied to any type of polarimetric SAR data, so it is often applied to the H-Alpha decompositions (dual or quad) [1]. Nine different classes were created using the Wishart classifier on the H-Alpha decomposition. These classes were visually inspected to determine which classes could be re-clustered to represent the desired three classes; (i) open water, (ii) flooded vegetation, and (iii) other. This processing method was applied to all four TSX images, producing four models each with three classes (Figure 3).

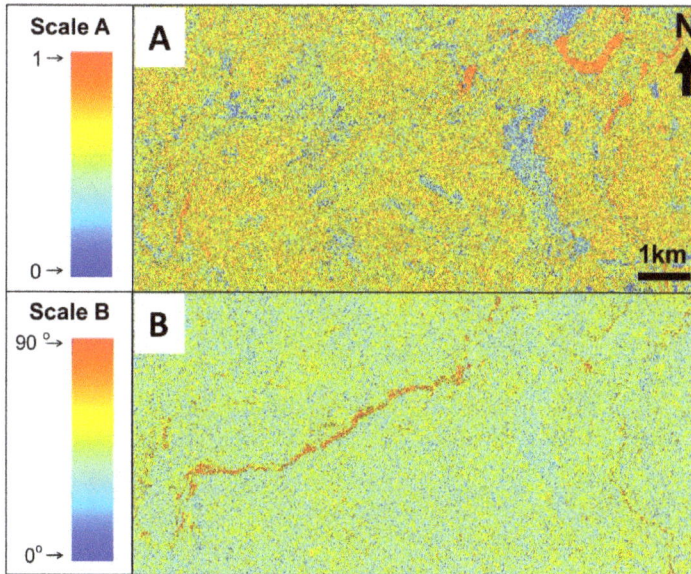

Figure 4. Images of parameters entropy (**A**) and alpha (**B**) for the TerraSAR-X scene from 2 April 2016.

The Kennaugh Element Framework developed by Schmitt et al. [27] linearly transforms the four-dimensional Stokes vector into a four-by-four scattering matrix, called the Kennaugh matrix [K]. Four normalized Kennaugh elements from this matrix K can be computed using dual-co-polarization data. In the case of HH/VV data, the elements are K_0, K_3, K_4, and K_7. K_0 represents the total intensity as a sum of HH and VV intensity; K_3 represents the difference between double-bounce and surface intensity; K_4 represents the difference between HH and VV intensity; and K_7 represents the phase shift between double-bounce and surface scattering mechanisms. These four elements have been shown to be very useful for identifying flooded vegetation. A study by Moser et al. [25] has demonstrated that open water has low values of K_0, because of the specular scattering nature of calm water causing a low backscatter signature. These areas generally form clusters that can be distinctly separated from the other classes. Flooded vegetation is characterized by high values of K_4, medium values of K_3 and lower values of K_0. Another study identified the significant difference between HH and VV intensity over flooded areas and inundated forests, emphasizing the importance of K_4 [36]. Using a pre-process Kennaugh chain, the Kennaugh elements were geocoded, calibrated, and enhanced using a multi-scale and multi-looking technique developed by Schmitt [37] (Figure 3). An example of the four Kennaugh elements produced for the 2 April 2016 scene can be seen in Figure 5. This scene was selected as it exhibits open water, ice cover, inundated vegetation, and other land cover. Open water is represented by the lowest values of K_0 (dark blue in Figure 5A) and medium values of K_3 (grey in Figure 5B). Flooded vegetation is represented by medium values of K_3 (grey and yellow in Figure 5B) and high values of K_4 (red and yellow in Figure 5C). Additionally, K_7 (Figure 5D), representing the phase shift between double-bounce and surface scattering, shows sensitivity to inundated vegetation, although not as strong as K_4, which was already found by Zalite et al. [36]. Ice cover is represented by low values of K_0 (light blue in Figure 5A) and low values of K_3 (dark blue in Figure 5B). These elements were processed using an unsupervised k-means classifier. The k-means clustering algorithm is one of the simplest forms of clustering techniques that aims at minimizing the Euclidean distance between points. The advantages of this technique are that it is fast and robust, and thus works well when applied to large, linear data sets, such as the Kennaugh elements. The k-means classifier produced 11 classes that were then visualized and analyzed to determine which classes represented (i) open water, (ii) flooded vegetation, and (iii) other. This processing method was applied to all four TSX images, producing four models each with three classes.

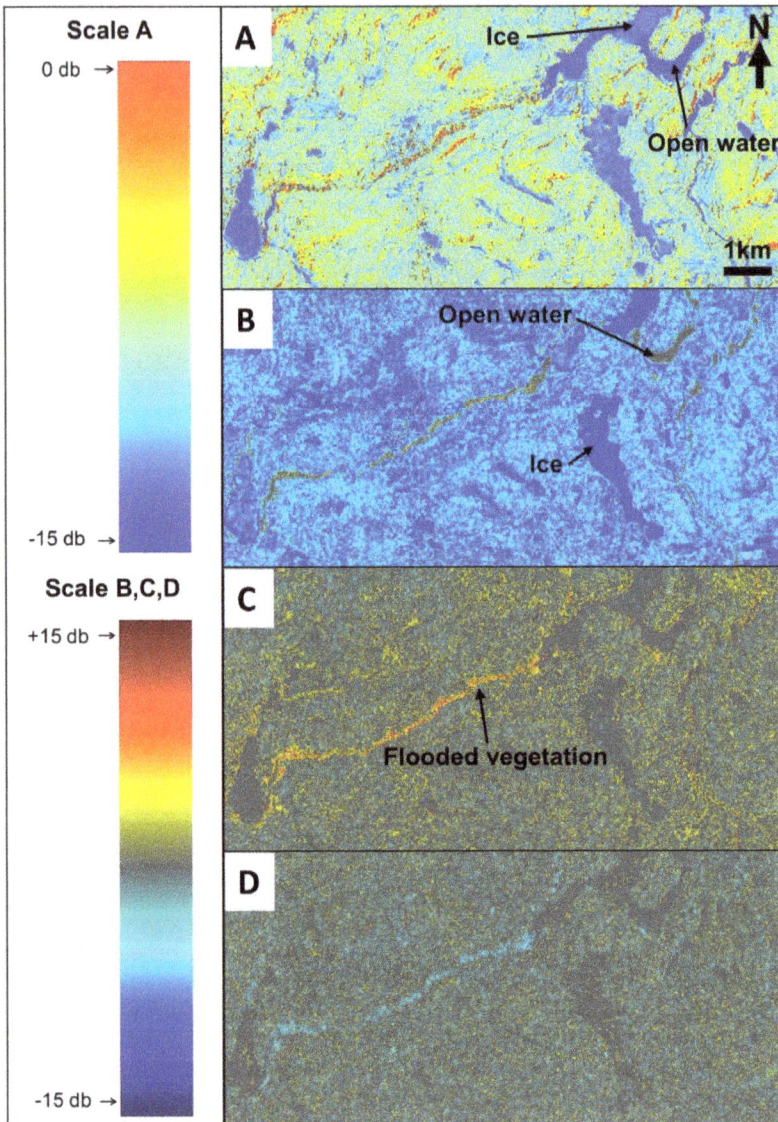

Figure 5. Four Kennaugh elements derived from the dual-pol TerraSAR-X image from 2 April 2016. (**A**) K_0—the total intensity sum of HH plus VV; (**B**) K_3—difference double-bounce minus surface scattering; (**C**) K_4—difference HH minus VV intensity; (**D**) K_7—phase shift between double-bounce and surface scattering mechanisms. Open water is represented by dark blue in (**A**) and grey in (**B**). Flooded vegetation is represented by grey and yellow in (**B**) and red and yellow in (**C**). Ice cover is represented by light blue in (**A**) and dark blue in (**B**). Inundated vegetation is shown in yellow/red in (**C**) and cyan in (**D**).

3. Results

3.1. Single Polarization Classification

The four water classification models created using grey-level thresholding can be seen in Figure 6. Areas of black represent water and grey areas represent non-water. Table 2 outlines the percentages of the two classes through time, as well as the threshold value used. Water classification changes from 8% to 10% throughout the four scenes, disagreeing with seasonal trends, which should show an increase in temperature causing a decrease in water. However, several other processes are occurring to account for this change. Ice can be seen in only the first scene (blue box in Figure 6A), as it is classified as both other and water, and decreases the amount of total water classified. The marsh land shown in the red box in Figure 6 is flooded in the first scene (A), but dried out by the last scene (D). This change agrees with seasonal change, despite the overall trend of water classification showing an increase. Counteracting this seasonal drying is an increase in overall misclassification in the first scene (A), as a result of ice and snow, and last scene (D), as a result of vegetation growth (shown in the yellow box in Figure 6). Although the single-polarization methodology was able to see seasonal changes in some wetlands, flooded vegetation was not classified, and misclassification errors occurred as a result of ice and tall vegetation causing an incorrect interpretation of the total surface water change in the area, a clear limitation of using single-polarization data only.

Table 2. Percentages of each class and threshold values used in each TerraSAR-X (TSX) scene using grey-level thresholding.

Date (2016)	Threshold Value (dB)	Water (%)	Other (%)
2 April	−17.38	8	92
24 April	−19.68	9	91
5 May	−18.56	10	90
18 June	−18.87	10	90

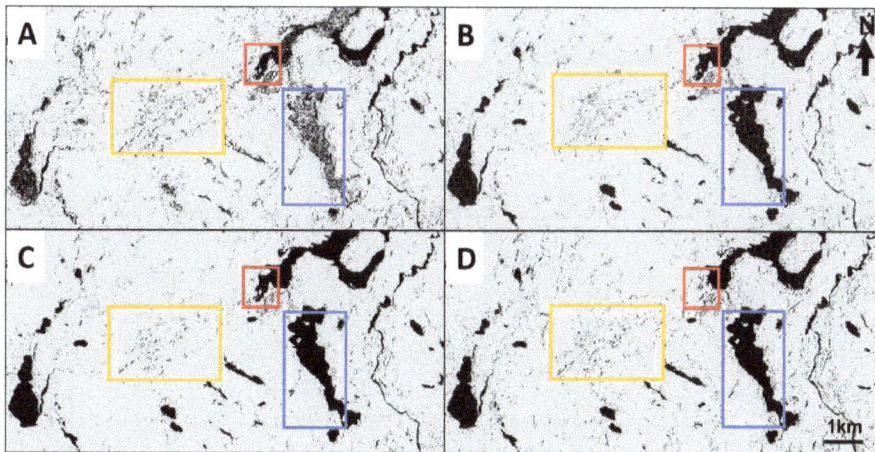

Figure 6. Grey-level thresholding classified models showing water (black) and other (grey) for (**A**) 2 April 2016; (**B**) 24 April 2016; (**C**) 5 May 2016; and (**D**) 18 June 2016. Coloured boxes indicate example areas of temporal change: blue—ice melting; red—marshland dries out; yellow—areas of misclassification due to ice (**A**) and vegetation (**C**).

3.2. Dual-Polarization Classification: H-Alpha–Wishart

The water classification models created using the unsupervised H-Alpha–Wishart classification can be seen in Figure 7, showing three classes: water (black), flooded vegetation (blue), and other (grey). Of the original nine classes, each model consistently identified Class 3 (blue) as flooded vegetation and Class 7 (black) as open water.

Table 3 display the results for each scene classified using the H-Alpha–Wishart method. The percent of water classified gradually decreases from the first scene until the last scene. This is expected as the seasonal changes from wet to dry occur during the study time period. However, some areas are misclassified as water, including a golf course located south of the Baie-de-l'Ours present in the optical imagery (blue box in Figure 7). Some fields identifiable in the optical imagery are classified as water as well (black areas in yellow box in Figure 7). The percent of flooded vegetation stays around approximately 5%, which is inconsistent with season trends. However, in the April scenes (A and B), there is flooded vegetation centered on the river, whereas this disappears in the later scenes (yellow box in Figure 7). The marsh area seen in the red box is shown to dry up in the final scene (D), which is consistent with the single-polarization SAR observations. Ice is not differentiable from open water in the first scene. The sum of water and flooded vegetation (total surface water) decreases and is consistent with the expected seasonal change. Some areas (yellow box in Figure 7) that are classified as water on 5 May (C) are classified as flooded vegetation on 18 June (D). This is an indication that vegetation changes may lead to a change in class, but not a change from water to non-water. Hence, the classification of total surface water seems more robust than the discrimination of open water and flooded vegetation. As the vegetation starts developing in May and is fully developed in June, the change in class could be a consequence of that. It is worth noting that snow fall occurred in the area throughout April, which may have led to snow and ice being misclassified as flooded vegetation in both April scenes. A rise in temperature by 15–20 degrees in early May may have melted the snow and ice, which became classified as water, considering that leafs have not developed yet. By June, vegetation has developed in and around water bodies, indicated by the increasing areas of classified flooded vegetation in the scene.

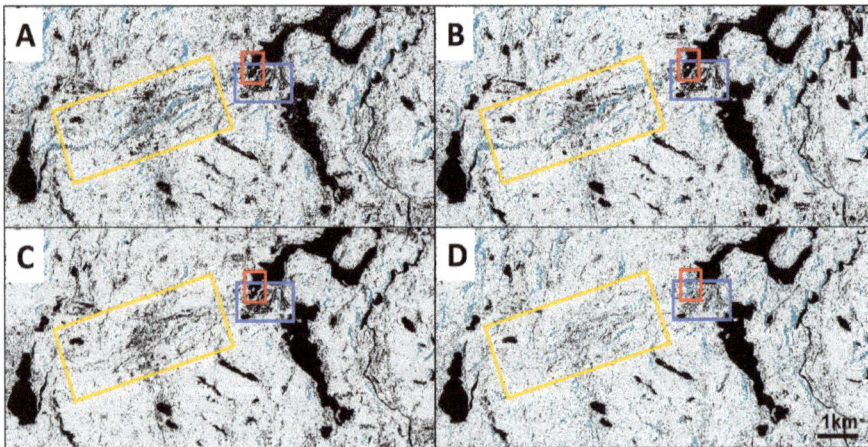

Figure 7. H-Alpha–Wishart classified models showing water (black), flooded vegetation (blue), and other (grey) for (**A**) 2 April 2016; (**B**) 24 April 2016; (**C**) 5 May 2016; and (**D**) 18 June 2016. Coloured boxes indicate example areas of change: blue—golf course misclassified as water; red—marshland dries out; yellow—flooded vegetation decreases, and misclassification of fields.

Table 3. Percentages of each class identified for each H-Alpha–Wishart model through time.

Date (2016)	Water (%)	Flooded Vegetation (%)	Other (%)
2 April	17	5	78
24 April	15	6	79
5 May	16	2	82
18 June	12	6	82

3.3. Dual-Polarization Classification: Kennaugh Element Framework

A time series of the four false colour composites of the processed Kennaugh elements, K_3-K_0-K_4, are shown in Figure 8. Colours are used to clearly differentiate the four classes. Bright pink represents open water; white/light pink colour represents flooded vegetation; green and blue represent the 'other' class, which includes forest, urban areas, agricultural land, or other classifications of land. The dark purple colour seen in the 2 April scene represents ice, which corresponds to the optical imagery.

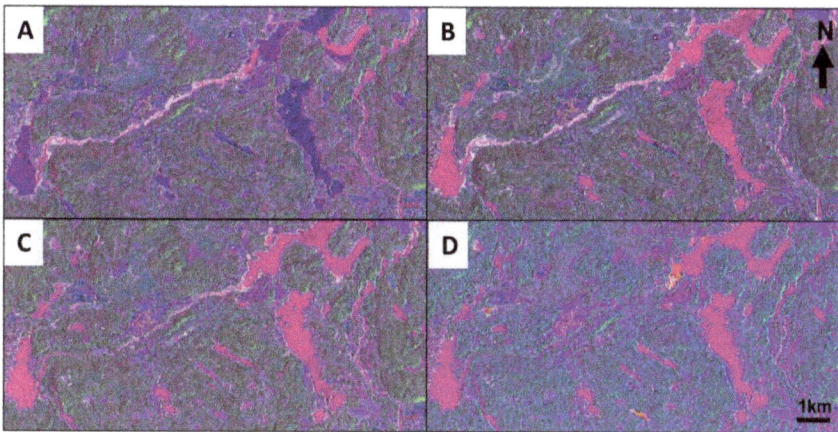

Figure 8. False colour composites of the processed Kennaugh elements, K_3-K_0-K_4, from (**A**) 2 April, (**B**) 24 April, (**C**) 5 May, and (**D**) 18 June 2016. Open water appears in pink, ice in dark purple, flooded vegetation in white/light pink, and 'other' in green and blue.

The Kennaugh element technique agrees with the optical imagery. The ice can be seen to disappear by the end of April and the water bodies are consistent. The Kennaugh elements were able to observe the flooded vegetation during melting, especially in the first two scenes, which the optical imagery was not able to identify. The Kennaugh elements were then classified using an unsupervised k-means classifier. This developed 11 classes for each scene. The average of each class of each Kennaugh element was analyzed and results are shown in Figure 9. It shows three plots, K_0 versus K_4, K_3 versus K_4, and K_0 versus K_3, for each of the four scenes. Using this technique, each point was assigned one class of three possible classes; open water, flooded vegetation, and other. It can be seen that the points assigned as water were consistently the lowest value of K_0. Flooded vegetation was consistently identified as the highest value in K_4. These findings are consistent with Moser et al. [25].

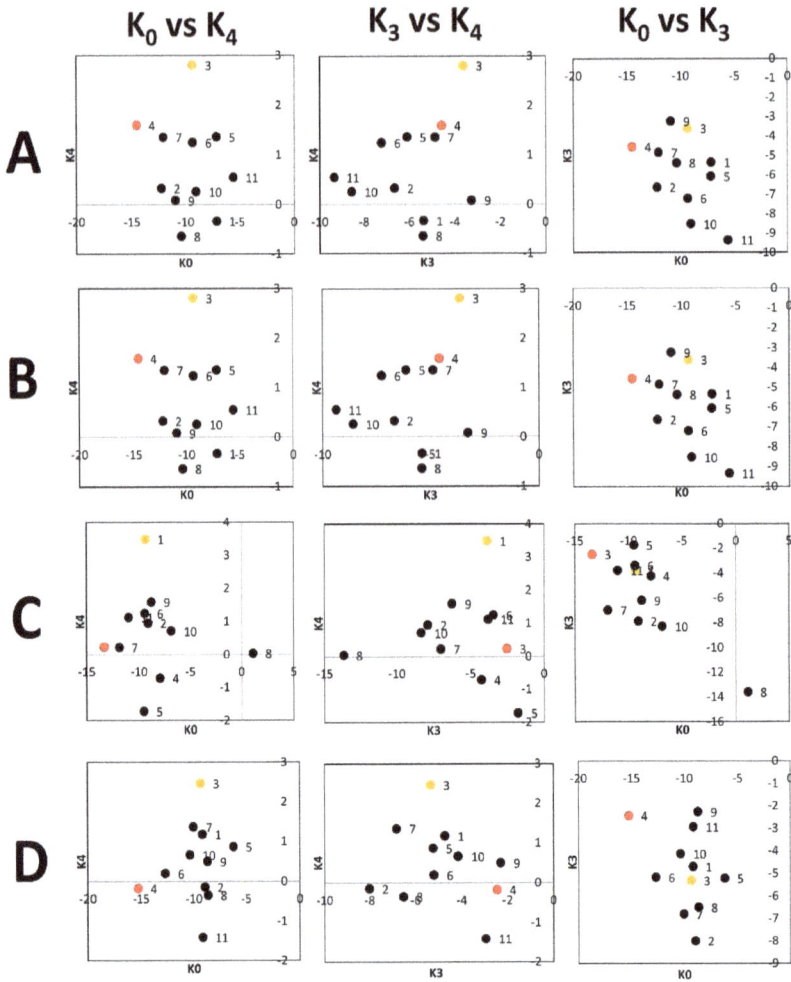

Figure 9. Graphs of the average of each class for the Kennaugh elements used to classify water (red), flooded vegetation (yellow), and other (black). (**A**) 2 April 2016; (**B**) 24 April 2016; (**C**) 5 May 2016; and (**D**) 18 June 2016.

The water classification models for each of the four scenes can be seen in Figure 10. Table 4 outlines the percentages of each class identified through time. The black areas represent open water and are shown to decrease from the first scene to the last scene from 16% to approximately 12%. However, in the first scene (green box in Figure 10A), the water is distributed and lakes are classified as both other and water, indicating the presence of ice. The flooded vegetation (blue) decreases from 13% to 5% with time. This corresponds with the seasonal melt that would occur in the first two scenes in early spring, and the seasonal drying that could occur during late spring and early summer of the last two scenes. Similar to H-Alpha–Wishart, misclassification of flooded vegetation can be seen distributed throughout the scene. However, the flooding around the river is evident in the first two scenes and not the last two (yellow box in Figure 10). A golf course south of Baie-de-l'Ours is misclassified as water (see Figure 1B), which is not seen in the optical imagery (blue box in Figure 10). Finally, the marsh area seen in the red box shows a drying out in the last scene similar to the other two classification methodologies.

Table 4. Percentages of each class identified for each Kennaugh Element model through time.

Date (2016)	Water (%)	Flooded Vegetation (%)	Other (%)
2 April	16	13	72
24 April	12	13	75
5 May	13	5	82
18 June	12	5	83

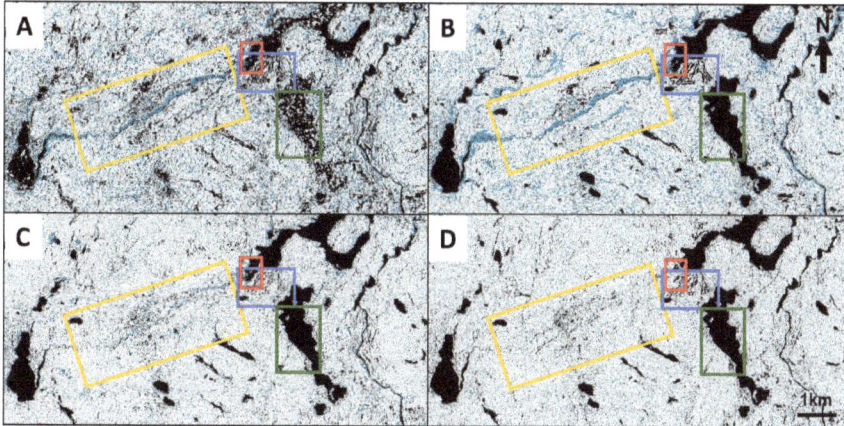

Figure 10. Kennaugh Element models classified showing water (black), flooded vegetation (blue), and other (grey) for (**A**) 2 April 2016; (**B**) 24 April 2016; (**C**) 5 May 2016; and (**D**) 18 June 2016. Coloured boxes indicate example areas of change: blue—golf course misclassified as water; red—marshland dries with time; yellow—flooded vegetation decrease, and misclassification of field areas; green—ice melting.

Differential Kennaugh elements [26] use the differences between the Kennaugh elements of two scenes to understand how the landscape has changed with time. This technique was applied to the first (2 April) and last (18 June) scene and can be seen in Figure 11. Each colour represents a type of change. Green represents the change from ice to open water, which is mainly reflected as a change in K_0 showing a decrease in total intensity. Dark red represents the change from flooded vegetation to land and accounts for a decrease in the difference between double-bounce and surface scattering or K_3. Yellow represents the change from open water to flooded vegetation, which is shown as an increase in K_3. This technique proves extremely useful, especially when comparing open water and flooded vegetation, because a clear distinction can be seen. The ability to identify the change from ice cover to open water is also important as it can indicate when the first seasonal melt and flooding occurs.

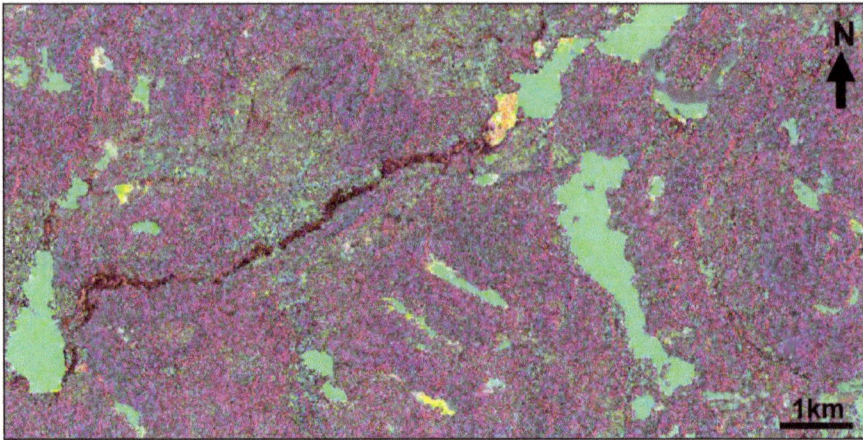

Figure 11. False colour composite of the differential Kennaugh elements, K_0-K_3-K_4, differenced between the 18 June 2016 scene and the 2 April 2016 scene. Red represents the change from flooded vegetation to land. Green represents the change from ice to open water. Yellow represents the change from open water to marshland.

4. Discussion

A comparison of the three methods, k-means performed on the Kennaugh elements, unsupervised H-Alpha–Wishart classification, and grey-level thresholding, can be seen in Figure 12. All methods were able to classify open water, however, the dual-polarization methods consistently classified more water than the grey-level thresholding technique. This was unexpected as the single-polarization data often have errors of commission from shadow zones classified as water. However, more misclassification occurred in the dual-polarization methods, including classifying fields and a golf course as water. Both of the dual-polarization methods show the correct decreasing trend of water over the study period from April to June 2016 that one would expect to see, including seasonal melting. The results from thresholding disagree with this and show a slight increase over time from 8% to 10%. No classification method was able to classify ice in a single class in the 2 April scene, however, grey-level thresholding and the Kennaugh Element k-means classifier were able to classify some portion of the ice covered lakes as other. Both dual-polarization methods were able to identify flooded vegetation. However, the Kennaugh elements classifier sees a decrease in flooded vegetation with time, which agrees better with seasonal trends. Both methods were able to classify the river flooding in the April scenes and no flooding occurring in the later scenes. The single-polarization method was unable to classify flooded vegetation, and instead classified it as other or non-water. The findings presented here are in line with the results of Mleczko and Mroz [38], who compared Sentinel-1 and Tandem-X multi-polarization data and concluded that dual polarization of TanDEM-X achieves the best results, while full polarization shows only marginally better performance for wetlands and flooded vegetation. Hence, full polarimetric acquisitions may not be efficient or needed for wetland mapping. Alternative classification procedures, such as Shannon Entropy [23] and interferometric coherence, have not shown to yield better results, unless full polarimetric coherence is used for simultaneous image acquisition by TerraSAR-X and Tandem-X [38]. Other studies using dual-polarimetric TerraSAR-X acquisitions and a variety of polarimetric indices demonstrate the strong dependence of the indices on vegetation conditions [39]. The performance of polarimetric indices thus depends on seasonal conditions, which mandates a classification technique such as differential Kennaugh elements, which is able to determine the change between different land cover classes throughout the available acquisition time period. An additional valuable parameter for water classification would be water levels, which have successfully been determined from wetland in SAR [40]. Multi-frequency acquisitions from different satellites, in combination with Kennaugh element decomposition, has also shown great potential for discriminating different vegetation types (e.g., [41]).

Model Classification through Time

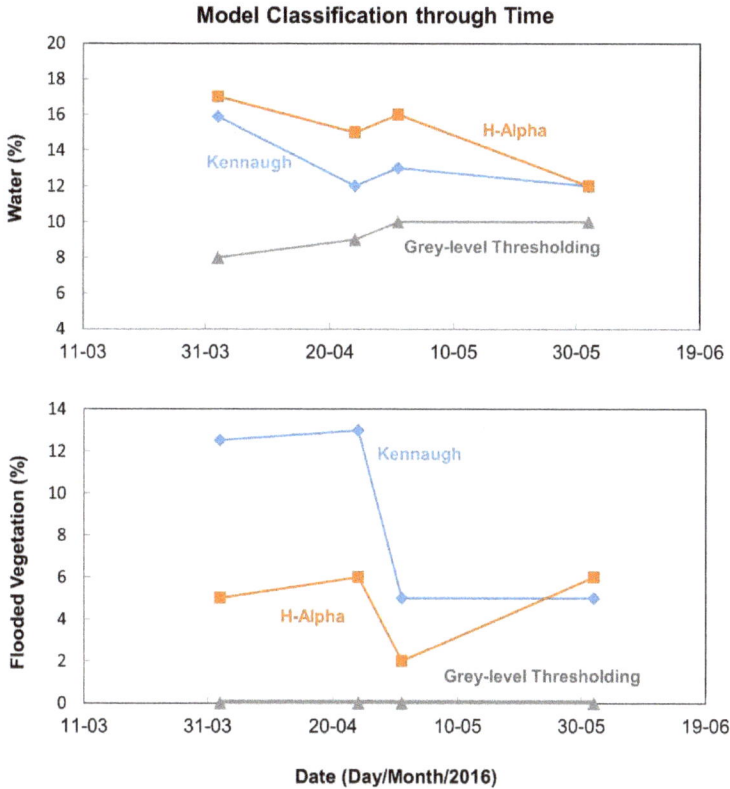

Figure 12. Graphs showing percent of water and percent of flooded vegetation classified through time for each of the classification methods. Lines: Blue diamond—unsupervised k-means classification on Kennaugh Elements, orange square—H-Alpha–Wishart unsupervised classification, and grey triangle—grey-level thresholding.

5. Conclusions

In this study, three different SAR classification methods were used to identify areas of water, ice, and flooded vegetation in four TSX scenes over a principally natural landscape. A river, which is seasonally flooded in the centre of the study area, serves as a test-bed for these methodologies. Single-polarization grey-level thresholding is an established technique for surface water monitoring and has the capability to classify areas of water and non-water. Using the techniques of H-Alpha–Wishart and the Kennaugh Element Framework applied to dual polarization data, the ability to analyze scattering mechanisms and classify water and flooded vegetation was tested and compared with the single-polarization method.

The H-Alpha–Wishart unsupervised classification for dual-polarization data was able to classify areas of open water and flooded vegetation. Flooded vegetation was classified surrounding the river in the first two scenes acquired in April, and not in the last two scenes acquired in May and

June. This corresponds with optical imagery of the same time period. This classifier was not able to distinguish areas of ice in the first scene, and misclassified them as open water.

The Kennaugh Element Framework was able to classify areas of open water, flooded vegetation, and ice. K_4 was used to distinguish areas of flooded vegetation and K_0 was used to identify areas of open water. Similar conclusions were found by Moser et al. [25]. The differential Kennaugh analysis on all four elements, looking at the difference from the first scene to the last scene, was able to indicate where changes occurred, as well as what changes occurred. The change between ice and open water, open water and marshland, and flooded vegetation and land were clearly identified using this method. Applying the k-means classifier allowed for the classification of open water and flooded vegetation which agreed with seasonality, but the areas of ice cover are less well-defined.

Finally, the single-polarization grey-level thresholding method proved to identify open water well. Part of the lakes were classified as other in the first scene, which indicates the potential to classify ice (or rather not misclassify water). However, the total surface water in each scene shows little change and no seasonal variation compared to the result of the Kennaugh Element Framework, which shows a decrease in flooded vegetation from 12% to 5%, indicating a seasonal change from flooding to drying. These conclusions could not have been drawn from the single-polarization data and, therefore, a clear advantage to dual-polarization data is the ability to show seasonal fluctuations. In addition, dual-polarization is able to distinguish open water and flooded vegetation.

The findings of this study confirm the expected advantages of dual-polarization observations, however, single-polarization observations are still useful in classifying water, albeit not sufficiently for identifying seasonal changes in vegetated areas. The applicability of single-polarization SAR for landscape dynamics is thus limited. Considering potential applications in earth system monitoring and process understanding, where not only the land cover type, but also the spatio-temporal transition from one type to another is highly relevant, the use of dual-polarization (or multi-polarization for specific land cover types) SAR data is a necessity.

Author Contributions: The research conducted and described in this manuscript was performed by all co-authors. Specifically, the following contributions were made: "Conceptualization, K.I., A.B., A.R.; Methodology, K.I., B.W., A.R.; Validation, K.I., A.R.; Resources, G.F., A.B., A.R.; Writing—Original Draft Preparation, K.I., G.F.; Writing—Review & Editing, K.I., G.F., A.B., A.R., B.W.; Visualization, K.I., A.B.; Supervision, G.F., A.B., A.R."

Funding: The German Space Agency (DLR) is acknowledged for providing TerraSAR-X data. NSERC is acknowledged for partially funding this research.

Conflicts of Interest: The authors declare no conflict of interest.

References

1. Lee, J.-S.; Pottier, E. *Polarimetric Radar Imaging: From Basics to Applications;* CRC Press: Boca Raton, FL, USA, 2009; 422p. [CrossRef]
2. Lee, J.-S.; Grunes, M.R.; Pottier, E. Quantitative Comparison of Classification Capability: Fully Polarimetric versus Dual and Single-Polarization SAR. *IEEE Trans. Geosci. Remote Sens.* **2001**, *39*, 2343–2351. [CrossRef]
3. Brisco, B.; Touzi, R.; van der Sanden, J.J.; Charbonneau, F.; Pultz, T.J.; D'Iorio, M. Water Resource Applications with RADARSAT-2—A Preview. *Int. J. Digit. Earth* **2008**, *1*, 130–147. [CrossRef]
4. White, L.; Brisco, B.; Dabboor, M.; Schmitt, A.; Pratt, A. A Collection of SAR Methodologies for Monitoring Wetlands. *Remote Sens.* **2015**, *7*, 7615–7645. [CrossRef]
5. Kuenzer, C.; Guo, H.; Huth, J.; Leinenkugel, P.; Li, X.; Dech, S. Flood Mapping and Flood Dynamics of the Mekong Delta: ENVISAT-ASAR-WSM Based Time Series Analyses. *Remote Sens.* **2013**, *5*, 687–715. [CrossRef]
6. Schlaffer, S.; Matgen, P.; Hollaus, M.; Wagner, W. Flood Detection from Multi-Temporal SAR Data Using Harmonic Analysis and Change Detection. *Int. J. Appl. Earth Obs. Geoinf.* **2015**, *38*, 15–24. [CrossRef]
7. White, L.; Brisco, B.; Pregitzer, M.; Tedford, B.; Boychuk, L. RADARSAT-2 Beam Mode Selection for Surface Water and Flooded Vegetation Mapping. *Can. J. Remote Sens.* **2014**, *40*, 135–151. [CrossRef]
8. Martinis, S.; Kersten, J.; Twele, A. A Fully Automated TerraSAR-X Based Flood Service. *ISPRS J. Photogramm. Remote Sens.* **2015**, *104*, 203–212. [CrossRef]

9. Mason, D.C.; Speck, R.; Devereux, B.; Schumann, G.; Neal, J.C.; Bates, P.D. Flood Detection in Urban Areas Using TerraSAR-X. *IEEE Trans. Geosci. Remote Sens.* **2010**, *48*, 882–894. [CrossRef]

10. Matgen, P.; Schumann, G.; Henry, J.B.; Hoffmann, L.; Pfister, L. Integration of SAR-Derived River Inundation Areas, High-Precision Topographic Data and a River Flow Model toward near Real-Time Flood Management. *Int. J. Appl. Earth Obs. Geoinf.* **2007**, *9*, 247–263. [CrossRef]

11. Liu, H.; Jezek, K.C. Automated Extraction of Coastline from Satellite Imagery by Integrating Canny Edge Detection and Locally Adaptive Thresholding Methods. *Int. J. Remote Sens.* **2004**, *25*, 937–958. [CrossRef]

12. Imhoff, M.L.; Vermillion, L.; Story, M.H.; Choudhury, A.M.; Gafoor, A.; Polcyn, F. Monsoon Flood Boundary Delineation and Damage Assessment Using Space Borne Imaging Radar and Landsat Data. *Photogramm. Eng. Remote Sens.* **1987**, *53*, 405–413.

13. Mason, D.C.; Horritt, M.S.; Dall'Amico, J.T.; Scott, T.R.; Bates, P.D. Improving River Flood Extent Delineation from Synthetic Aperture Radar Using Airborne Laser Altimetry. *IEEE Trans. Geosci. Remote Sens.* **2007**, *45*, 3932–3943. [CrossRef]

14. Martinis, S.; Twele, A.; Voigt, S. Towards Operational near Real-Time Flood Detection Using a Split-Based Automatic Thresholding Procedure on High Resolution TerraSAR-X Data. *Nat. Hazards Earth Syst. Sci.* **2009**, *9*, 303–314. [CrossRef]

15. Giustarini, L.; Chini, M.; Hostache, R.; Pappenberger, F.; Matgen, P. Flood Hazard Mapping Combining Hydrodynamic Modeling and Multi Annual Remote Sensing Data. *Remote Sens.* **2015**, *7*, 14200–14226. [CrossRef]

16. Irwin, K.; Beaulne, D.; Braun, A.; Fotopoulos, G. Fusion of SAR, Optical Imagery and Airborne LiDAR for Surface Water Detection. *Remote Sens.* **2017**, *9*, 890. [CrossRef]

17. Gstaiger, V.; Huth, J.; Gebhardt, S.; Wehrmann, T.; Kuenzer, C. Multi-Sensoral and Automated Derivation of Inundated Areas Using TerraSAR-X and ENVISAT ASAR Data. *Int. J. Remote Sens.* **2012**, *33*, 7291–7304. [CrossRef]

18. Schmitt, A.; Brisco, B. Wetland Monitoring Using the Curvelet-Based Change Detection Method on Polarimetric SAR Imagery. *Water* **2013**, *5*, 1036–1051. [CrossRef]

19. Mermoz, S.; Allain-Bailhache, S.; Bernier, M.; Pottier, E.; Van Der Sanden, J.J.; Chokmani, K. Retrieval of River Ice Thickness from C-Band PolSAR Data. *IEEE Trans. Geosci. Remote Sens.* **2014**, *52*, 3052–3062. [CrossRef]

20. Brisco, B.; Schmitt, A.; Murnaghan, K.; Kaya, S.; Roth, A. SAR Polarimetric Change Detection for Flooded Vegetation. *Int. J. Digit. Earth* **2011**, *6*, 103–114. [CrossRef]

21. Gallant, A.L.; Kaya, S.G.; White, L.; Brisco, B.; Roth, M.F.; Sadinski, W.; Rover, J. Detecting Emergence, Growth, and Senescence of Wetland Vegetation with Polarimetric Synthetic Aperture Radar (SAR) Data. *Water* **2014**, *6*, 694–722. [CrossRef]

22. Hong, S.; Jang, H.; Kim, N.; Sohn, H.G. Water Area Extraction Using RADARSAT SAR Imagery Combined with Landsat Imagery and Terrain Information. *Sensors* **2015**, *15*, 6652–6667. [CrossRef] [PubMed]

23. Betbeder, J.; Rapinel, S.; Corgne, S.; Pottier, E.; Hubert-Moy, L. TerraSAR-X Dual-Pol Time-Series for Mapping of Wetland Vegetation. *ISPRS J. Photogramm. Remote Sens.* **2015**, *107*, 90–98. [CrossRef]

24. Dabboor, M.; White, L.; Brisco, B.; Charbonneau, F. Change Detection with Compact Polarimetric SAR for Monitoring Wetlands. *Can. J. Remote Sens.* **2015**, *41*, 408–417. [CrossRef]

25. Moser, L.; Schmitt, A.; Wendleder, A.; Roth, A. Monitoring of the Lac Bam Wetland Extent Using Dual-Polarized X-Band SAR Data. *Remote Sens.* **2016**, *8*, 302. [CrossRef]

26. Schmitt, A.; Leichtle, T.; Huber, M.; Roth, A. On the Use of Dual-Co-Polarized TerraSAR-X Data for Wetland Monitoring. *Int. Arch. Photogramm. Remote Sens. Spat. Inf. Sci. ISPRS Arch.* **2012**, *39*, 341–344. [CrossRef]

27. Schmitt, A.; Wendleder, A.; Hinz, S. The Kennaugh Element Framework for Multi-Scale, Multi-Polarized, Multi-Temporal and Multi-Frequency SAR Image Preparation. *ISPRS J. Photogramm. Remote Sens.* **2015**, *102*, 122–139. [CrossRef]

28. Hess, L.L.; Melack, J.M.; Filoso, S.; Wang, Y. Delineation of Inundated Area and Vegetation along the Amazon Floodplain with the SIR-C Synthetic Aperture Radar. *IEEE Trans. Geosci. Remote Sens.* **1995**, *33*, 896–904. [CrossRef]

29. Bourgeau-Chavez, L.L.; Kasischke, E.S.; Brunzell, S.M.; Mudd, J.P.; Smith, K.B.; Frick, L. Analysis of Space-Borne SAR Data for Wetland Mapping in Virginia Riparian Ecosystems. *Int. J. Remote Sens.* **2001**, *22*, 3665–3687. [CrossRef]

Remote Sens. **2018**, *10*, 949

30. Townsend, P. A Synthetic Aperture Radar–based Model to Assess Historical Changes in Lowland Floodplain Hydroperiod. *Water Resour. Res.* **2002**, *38*. [CrossRef]
31. Manjusree, P.; Kumar, L.P.; Bhatt, C.M.; Rao, G.S.; Bhanumurthy, V. Optimization of Threshold Ranges for Rapid Flood Inundation Mapping by Evaluating Backscatter Profiles of High Incidence Angle SAR Images. *Int. J. Disaster Risk Sci.* **2012**, *3*, 113–122. [CrossRef]
32. Brisco, B.; Short, N.; Van Der Sanden, J.; Landry, R.; Raymond, D. A Semi-Automated Tool for Surface Water Mapping with RADARSAT-1. *Can. J. Remote Sens.* **2009**, *35*, 336–344. [CrossRef]
33. Kasischke, E.S.; Melack, J.M.; Dobson, M.C. The Use of Imaging Radars for Ecological Applications—A Review. *Remote Sens. Environ.* **1997**, *59*, 141–156. [CrossRef]
34. Cloude, S.R. The Dual Polarisation Entropy/Alpha Decomposition. In Proceedings of the 3rd International Workshop on Science and Applications of SAR Polarimetry and Polarimetric Interferometry, Frascati, Italy, 22–26 January 2007; pp. 1–6.
35. Cloude, S.R.; Pottier, E. A Review of Target Decomposition Theorems in Radar Polarimetry. *IEEE Trans. Geosci. Remote Sens.* **1996**, *34*, 498–518. [CrossRef]
36. Zalite, K.; Voormansik, K.; Olesk, A.; Noorma, M.; Reinart, A. Effects of Inundated Vegetation on X-Band HH-VV Backscatter and Phase Difference. *IEEE J. Sel. Top. Appl. Earth Obs. Remote Sens.* **2014**, *7*, 1402–1406. [CrossRef]
37. Schmitt, A. Multiscale and Multidirectional Multilooking for SAR Image Enhancement. *IEEE Trans. Geosci. Remote Sens.* **2016**, *54*, 5117–5134. [CrossRef]
38. Mleczko, M.; Mróz, M. Wetland Mapping Using SAR Data from the Sentinel-1A and TanDEM-X Missions: A Comparative Study in the Biebrza Floodplain (Poland). *Remote Sens.* **2018**, *10*, 78. [CrossRef]
39. Heine, I.; Jagdhuber, T.; Itzerott, S. Classification and Monitoring of Reed Belts Using Dual-Polarimetric TerraSAR-X Time Series. *Remote Sens.* **2016**, *8*, 552. [CrossRef]
40. Wdowinski, S.; Hong, S.H. Wetland InSAR: A review of the technique and applications. In *Remote Sensing of Wetlands Applications and Advances*; Tiner, R.W., Lang, M.W., Klemas, V.V., Eds.; CRC Press: Boca Raton, FL, USA, 2015; pp. 137–154. [CrossRef]
41. Ullmann, T.; Banks, S.N.; Schmitt, A.; Jagdhuber, T. Scattering characteristics of X-, C- and L-band PolSAR data examined for the tundra environment of the Tuktoyaktuk Peninsula, Canada. *Appl. Sci.* **2017**, *7*, 595. [CrossRef]

remote sensing

MDPI

Article

Internal Solitary Waves in the Andaman Sea: New Insights from SAR Imagery

Jorge M. Magalhaes [1,2] and José C. B. da Silva [1,2,*]

[1] Department of Geosciences, Environment and Spatial Planning (DGAOT), Faculty of Sciences, University of Porto, 4169-007 Porto, Portugal; jmagalhaes@fc.ul.pt
[2] Interdisciplinary Centre of Marine and Environmental Research (CIIMAR), 4450-208 Matosinhos, Portugal
* Correspondence: jdasilva@fc.up.pt; Tel.: +351-220-402-476

Received: 30 April 2018; Accepted: 28 May 2018; Published: 1 June 2018

Abstract: The Andaman Sea in the Indian Ocean has been a classical study region for Internal Solitary Waves (ISWs) for several decades. Papers such as Osborne and Burch (1980) usually describe mode-1 packets of ISWs propagating eastwards, separated by distances of around 100 km. In this paper, we report on shorter period solitary-like waves that are consistent with a mode-2 vertical structure, which are observed along the Ten Degree Channel, and propagate side-by-side the usual large mode-1 solitary wave packets. The mode-2 waves are identified in TerraSAR-X images because of their distinct surface signatures, which are reversed when compared to those that are typical of mode-1 ISWs in the ocean. These newly observed regularly-spaced packets of ISW-like waves are characterized by average separations of roughly 30 km, which are far from the nominal mode-1 or even the mode-2 internal tidal wavelengths. On some occasions, five consecutive and regularly spaced mode-2 ISW-like wave envelopes were observed simultaneously in the same TerraSAR-X image. This fact points to a tidal generation mechanism somewhere in the west shallow ridges, south of the Nicobar Islands. Furthermore, it implies that unusually long-lived mode-2 waves can be found throughout the majority of the fortnightly tidal cycle. Ray tracing techniques are used to identify internal tidal beams as a possible explanation for the generation of the mode-2 solitary-like waves when the internal tidal beam interacts with the ocean pycnocline. Linear theory suggests that resonant coupling with long internal waves of higher-mode could explain the longevity of the mode-2 waves, which propagate for more than 100 km. Owing to their small-scale dimensions, the mode-2 waves may have been overlooked in previous remote sensing images. The enhanced radiometric resolution of the TerraSAR-X, alongside its wide coverage and detailed spatial resolutions, make it an ideal observational tool for the present study.

Keywords: SAR; internal waves; Andaman Sea

1. Introduction

Efficiently monitoring the ocean by means of satellite imagery is becoming an increasingly valuable tool, as its surveying ability on a global basis is far beyond that of traditional in situ measurements. Satellite remote sensing especially excels in acquiring and integrating multiple views of the ocean including in synergy with numerical modelling and in situ data, and hence it leverages our ability to document, understand, and predict a wide range of ocean phenomena. In particular, orbiting Synthetic Aperture Radars (SARs) have been acquiring high-resolution radar images in all-weather conditions for nearly four decades—since Seasat first flew in 1978. Their unique views of the ocean surface have been providing the ocean sciences communities with unprecedented insights into numerous processes and ocean applications (for more details see [1]). For instance, despite being a classical subject in physical oceanography, Internal Wave (IW) research has greatly benefited from SAR evidence in the last few decades, especially when examining their generation, propagation, and dissipation mechanisms (see e.g., [2]). IWs may be regarded as density oscillations propagating along the ocean stratification, and it is now acknowledged that they impact a series of ocean phenomena ranging from fundamental oceanography to multiple other fields of research [3]. For instance, recent studies have revealed that they are increasingly important in quantifying and understanding ocean energetics and mixing [4,5]. Furthermore, recently quantified organic carbon fluxes that are associated with large IWs have been shown by Li et al. [6] to increase by as much as a factor of 3.

IW generation mechanisms in the ocean are of different kinds, and are still the subject of active discussion and research. An extensive review is beyond the scope of this paper, however it is important to note that IW observations in SAR are mostly tidal in origin—the reader is referred to the works by Jackson et al. [7] and da Silva et al. [2] for a comprehensive review of IW generation mechanisms. In this case, some of the IW generation details may be obtained via parameterized generation regimes, which are in turn based on characteristic measures of stratification, tidal currents, and bathymetry dimensions [8,9]. The topographic Froude number (Fr_t) and the normalized tidal excursion length (δ) are set as two major parameters governing the different IW generation regimes—defined as $\frac{N_{max}H}{U_{max}}$ and $\frac{U_{max}}{L\omega}$ (respectively), where N_{max} is a representative maximum of the buoyancy frequency, U_{max} is a typical value of the maximum flow velocity over some topography with characteristic width L and height H, and ω is usually the semi-diurnal tidal frequency. These non-dimensional parameters indicate that, for example, increasing or decreasing Fr_t means more or less blocking from the bathymetry on the incoming flow, or that the tidal excursion length is simply a measure of how much of the topography is traversed during a tidal excursion. A third parameter is also particularly important in the discussions to come. The internal or densimetric Froude number (Fr) is usually defined as U/c, and essentially describes the hydraulic state of a stratified flow with magnitude U (i.e., subcritical, critical, or supercritical) with respect to a characteristic IW phase speed (c). While these parameters are useful in describing IW generation in the vicinities of some bathymetry feature, they also define the upstream or downstream locations of the exact isopycnal perturbations from which IW trains eventually originate (also referred to as a lee wave or upstream influence). However, these details need not concern us here, especially since they are hard to determine from SAR imagery alone (i.e., without additional in situ or modelling data). What is particularly important for the observations discussed in this work is the resonant coupling between IWs of different vertical modes that has recently come to light from SAR imagery. This resonant coupling involves sea surface signatures of small-scale IWs that appear in between the larger IW packets usually observed in SAR, and it has been documented in the South China Sea [10] and in the Mascarene Ridge of the Indian Ocean [2]. In these independent studies, coupled IW systems propagating with similar phase speeds were documented by means of SAR and were investigated with numerical modelling. They are then said to be in resonant coupling in the sense that energy transfer between both IW systems is possible, with the larger-scale waves travelling deeper and the short-scale waves being trapped within the upper layers close to the ocean surface. In this paper, similar wave features will also be shown in the Andaman Sea.

The Andaman Sea in the Indian Ocean, located between the Malay Peninsula and the Andaman and Nicobar Islands (see Figure 1 for location), has been a classical study region for IWs for more than 50 years. However, observations of extraordinarily large IWs in this study region were reported long before their first scientific accounts. In fact, in 1861 the oceanographer Matthew F. Maury had already documented some large bands of choppy water or ripplings stretching from horizon to horizon, which were commonly observed by mariners during their journeys in the Andaman Sea (see p. 389 in [11]). More than a century later, these descriptions were indeed recognized as the surface manifestation of large-scale and fast propagating IWs—when Perry and Schimke [12] conducted the first IW oceanographic measurements in this region. While aiming to determine the impact of IWs in underwater oil drilling structures, a series of measurements were also conducted by Osborne and Burch [13] in the southern edge of the Andaman Sea. Their detailed measurements revealed that these waves appeared in packets with ranking-ordered amplitudes, beginning at approximately 60 m in the leading wave, and then decreasing towards the rear. These are characteristic features of Internal

Solitary Waves (ISWs)—a special class of IWs which retain their shape and speed for considerable distances—which are usually described under the solitary wave theory of Korteweg and deVries (KdV). In fact, their measurements showed that these waves could be better predicted and interpreted under the shallow water KdV equation, and hence the term ISWs is usually referred to when describing the IWs propagating in the Andaman Sea.

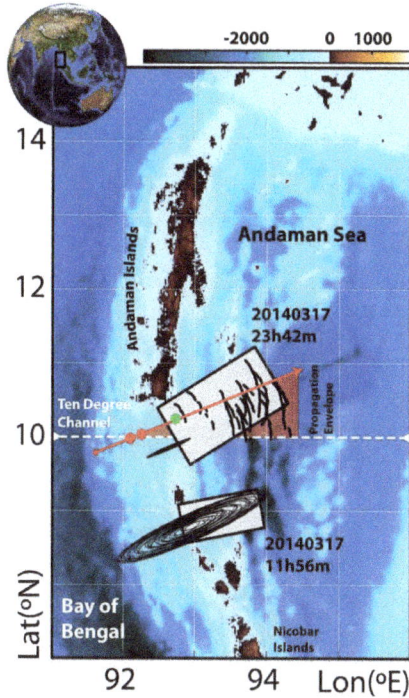

Figure 1. Bathymetry of the Andaman Sea corresponding to the inset on the top-left corner (colour-scaled on top). Grey rectangles depict subsets corresponding to TerraSAR-X acquisitions shown in Figure 3. Representative tidal ellipses are shown in black for a full spring-neap tidal cycle (the largest having a semi-major axis around 0.9 m/s). The red transect and colour-filled circles are reference locations discussed for a particular bathymetry profile shown in Figure 5.

Few other studies have emerged since. Particularly important are those by Alpers et al. [14] and Jackson [15], showing long SAR strips with exceptionally strong sea surface manifestations of ISW packets propagating onto the shelf of the Malayan Peninsula, suggesting tidally-driven generation sources in the western sections of the Andaman Sea. Acoustic Doppler Current Profiles presented by Neng et al. [16] also suggest IW sources near the shallow ridges separating the Andaman Sea from the Bay of Bengal, which were further confirmed in the high-resolution 3D numerical simulations shown in Shimizu and Nakayama [17]. Nonetheless, comprehensive descriptions of the IW fields are yet to be documented in this region—especially by means of SARs, which has been done in other equally important IW hotspots such as the South China Sea (e.g., [18]), the Mascarene Ridge in the Indian Ocean [2], and off the Amazon shelf [19].

Interestingly, however, global ocean models reveal IW energetics in this region to be at least of the same order as (if not greater than) those in other intensively studied regions (see e.g., [20]). In fact, in the particular case of the Andaman Sea (see Figure 1), a series of shallow ridges separate this marginal sea from the Bay of Bengal along more than 1000 km—all of which are potential IW generation sites.

Likewise, satellite views in this region typically reveal at least half a dozen distinct generation sites, almost resembling a series of ISW hotspots stacked meridionally—for example, shown in Jackson [15] and at https://go.nasa.gov/2ow8eSK. Furthermore, different characteristics can be found across these multiple generation sites involving important parameters such as bottom depth, bottom slopes, and tidal currents. This is especially interesting since direct comparisons can be made between different IW generation regimes while surveying them simultaneously in SAR imagery or other means of remote sensing.

Of particular interest to this study are two neighbouring and yet distinct stretches of the Andaman Sea—one approximately between 8° and 9°N, and the other along what is commonly referred to as the Ten Degree (10°) Channel (see Figure 1 for locations). In this study, we will show that these contiguous sections feature two very different ISW regimes. In fact, a recent survey using TerraSAR-X acquisitions—which are especially suited for ISW surveys as they deliver considerable wide swaths at very high spatial resolutions—from the Andaman Sea revealed a much more intricate structure than previously thought. In particular, evidence for resonant coupling between larger Internal Tides (ITs, i.e., IWs of tidal period) and shorter-scale mode-1 ISWs will be shown to be at work in the Andaman Sea, similar to recently documented cases in the South China Sea [10,21] and in the Mascarene Ridge [2]. Moreover, unprecedented SAR evidence will be shown that features short ISW envelopes with characteristic mode-2 sea surface signatures that are regularly spaced by higher-mode IWs.

In this paper, we seek to document the 2D horizontal structure on these particular sections of Andaman Sea (approximately between 8° and 11°N). The rest of the paper is organized as follows. Section 2 will describe the methods concerning satellite SAR imagery and important methodologies thereof. Section 3 will present key acquisitions that are representative of the study region, and a detailed analysis will follow in Section 4. Finally, Section 5 will summarise our main findings, along with some concluding remarks.

2. Methodology

Six SAR images that were acquired and processed from TerraSAR-X were selected as representative of the ISW activity in the Andaman Sea between 8° and 11°N. These are listed chronologically in Table 1, together with the Envisat-ASAR acquisition used in da Silva and Magalhaes [22], which is also especially important in this study (to be discussed in the next sections). All TerraSAR-X acquisitions were in ScanSAR-mode (with a nominal spatial resolution of 18 m), which means that the viewing strips were approximately 200 km wide and as long as 1000 km long in some cases. This acquisition mode is especially suited when surveying ISWs since wide swaths can be delivered with high spatial resolutions around tens of meters, allowing extended views of the waves along their propagation path, while preserving their detailed spatial structure.

ISWs are particularly well observed in SAR images, essentially owing to their characteristic sea surface manifestations (or signatures) in the short-scale (i.e., centimetre to decimetre) sea surface roughness, which is in turn proportional to the waves' surface current gradients [23–25]. An illustration is given in the schematic representation of Figure 2, which shows how the normalized radar cross section (σ_0) is modulated in the presence of mode-1 ISWs of depression (left panel in Figure 2)—as these are the most commonly observed in the ocean and in SAR images (see e.g., [15]). In this case, isopycnals are depressed as the wave propagates towards the right, while the wave's orbital velocity field creates a surface gradient with consecutive convergence and divergence sections. Consequently, sea surface roughness will increase and decrease in comparison with the unperturbed background. In a SAR image, when considering the wave direction of propagation (in agreement with [23]), this means mode-1 ISWs appear first as a bright and enhanced backscatter section, and after this a trailing dark and reduced backscatter section follows.

Table 1. List of images used in this study and in the composite map in Figure 1 (solid black lines). Tidal ranges within spring-neap cycles are given together with tidal currents (along the waves' direction of propagation) at time of acquisition. The number of wave-tail envelopes (N_{wt}) in each image is also shown with an averaged value for the length between consecutive wave-tails ($\bar{\lambda}$).

Acquisition Date/Time	Satellite Sensor & Mode	Tidal Range & Current	N_{wt} & $\bar{\lambda}$
2 January 2007 15h49m UTC	Envisat ASAR, Image Mode	+23 to −21 cm/s −5 cm/s	N_{wt} = 4 $\bar{\lambda}$ = 29 km
17 March 2014 11h56m UTC (in Figure 3a)	TerraSAR-X TSX-SAR, ScanSAR	−22 to 23 cm/s +15 cm/s	N_{wt} = 2 $\bar{\lambda}$ = 30 km
17 March 2014 23h42m UTC (in Figure 3b)	TerraSAR-X TSX-SAR, ScanSAR	−23 to +24 cm/s +14 cm/s	N_{wt} = 5 $\bar{\lambda}$ = 32 km
1 April 2015 11h48m UTC	TerraSAR-X TSX-SAR, ScanSAR	−14 to +15 cm/s +15 cm/s	N_{wt} = 3 $\bar{\lambda}$ = 24 km
23 April 2015 11h48m UTC	TerraSAR-X TSX-SAR, ScanSAR	−18 to +14 cm/s −9 cm/s	N_{wt} = 2 $\bar{\lambda}$ = 28 km
4 May 2015 11h48m UTC	TerraSAR-X TSX-SAR, ScanSAR	−22 to +20 cm/s +15 cm/s	N_{wt} = 1 $\bar{\lambda}$ = NA
4 May 2015 23h34m UTC	TerraSAR-X TSX-SAR, ScanSAR	−20 to 22 cm/s +17 cm/s	N_{wt} = 3 $\bar{\lambda}$ = 31 km

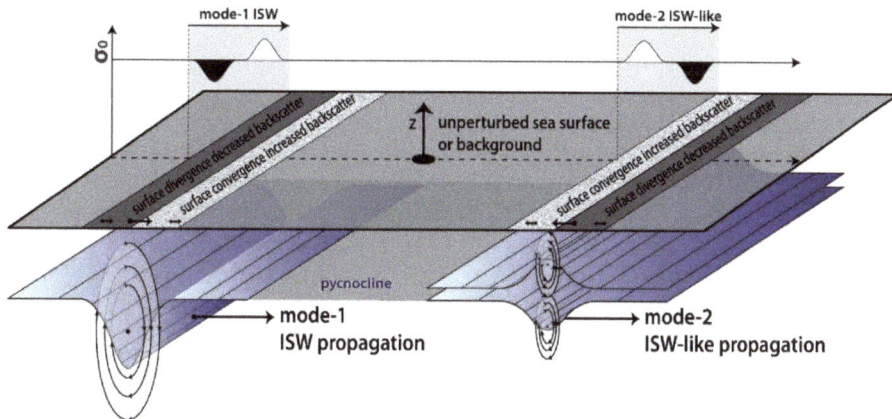

Figure 2. Schematic representation of Internal Solitary Wave (ISW) sea surface manifestations in Synthetic Aperture Radar (SAR) images with respect to an unperturbed background. A comparison between the mode-1 (**left**) and mode-2 (**right**) vertical structures is illustrated for waves propagating rightwards along the ocean pycnocline. From top to bottom, a measure for σ_o is seen to depart from the background owing to the convergence and divergence patterns induced by the waves' surface currents—all of which reverse between mode-1 and mode-2 waves.

However, an increasing number of ISW observations have recently been documented in which SAR signatures are not consistent with this picture, and are instead reversed in relation to mode-1 ISWs. These waves are often discussed as resulting from a mode-2 internal vertical structure, and hence are termed mode-2 ISW-like waves, since they too usually appear in rank-ordered packets. According to Figure 2 (right panel), a convex mode-2 internal structure is best thought of as a bulge propagating along horizontally which displaces isopycnals upward and downward in the upper and lower layers, respectively. It is as if two individual ISWs are placed vertically on top of each other, and hence two orbital velocity fields appear instead. Only this time, the upper velocity field is reversed in relation to the lower velocities, and hence in relation to typical mode-1 ISWs. In turn, this eventually means that the polarity of a mode-2 ISW-like wave signature at the surface is also reversed, meaning a dark reduced backscatter will precede bright enhanced backscatter in the propagation direction of the wave (see e.g., [2,26] for more details).

The information retrieved from SAR imagery can be analysed together with in situ data and linear theory to provide further insights into the generation and propagation details. In particular, if ISW packets are found regularly spaced in the images, then they are usually of tidal origin and their separations are interpreted as tidal wavelengths. In this case, travel-time plots (in the same fashion as in [2,27]) can be used to determine possible proxies for the waves' generation position and tidal phase. These can then be compared with the Baines [28] barotropic tidal forcing to determine the most likely ISW hotspots. The barotropic tidal forcing has been found to be a useful indicator for generation sites in other independent studies, including in the Iberian Shelf [29–32] and in the Mozambique Channel [33]. Following Baines [28], the barotropic forcing term (F) for ITs resulting from tidal currents and bottom bathymetry, is a scalar quantity found at each depth level as:

$$F(z) = zN^2(z) \int Q(x,y) \cdot \nabla \left(\frac{1}{h(x,y)} \right) dt \tag{1}$$

where z is the vertical coordinate (positive upwards), N is the buoyancy frequency, Q is the barotropic mass flux vector $Q = (uh, vh)$ with u and v being the zonal and meridional components of the barotropic velocity, and h the bottom depth. This tidal forcing is usually integrated analytically along a tidal cycle, provided that tidal currents are previously known and best-fitted to tidal ellipses. In this study, tidal currents were estimated from a regional solution for the Andaman Sea provided from the Oregon State University Tidal Inversion Software [34]), which runs at $1/30°$ and includes 4 semidiurnal and 4 diurnal constituents (M_2, S_2, N_2, K_2, K_1, O_1, P_1, Q_1). The bathymetry data is part of the one minute global bathymetry from Smith and Sandwell [35], and the stratification was obtained from climatological monthly means (available at www.esrl.noaa.gov), which is assumed constant in the x (i.e., zonal) and y (i.e., meridional) directions in Equation (1).

According to this formulation, elevated barotropic forcing sites where ITs are being generated are usually found where large tidal currents oscillate across shallow steep bathymetry. This is certainly the case in the shallow ridges along the Andaman Sea (see Figure 1). Two distinct cases then follow, depending on how local stratification, bottom bathymetry, and tidal currents interact. Either the vertical oscillations induced in the water column radiate away horizontally in the form of interfacial waves directly along the nearby pycnocline, or IW beams (or rays) form and propagate coherently along an oblique direction to the horizontal [36]. In the first case, ocean ridges feature large interfacial ITs forming directly above steep bottom bathymetry, which typically evolve (through nonlinear processes) to higher-frequency ISW packets (see e.g., Figure 2 in [2]). Otherwise, when critical (or near-critical) slopes exist, meaning the bottom topographic slope matches the slope of the characteristic IT propagation paths, the IT energy propagates at an angle (θ) to the horizontal such that,

$$\tan \theta = \left(\frac{\omega^2 - f^2}{N^2 - \omega^2} \right)^{\frac{1}{2}} \tag{2}$$

where ω is the tidal frequency, f is the Coriolis parameter, and N is, again, the buoyancy frequency [36]. In other words, when the forcing barotropic flow is coincident with the motion plane for free IWs, resonant conditions usually result in enhanced IW beam generation. In continuously stratified fluids (such as the ocean), Equation (2) is often used in a ray tracing technique that follows IT beams, as these follow characteristic pathways along which the internal tide energy can propagate. These rays can then reflect from the seafloor or the sea surface, and interact afterwards with the ocean's pycnocline to generate ISWs by means of a different mechanism. Note that these ISWs do not originate from nearby bathymetry, but from IT beams instead, as these interact at an oblique angle with the ocean pycnocline, generating interfacial waves there that may evolve through nonlinear steepening into ISW packets. This mechanism is therefore usually referred to as *local generation* and it has been measured in situ [37,38], modelled numerically [39–41], and observed in laboratory experiments [42]. SAR imagery has confirmed its effectiveness as well, including in the Bay of Biscay, the Iberian shelf-break, and the Mozambique Channel [30,31,33,43].

More recently, in the Mascarene Ridge of the Indian Ocean, da Silva et al. [2] also found consistent evidence between SAR and fully nonlinear and non-hydrostatic simulations, showing large interfacial mode-2 IWs that originated from IT beams, interacting with the pycnocline from above after reflecting from the sea surface (see their Figure 9). It is noteworthy that these mode-2 features were found in resonant coupling with very short-scale mode-1 ISWs (meaning they both travelled at the same speed) which were clearly observed in SAR and referred to as wave-tails. This is especially important to the present study, since similar features will also be shown in this paper to be at work in the Andaman Sea. However, unlike the Mascarene Ridge, SAR observations of both mode-1 and mode-2 ISW-like waves will be shown to be recurring features in this study region—but they too will be consistent with resonance with higher-mode IWs.

3. SAR Observations

We now present two selected case studies that are representative of the ISW field in the Andaman Sea between 8° and 11°N. These images show typical views along two adjacent stretches (see Figure 1 for locations), which will nonetheless appear very different. The first case study is presented in Figure 3a, and shows a TerraSAR-X image dated 17 March 2014, acquired as the satellite flew over the Andaman Sea (see also Supplemental Material S1). In this image, as in many others in this study region, two large mode-1 ISW packets may be seen, which are typically separated by 100 km along their propagation direction. A detailed inspection of the SAR signatures, in the same fashion as [44], reveals that the ISW packets propagate eastwards into the Andaman Sea and away from the shallow ridges near 9°N (see also the grey rectangle in Figure 1). Their backscatter modulations in the SAR consist essentially of very bright bands, behind which some less pronounced darker sections follow (especially in the easternmost wave). Considering that the ocean depths here are greater than 1000 m, this indicates that the sea surface signatures are consistent with mode-1 ISWs of depression. These large-scale packets are actually a typical view in the Andaman Sea, and are in fact very similar to those reported in other studies reporting similar IW activity in this region (e.g., [14,15]).

Careful examination reveals two other wave features in between the larger mode-1 ISW trains, which do not appear to belong to the same family of those. These are labelled as wave-tail envelopes in Figure 3a, owing to their wavy and clustered quasi-linear characteristics, whose total length along the propagation direction is at least 20 km. Their backscatter intensities are comparable to the larger ISWs, while still being characteristic of short-scale mode-1 ISWs of depression. Similar features have previously been reported in the South China Sea [10,21] and in the Mascarene Ridge [2]. In both cases, the authors also documented similar short-scale mode-1 ISWs, whose envelopes were seen to propagate coupled with larger mode-2 IWs (with a semidiurnal periodicity). Interestingly, the two wave-tails in Figure 3a were also separated by approximately half their larger mode-1 companions, and hence were consistent with a wavelength of a mode-2 IT. We also note in passing that these wave-tails are commonly observed in the Andaman Sea, including in other means of remote sensing as shown in Supplemental Material S2. In this case, an RGB composite (from MODIS) was acquired two days later in cloud free conditions, and again shows a wave-tail feature, this time a bit farther ahead its propagation path and in between two large mode-1 ISWs (almost identical to Figure 8 in [2]). As these features appear to be observed frequently in yet another region of the world's oceans, we feel that they add to the case that higher-mode IWs may be commonly operative in many other ISW hotspots.

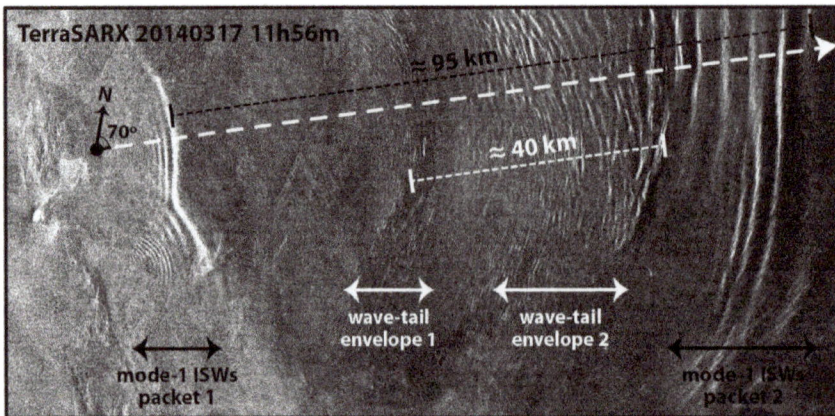

(a)

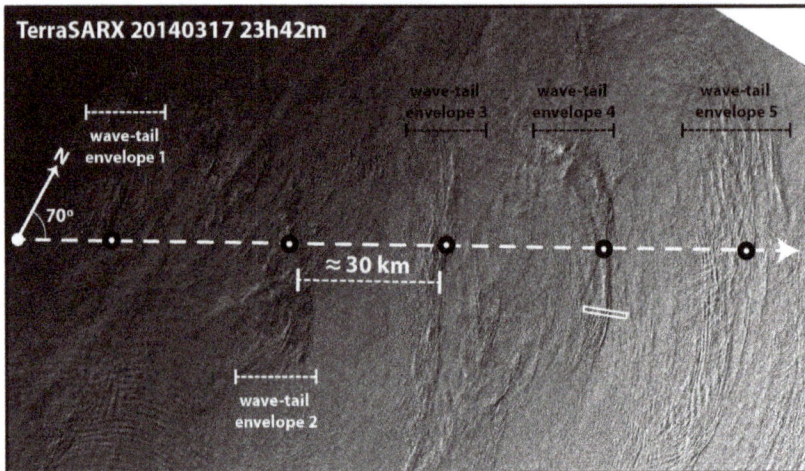

(b)

Figure 3. (**a**) Subset of a TerraSAR-X image, approximately 123×58 km^2, showing a typical view of the ISW field along 9°N in the Andaman Sea (see Figure 1 and Supplemental Material S1 for location). ISWs are propagating approximately towards 70°TN (white dashed arrow), and two mode-1 packets may be identified with a separating distance of roughly 100 km. In between these, two "wave-tail" envelopes are also depicted by white arrows and are spaced by about half of the previous wavelength; (**b**) Subset of a TerraSAR-X image, approximately 167×94 km^2, showing a typical view of the ISW field along 10°N in the Andaman Sea (see Figure 1 for location). The wave field is propagating approximately towards 70°TN (white dashed arrow), but no mode-1 packets are observed. Instead, five regularly-spaced "wave-tail" envelopes are depicted by dashed borderlines. The white rectangle depicts the location of an ISW transect shown in Figure 4, and is consistent with a mode-2 ISW-like wave.

The second case study addressed in this paper presents an entirely different view. In this example, Figure 3b shows sea surface signatures that are typically observed along what is commonly referred to as the Ten Degree Channel of the Andaman Sea (see Figure 1 for location), corresponding to a TerraSAR-X image acquired approximately a semidiurnal tidal cycle later than Figure 3a. Nonetheless, unlike the previous case, the usual bright and large-scale mode-1 ISWs packets are all but missing.

Instead, the major features in this image are actually consistent with mode-2 short-period ISW-like waves. Indeed, while the IW field is also seen to propagate eastwards and away from the shallow ridges south of the Andaman Islands (c.f. Supplemental Material S3), the majority of the individual waves in this case have much shorter scales and their sea surface signatures are reversed when compared with those shown in Figure 3a. A detailed inspection of these sea surface signatures is presented in Figure 4. Recalling Figure 2, we note that the backscatter modulations along the waves in this case, reveal dark bands (i.e., decreased backscatter) preceding bright bands (i.e., increased backscatter) in the propagation direction, which for those deep-ocean conditions, are consistent with the sea surface manifestations of mode-2 ISW-like waves. Interestingly, despite the much shorter-scales and reduced packet coherency (comparing with larger mode-1 ISW packets), the profile in Figure 4 still suggests that these waves may have rank ordered amplitudes, at least in the leading and more prominent waves.

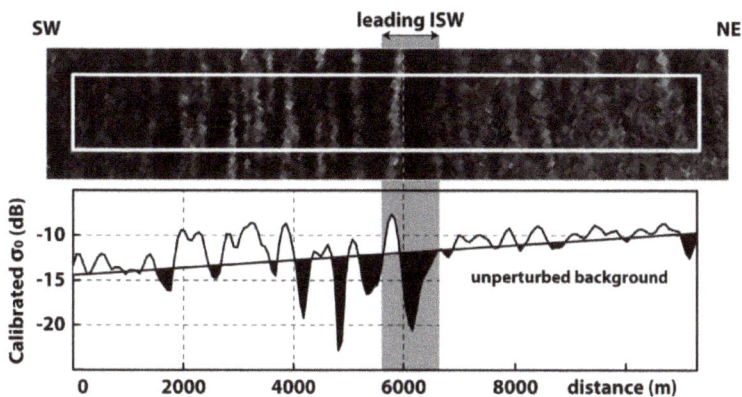

Figure 4. The top panel is a zoom in of the white rectangle in Figure 3b. The bottom panel is a transect along the white rectangle on top (extending approximately from the southwest to the northeast), showing a calibrated backscatter cross-section across a mode-2 ISW-like feature (leading wave highlighted in a grey background). The black line is a linear best fit used to illustrate signal modulation above (in white) and below (in black) a representative local average.

The IW field in Figure 3b is also very different from the traditional views of the Andaman Sea. Despite being acquired just north of Figure 3a (and a tidal cycle later), it is remarkable to see that no sign of large mode-1 ISW packets can be observed. Regularly-spaced envelopes still exist, but this time their average separations are roughly around 30 km, which are far from nominal mode-1 or even mode-2 IT wavelengths. Furthermore, five distinct envelopes may be identified, which means that the longevity of these wave-packets is at least 48 h (assuming semidiurnal periodicity). In fact, scrutiny suggests that the envelopes' characteristic dimensions increase from about 10 to more than 20 km as the waves develop along their direction of propagation. We stress that this represents an unprecedented view in the Andaman Sea and everywhere else in the ocean, in which five consecutive and regularly spaced mode-2 ISW-like wave envelopes were observed simultaneously in the same image.

Interestingly however, it is not an isolated event, but has instead been found in a number of images along this particular stretch of the Andaman Sea. Owing to their small-scale dimensions, these waves may have been overlooked in other remote sensing images. For instance, it is very hard to observe them in standard 250 m resolution sunglint images, where the larger mode-1 ISW packets are easily seen (see e.g., Supplemental Material S2). On the other hand, the enhanced radiometric resolution of the TerraSAR-X, alongside its wide coverage and detailed spatial resolutions, make it an ideal observational tool for these waves. In fact, after surveying the entire record, five more cases were identified with similar surface signatures consistent with mode-2 ISW-like waves—all of which showed no signs of the larger mode1 ISW packets. These are listed in Table 1 in chronological order, and span between 2014 and 2015. We note in passing that TerraSAR-X images were acquired only on 11 days throughout its entire record, which means that waves along the Ten Degree Chanel were found on nearly 50% of these occasions. Furthermore, the tidal currents at the time of each acquisition are also listed in Table 1, and suggest that these mode-2 waves are found throughout the majority of the fortnightly tidal cycle. This is somewhat unexpected since the larger mode-1 ISW packets commonly observed in the Andaman Sea are known to be more frequently observed close to spring tides (see e.g., [13]). For consistency, one other image which is presented in da Silva and Magalhaes [22] is also presented in Table 1, and contains a very similar picture to that in Figure 3b.

A composite map consisting of all of the images listed in Table 1 is presented in Figure 1 (see solid black curves within the light red propagation envelope). In this map, only the leading and strongest observations were accounted for, and the mean distance between mode-2 ISW-like packet envelopes in each image $(\overline{\lambda})$ was also listed in Table 1, along with their corresponding number of observations (N_{WT}). It is worthwhile noting that these features consistently present interpacket distances of around 30 km, just as is observed in our case study in Figure 3b and in da Silva and Magalhaes [22]. Also, the composite map in Figure 1 reveals a well-organized IW field whose propagation envelope (in light red in Figure 1) is fairly along the tidal currents there, and suggests an origin along the shallow ridges to the west.

4. ISW Generation, Propagation, and Resonant Coupling

In general, the exact IW generation and propagation details cannot be confidently resolved without dedicated in situ measurements, modelling results, or a sufficiently large number of satellite images. Nonetheless, further evidence may be obtained from carefully inspecting the present dataset, especially in light of the methodologies described in the previous sections. The overall impression from the SAR is that substantial differences exist between the two IW fields depicted in Figure 3a,b (see Figure 1 for locations). While farther south, the SAR shows the usual mode-1 ISW packets propagating eastwards along with short-scale mode-1 wave-tails, the picture along the Ten Degree Channel is very different, since in this stretch only regularly spaced mode-2 ISW-like waves exist.

These short mode-2 ISW-like wave envelopes are an unprecedented view in this study region, and we therefore turn to the Ten Degree Channel since it is there that they are frequently observed. The observations in Figure 3b along this stretch reveal that the waves appear regularly spaced at about 30 km, which points to a tidal generation mechanism somewhere in the west shallow ridges south of the Nicobar Islands. A bathymetry profile along the waves propagation path (red transect in Figure 1) is shown in Figure 5a, and confirms a series of shallow ridges, all of which are potential IW generation sites. These can be further analysed by taking into consideration the methodologies discussed in the first sections. We begin by analysing the barotropic tidal forcing (F) along these shallow ridges, which is shown as an inset in Figure 5a. Recalling Section 2, we note that stratification in these calculations was assumed constant and was taken from a climatological mean for March (see Figure 5d)—taken as representative for the observed waves (see Table 1). The inset in Figure 5 then essentially shows the maxima depth integrated F along a full fortnightly cycle. The tidal forcing at the East Summit (ES) is especially strong, being above 1 m^2 s^{-2} for more than 10 km, and is in fact of the same order of other ISW hotspots in the Mozambique Channel (see Figure 4 in [33]). This indicates that energy is being converted from the barotropic tide into baroclinic motions at ES, which may then generate internal motions directly above the bathymetry, or evolve into IT beams (as discussed in the previous sections).

Both possibilities may be investigated by means of the parameters governing typical tidal generation regimes. Interestingly, in the case of ES and the mode-2 ISW-like waves observed in Figure 3b, both Fr and Fr_t are quite low, which is essentially a consequence of weak tidal currents (of the order of 10 cm/s, see Table 1) interacting with ES at depths where stratification itself is outside the bulk of the pycnocline. Namely, Fr is around 0.3 at most and Fr_t of the order of 100, when considering typical tidal currents and stratification, and characteristic phase speeds between 0.6 and 0.7 m/s—as computed from the interpacket distances separating the mode-2 ISW-like waves in Figure 3b. Low internal and topographic Froude numbers suggest that it is unlikely that the waves seen in Figure 3b are being generated directly in the vicinities of ES (see [2,9]).

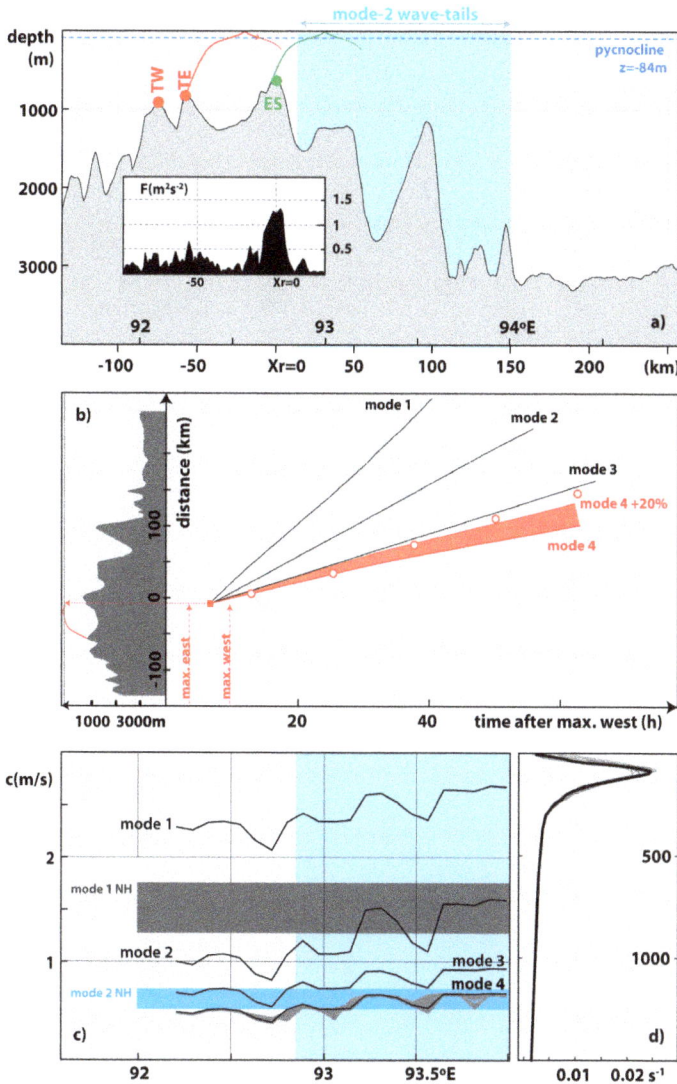

Figure 5. (a) Bathymetry profile along the red transect in Figure 1, including references to the East Summit (ES), Twin East (TE), and West (TW) summits. IT beams are also shown in corresponding colours, along with a representative pycnocline depth. The light blue shaded area depicts where the majority of the mode-2 observations are seen in the SAR; (b) Travel-time plot along the bathymetry profile in the top panel (see inset on left side); (c) Linear IW phase speeds for hydrostatic vertical modes 1–4 (solid black lines). Envelopes for the non-hydrostatic (NH) waves with wavelengths between 500 and 1500 m are also shown for modes 1 (grey) and 2 (blue); (d) March climatological mean stratification (solid black line), and remaining monthly means (in grey). See text for more details.

Alternatively, IT beams forming on critical slopes along these shallow features may be a suitable hypothesis instead, since tidal excursion lengths (δ) along these ridges appear to be of the order of 0.1. In the framework of IW generation regimes, this essentially translates to an optimum IT beam generation regime in the presence of critical (or near-critical) slopes (see e.g., [8] and Figure 6.36 in [45]). However, this means acknowledging the possibility that mode-2 ISW-like waves (rather than the usual mode-1 ISWs) could result from an IT beam impinging on the pycnocline and locally generating the waves there. Several studies have already documented the generation of waves with higher mode vertical structures in the vicinity of an IW beam impact, including in numerical models [41], laboratory experiments [42], and SAR imagery [2].

The above-mentioned studies indicate that mode-2 ISW-like waves may result from IT beams, provided that critical bathymetry exists from which IT beams can propagate and interact with the seasonal pycnocline in a geometric configuration consistent with the SAR observations—i.e., the waves observed in the SAR need to be ahead of the beam impact. Representative IT beams emanating from critical bathymetry in the vicinities of ES are depicted in green in Figure 5a as computed according to Equation (2), with the same mean stratification for March used in the barotropic tidal forcing calculations. However, interactions with the pycnocline in this case would be beyond (i.e., to the East of) the earliest SAR observations (see blue rectangle in Figure 5a), rendering this hypothesis less likely to be at work. A more promising possibility comes from ray tracing an IT beam from critical slopes farther west, in particular from TE (in red in Figure 5a) since interactions with the pycnocline there precede the SAR observations while still having a significant F around 0.5 m^2 s^{-2}.

To verify consistency with the SAR ground truth, the summit location (TE in Figure 5a) can be investigated by means of a travel-time plot assembled in the same fashion as in [2,27], which allows for a proxy of the waves' propagation history to be reconstructed back to their point of origin. This is presented in Figure 5b for the five observations in Figure 3b (depicted as red circles). The data suggest a generation somewhat close to where the downward beam interacts with the pycnocline after it reflects from the surface at a time when the tide transitions from maximum east to westward flow (see also inset on the left-hand side of Figure 5b). This would mean that after maximum eastward flow, which is consistent with IT beams forming on the western slopes and leaning upwards to the east, mode-2 ISW-like waves result from the beam impact some 20 km before (i.e., to the west) the first observations are detected in the SAR—very similar to Figures 8 and 9 in [2].

However, it is important to note that the waves in Figure 5a are propagating over shallow and variable-depth bathymetry, which will alter their propagation speeds. Taking this into account means estimating phase speeds along variable bathymetry, which can be integrated numerically over time to yield an estimate for the waves' trajectories in a travel-time plot. In turn, this allows the waves in the SAR to be compared with the propagation predictions computed from linear theory as given by a standard boundary value problem, assuming the Taylor-Goldstein (TG) equation with appropriate boundary conditions as follows:

$$\frac{d^2\phi}{dz^2} + \left[\frac{N(z)^2}{(U-c)^2} - \frac{\frac{d^2U}{dz^2}}{U-c} - k^2\right]\phi = 0, \ \phi(0) = \phi(-H) = 0 \tag{3}$$

In this formulation, ϕ is the modal structure, k is the wavenumber representing the nonhydrostatic term, U is the current velocity along the waves' direction of propagation, c is a characteristic phase speed, and H is the bottom depth. In Figure 5b, distance versus time simulations are presented for the first four vertical modes assuming a semidiurnal tidal frequency and wavenumbers as observed from the SAR (set nominally to 30 km). The linear predicted trajectories presented as solid lines in Figure 5b reveal that the first three modes are too steep when compared with the observations, whereas their best fit is along a mode-4 long IW (solid line in red). Note that an increased wave speed is expected for the SAR observations when compared with those obtained from Equation (3), because of an additional contribution from nonlinearity. This effective wave speed increase due to nonlinearity is usually around 20%, as documented in similar studies in other ISW hotspots (e.g., [2,27]), and it is indicated in Figure 5b as light red shades departing from the mode-4 solid red line. This corrected trajectory is quite consistent with the observations and propagation speeds around 0.7 m/s, just as estimated from the average spacing between the observations in Figure 3b. Interestingly, the solid red line representing a propagation trajectory consistent with a long mode-4 IW can be traced back in agreement with a generation close to where the downward beam interacts with the pycnocline, at a time when the tide transitions from maximum east to westward flow. We therefore believe this hypothesis is more consistent in light of the SAR, the in situ data, and the methodology described in Sections 1 and 2. We also note in passing that TW in Figure 5a yields very similar results (not shown). However, the Baines [28] body force (F) is slightly smaller around TW compared to TE, and we therefore do not pursue a generation hypothesis at TW here, leaving its detailed analysis for forthcoming investigations.

A closer inspection of Figure 5b indicates that the observations of mode-2 ISW-like waves extend for at least 100 km to the east, while enduring at least 48 h and 5 tidal cycles (assuming that the semidiurnal tide is at work). This is an unusually large time span when comparing with other mode-2 observations documented in the literature [2,46]. For instance, in da Silva et al. [2] some mode-2 ISW-like waves were found to be short-lived, not exceeding a single tidal-cycle, as were those documented off the coast of New Jersey with lifetimes typically less than a few hours [46]. The extended longevity observed in Figure 3b, and other images alike, could come from resonant conditions between the observed mode-2 wave-tails and larger mode-4 IWs (as indicated in Figure 5b). This hypothesis is inspired in recent studies such as Guo et al. [10] for the mode-1 wave-tails observed in the SAR, and larger mode-2 IWs that propagate along deeper water layers below in the South China Sea. In this particular stretch of the Andaman Sea, the effectiveness of this resonant coupling involving mode-2 short internal waves can be further investigated, even though the existence of a hypothetical mode-4 IW cannot be confirmed without dedicated in situ measurements or advanced numerical modelling.

Resonant conditions are usually discussed in the literature by comparing the phase speeds predicted according to Equation (3) for large IWs in hydrostatic mode (i.e., $k \approx 0$), with those predicted in a non-hydrostatic regime characteristic of the short-scale ISWs observed in the wave-tails (see [2,10,47]). In the case of Figure 3b, this means comparing large IWs with typical mode-4 wavelengths and mode-2 ISW-like wave-tails, which propagate along the direction of the red transect in Figure 1. Because the mode-2 ISW-like waves are long-lived, we choose to investigate resonant conditions along the waves' propagation path by comparing the characteristic phase speeds along the full extent of the IW field, as observed in the bottom profile in Figure 5a. This is done in Figure 5c for the same March stratification (see Figure 5d) by presenting phase speeds for the first four hydrostatic vertical modes (in solid black lines) along the bottom profile shown in Figure 5a, and the corresponding non-hydrostatic estimates assuming wavelengths between 500 and 1500 m, as observed in Figure 4 and in the SAR imagery listed in Table 1. Note that, while the hydrostatic case is somewhat sensitive to bathymetry changes, the shorter wavelengths in the non-hydrostatic cases are not—as expected based on linear IW theory. According to these results, the mode-2 ISW-like waves observed in the SAR have linear phase speed estimates (computed from Equation (3) and depicted in a light blue rectangle) close to both mode-3 and mode-4 long IWs (recalling their observed wavelengths is around 30 km in the SAR). However, they compare better with mode-4 when considering the full extent of the propagation path (especially to the East of 93°E). This means that the observations in the SAR are consistent with a resonant coupling predicted by linear theory between large mode-4 IWs and short-scale mode-2 ISW-like wave-tails. We also note that resonant conditions between mode-2 IWs and short mode-1 wave-tails (see grey shaded rectangle) exist, both along the Ten Degree Channel and farther south along the propagation path of the waves in Figure 3a. However, we reiterate that the SAR ground truth shows neither the usual large mode-1 ISW packets nor the mode-1 wave-tails along the Ten Degree Channel of the Andaman Sea.

According to the composite map in Figure 1, the waves' propagation envelope (depicted in light red) includes increasing eastward components than that corresponding to the profile in Figure 5a. For consistency then, ten more profiles corresponding to waves with more eastward components are analysed in Figure 5c, and correspond to the grey lines clustering around the hydrostatic mode-4 black curve which show that these too are fairly in resonant coupling with mode-2 ISW-like wave-tails. Furthermore, SAR shows these wave-tails to be observed, at least, from January to May in the Ten Degree Channel, and therefore similar plots were investigated for the remaining density profiles (climatological monthly means). In practice, this means having the pycnocline shifting some 50 m in the vertical and some 30% in the horizontal (see Figure 5d), while still yielding similar results to those in Figure 5c. This therefore adds to the robustness of the resonant coupling conditions in this region.

5. Summary and Conclusions

SAR imagery from TerraSAR-X is used to document the 2D horizontal structure of ISWs in the Andaman Sea between 8° and 11°N. According to the satellite imagery, two very distinct IW systems appear to be operating simultaneously in nearby stretches in this region. On the one hand, farther south the usual large mode-1 ISW packets are regularly observed, whereas mode-1 ISW wave-tails separated by long mode-2 wavelengths are also frequently observed, just as in other regions of the world's oceans. On the other hand, sea surface signatures in the SAR are very different along the Ten Degree Channel. In this case, no mode-1 ISW packets are apparent, while long-lived and regularly spaced mode-2 ISW-like wave-tails are frequently observed. These waves were analysed in light of additional in situ data and linear theory. While semidiurnal in nature, their generation appears to be more consistent with IT beams interacting with the local pycnocline, rather than resulting directly from the barotropic tide as it flows over the shallow ridges south of the Nicobar Islands. The unusually long-lived mode-2 wave-tails were also found consistent with a resonant coupling with larger mode-4 IWs. Altogether, we believe that this study region offers unique prospects as far as IW dynamics is concerned. In particular, several generation mechanisms could be operating simultaneously in close proximity, therefore providing SAR ground truth for forthcoming in situ measurements and/or numerical modelling. Evidence consistent with resonant coupling between different IW vertical structures in the Andaman Sea adds to an increasing awareness of the phenomenon in SAR observations, especially since in the Andaman Sea, they include simultaneous, multiple modes.

Supplementary Materials: The following are available online at http://www.mdpi.com/2072-4292/10/6/861/s1. S1: TerraSAR-X image shown in Figure 3a. S2: RGB composite from a MODIS-Terra image dated 2014/03/19. S3: TerraSAR-X image shown in Figure 3b.

Author Contributions: Both authors have equally contributed during all stages of this work.

Acknowledgments: The Authors would like to thank DLR projects OCE3154 and OCE2254 for providing the SAR images. This work was supported in part by FCT under Grant SFRH/BPD/84420/2012.

Conflicts of Interest: The authors declare no conflicts of interest.

References

1. Jackson, C.R.; Apel, J.R. *Synthetic Aperture Radar Marine User's Manual*; U.S. Department of Commerce: Washington, DC, USA, 2004; p. 464.
2. Da Silva, J.C.B.; Buijsman, M.C.; Magalhaes, J.M. Internal waves on the upstream side of a large sill of the Mascarene Ridge: A comprehensive view of their generation mechanisms and evolution. *Deep Sea Res.* **2015**, *99*, 87–104. [CrossRef]
3. Alford, M.H.; Peacock, T.; MacKinnon, J.A.; Nash, J.D.; Buijsman, M.C.; Centurioni, L.R.; Chao, S.Y.; Chang, M.H.; Farmer, D.M.; Fringer, O.B.; et al. The formation and fate of internal waves in the South China Sea. *Nature* **2015**, *521*, 65–69. [CrossRef] [PubMed]
4. Waterhouse, A.F.; MacKinnon, J.A.; Nash, J.D.; Alford, M.H.; Kunze, E.; Simmons, H.L.; Polzin, K.L.; St Laurent, L.C.; Sun, O.M.; Pinkel, R.; et al. Global patterns of mixing from measurements of the turbulent dissipation rate. *J. Phys. Oceanogr.* **2014**, *44*, 1854–1872. [CrossRef]

5. Kunze, E. Internal-Wave-Driven Mixing: Global Geography and Budgets. *J. Phys. Oceanogr.* **2017**, *47*, 1325–1345. [CrossRef]

6. Li, D.; Chou, W.-C.; Shih, Y.-Y.; Chen, G.-Y.; Chang, Y.; Chow, C.H.; Lin, T.-Y.; Hung, C.-C. Elevated particulate organic carbon export flux induced by internal waves in the oligotrophic northern South China Sea. *Nature* **2018**, *8*, 1–7. [CrossRef] [PubMed]

7. Jackson, C.R.; da Silva, J.C.B.; Jeans, G. The generation of nonlinear internal waves. *Oceanography* **2012**, *25*, 108–123. [CrossRef]

8. Nakamura, T.; Awaji, T.; Hatayama, T.; Akitomo, K.; Takizawa, T.; Kono, T.; Kawasaki, Y.; Fukasawa, M. The generation of large-amplitude unsteady lee waves by subinertial K1 tidal flow: A possible vertical mixing mechanism in the Kuril Straits. *J. Phys. Oceanogr.* **2000**, *30*, 1601–1621. [CrossRef]

9. Garrett, C.; Kunze, E. Internal tide generation in the deep ocean. *Annu. Rev. Fluid Mech.* **2007**, *39*, 57–87. [CrossRef]

10. Guo, C.; Vlasenko, V.; Alpers, W.; Stashchuk, N.; Chen, X. Evidence of short internal waves trailing strong internal solitary waves in the northern South China Sea from synthetic aperture radar observations. *Remote Sens. Environ.* **2012**, *124*, 542–550. [CrossRef]

11. Maury, M.F. *The Physical Geography of the Sea and Its Meteorology*; Harper: New York, NY, USA, 1861; p. 457.

12. Perry, R.B.; Schimke, G.R. Large-Amplitude Internal Waves Observed off the Northwest Coast of Sumatra. *J. Geophys. Res.* **1965**, *70*, 2319–2324. [CrossRef]

13. Osborne, A.R.; Burch, T.L. Internal Solitons in the Andaman Sea. *Science* **1980**, *208*, 451–460. [CrossRef] [PubMed]

14. Alpers, W.; Wang-Chen, H.; Hock, L. Observation of Internal Waves in the Andaman Sea by ERS SAR. IGARSS 97. *Remote Sens. Sci. Vis. Sustain. Dev.* **1997**, *4*, 1518–1520.

15. Jackson, C.R. *An Atlas of Internal Solitary-like Waves and Their Properties*, 2nd ed.; Global Ocean Associates: Alexandria, VA, USA, 2004; p. 560.

16. Yi-Neng, L.; Shi-Qiu, P.; Xue-Zhi, Z. Observations and Simulations of the Circulation and Mixing around the Andaman-Nicobar Submarine Ridge. *Atmos. Ocean. Sci. Lett.* **2012**, *5*, 319–323. [CrossRef]

17. Shimizu, K.; Nakayama, K. Effects of topography and Earth's rotation on the oblique interaction of internal solitary-like waves in the Andaman Sea. *J. Geophys. Res.* **2017**, *122*, 1–17. [CrossRef]

18. Jackson, C.R. An empirical model for estimating the geographic location of nonlinear internal solitary waves. *J. Atmos. Ocean. Technol.* **2009**, *26*, 2243–2255. [CrossRef]

19. Magalhaes, J.M.; da Silva, J.C.B.; Buijsman, M.C.; Garcia, C.A.E. Effect of the North Equatorial Counter Current on the generation and propagation of internal solitary waves off the Amazon shelf (SAR observations). *Ocean Sci.* **2016**, *12*, 243–255. [CrossRef]

20. Buijsman, M.C.; Ansong, J.K.; Arbic, B.K.; Richman, J.G.; Shriver, J.F.; Timko, P.G.; Wallcraft, A.J.; Whalen, C.B.; Zhao, Z.X. Impact of parameterized internal wave drag on the semidiurnal energy balance in a global ocean circulation model. *J. Phys. Oceanogr.* **2016**, *46*, 1399–1419. [CrossRef]

21. Vlasenko, V.; Stashchuk, N.; Guo, C.; Chen, X. Multimodal structure of baroclinic tides in the South China Sea. *Nonlinear Process. Geophys.* **2010**, *17*, 529–543. [CrossRef]

22. Da Silva, J.C.B.; Magalhaes, J.M. Internal solitons in the Andaman Sea: A new look at an old problem. *Proc. SPIE* **2016**. [CrossRef]

23. Alpers, W. Theory of radar imaging of internal waves. *Nature* **1985**, *314*, 245–247. [CrossRef]

24. Da Silva, J.C.B.; Ermakov, S.A.; Robinson, I.S.; Jeans, D.R.G.; Kijashko, S.V. Role of surface films in ERS SAR signatures of internal waves on the shelf: 1. Short-period internal waves. *J. Geophys. Res.* **1998**, *103*, 8009–8031. [CrossRef]

25. Da Silva, J.C.B.; Ermakov, S.A.; Robinson, I.S. Role of surface films in ERS SAR signatures of internal waves on the shelf: 3. Mode transitions. *J. Geophys. Res.* **2000**, *105*, 24089–24104. [CrossRef]

26. Jackson, C.R.; da Silva, J.C.B.; Jeans, G.; Alpers, W.; Caruso, M.J. Nonlinear Internal Waves in Synthetic Aperture Radar Imagery. *Oceanography* **2013**, *26*, 68–79. [CrossRef]

27. Da Silva, J.C.B.; New, A.L.; Magalhaes, J.M. On the structure and propagation of internal solitary waves generated at the Mascarene Plateau in the Indian Ocean. *Deep Sea Res.* **2011**, *58*, 229–240. [CrossRef]

28. Baines, P.G. On internal tide generation models. *Deep Sea Res.* **1982**, *29*, 307–338. [CrossRef]

29. Sherwin, T.J.; Vlasenko, V.I.; Stashchuk, N.; Jeans, D.R.G.; Jones, B. Along-slope generation as an explanation for some unusually large internal tides. *Deep Sea Res.* **2002**, *49*, 1787–1799. [CrossRef]

30. Azevedo, A.; da Silva, J.C.B.; New, A.L. On the generation and propagation of internal solitary waves in the southern Bay of Biscay. *Deep Sea Res.* **2006**, *53*, 927–941. [CrossRef]
31. Da Silva, J.C.B.; New, A.L.; Azevedo, A. On the role of SAR for observing local generation of internal solitary waves off the Iberian Peninsula. *Can. J. Remote Sens.* **2007**, *33*, 388–403. [CrossRef]
32. Magalhaes, J.M.; da Silva, J.C.B. SAR observations of internal solitary waves generated at the Estremadura Promontory off the west Iberian coast. *Deep Sea Res.* **2012**, *69*, 12–24. [CrossRef]
33. Da Silva, J.C.B.; New, A.L.; Magalhaes, J.M. Internal solitary waves in the Mozambique Channel: Observations and interpretation. *J. Geophys. Res.* **2009**, *114*, C05001. [CrossRef]
34. Egbert, G.D.; Erofeeva, S.Y. Efficient inverse modeling of barotropic ocean tides. *J. Ocean. Atmos. Technol.* **2002**, *19*, 183–204. [CrossRef]
35. Smith, W.H.F.; Sandwell, D.T. Global sea floor topography from satellite altimetry and ship depth soundings. *Science* **1997**, *277*, 1956–1962. [CrossRef]
36. Leblond, P.H.; Mysak, L.A. *Waves in the Ocean*; Elsevier: New York, NY, USA, 1978; Volume 20, p. 602.
37. New, A.L.; Pingree, R.D. Large-amplitude internal soliton packets in the central Bay of Biscay. *Deep Sea Res.* **1990**, *37*, 513–524. [CrossRef]
38. New, A.L.; Pingree, R.D. Local generation of internal soliton packets in the central Bay of Biscay. *Deep Sea Res.* **1992**, *39*, 1521–1534. [CrossRef]
39. Gerkema, T. Internal and interfacial tides: Beam scattering and local generation of solitary waves. *J. Mar. Res.* **2001**, *59*, 227–255. [CrossRef]
40. Akylas, T.R.; Grimshaw, R.H.J.; Clark, S.R.; Tabaei, A. Reflecting tidal wave beams and local generation of solitary waves in the ocean thermocline. *J. Fluid Mech.* **2007**, *593*, 297–313. [CrossRef]
41. Grisouard, N.; Staquet, C.; Gerkema, T. Generation of internal solitary waves in a pycnocline by an internal wave beam: A numerical study. *J. Fluid Mech.* **2011**, *676*, 491–513. [CrossRef]
42. Mercier, M.J.; Mathur, M.; Gostiaux, L.; Gerkema, T.; Magalhaes, J.M.; da Silva, J.C.B.; Dauxois, T. Soliton generation by internal tidal beams impinging on a pycnocline: Laboratory experiments. *J. Fluid Mech.* **2012**, *704*, 37–60. [CrossRef]
43. New, A.L.; da Silva, J.C.B. Remote-sensing evidence for the local generation of internal soliton packets in the central Bay of Biscay. *Deep Sea Res.* **2002**, *49*, 915–934. [CrossRef]
44. Thompson, D.R.; Gasparovic, R.F. Intensity modulation in SAR images of internal waves. *Nature* **1986**, *320*, 345–348. [CrossRef]
45. Vlasenko, V.; Stashchuk, N.; Hutter, K. *Baroclinic Tides: Theoretical Modeling and Observational Evidence*; Cambridge University Press: New York, NY, USA, 2005; p. 351.
46. Shroyer, E.L.; Moum, J.N.; Nash, J.D. Mode 2 waves on the continental shelf: Ephemeral components of the nonlinear internal wave field. *J. Geophys. Res.* **2010**, *115*, C07001. [CrossRef]
47. Pingree, R.D.; Mardell, J.T.; New, A.L. Propagation of internal tides from the upper slopes of the Bay of Biscay. *Nature* **1986**, *321*, 154–158. [CrossRef]

remote sensing

MDPI

Review

TerraSAR-X and Wetlands: A Review

Christian Wohlfart [1,*], Karina Winkler [1,2,3], Anna Wendleder [1] and Achim Roth [1]

1   German Remote Sensing Data Center (DFD), German Aerospace Center (DLR), Muenchener Strasse 20,
    82234 Wessling, Germany; karina.winkler@dlr.de (K.W.); anna.wendleder@dlr.de (A.W.);
    achim.roth@dlr.de (A.R.)
2   Company for Remote Sensing and Environmental Research (SLU), Kohlsteiner Strasse 5,
    81243 Munich, Germany
3   Institute of Meteorology and Climate Research (IMK-IFU), Karlsruhe Institute of Technology (KIT),
    Kreuzeckbahnstr. 19, 82467 Garmisch-Partenkirchen, Germany
*   Correspondence: christian.wohlfart@dlr.de; Tel.: +49-8153-28-3418

Received: 17 April 2018; Accepted: 8 June 2018; Published: 10 June 2018

Abstract: Since its launch in 2007, TerraSAR-X observations have been widely used in a broad range of scientific applications. Particularly in wetland research, TerraSAR-X's shortwave X-band synthetic aperture radar (SAR) possesses unique capabilities, such as high spatial and temporal resolution, for delineating and characterizing the inherent spatially and temporally complex and heterogeneous structure of wetland ecosystems and their dynamics. As transitional areas, wetlands comprise characteristics of both terrestrial and aquatic features, forming a large diversity of wetland types. This study reviews all published articles incorporating TerraSAR-X information into wetland research to provide a comprehensive study of how this sensor has been used with regard to polarization, and the function of the data, time-series analyses, or the assessment of specific wetland ecosystem types. What is evident throughout this literature review is the synergistic fusion of multi-frequency and multi-polarization SAR sensors, sometimes optical sensors, in almost all investigated studies to attain improved wetland classification results. Due to the short revisiting time of the TerraSAR-X sensor, it is possible to compute dense SAR time-series, allowing for a more precise observation of the seasonality in dynamic wetland areas as demonstrated in many of the reviewed studies.

Keywords: synthetic aperture radar; X-band; marine; estuarine; lacustrine; riverine; palustrine; time-series; SAR applications; vegetation; remote sensing data

1. Introduction

Wetlands are among the most biologically diverse and productive ecosystems in the world. Home to a large variety of floral and faunal communities, wetlands provide essential ecosystem services and goods of great ecological and economic value to millions of people [1–4]. Their services include water quality preservation, erosion control and sediment retention, groundwater recharge, reducing risks of storm surges and flooding, and climate change mitigation [5–7]. Covering approximately 6% of the Earth's surface, wetland soils hold a disproportionate amount of carbon, accounting for around 12% of global carbon stocks [6,8], and therefore play a critical role in global carbon dynamics [9].

Despite their recognition as one of the most valuable natural resources, major wetland areas have irreversibly diminished or deteriorated and subsequently lost their functional properties with significant repercussions on ecological, economic, social, and cultural benefits [10]. During the 20th and 21st centuries, around 64% of the natural wetlands have been vanished and continue to disappear at an annual loss rate of 1%, although this pace is slowing [11]. Anthropogenic factors are the main reason for the disappearing wetland resources, including the conversion of wetlands to urban or agricultural land, damming, or pollution, being more pronounced in inland ecosystems in light of the aforementioned benefits. To counteract the continued decline, there is an increasing need to detect and monitor the ecosystem remnants. Precise spatio-temporal information about wetland cover and their dynamics are central components for establishing effective conservation strategies, and help to understand the underlying drivers of change [12,13].

The spatial diversity and complexity of wetland ecosystems around the world make wetland identification and mapping a challenging task. Over the past several decades, the domain of remote sensing techniques has greatly advanced in terms of sensor variety and derived products to exploit new fields of application to understand Earth surface cover and dynamics [14]. A fleet of various remote sensing sensors, including both optical and radar, has been used to identify, delineate, and classify the extent and quality of wetland ecosystems on different spatial scales, given their ability to monitor large areas in a comparatively short period of time [15–20]. So far, scientific studies have put more emphasis on optical sensors (Figure 1), although radar systems offer great potential in delineating unique wetland characteristics, including species composition, water surface detection, vegetation health, or structural properties [17,21,22].

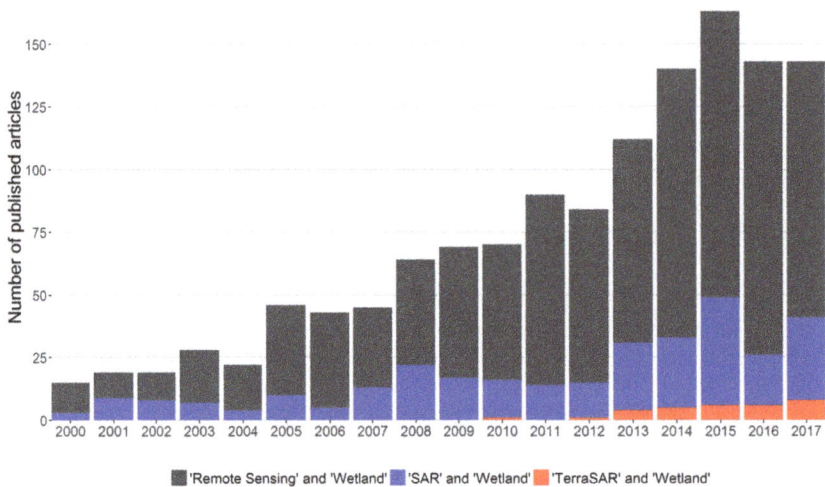

Figure 1. Number of wetland publications using remote sensing (gray), synthetic aperture radar (SAR, blue), and TerraSAR-X (red) data from 2000 to 2017.

Synthetic aperture radar (SAR) technology provides complementary and supplementary information to multi- and hyperspectral or thermal sensors, since, in the microwave spectrum, radar frequently monitors at the increased spatial resolution necessary for mapping wetlands. In contrast to sensors operating in the optical and thermal range, pulsed SAR signals can penetrate the clouds and haze that are very common in cloud-prone wetland regions, as well as image surfaces during low-light conditions in near-real-time [17,23]. Furthermore, SAR has the ability, to varying degrees, to penetrate vegetation canopies for the delineation of understory information, and is sensitive to dielectric properties such as soil moisture and inundation level. Evidently, scientific reviews about the radar remote sensing of wetland ecosystems began appearing in the literature in the 1990s [17,22,24–28]. These published studies published describe the optimal relationship between SAR specifications (wavelength, polarization, incidence angle, spatial resolution) and specific ground features of wetlands (e.g., structure, roughness, dielectric constant). Since the launch of the first spaceborne SAR missions including NASA's Spaceborne Imaging Radar (SIR) and ESA's European Remote Sensing (ERS) satellites [29], many studies have used this new source of data for wetland applications. With the recent development of very-high-resolution SAR sensors such as RADARSAT-2 and TerraSAR-X, new possibilities in wetland research have emerged. TerraSAR-X provides unprecedented opportunities for wetland assessment, since its X-band signal (3.1 cm wavelength) has proven suitable for observing tidal flats [30] or herbaceous wetlands with higher levels of detail and resolutions than other SAR sensors [31].

To date, there is no comprehensive literature review about the potential and range of applications TerraSAR-X provides with regard to wetland analyses. This paper aims to close this gap by providing an update to existing wetland-radar review studies explicitly describing wetland-related research with a specific focus on high-resolution and high-frequency TerraSAR-X data. The next section briefly introduces the wetland types and characterization we used in this study followed by a section about the methodology in the review process. Then, SAR principles and backscatter mechanisms in mapping wetland characteristics are pictured. Section 5 provides the reader with a detailed and comprehensive overview of how TerraSAR-X data have been used in wetland-related research, including sensor specifications, function of TerraSAR-X data and wetland categories. The paper finishes with some concluding remarks.

2. Wetland Concept, Types, and Characteristics

Wetlands are often characterized as transitional ecotones forming at the interface of terrestrial and aquatic ecosystems and, hence, holding features of both [32]. While variations in appearance and species composition may be high, temporary or permanent inundation by water is a common feature that shapes plant adaption, soil processes, as well as animal life [33,34]. The Ramsar Convention established a broad concept of wetlands in 1971, which represents the most commonly used, internationally accepted wetland definition. According to this, wetlands are described as

> "areas of marsh, fen, peatland or water, whether natural or artificial, permanent or temporary, with water that is static or flowing, fresh, brackish or salt, including areas of marine water the depth of which at low tide does not exceed six metres" [35].

Table 1. Overview of characteristics and functions of major natural wetland types according to the Ramsar international wetland classification [36,37] and TerraSAR-X-related studies, respectively (based on: [28,33,34]).

Wetland Type	Characteristics	Functions	Selected Studies Using TerraSAR-X
Marine and Coastal Wetlands (Mostly Marine and Estuarine)			
G: Intertidal (salt) flats	Non-vegetated shore with tidal flooding (salt restricting plant growth)	Shoreline stabilization; Nutrient and sediment retention; Vital habitats for clams, crabs, and juvenile fish	Gade et al. (2014 [38], 2018 [39]); Gade & Melchionna (2016) [40]; Kim et al. (2011) [41]; Jung et al. (2015) [42]
H: Intertidal (salt) marshes, freshwater marshes	Herb-dominated shores with tidal/long-term flooding during the year (often with halophytic plants) and hydric soils	Carbon sequestration; Nutrient and sediment retention; Water-quality renovation; Wildlife habitat	Lee et al. (2011) [43]; Da Lio et al. (2018) [44]
I: Intertidal forested wetlands (mangrove and tidal freshwater swamps)	Tree-dominated shores with tidal/long-term flooding during the year and hydric soils	Carbon sequestration; Valuable habitat for fish and crustacean species; Coastal protection; Water-quality renovation; Production of timber	Hong et al. (2010 [45], 2015 [46]); Lagomasino et al. (2016) [47]
Inland Wetlands (Lacustrine, Riverine, Palustrine)			
Peatlands	Peat-forming soils	Carbon sequestration in the form of peat; Provision of substrate; Bird breeding and wildlife habitat	
U: Non-forested peatlands (bogs, swamps, fens)	Dominated by moss, sedges, and grasses		Morishita & Hansen (2015) [48]; Mahdianpari et al. (2017) [49]; Regmi et al. (2012) [50]
Xp: Forested peatlands (peatswamp forest)	Tree-dominated		Wijaya et al. (2010) [51]; Englhart et al. (2011) [52];
Marshes and swamps	Inorganic soils	Carbon sequestration; Water-quality renovation; Bird breeding and wildlife habitat; Production of timber	Irwin et al. (2017) [53]; Betbeder et al. (2014) [31]; Parker et al. (2017) [54]
S/T: Marshes/pools	saline/brackish/alkaline or freshwater, herb-dominated		
W: Shrub-dominated wetlands			
Xf: Tree-dominated wetlands			

Due to the diversity of wetlands that exist around the globe, they can be categorized according to numerous classification schemes relying on biological, physical, chemical, hydrogeomorphological variables or functions [34]. Although many wetland classification systems have been developed by scientists at regional and national levels, the Ramsar Convention on Wetlands established the first globally consistent classification framework [37] in order to characterize wetlands of international importance [34,55]. Generally, five major wetland types are distinguished according to their landscape features and formation [56].

- **marine** (coastal wetlands such as lagoons, shores, and coral reefs)
- **estuarine** (deltas, tidal marshes, mudflats, and mangrove swamps)
- **lacustrine** (wetlands associated with lakes)
- **riverine** (wetlands along rivers and streams)
- **palustrine** ("marshy" wetlands like marshes, swamps, and bogs)

Additionally, lhuman-made wetlands such as aquaculture ponds, salt pans, dams, and reservoirs are recognized. The Convention has adopted these types in its Ramsar Classification of Wetland Types, which comprises 42 types grouped into the categories Marine and Coastal Wetlands, Inland Wetlands, and Human-Made Wetlands [56]. This has gained widespread acceptance, and thus is used as a foundation for categorizing wetland types in this study. Table 1 displays a summarized representation of major wetland types, their description, functions, and TerraSAR-X related studies, respectively. Since the focus was laid on natural wetlands, studies focusing explicitly on human-made wetlands such as aquacultures and irrigated areas were excluded from the analysis.

3. Data and Methodology

Both the comprehensive Scientific Citation Index Extented (SCIE) database from Web of Science and Elsevier's Scopus database were used for the literature review. To get an overview of remote sensing-based studies related to wetlands as well as those using SAR and explicitly TerraSAR-X data, we applied a search based on the keyword string combinations of "wetland" with "remote sensing", "SAR", and "TerraSAR-X", respectively.

The scientific literature related to wetlands using remote sensing applications is rapidly growing. Studies using radar remote sensing still represent a minor part of all remote sensing-based publications on wetlands. However, numbers have continuously increased since 2000. Especially in the last decade, the usage of newly available sensors such as TerraSAR-X has steadily grown and contributed to this development (see Figure 1).

In order to obtain a consistent database of published articles on the remote sensing of wetlands using TerraSAR-X data, we used a combination of "TerraSAR-X" with of one of the following keywords: "wetland", "mangrove", "peat", "bog", "fen", "marsh", and "swamp". After revising the entries and excluding papers without the usage of TerraSAR-X images or without a focus on (semi-)natural wetlands, a total of 32 articles were obtained. Table A1 in the Appendix summarizes all selected studies for this review.

Figure 2 shows the spatial distribution of study sites from the selected publications and major wetland areas.

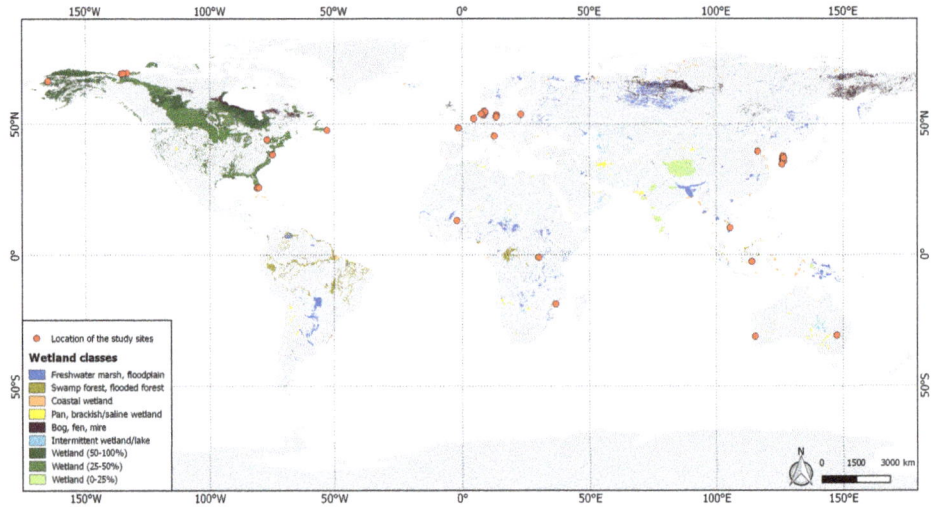

Figure 2. Generalized global distribution of wetland types based on Lehner et al. [57] and location of the study sites of published TerraSAR-X-related articles.

4. SAR Principles for Mapping Wetland Characteristics

SAR is a common imaging radar technology, and has been widely used in wetland-related studies. As an active sensor, SAR measures the receipt (backscatter) of an actively transmitted pulse, and therefore offers unique characteristics that make it ideal for mapping and monitoring wetland features. The electromagnetic spectrum is sampled at much longer wavelengths compared to optical sensors, which usually operate in the visible and infrared spectrum. The ability of radar to penetrate through clouds makes this sensor particularly valuable for the detection of wetlands, which are usually distributed across areas with frequent cloud coverage. Furthermore, radar backscatter is sensitive to dielectric properties including inundation level, soil moisture, or salinity, which are common features of wetland ecosystems [17]. SAR data are often categorized by their wavelength (e.g., X-band with 3.1 cm, C-band with 5.6 cm, and L-band with 23.6 cm) or polarization (e.g., HH = horizontally transmitted and received and VV = vertically transmitted and received). The pulsed radar beam is transmitted at different angles relative to the surface (look angles). Thus, a radar image provides structural and textural information about the land surface [58]. The relationship of these sensor parameters to the scattering mechanisms of wetland vegetation types have been shown in many scientific reviews [17,22,24–26,28]. Figure 3 provides a simplified illustration of the basic backscatter mechanisms related to different wetland characteristics for the most commonly used SAR bands. Four general backscattering types can be distinguished:

- **surface scattering** (rough surface reflects transmitted radiation in all directions)
- **specular scattering** (smooth surface reflects radiation away from the sensor)
- **volume scattering** (radiation is both reflected and refracted/transmitted in an object or volume)
- **double-bounce scattering** (radiation is reflected on two perpendicular planes)

When there is dry soil cover accompanied by herbaceous vegetation or small shrubs which create a rougher surface, surface reflection is the dominant scattering mechanism at both short and long wavelengths, whereas longer wavelengths show a greater penetration into the dry soil. Denser vegetation can also result in volume scattering. On smoother surfaces, the reflection becomes specular for shorter and longer wavelengths. Smooth surfaces are calm water where the signal is reflected away from the sensor due to the dielectric constant or bare soils where the surface roughness is smaller than the wavelength of the sensor. For flooded vegetation such as reed, short wavelength scatter is dominated by volume scattering in dense vegetation structure. For example, X-band backscatter mainly interacts with leaves, twigs, and small branches, and thus is sensitive to herbaceous vegetation [52]. In addition, double-bounce can occur between two smooth surfaces, such as water and vertical emergent vegetation. In contrast, longer wavelengths penetrate through herbaceous vegetation, resulting in specular scattering, assuming little water roughness. Forests cause volume scattering within dense canopies at shorter wavelengths, because the electromagnetic signal is not able to penetrate the crown cover. Longer wavelengths appear to scatter from branches and tree structures in combination with double-bounce backscatter, whereas inundated forests amplify the double-bounce scatter. Another important parameter which influences the scattering mechanism is the incidence angle, which describes the angle between the radar beam and the vertical (normal) direction to the intercepting surface. The selection of an appropriate incidence angle depends on the research goal [59]. Smaller incidence angles (less than 28 degrees) are able to better penetrate vegetation and can potentially map sub-canopy water. Furthermore, smaller incidence angles reduce the effect of shadow, particularly in regions with distinct topography [60,61].

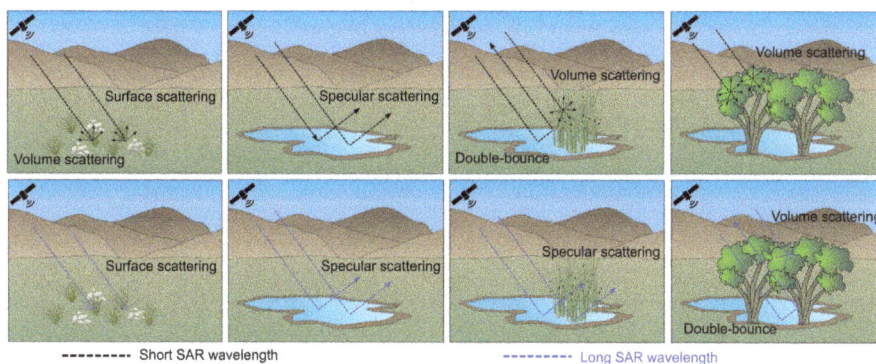

Figure 3. Simplified schematic illustration of the dominant backscatter mechanisms for short (e.g., X, 3.1 cm) and long (e.g., L, 23.6 cm) wavelengths for various surface cover conditions related to wetland.

5. TerraSAR-X and Wetlands

TerraSAR-X was launched into orbit on 15 June 2007, has been fully operational since January 2008, and is the first German operational radar satellite mission [62]. The X-band radar sensor offers a wide range of beam modes, allowing it to record images with different swath widths, resolutions, polarizations, and incidence angles. Using its active radar antenna, TerraSAR-X is able to produce image data ranging from medium-scale imaging in ScanSAR and StripMap mode to high-resolution small-scale application in the SpotLight mode, with a spatial resolution of down to one meter (see Figure 4), allowing for new research perspectives in various fields, including hydrology, geology, oceanography, or ecology [63].

Particularly in the field of wetlands, research requires the highest spatial resolution to characterize the small-scale heterogeneous nature of wetland ecosystems. Additionally, due to its repeat cycle of 11 days, TerraSAR-X provides the opportunity of acquiring high-resolution time-series of radar images. This holds the potential to strongly improve the accuracy of wetland assessments, in which data acquisition timing is a critical issue, given the significant temporal variability of most wetland types [34].

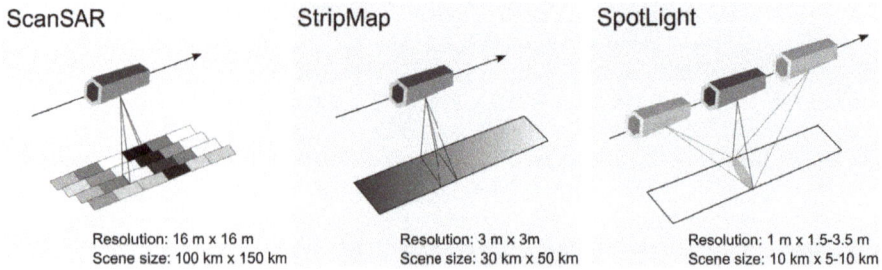

ScanSAR **StripMap** **SpotLight**

Resolution: 16 m x 16 m
Scene size: 100 km x 150 km

Resolution: 3 m x 3m
Scene size: 30 km x 50 km

Resolution: 1 m x 1.5-3.5 m
Scene size: 10 km x 5-10 km

Figure 4. TerraSAR-X basic imaging modes and main configurations [62,63].

5.1. How Has TerraSAR-X Been Used in Wetland Studies?

It is important to appreciate differences in these instrument specifications because the selection of optimal specifications must be tailored to different applications to help ensure project success. Each study design and research question related to wetlands requires particular image geometries, polarimetric information, and extent of data on spatial and temporal scale. The following subsections describe the different usage of TerraSAR-X images, focusing on sensor specifications such as imaging modes, polarization, incidence angle, and amount of data sets, acquisition, and the respective function of images used in wetland studies.

5.1.1. Sensor Specifications

The majority of the analyzed studies (21 from 32 articles, see Figure 5) used TerraSAR-X images in StripMap mode, which provides a good trade-off between resolution and coverage. As an advantage, even very small land cover patches can be detected in classification approaches to enable an accurate estimation of wetland extent in large heterogeneous areas. The SpotLight imaging mode was utilized in seven studies, indicating that the advantages of high spatial resolution data, especially for the classification of intertidal areas such as the detection of mussel beds, were investigated [38,40]. ScanSAR images with a wide area coverage and a reduced resolution were used least amongst the selected TerraSAR-X-related articles, and were mainly acquired for regional-scale analysis [64].

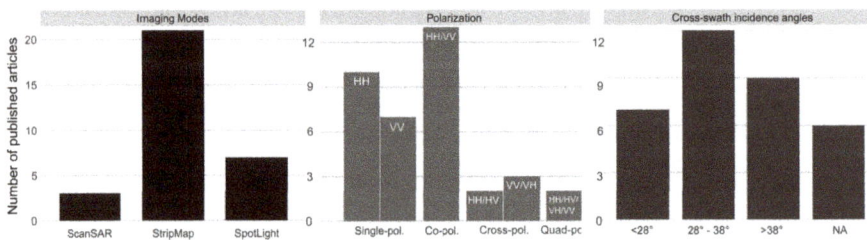

Figure 5. Sensor specifications (imaging modes, polarization, cross-swath incidence angles) of TerraSAR-X data for wetland studies.

The majority of all wetlands publications (20) used dual-polarized TerraSAR-X images, with emphasis on HH/VV co-polarizations. Wetlands in the form of flooded vegetation are expected to show a strong double-bounce backscattering. This mechanism can only be identified by analyzing dual (HH/VV, HH/VH, or VV/VH) or quad polarized (all four polarizations, quad-pol) datasets. Multi-polarized images enable the exploitation of polarimetric parameters as an additional input for wetland mapping [31,65,66]. The added value of multi-polarizations stands out in the distribution of polarizations used in TerraSAR-X-related wetland studies (see Figure 5). The enhanced backscatter interactions of the horizontally (HH) with little increase in the vertically polarized (VV) between flooded surface and wetland vegetation allows the identification of inundation, and additionally ensures a stable phase relation [66,67]. Thus, the usage of co-polarized images (13) predominates within wetland studies. With respect to single-polarized data, HH (10) slightly outweighed VV (7) polarization. HH-polarized SAR data are less sensitive to wind-induced surface roughness in open water and show a stronger contrast between flooded and non-flooded areas, and were consequently favored over VV [68]. Quad-polarization data were least used (2), and play a minor role in wetland studies, since the quad-polarimetric mode was only a short-term experimental phase for TerraSAR-X and the availability of acquisitions is limited [69,70]. Still, exploiting quad-polarization parameters for wetland classification showed promising results for wetland mapping in the Biebrza Floodplain, Poland, compared to other sensors and polarizations [71]. Notably, 9 from 33 wetland studies used a combination of different polarizations for a comparative analysis of input parameters for wetland classification.

As already stated in the previous chapter, the incidence angle can have a significant effect on the radar signal related to wetland research. We categorized the incidence angles in three main categories according to their degrees across all swath modes: small (lower than 28 degrees); intermediate (between 28 and 38 degrees); and large (above 38 degrees). Six studies did not provide any information about the incidence angle. The majority of all publications (12) applied the intermediate incidence angle range, which is well-suited for land applications with no major relief. Studies using large (9) or small (8) incidence angles tend to be balanced.

However, in many research projects, the configuration used additionally depends on the frequency of the acquired mode, and thus does not necessarily constitute the usefulness of that particular sensor specification. This constrains a general comparison of sensor configurations in terms of their benefits.

5.1.2. Usage of TerraSAR-X Data

The number of TerraSAR-X acquisitions used for wetland studies differed depending on the extent of the study area and the analyzed time frame. Most studies (13) used between two and five images for wetland mapping. However, 11 articles utilized more than 20 TerraSAR-X images, most frequently for multi-temporal analyses. With a total of 143 scenes, the study of Da Lio et al. [44] used the highest number of TerraSAR-X acquisitions within a single study.

The majority of wetland articles (19 studies) exploited time-series of TerraSAR-X data (see Figure 6). Multi-temporal images are an ideal tool set for capturing changing wetland dynamics, classifying and differentiating seasonally dependent vegetation [67]. Given the high repetition time of TerraSAR-X (11 days), high-resolution time-series of images are potentially available for any desired region, in contrast to other high-resolution SAR sensors.

Often, TerraSAR-X acquisitions were combined or fused with other remote sensing sensors. Optical data from Landsat (seven articles) and very-high-resolution (VHR) optical sensors (six articles) were used most frequently as an additional source of information or for validation purposes (see Figure 6). Likewise, images from other radar sensors such as RADARSAT-2, ALOS-PALSAR (six articles each), and Sentinel-1 (three articles) found wide use.

Regarding the methods used in TerraSAR-X-based wetland studies, the direct analysis of backscatter intensity (amplitude) and the application of polarimetric SAR (PolSAR) were in balance (13 articles each, see Figure 6). On the one hand, the intensity of radar backscatter (amplitude) was frequently taken as an input parameter for wetland classification, in most cases by applying a certain threshold for class discrimination (e.g., [72]). On the other hand, multiple articles took advantage of PolSAR techniques as a sophisticated approach for discriminating different land cover and wetland types according to its respective dominant scattering mechanisms [46,73]. Interferometric SAR (InSAR) is another method applied on TerraSAR-X images (six articles). For wetland environments, it is mostly utilized to detect water-level changes or ground deformation processes due to a lowering groundwater table (e.g., [74]).

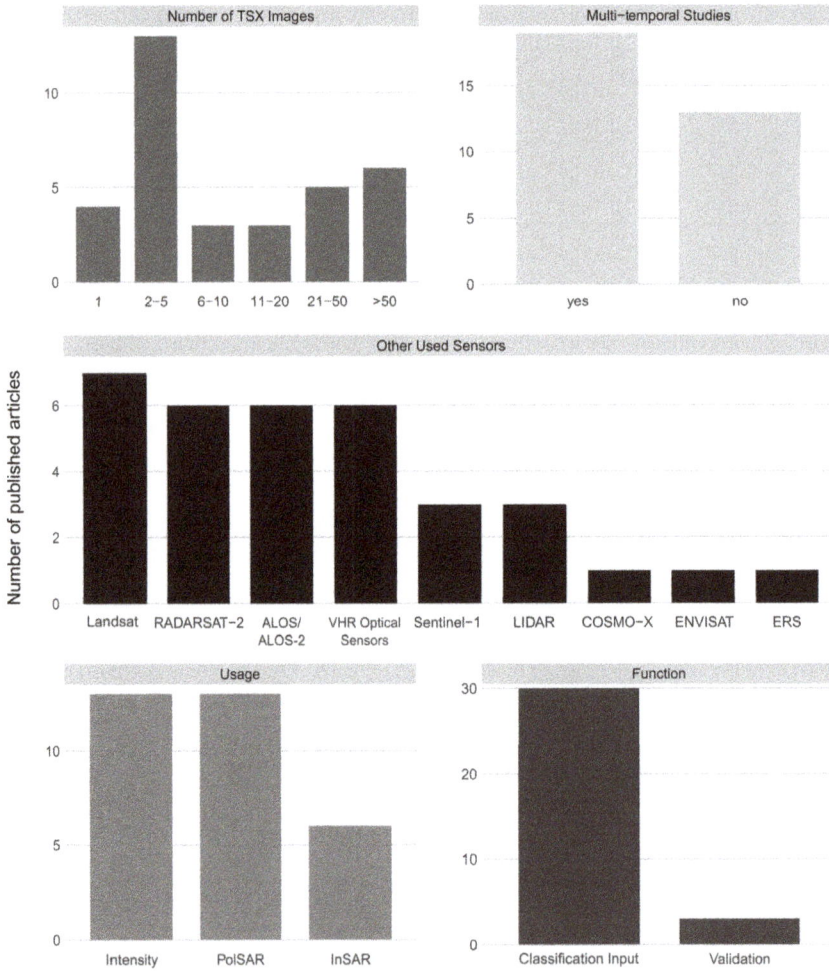

Figure 6. Usage of TerraSAR-X (TSX) data for wetland studies: Amount of images, multi-temporal analysis, other sensors combined, methodological approach, and function of the data. InSAR: interferometric SAR: PolSAR: polarimetric SAR.

Almost all of the selected wetland studies used TerraSAR-X images as a main input for land cover or wetland classification (30 articles). Only three papers utilized them as reference data for validation.

5.2. Delineation and Characterization of Various Wetland Types

As previously stated in Section 2, wetland ecosystems comprise spatially and temporally heterogeneous structures, which according to the Ramsar Classification Scheme can be categorized into marine, estuarine, lacustrine, riverine, and palustrine [36,37]. They share similar hydrologic, hydrogeomorphologic, chemical, or biological features. We classified the selected studies according to these generalized classes. The number of published articles ranked by wetland category is displayed in Figure 7. Some studies contained multiple wetland classes, which have been assigned to the predominant category within the study. Marine and estuarine ecosystems received the most scientific attention, with 11 and 10 published articles, respectively. There were only six and four TerraSAR-X articles incorporating the lacustrine and palustrine wetland classes, respectively.

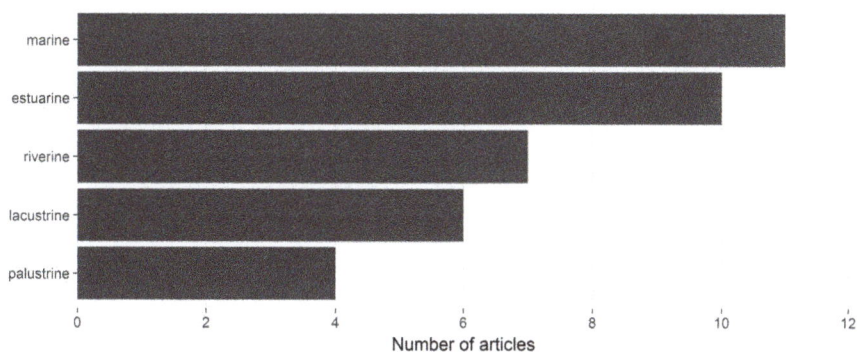

Figure 7. Number of studies categorized by wetland type.

5.2.1. Palustrine Wetlands

Palustrine ecosystems represent the "traditional" image of wetlands including marshes, swamps, and bogs. They comprise all non-tidal and non-riverine wetlands that are vegetated by trees, shrubs, or emergent herbaceous plants, as well as all tidal wetlands where the concentration of ocean-derived salts is less than 0.5 ppt [75]. In the following, all published articles related to palustrine wetland ecosystems are introduced in chronological order.

The synergistic use of TerraSAR-X and optical Landsat ETM+ data for delineating heterogeneous landscape features in a cloud-prone area (Uganda) was demonstrated by the study of Otukei et al. [72]. The authors fused a single-polarized (HH) TerraSAR-X band and all spectral Landsat bands acquired in 2009 by applying high-pass filtering (HPF) and principal component analysis (PCA). The latter fusion technique was further divided into a PCA with band substitution and a wavelet transform. The mono-temporal classification was based on a C 4.5 decision tree to predict a total of 13 land cover classes, including 3 wetland classes. Accuracy assessment was performed using optical high-resolution IKONOS images. Overall, the PCA with wavelet transform showed highest accuracies reaching approximately 85%, while the wetland classes ranged from 68% for degraded wetlands to 80% for papyrus vegetation. The authors suggested the exploration of object-based methods for further analysis.

A comparison of dense intra-annual time-series (from 2010) of high-resolution spatial data from multi-spectral RapidEye and HH-polarized TerraSAR-X data to classify various small-scale wetland vegetation habitats in Germany was conducted by Schuster et al. [76]. The high temporal resolution of the remote sensing data is able to identify different vegetation classes using the inherent phenological properties. In general, both sensors attained very high overall accuracies, greater than 90%. TerraSAR-X time series showed better classification results particularly for *Arrhenaterion elatioris*, *Solidago canadensis*, and *Phragmitetum australis*. Due to their distinct and fast-growing vertical structure, these plant communities return a clear trajectory in TerraSAR's X-band. In contrast, the low structural growth of *Caricion elatae* and *Calthion palustris* resulted in lower accuracies compared to the optical data. In summary, this study demonstrated the reliable application of very high spatial and temporal resolution optical or X-band SAR data for a wide range of different grassland ecosystems.

Dabrowska-Zielinska et al. [64] explored the potential of interferometric Sentinel-1 data to assess carbon flux and soil moisture over the Biebrza Wetland in north-eastern Poland. In this context, three ScanSAR TerraSAR-X VV scenes combined with one Landsat 8 OLI acquisition served as classification input to delineate various wetland habitat types, such as reeds, sedge-moss, sedges, or grassland. The supervised maximum likelihood classification yielded accuracies of around 87%. Furthermore, class-specific CO_2 flux and soil moisture were modeled based on interferometric Sentinel-1 data.

A multi-hierarchical classification framework for discriminating wetland types on the Avalon Peninsula (Canada) based on multi-polarization and multi-frequency SAR data was developed by Mahdianpari et al. [49]. They integrated single-polarized TerraSAR-X (HH), dual-polarized ALOS-PALSAR-2 (HH/HV), and quad-polarized RADARSAT-2 images. As a first step, a SAR backscatter analysis separated water from non-water classes (Level-I), followed by a more detailed Level-II classification, in which the water class was further divided into shallow and deep-water, and non-water separated into herbaceous and non-herbaceous vegetation classes. The final Level-III classification split the herbaceous class into bog, fen, and marsh classes. For the Level-II and Level-III classifications, different polarimetric features were extracted as input variables for the object-based random forest classification. These included covariance, coherency, Cloude–Pottier, Freeman–Durden, Yamaguchi decompositions, and Kennaugh matrices computed from dual- and quad-polarimetric SAR data, as well as the single-pol TerraSAR-X intensity. An overall accuracy of around 94% was yielded, whereas wetland classes reached 81%. The detailed backscatter analysis revealed the highest backscatter values for HH polarized TerraSAR-X images. When comparing wavelengths, the X-band possessed higher backscatter for herbaceous classes, while L-band represented the higher backscatter in the swamp class.

5.2.2. Lacustrine Wetlands

Lacustrine wetlands are associated with lakes, ponds, and their margins. They are characterized by non-flowing water and non-saline conditions with a depth of more than 2 m at low water level [75]. They are dominated by open water areas and beds of submerged, floating, or non-persistent vegetation, and are usually lacking trees and shrubs. Edge vegetation includes tall reed-like plants, such as rush. Lakes can also be associated with peat. Shallower water bodies are assigned to the palustrine ecosystem. This subsection chronologically presents all articles dealing with lacustrine ecosystems.

The study of Regmi et al. [50] investigated the possible application of TerraSAR-X's X-band to determine the times where thermokarst lakes on the Alaska Arctic Coastal Plain drained in order to understand post-drainage succession. In addition, the Normalized Difference Vegetation Index (NDVI) derived from a multi-spectral Landsat-5 image was utilized to better interpret the backscatter signal and thus to identify major biophysical parameters. A total of six single-polarized TerraSAR-X HH (StripMap) images were acquired during the growing season in 2009 covering 14 thermokarst lakes, previously radio-carbon-dated by Jones et al. [77], with a drainage age ranging from 225 to around 9500 years BP. Younger drained lakes (1950–2010) were determined by aerial and satellite sensors. The authors performed regression analysis between SAR backscatter and the NDVI of all selected lakes, which were grouped into three temporal subsets (0–50 years, 50–10,000 years, and 0–10,000 years). The regression analyses indicated a significant relationship between X-band backscatter/NDVI and drainage age for the long-term subsets (0–10,000 and 50–10,000). Surprisingly, no clear relationship was found for the modern 50 year time-scale. The study revealed an overall decreasing trend in backscatter and NDVI with increasing age of the the drained lake system, due to changing surface properties related to vegetation succession, soils, and hydrology.

Using TerraSAR-X time-series of 19 dual-polarized StripMap data, Heine et al. [70] identified reed belts of a lake in northeastern Germany and monitored their phenological changes. The authors calculated 16 different polarimetric indices based on HH/VV intensities serving a random forest classifier to delineate five major landscape features. Results showed that intensity difference HH-VV, the mean alpha scattering angle, intensity ratio HH/VV, and the coherence (phase) tend to be the most explaining variables for reed classification. The study also confirmed a better suitability of winter acquisitions to separate reed plants from other vegetation types due to an increase of double-bounce scattering when vegetation defoliates. In summer, volume scattering is the dominating scattering mechanism. The overall accuracy of all five classes reached more than 90%, whereas reeds only achieved around 50%. The authors suggest a denser temporal data stack to make use of the inherent and fast-changing phenology in order to improve the classification accuracy.

Another study involving TerraSAR-X time-series was undertaken by Moser et al. [68]. They mapped the spatio-temporal dynamics of prevailing land cover types of the largest freshwater lake in Burkina Faso, Lac Bam. The dense dual-polarized HH-VV StripMap time-series data (21 images) were acquired between 2013 and 2014 covering both rainy and dry seasons. Apart from the frequently applied decomposition techniques developed by Cloude and Pottier [78] or Pauli [79], the normalized Kennaugh elements developed by Schmitt et al. [65] were chosen here as the preferred decomposition technique, as they can be applied on dual- and quad-polarized SAR data of any wavelength. The authors carried out two supervised classification approaches. In the first case, called the mono-temporal approach, four static land cover categories (including flooded vegetation) were defined, and every Kennaugh image was classified. The second or multi-temporal case included dynamic processes by defining change classes through visual inspection of very-high-resolution optical RapidEye data. The overall accuracy of dual-polarimetric SAR data reached 88.5%, whereas a single-pol classification only using K_0 showed lower accuracy (82.2%). Overall, the static classes performed better than the dynamic ones. Flooded vegetation held class-specific accuracies ranging from 84.2 to 94.9%. This study specifically demonstrated the out-performance of dual-pol SAR acquisitions in wetland mapping compared to single polarized imagery.

A data fusion approach combining the TerraSAR-X sensor, optical multi-spectral data and airborne laser scanning light detection and ranging (LIDAR) for surface water detection was conducted by Irwin et al. [53]. Five single-polarization HH SpotLight acquisitions from April to September 2016 were processed to create water masks for each time step. A gray level thresholding analysis revealed an intensity value of -25 dB to clearly separate water from non-water areas. Based on the temporal complementary LIDAR and RapidEye datasets, individual water masks were created. To take advantage of each sensor's strength and compensate for the weaknesses of SAR, LIDAR, and optical systems, all water masks were combined into a fused water mask model. In doing so, the final fused model showed higher accuracies compared to the classification of the single-polarized TerraSAR-X water mask, which was 17–23% lower. The uncertainty of the final fusion water masks ranged between 4% and 9%.

Interferometric synthetic aperture radar (InSAR) is one of the most popular techniques for measuring ground deformation and subsidence, which is often associated with changes in groundwater levels. Parker et al. [54] used Sentinel-1A (C-Band) to delineate ground displacements over a wetland area in the Perth Basin (Australia). In this study, TerraSAR-X InSAR time-series (21 acquisitions) served as an independent validation dataset together with a continuous GPS measurement. Both satellites are able to detect small-magnitude deformation, and showed close agreement between the vertical displacement and the measured GPS data. The average deviations for Sentinel-1A and TerraSAR-X are 9 mm and 5 mm, respectively.

Ullmann et al. [73] characterized a heterogeneous tundra landscape with abundant thermokarst lakes and associated wetlands at the Tuktoyaktuk Peninsula in Canada. They calculated and compared several decomposition features of quad-, co-, and cross-polarized data acquired from different sensors, including X-band from TerraSAR-X, C-band (RADARSAT-2), and long-wave L-band from ALOS-PALSAR and ALOS-PALSAR-2. Shorter wavelength (X- and C-band) tend to be more efficient in separating wetland and tundra vegetation, while the L-band can better characterize bare ground classes. In general, the Kennaugh matrix decomposition [65] was found to be the most crucial variable for class discrimination, of which the intensity-based elements of the quad-polarized data offered the best separation. For the wetland class, best class-specific separability could be attributed to Kennaugh features of quad-polarized data (ALOS-PALSAR, RADARSAT-2) and dual- as well as cross-polarized data from TerraSAR-X. Correlation analyses also revealed high correlation between quad-polarized Kennaugh matrix elements (K_0, K_1, K_3, K_4, K_7) and dual-polarized elements, which could therefore be a substitute for those data.

5.2.3. Riverine Wetlands

Riverine wetlands are associated with flowing water in rivers or streams and their morphology. The water levels and volumes can be highly variable from season to season. Wetlands may contain flowing water permanently or may remain periodically dry for long intervals. As a result, riverine wetland ecosystems support a wide range of variable species. A total of six articles related to riverine wetlands have been published (Figure 7).

The study of Englhart et al. [52] explored the capabilities of X- (TerraSAR-X) and L-band (ALOS-PALSAR) to estimate aboveground biomass (AGB) in intact and degraded swamp forests in Borneo (Indonesia). Based on full-waveform LIDAR combined with field inventory data, an AGB model (multiple linear regression) was developed, which served as the reference data set. TerraSAR-X images acquired in the ScanSAR mode at VV polarization were considered. The long-wave SAR counterpart L-band ALOS-PALSAR was acquired at HH/HV polarization. The LIDAR-based estimated AGB were used to find the relationship between biomass and SAR backscatter signal and were subsequently upscaled across a larger area to provide multi-temporal information about various forest types and their condition. Results revealed that ALOS-PALSAR L-band backscatter is more sensitive to high biomass conditions (more than 100 t/ha), whereas TerraSAR-X performed better in low biomass ranges depicting degraded swamp forests. The most reliable results were achieved with the combined X- and L-band model using multi-temporal information. The overall model achieved accuracies of 53%. This work again highlighted the beneficial use of multi-frequent SAR information to delineate intact swamp forests with L-band, whereas X-band was more explanatory for degraded ecosystems.

Two studies from Betbeder et al. [31,80] presented another multi-temporal approach for wetland mapping based on TerraSAR-X time-series. In both studies, the authors utilized dual-co-polarized HH/VV images to calculate one polarimetric parameter, the Shannon entropy, and two intensity parameters to characterize wetland vegetation from six [31] and eight [80] images, respectively. The latter study also determined the optimal minimum number of required SAR images, the best timing of acquisitions, as well as best parameter combinations to achieve sufficient classification accuracy. Both chose support vector machine techniques for the final classification, attaining very high overall accuracies of 84% in the first study [31] and even 94% in the second one [80]. They also revealed the importance of polarimetric features for mapping wetland formations, rather than only using backscattering coefficients. Betbeder et al. [80] also indicated late winter (February), spring (April and May), and the beginning of summer (June and July) as the most relevant periods for wetland mapping covering the inherent phenological trajectories of each land cover class. Sensitivity analysis revealed that five TerraSAR-X acquisitions saturated almost at the final classification accuracy level (95%), hence being sufficient to balance the trade-off between the number of acquisitions and overall accuracy.

The effect of inundated vegetation on HH/VV backscatter and phase difference for TerraSAR's X-band was examined by Zalite et al. [81]. The authors ordered two TerraSAR-X StripMaps at HH/VV polarization, representing both flooded and non-flooded conditions, covering the Soomaa National Park located in south-western Estonia. This landscape encompasses two transiting rivers with adjacent floodplains and forests. In this study, coniferous and deciduous forest stands with different tree heights were investigated. As a first step, complex HH/VV images were calculated by subtracting the VV from the HH channel to gain information about backscatter mechanism (e.g., double-bounce) and calibrated to σ^0. To cover the prevailing land cover characteristics, 12 patches of 115 m × 115 m were selected. For each patch, the differences of backscatter and phase between the two time-steps were calculated and compared. Results revealed an increase in HH backscatter from 2.3 to 8.2 dB for the complex HH/VV and from 3 to 9.8 dB over all patches, when vegetation became inundated. The study also showed that the dual-co-polarization HH/VV channel separated flooded from non-flooded vegetation better than the single-polarized band. The considerable increase in phase shift suggests a stronger double-bounce backscatter mechanism. When comparing evergreen and deciduous tree species (leaf-off season), the latter tend to have higher phase differences due to stronger double-bounce backscatter.

A multi-frequent backscatter analysis for mapping and monitoring wetland extent and inundation pattern was conducted by Mitchell et al. [67] in the Macquarie River Catchment in north-central Australia. This area consists of various riverine wetland types, including numerous herbaceous and woody vegetation communities. The authors acquired L-band ALOS-PALSAR and TerraSAR-X time-series data during dry, wet, and transitional wetland conditions. However, two TerraSAR-X cross-polarized HH and HV polarized StripMap scenes were exclusively applied on the transitional phase. For water body detection, only the longwave L-band data was used. The synergy of both sensors was explored for discriminating different vegetation communities. Two statistical approaches were tested. In the first, the mean radar spectra for each land cover class were calculated and spectrally separated by using two separability measures (Jeffries–Matusita and the Transformed Divergence). The second approach was only applied to the ALOS-PALSAR data. A canonical variate analysis (CVA) was used to determine the best single band and band combination for wetland discrimination. The results of this study once again showed the improved wetland discrimination when multi-frequency SAR data are used in a complex environment. Due to the occurrence of woody vegetation, L-band performed better in mapping sub-canopy inundation, whereas the X-band was able to detect flushes in growth of vegetation in response to higher soil moisture from flooding.

A very recent study was published by Mleczko and Mrôz [71] investigating the desiccation of wetlands and biodiversity loss in the floodplains of the Biebrza River in Poland. This comparative study applied C-band from Sentinel-1 and X-band from TerraSAR-X to map different vegetation types (wet grasslands, reed, flooded deciduous forest) and flooding conditions (permanent and temporarily flooded). The six TerraSAR-X images were acquired in quad-pol during the quad-polarimetric experimental mode [69]. For the wetland classification, σ^0, coherence, Shannon entropy, and the Yamaguchi decomposition parameter (quad-pol data only) were computed. The best classification results were achieved with the inclusion of the TerraSAR-X quad-pol data with overall accuracies of 79%, whereas the Sentinel-1 time-series alone resulted in 65%. Comparing the different TerraSAR-X polarimetric modes, the dual-polarized dataset performed with an overall accuracy of 76% in a similar range as the quad-pol accuracies (79%). As descriptor for characterizing wetland types, the Shannon entropy yielded lower accuracies ranging from 47% to 58%. The integration of the interferometric coherence in the classification scheme tends to have less importance in distinguishing between wetland classes.

5.2.4. Estuarine Wetlands

Estuaries are located in marine deltas, where a river meets the sea. Lying along an estuary with a mix of saltwater coming from the sea and freshwater from the river and being subject to tidal flooding, these wetlands are associated with estuarine marshes and salt marshes, intertidal flats, swamps, and mangrove forests [82]. Estuarine wetlands are valuable coastal ecosystems that not only provide flood control and shoreline stabilization, but also offer an important habitat for juvenile marine organisms and migrating birds. They support a wide range of threatened plant species. Regarding the selected wetland studies, estuarine wetlands cannot be easily distinguished from marine wetland applications, since both wetland types often coincide. In the following, TerraSAR-X-related studies primarily dealing with estuarine wetlands are presented.

Hong et al. [45] examined water-level changes over South Florida's Everglades wetlands during an eight month period in 2008 based on the interferometric analysis of dual-polarized TerraSAR-X observations. Interferometric coherence of all four polarizations was evaluated in order to determine the dominant scattering mechanisms in each of the different wetland environments within the study area, containing both managed and natural flow fresh- and salt-water wetlands. The authors revealed high coherence in all polarizations, with the highest response in HH. Fringe patterns of interferograms derived from multi-polarized data did not differ significantly among the polarization types, indicating that the phase information reflects water-level changes, and volume scattering was ruled out as the dominant scattering mechanism. Overall, the applied wetland InSAR approach worked well with X-band TerraSAR-X data.

In 2015, Hong et al. [46] published another study related to the Florida Everglades. The authors examined the applicability of TerraSAR-X quad-polarization data for wetland vegetation mapping. In doing so, different scattering mechanisms were extracted using both Hong & Wdowinsky four-component decomposition as well as Yamaguchi's decomposition, which were utilized for classifying vegetation types with an object-based image analysis approach. High-resolution optical RapidEye imagery was used for validation. Good accuracies of over 93% were achieved by classifying with the SAR image features. Overall, quad-pol X-band PolSAR products performed as good indicators for wetland vegetation mapping, especially for mangrove forests. According to the authors, the availability of operational X-band quad-pol images remains as one of the main limitations of PolSAR-based wetland classification.

With the aim of evaluating the potential of dual-polarized TerraSAR-X data for mapping different peatland forests, Wijaya et al. [51] used $H/\alpha/A$ (entropy/alpha/anisotropy) decomposition of the covariance matrix and canonical discrimination analysis for determining the importance of SAR parameters for peatland discrimination in tropical swamp and mangrove forests within the Tanjung Putig National Park in Central Kalimantan, Indonesia. Although anisotropy and entropy proved to be more important than other SAR features for peatland classification, X-band SAR alone could not provide satisfactory results. The distinction of shallow and deep peat soils under relatively closed canopies remains challenging. However, a supervised maximum likelihood classification based on TerraSAR-X fused with optical Landsat ETM data yielded up to 87% accuracy.

Lagomasino et al. [47] compared different remote sensing techniques to measure mangrove forest canopy heights in the Zambezi River delta from airborne and space-borne platforms using LIDAR, stereophotogrammetry, and radar interferometry. The authors derived four canopy height models based on VHR stereophotogrammetry from WorldView-1, TerraSAR-X add-on for Digital Elevation Measurements (TanDEM-X), and the Shuttle Radar Topography Mission (SRTM). For this, the spatiotemporal variability between the sensors as well as a comparison of model efficiency for canopy height models were of primary importance. SRTM overlooks fine-scale forest canopy and poorly represents the top of the canopy, but provides temporal stability and is an accurate indicator for the mean height distribution of mature, intact mangrove forests. Taller canopies are accurately represented by high-resolution TanDEM-X models, but shorter canopies are not estimated in detail. Lastly, VHR canopy height models derived from stereophotogrammetry showed good accordance with LIDAR as "the gold standard" of mangrove forest canopy models.

Furthermore, Ullmann et al. [83] applied a two-component polarimetric decomposition technique for quad-pol RADARSAT-2 (C-band) and dual-co-pol HH/VV TerraSAR-X (X-band) in order to investigate the role of volume scattering from vegetation in a tundra environment across the Mackenzie Delta region in Canada. The approach decomposed total backscatter into surface and double-bounce scattering, and was compared to a three-component decomposition (quad-pol RADARSAT-2). The revealed PolSAR features were analyzed for each land cover class and correlated with each other based on scattering characteristics derived from in-situ ground truth data on land cover. Generally, little influence of volume scattering was revealed for land cover classes of low vegetation height, whereas wetland vegetation showed a high degree of volume scattering. The quad-pol RADARSAT-2 data provided the best separation of tundra land cover classes. Nevertheless, wetlands showed a clear position in both the C- and the X-band HH/VV-PolSAR feature spaces, indicating an accurate classification.

5.2.5. Marine Wetlands

Marine wetlands are located along coasts, in offshore reefs, dune hollows, or sand plains. They include rocky shores, coastal lagoons, coral reefs, and subtidal seagrass beds with a depth of up to six meters (at low tide). Influenced by saltwater, marine wetlands are exposed to ocean currents, waves, and tides. Further, they are dominated by saltwater-tolerant plants and provide not only an essential nursery and feeding space for a wide range of fish and sea turtles, and also represent a valuable habitat for migratory birds [56,82]. Related to TerraSAR-X, 11 studies with a focus on marine wetlands have been released, indicating the most-studied wetland type of our review.

With the aim of investigating the extent of submarine groundwater discharge (SGD) to large tidal flats along the Korean coast, Kim et al. [41] compared polarimetric backscattering coefficients obtained from both TerraSAR-X (X-band) and RADARSAT-2 (C-band) data with theoretical scattering models to detect water puddles. Based on in-situ measurements of surface roughness, soil moisture, and water salinity, two different scattering models—an updated integral equation model (IEM) and Oh's empirical model—were applied. Compared to the models, the SAR data showed lower backscattering over areas with water puddles. The authors revealed that both X- and C-band SAR are suitable tools for identifying water puddles in tidal flats. Eventually, these puddles can be associated with SGD according to their shape and location.

Another study dedicated to Korean tidal flats was published by Lee et al. [84], who used a backscattering analysis of 16 TerraSAR-X images, HH single, and VH/VV dual-cross polarizations for mapping halophytes of the Ganghwa tidal flat. Based on a decision tree classifier, a rule-based salt marsh mapping approach was designed in order to examine the ecological status, distribution changes, and optimum season for the mapping of halophytes. Winter was revealed as the best season for distinction between an annual plant, *Suaeda japonica*, and a typical perennial salt marsh, *Phragmites australis*, based on HH polarization backscatter. It was shown that TerraSAR-X is an effective instrument for mapping and monitoring seasonal variations of halophyte species in tidal flats.

A PolSAR-based land cover classification of an arctic tundra environment on Richards Island, Canada, was carried out by the authors of Ullmann et al. [85]. Here, a combination of dual-co-polarized TerraSAR-X (HH/VV), quad-polarized RADARSAT-2, and multispectral Landsat 8 OLI data was used for unsupervised and supervised classification of tundra land cover types. In-situ measurements and high-resolution aerial images were obtained during field campaigns, and served as reference data. The study did not exclusively focus on marine wetlands. However, a large part of the coastal wetlands was located along the shore and was shaped by an ocean setting. It was shown that combination of optical data with PolSAR features (entropy/alpha, Yamaguchi, two-component decomposition) significantly improved the classification results. Decomposition features of the dual and quad-polarized data showed a high sensitivity for double-bounce backscattering in wetland areas.

Several studies investigating intertidal flats have been published by Gade et al. (2014) [38], Jung et al. [42], and Gade et al. [39]. The first study of Gade et al. [38] revealed the high potential of SAR to complement and improve the monitoring of intertidal flats in the German Wadden Sea. For this, multiple SAR sensors including ALOS-PALSAR (L-band), ERS-1/2, RADARSAT-2, ENVISAT ASAR (C-band each), and TerraSAR-X (X-band) were utilized in order to detect benthic fauna such as mussel or oyster beds based on surface roughness parameters. Here, single-sensor performance was compared with multiple-sensor performance. As a result, statistical analysis of the multi-temporal TerraSAR-X data significantly improved the classification of mussel beds through a synergistic monitoring approach.

Jung et al. [42] also published a multi-sensor methodology for classifying intertidal areas in the Wadden Sea. Therefore, RapidEye and TerraSAR-X satellite data were combined. A hierarchical decision-tree classifier was applied to firstly separate water from tidal flat areas and secondly delineate the shellfish beds based on TerraSAR-X-derived textural features and morphological filters. Thirdly, vegetation was classified by means of a vegetation index and object-based feature extraction. The authors highlighted the great value of TerraSAR-X data for a detailed fish bed classification that reached an overall accuracy of 98%. For this purpose, the usage of high-resolution radar sensors was regarded as indispensable.

A further study related to intertidal flats in the German Wadden Sea was published by Gade et al. (2016) [40], who utilized combined pairs of single-polarized (VV) TerraSAR-X and RADARSAT-2 as well as HH and VV-polarized TerraSAR-X images between 2008 and 2013 in order to image sediments, macrophytes, and mussels. The focus was laid on the SAR-based detection of morphological changes and bivalve beds of dry-fallen intertidal flats in order to improve existing classification systems. The results demonstrated that multi-temporal SAR imagery was a valuable input for classification due to its strong backscatter in bivalve beds. In particular, the polarization coefficient of SAR data based on both co-polarizations allowed the detection of mussel or oyster habitats. Further, with additional information about the water level available, multi-temporal and multi-frequency SAR proved to be a useful instrument for a change detection of intertidal flats.

Another more recent publication with regard to tidal flats on the German North Sea coast was released by Gade et al. [39] in 2018. The authors investigated the influence of imaging geometry and environmental parameters such as wind speed, water level, and vegetation period on the radar backscatter from exposed tidal flats using RADARSAT-2 and TerraSAR-X single- and dual-co-polarized SAR images. Further, TerraSAR-X images were utilized for polarimetric analyses (PolSAR) based on the decomposition of the Kennaugh matrix. As a result, VV-polarized TerraSAR-X data was most sensitive to surface roughness as one of the most important environmental influencing factors. Other essential variables shaping the radar return are the local incidence angle depending on local topography, SAR look angle, and direction. Water-level history (e.g., low or high tide, level minimum at low tide) was also considered important for the backscatter intensity. Altogether, the potential of polarimetric decomposition for identifying surface types of intertidal flats was confirmed.

With the objective of investigating the linkage of surge extent and persistence after Hurricane Sandy landfall and condition change of coastal marshes along the New Jersey Atlantic coast, Rangoonwala et al. [86] used SAR time-series for surge mapping and optical satellite images for assessing changes in marsh condition. With a combination of TerraSAR-X and COSMO-SkyMed SAR data (HH polarization), the spatial and temporal extent of post-hurricane flooding in back-barrier lagoon marshes could be captured based on relative horizontal backscatter attenuation. Here, the ability of X-band to perceive marsh flooding at all depths was exploited. Linked with an analysis of marsh condition based on optical sensors (SPOT, MODIS), high surge-related marsh impacts in back-barrier lagoons north of the hurricane landfall track could be identified.

Recently in 2018, Da Lio et al. [44] presented an advanced persistent scatterer interferometry (PSI) technique on a 5-year long stack of TerraSAR-X acquisitions of the Venice Lagoon salt marshes for quantifying land subsidence between 2008 and 2013. Thousands of measurable persistent targets could be identified due to the short revisiting time of 11 days and the high spatial resolution of 3 m. By integrating the SAR resolution with ground-based multi-depth displacement records, a quantitative conceptual model of salt marsh vertical dynamics could be established. Results showed a different subsidence behavior between natural and man-made salt marshes based on soil consolidation stages. Overall, the high heterogeneity of the superficial-subsoil dynamics were revealed, mirroring the diverse bio-morphological and geological setting at the salt marsh-scale.

6. Conclusions

This review demonstrates the wide-ranging application of TerraSAR-X data regarding mapping various types of wetlands, sensor specifications, temporal and spatial scales, as well as the applied methodology. Notably, marine and estuarine wetlands were amongst the most studied wetland types within the selected studies. In this context, the benefit of a synergy between the ability of X-band radar to capture non-vegetated or low herbaceous wetland vegetation along the coast without penetrating the ground surface and the high spatial and temporal resolution provided by TerraSAR-X has been utilized in many scientific publications. Furthermore, the distribution of wetland-related studies across the globe indicates an increased use of TerraSAR-X in cloud-prone areas, which mirrors the penetration of clouds as one of the main advantages of radar-based applications. Dual-polarized data were the most used among the selected studies, showing a peak in HH/VV co-polarization based on the strong double-bounce backscattering signal from flooded vegetation as well as the possibility of exploiting polarimetric parameters. In the case of the single-polarization data, HH was preferred over VV. Few studies acquired TerraSAR-X images during the experimental qual-polarization phase, providing promising results.

In this review, it was shown that polarimetric decomposition is a frequently applied technique for revealing the physical characterizations of the scattering scenario and the involved scatterers, and thus allows a more detailed view of the complex wetland environments. In general, PolSAR is a rapidly growing research field and, especially when combined with X-band radar, provides valuable information for wetland analyses due to the dominant double-bounce/surface scattering mechanisms to discriminate between different wetland characteristics. However, the limited availability of quad-pol data restricts its wider usage with TerraSAR-X. The high amount of wetland studies applying multi-temporal analyses indicates that the short revisiting time of TerraSAR-X (11 days) is an advantage of time-series analysis, which can provide crucial information about the seasonality of vegetation in wetland areas or tidal dynamics. However, the selection of the optimal timing is more important than the amount of data. Throughout all studies, results reveal that single-date acquisitions are not adequate for mapping wetland accurately.

Finally, the strengths of TerraSAR-X for wetland assessments lie in its high sensitivity to soil moisture and surface roughness on low-biomass surfaces. In particular, a good performance of TerraSAR-X for mapping salt marshes, including fish or mussel bed classification, as well as the depiction of degraded swamp forests and herbaceous vegetation, was highlighted in the reviewed studies. With respect to multi-temporal analyses, vegetation species showing fast growing vertical structures could be best distinguished by using X-band. However, the usage of TerraSAR-X data for areas with high-biomass conditions such as swamp forests with dense canopies showed limitations. Here, a better performance of L-band radar was discovered in several studies. On that basis, the combined usage of TerraSAR-X with other sensors yielded promising results for wetland mapping.

However, while SAR can improve the accuracy of wetland classifications in inundated areas, the partially limited availability and often high cost of radar imagery still limits its broader use [58]. Despite being free of charge for scientific and non-commercial users, the access to TerraSAR-X products is not straightforward. As a first step, data need to be ordered via a detailed science proposal describing research objectives, work plan, data requirements, and detailed information about the "scientific use" criterion. Each proposal is evaluated by a scientific and technical committee, which judge the scientific priority. If accepted, the data will be delivered electronically. However, depending on the project, commercial users sometimes receive higher priority, conflicting with orders from scientific users, which could result in a reduction of the data or even in a complete rejection. This may change in the near future due to transitions in data policies, indicating a trend towards freely available remote sensing data. Moreover, the synergistic use of different SAR data (e.g., TerraSAR-X with L-band radar sensors such as ALOS-PALSAR-2) is a promising approach in which both the advantage of high-resolution X-band radar for observing herbaceous wetland vegetation and the suitability of L-band radar to map forested wetlands by penetrating canopies are combined. Further, data fusion of TerraSAR-X with high-resolution optical data (e.g., RapidEye) proved to be extremely valuable for a detailed wetland classification, as high-resolution SAR data might be influenced by speckle noise, which degrades the true spatial resolution. In the future, multi-sensor approaches may become increasingly important for wetland assessment due to the growing availability of different sensors with the advantage of complementing each other.

Author Contributions: C.W. and K.W. developed the concept of this study, reviewed the literature, and wrote the majority of this paper. C.W. designed and prepared all figures. A.W. and A.R. provided written input and feedback on the content of this manuscript during the initial preparation and revision.

Funding: The TanDEM-X project is partly funded by the German Federal Ministry for Economics Affairs and Energy (Förderkennzeichen 50 EE 1035).

Acknowledgments: The authors would like to express their gratitudes to Isabel Georg for reviewing the English language.

Conflicts of Interest: The authors declare no conflict of interest.

Appendix A

Table A1. Summary of all considered studies in this review using TerraSAR-X data in the domain of wetland research (in chronological order).

Study (Year Published)	Study Area (Country)	Wetland Type	Research Objectives
Da Lio et al. (2018) [44]	Venice Laggon (Italy)	marine, estuarine	land subsidence
Gade et al. (2018) [39]	Wadden Sea (Germany)	marine	monitoring tidal flats
Mleczko et al. (2018) [71]	Biebrza Valley (Poland)	riverine	wetland classification
Irwin et al. (2018) [53]	Ontario (Canada)	lacustrine	wetland classification
Mahdianpari et al. (2017) [49]	Avalon Peninsula (Canada)	palustrine	wetland classification
Parker et al. (2017) [54]	Perth Basin (Australia)	lacustrine	land subsidence
Ullmann et al. (2017) [73]	Tuktoyaktuk Peninsula (Canada)	lacustrine	tundra classification
Zhou et al. (2017) [87]	Eastern Beijing Plain (China)	palustrine	land subsidence
Dabrowska-Zielinska et al. (2016) [64]	Biebrza Valley (Poland)	palustrine, riverine	carbon fluxes
Gade et al. (2016) [40]	North Sea (Germany)	marine	detection of bivalve beds
Heine et al. (2016) [70]	Lake Fürstenseer (Germany)	lacustrine	reed classification
Lagomasino et al. (2016) [47]	Zambezi River Shed (Mozambique)	estuarine	mangrove height classification
Moser et al. (2016) [68]	Lac Bam (Burkina Faso)	lacustrine	wetland classification
Ullmann et al. (2016) [73]	Mackenzie Delta (Canada)	esturine	wetland classification
Rangoonwala et al. (2016) [86]	New Yersey Bay (USA)	marine, esturine	coastal wetland classification
Betbeder et al. (2015) [80]	Mont-Saint-Michel (France)	riverine	wetland classification
Jung et al. (2015) [42]	Norderney (Germany)	marine	tidal flats classification
Mitchell et al. (2015) [67]	Macqarie Marshes (Australia)	riverine	wetland classification
Hong et al. (2015) [46]	Florida Everglades (USA)	estuarine	wetland classification
Morishita et al. (2015) [48]	Delft (Netherlands)	palustrine	peatland classification
Schuster et al. (2015) [76]	Döberitzer Heide (Germany)	palustrine	wetland classification
Otukei et al. (2015) [76]	Bwindi (Uganda)	palustrine	wetland classification
Ullmann et al. (2014) [85]	Richard Island (Canada)	marine, estuarine	wetland classification
Betbeder et al. (2014) [31]	Mont-Saint-Michel (France)	riverine	wetland classification
Gade et al. (2014) [38]	North Sea (Germany)	marine	intertidal flats classification
Zalite et al. (2014) [81]	Soomaa National Park (Estonia)	riverine	wetland characterization
Lee et al. (2012) [84]	Ganghwa (Korea)	marine, estuarine	tidal flats classification
Regmi et al. (2012) [50]	Seward Peninsula (USA)	lacustrine	peat age classification
Kim et al. (2011) [41]	Namyang Bay (Korea)	marine	tidal flats classification
Englhart et al. (2011) [52]	River Sebangau (Borneo)	riverine	swamp forest classification
Hong et al. (2010) [45]	Everglades (USA)	riverine	mangrove classification
Wijaya et al. (2010) [51]	Southern Kalimantan (Borneo)	riverine	swamp forest classification

References

1. Costanza, R.; d'Arge, R.; de Groot, R.; Farber, S.; Grasso, M.; Hannon, B.; Limburg, K.; Naeem, S.; O'Neill, R.V.; Paruelo, J.; et al. The value of the world's ecosystem services and natural capital. *Nature* **1997**, *387*, 253. [CrossRef]

2. Assessment, Millennium Ecosystem. *Ecosystems and Human Well-Being: General Synthesis*; Island Press: Washington, DC, USA, 2005.

3. Mitsch, W.J.; Bernal, B.; Hernandez, M.E. Ecosystem services of wetlands. *Int. J. Biodivers. Sci. Ecosyst. Serv. Manag.* **2015**, *11*, 1–4. [CrossRef]

4. Sharma, B.; Rasul, G.; Chettri, N. The economic value of wetland ecosystem services: Evidence from the Koshi Tappu Wildlife Reserve, Nepal. *Ecosyst. Serv.* **2015**, *12*, 84–93. [CrossRef]

5. Dodds, W.K.; Wilson, K.C.; Rehmeier, R.L.; Knight, G.L.; Wiggam, S.; Falke, J.A.; Dalgleish, H.J.; Bertrand, K.N. Comparing ecosystem goods and services provided by restored and native lands. *BioScience* **2008**, *58*, 837–845.[CrossRef]

6. Erwin, K.L. Wetlands and global climate change: The role of wetland restoration in a changing world. *Wetl. Ecol. Manag.* **2008**, *17*, 71. [CrossRef]

7. Shepard, C.C.; Crain, C.M.; Beck, M.W. The protective role of coastal marshes: A systematic review and meta-analysis. *PLoS ONE* **2011**, *6*, e27374. [CrossRef] [PubMed]

8. IPCC. Summary for Policymakers. In *Climate Change 2013: The Physical Science Basis. Contribution of Working Group I to the Fifth Assessment Report of the Intergovernmental Panel on Climate Change*; Stocker, T., Qin, D., Plattner, G.K., Tignor, M., Allen, S., Boschung, J., Nauels, A., Xia, Y., Bex, V., Midgley, P., Eds.; Book Section SPM; Cambridge University Press: Cambridge, UK; New York, NY, USA, 2013; pp. 1–30.

9. Mitsch, W.J.; Bernal, B.; Nahlik, A.M.; Mander, Ü.; Zhang, L.; Anderson, C.J.; Jørgensen, S.E.; Brix, H. Wetlands, carbon, and climate change. *Landsc. Ecol.* **2013**, *28*, 583–597. [CrossRef]

10. Hu, S.; Niu, Z.; Chen, Y.; Li, L.; Zhang, H. Global wetlands: Potential distribution, wetland loss, and status. *Sci. Total Environ.* **2017**, *586*, 319–327. [CrossRef] [PubMed]

11. Davidson, N.C. How much wetland has the world lost? Long-term and recent trends in global wetland area. *Mar. Freshw. Res.* **2014**, *65*, 934–941. [CrossRef]

12. Foley, J.A.; DeFries, R.; Asner, G.P.; Barford, C.; Bonan, G.; Carpenter, S.R.; Chapin, F.S.; Coe, M.T.; Daily, G.C.; Gibbs, H.K.; et al. Global consequences of land use. *science* **2005**, *309*, 570–574. [CrossRef] [PubMed]

13. Nagendra, H.; Lucas, R.; Honrado, J.P.; Jongman, R.H.G.; Tarantino, C.; Adamo, M.; Mairota, P. Remote sensing for conservation monitoring: Assessing protected areas, habitat extent, habitat condition, species diversity, and threats. *Ecol. Indic.* **2013**, *33*, 45–59. [CrossRef]

14. Kuenzer, C.; Ottinger, M.; Wegmann, M.; Guo, H.; Wang, C.; Zhang, J.; Dech, S.; Wikelski, M. Earth observation satellite sensors for biodiversity monitoring: Potentials and bottlenecks. *Int. J. Remote Sens.* **2014**, *35*, 6599–6647. [CrossRef]

15. Hess, L.L.; Melack, J.M.; Novo, E.M.; Barbosa, C.C.; Gastil, M. Dual-season mapping of wetland inundation and vegetation for the central Amazon basin. *Remote Sens. Environ.* **2003**, *87*, 404–428. [CrossRef]

16. Schmidt, K.; Skidmore, A. Spectral discrimination of vegetation types in a coastal wetland. *Remote Sens. Environ.* **2003**, *85*, 92–108. [CrossRef]

17. Henderson, F.M.; Lewis, A.J. Radar detection of wetland ecosystems: A review. *Int. J. Remote Sens.* **2008**, *29*, 5809–5835. [CrossRef]

18. Petus, C.; Lewis, M.; White, D. Monitoring temporal dynamics of Great Artesian Basin wetland vegetation, Australia, using MODIS NDVI. *Ecol. Indic.* **2013**, *34*, 41–52. [CrossRef]

19. Wohlfart, C.; Liu, G.; Huang, C.; Kuenzer, C. A river basin over the course of time: Multi-temporal analyses of land surface dynamics in the Yellow River Basin (China) based on medium resolution remote sensing data. *Remote Sens.* **2016**, *8*, 186. [CrossRef]

20. Klein, I.; Gessner, U.; Dietz, A.J.; Kuenzer, C. Global WaterPack—A 250 m resolution dataset revealing the daily dynamics of global inland water bodies. *Remote Sens. Environ.* **2017**, *198*, 345–362. [CrossRef]

21. Martinis, S.; Kuenzer, C.; Wendleder, A.; Huth, J.; Twele, A.; Roth, A.; Dech, S. Comparing four operational SAR-based water and flood detection approaches. *Int. J. Remote Sens.* **2015**, *36*, 3519–3543. [CrossRef]

22. White, L.; Brisco, B.; Dabboor, M.; Schmitt, A.; Pratt, A. A collection of SAR methodologies for monitoring wetlands. *Remote Sens.* **2015**, *7*, 7615–7645. [CrossRef]

23. Smith, L.C. Satellite remote sensing of river inundation area, stage, and discharge: A review. *Hydrol. Process.* **1997**, *11*, 1427–1439. [CrossRef]

24. Hess, L.L.; Melack, J.M.; Simonett, D.S. Radar detection of flooding beneath the forest canopy: A review. *Int. J. Remote Sens.* **1990**, *11*, 1313–1325. [CrossRef]

25. Schmullius, C.; Evans, D. Review article Synthetic aperture radar (SAR) frequency and polarization requirements for applications in ecology, geology, hydrology, and oceanography: A tabular status quo after SIR-C/X-SAR. *Int. J. Remote Sens.* **1997**, *18*, 2713–2722. [CrossRef]

26. Kuenzer, C.; Bluemel, A.; Gebhardt, S.; Quoc, T.V.; Dech, S. Remote sensing of mangrove ecosystems: A review. *Remote Sens.* **2011**, *3*, 878–928. [CrossRef]

27. Kuenzer, C.; Knauer, K. Remote sensing of rice crop areas. *Int. J. Remote Sens.* **2013**, *34*, 2101–2139. [CrossRef]

28. Guo, M.; Li, J.; Sheng, C.; Xu, J.; Wu, L. A review of wetland remote sensing. *Sensors* **2017**, *17*, 777. [CrossRef] [PubMed]

29. Moreira, A.; Prats-Iraola, P.; Younis, M.; Krieger, G.; Hajnsek, I.; Papathanassiou, K.P. A tutorial on synthetic aperture radar. *IEEE Geosci. Remote Sens. Mag.* **2013**, *1*, 6–43. [CrossRef]

30. Ryu, J.H.; Choi, J.K.; Lee, Y.K. Potential of remote sensing in management of tidal flats: A case study of thematic mapping in the Korean tidal flats. *Ocean Coast. Manag.* **2014**, *102*, 458–470. [CrossRef]

31. Betbeder, J.; Rapinel, S.; Corpetti, T.; Pottier, E.; Corgne, S.; Hubert-Moy, L. Multitemporal classification of TerraSAR-X data for wetland vegetation mapping. *J. Appl. Remote Sens.* **2014**, *8*, 083648. [CrossRef]

32. Mitsch, W.; Gosselink, J. *Wetlands*, 4th ed.; Wiley: Hoboken, NJ, USA, 2007.

33. Keddy, P.A. *Wetland Ecology: Principles and Conservation*; Cambridge University Press: Cambridge, UK, 2010.

34. Tiner, R.W. Wetlands. An Overview. In *Remote Sensing of Wetlands: Applications and Advances*; CRC Press: Boca Raton, FL, USA, 2015; pp. 1–18.

35. Ramsar, Iran. Convention on wetlands of international importance, especially as waterfowl habitat. *Ramsar (Iran)* **1971**, 1–3.

36. Frazier, S.; International, W. *An Overview of the World's Ramsar Sites*; Wetlands International: Wageningen, The Netherlands, 1996.
37. Ramsar, Iran, 1971. Recommendation 4.7: Mechanisms for improved application of the Ramsar Convention. In Proceedings of the Meeting of the Conference of the Contracting Parties, Montreux, Switzerland, 27 June–4 July 1990; Volume 4.
38. Gade, M.; Melchionna, S.; Stelzer, K.; Kohlus, J. Multi-frequency SAR data help improving the monitoring of intertidal flats on the German North Sea coast. *Estuar. Coast. Shelf Sci.* **2014**, *140*, 32–42. [CrossRef]
39. Gade, M.; Wang, W.; Kemme, L. On the imaging of exposed intertidal flats by single-and dual-co-polarization Synthetic Aperture Radar. *Remote Sens. Environ.* **2018**, *205*, 315–328. [CrossRef]
40. Gade, M.; Melchionna, S. Joint use of multiple Synthetic Aperture Radar imagery for the detection of bivalve beds and morphological changes on intertidal flats. *Estuar. Coast. Shelf Sci.* **2016**, *171*, 1–10. [CrossRef]
41. Kim, D.J.; Moon, W.M.; Kim, G.; Park, S.E.; Lee, H. Submarine groundwater discharge in tidal flats revealed by space-borne synthetic aperture radar. *Remote Sens. Environ.* **2011**, *115*, 793–800. [CrossRef]
42. Jung, R.; Adolph, W.; Ehlers, M.; Farke, H. A multi-sensor approach for detecting the different land covers of tidal flats in the German Wadden Sea—A case study at Norderney. *Remote Sens. Environ.* **2015**, *170*, 188–202. [CrossRef]
43. Lee, Y.K.; Park, W.; Choi, J.K.; Ryu, J.H.; Won, J.S. Assessment of TerraSAR-X for mapping salt marsh. In Proceedings of the Geoscience and Remote Sensing Symposium (IGARSS), Vancouver, BC, Canada, 24–29 July 2011; pp. 2330–2333.
44. Da Lio, C.; Teatini, P.; Strozzi, T.; Tosi, L. Understanding land subsidence in salt marshes of the Venice Lagoon from SAR Interferometry and ground-based investigations. *Remote Sens. Environ.* **2018**, *205*, 56–70. [CrossRef]
45. Hong, S.H.; Wdowinski, S.; Kim, S.W. Evaluation of TerraSAR-X observations for wetland InSAR application. *IEEE Trans. Geosci. Remote Sens.* **2010**, *48*, 864–873. [CrossRef]
46. Hong, S.H.; Kim, H.O.; Wdowinski, S.; Feliciano, E. Evaluation of polarimetric SAR decomposition for classifying wetland vegetation types. *Remote Sens.* **2015**, *7*, 8563–8585. [CrossRef]
47. Lagomasino, D.; Fatoyinbo, T.; Lee, S.; Feliciano, E.; Trettin, C.; Simard, M. A comparison of mangrove canopy height using multiple independent measurements from land, air, and space. *Remote Sens.* **2016**, *8*, 327. [CrossRef] [PubMed]
48. Morishita, Y.; Hanssen, R.F. Temporal decorrelation in L-, C-, and X-band satellite radar interferometry for pasture on drained peat soils. *IEEE Trans. Geosci. Remote Sens.* **2015**, *53*, 1096–1104. [CrossRef]
49. Mahdianpari, M.; Salehi, B.; Mohammadimanesh, F.; Motagh, M. Random forest wetland classification using ALOS-2 L-band, RADARSAT-2 C-band, and TerraSAR-X imagery. *ISPRS J. Photogramm. Remote Sens.* **2017**, *130*, 13–31. [CrossRef]
50. Regmi, P.; Grosse, G.; Jones, M.C.; Jones, B.M.; Anthony, K.W. Characterizing post-drainage succession in thermokarst lake basins on the Seward Peninsula, Alaska with TerraSAR-X backscatter and Landsat-based NDVI data. *Remote Sens.* **2012**, *4*, 3741–3765. [CrossRef]
51. Wijaya, A.; Reddy Marpu, P.; Gloaguen, R. Discrimination of peatlands in tropical swamp forests using dual-polarimetric SAR and Landsat ETM data. *Int. J. Image Data Fusion* **2010**, *1*, 257–270. [CrossRef]
52. Englhart, S.; Keuck, V.; Siegert, F. Aboveground biomass retrieval in tropical forests—The potential of combined X-and L-band SAR data use. *Remote Sens. Environ.* **2011**, *115*, 1260–1271. [CrossRef]
53. Irwin, K.; Beaulne, D.; Braun, A.; Fotopoulos, G. Fusion of SAR, optical imagery and airborne LiDAR for surface water detection. *Remote Sens.* **2017**, *9*, 890. [CrossRef]
54. Parker, A.L.; Filmer, M.S.; Featherstone, W.E. First results from Sentinel-1A InSAR over Australia: Application to the Perth Basin. *Remote Sens.* **2017**, *9*, 299. [CrossRef]
55. Scott, D.; Jones, T. Classification and inventory of wetlands: A global overview. *Vegetatio* **1995**, *118*, 3–16. [CrossRef]
56. Ramsar Convention Secretariat. *An introduction to the Convention on Wetlands (previously The Ramsar Convention Manual)*; CRC Press: Gland, Switzerland, 2016.
57. Lehner, B.; Döll, P. Development and validation of a global database of lakes, reservoirs and wetlands. *J. Hydrol.* **2004**, *296*, 1–22. [CrossRef]

58. Knight, J.F.; Corcoran, J.M.; Rampi, L.P.; Pelletier, K.C. Theory and applications of object-based image analysis and emerging methods in wetland mapping. In *Remote Sensing of Wetlands: Applications and Advances*; CRC Press: Boca Raton, FL, USA, 2015; pp. 175–194.
59. Baghdadi, N.; Bernier, M.; Gauthier, R.; Neeson, I. Evaluation of C-band SAR data for wetlands mapping. *Int. J. Remote Sens.* **2001**, *22*, 71–88. [CrossRef]
60. Töyrä, J.; Pietroniro, A. Towards operational monitoring of a northern wetland using geomatics-based techniques. *Remote Sens. Environ.* **2005**, *97*, 174–191. [CrossRef]
61. Huang, S.; Potter, C.; Crabtree, R.L.; Hager, S.; Gross, P. Fusing optical and radar data to estimate sagebrush, herbaceous, and bare ground cover in Yellowstone. *Remote Sens. Environ.* **2010**, *114*, 251–264. [CrossRef]
62. Werninghaus, R.; Buckreuss, S. The TerraSAR-X mission and system design. *IEEE Trans. Geosci. Remote Sens.* **2010**, *48*, 606–614. [CrossRef]
63. Buckreuss, S.; Schattler, B. The TerraSAR-X ground segment. *IEEE Trans. Geosci. Remote Sens.* **2010**, *48*, 623–632. [CrossRef]
64. Dabrowska-Zielinska, K.; Budzynska, M.; Tomaszewska, M.; Malinska, A.; Gatkowska, M.; Bartold, M.; Malek, I. Assessment of carbon flux and soil moisture in wetlands applying Sentinel-1 Data. *Remote Sens.* **2016**, *8*, 756.
65. Schmitt, A.; Wendleder, A.; Hinz, S. The Kennaugh element framework for multi-scale, multi-polarized, multi-temporal and multi-frequency SAR image preparation. *ISPRS J. Photogramm. Remote Sens.* **2015**, *102*, 122–139. [CrossRef]
66. Brisco, B. Mapping and monitoring surface water and wetlands with synthetic aperture radar. In *Remote Sensing of Wetlands: Applications and Advances*; CRC Press: Boca Raton, FL, USA, 2015; pp. 119–136.
67. Mitchell, A.; Milne, A.; Tapley, I. Towards an operational SAR monitoring system for monitoring environmental flows in the Macquarie Marshes. *Wetl. Ecol. Manag.* **2015**, *23*, 61–77. [CrossRef]
68. Moser, L.; Schmitt, A.; Wendleder, A.; Roth, A. Monitoring of the Lac Bam wetland extent using dual-polarized X-band SAR data. *Remote Sens.* **2016**, *8*, 302. [CrossRef]
69. Hajnsek, I.; Busche, T.; Krieger, G.; Zink, M.; Moreira, A. *Announcement of Opportunity: TanDEM-X Science Phase*; DLR Public Document TD-PD-PL-0032, Issue: 1.0, 19 May 2014; TanDEM-X Ground Segment, Microwaves and Radar Institute(DLR-HR), Wessling, Germany, pp. 1–27.
70. Heine, I.; Jagdhuber, T.; Itzerott, S. Classification and monitoring of reed belts using dual-polarimetric TerraSAR-X time series. *Remote Sens.* **2016**, *8*, 552. [CrossRef]
71. Mleczko, M.; Mróz, M. Wetland Mapping Using SAR Data from the Sentinel-1A and TanDEM-X Missions: A Comparative Study in the Biebrza Floodplain (Poland). *Remote Sens.* **2018**, *10*, 78. [CrossRef]
72. Otukei, J.R.; Blaschke, T.; Collins, M. Fusion of TerraSAR-x and Landsat ETM+ data for protected area mapping in Uganda. *Int. J. Appl. Earth Obs. Geoinform.* **2015**, *38*, 99–104. [CrossRef]
73. Ullmann, T.; Banks, S.N.; Schmitt, A.; Jagdhuber, T. Scattering characteristics of X-, C-and L-Band polsar data examined for the tundra environment of the Tuktoyaktuk Peninsula, Canada. *Appl. Sci.* **2017**, *7*, 595.
74. Wdowinski, S.; Amelung, F.; Kim, S.W.; Dixon, T. Wetland InSAR. In *Remote Sensing of Wetlands: Applications and Advances*; CRC Press: Boca Raton, FL, USA, 2015; p. 137.
75. Hammerson, G. *Connecticut Wildlife: Biodiversity, Natural History, and Conservation*; University Press of New England: Lebanon, NH, USA, 2004.
76. Schuster, C.; Schmidt, T.; Conrad, C.; Kleinschmit, B.; Förster, M. Grassland habitat mapping by intra-annual time series analysis–Comparison of RapidEye and TerraSAR-X satellite data. *Int. J. Appl. Earth Obs. Geoinform.* **2015**, *34*, 25–34. [CrossRef]
77. Jones, M.C.; Grosse, G.; Jones, B.M.; Walter Anthony, K. Peat accumulation in drained thermokarst lake basins in continuous, ice-rich permafrost, northern Seward Peninsula, Alaska. *J. Geophys. Res. Biogeosci.* **2012**, *117*. [CrossRef]
78. Cloude, S.R.; Pottier, E. An entropy based classification scheme for land applications of polarimetric SAR. *IEEE Trans. Geosci. Remote Sens.* **1997**, *35*, 68–78. [CrossRef]
79. Cloude, S.R.; Pottier, E. A review of target decomposition theorems in radar polarimetry. *IEEE Trans. Geosci. Remote Sens.* **1996**, *34*, 498–518. [CrossRef]
80. Betbeder, J.; Rapinel, S.; Corgne, S.; Pottier, E.; Hubert-Moy, L. TerraSAR-X dual-pol time-series for mapping of wetland vegetation. *ISPRS J. Photogramm. Remote Sens.* **2015**, *107*, 90–98. [CrossRef]

81. Zalite, K.; Voormansik, K.; Olesk, A.; Noorma, M.; Reinart, A. Effects of inundated vegetation on X-band HH–VV backscatter and phase difference. *IEEE J. Sel. Top. Appl. Earth Obs. Remote Sens.* **2014**, *7*, 1402–1406. [CrossRef]

82. Tiner, R.W. Classification of wetland types for mapping and large-scale inventories. In *Remote Sensing of Wetlands: Applications and Advances*; CRC Press: Boca Raton, FL, USA, 2015; pp. 19–42.

83. Ullmann, T.; Schmitt, A.; Jagdhuber, T. Two component decomposition of dual polarimetric HH/VV SAR data: Case study for the tundra environment of the Mackenzie Delta region, Canada. *Remote Sens.* **2016**, *8*, 1027. [CrossRef]

84. Lee, Y.K.; Park, J.W.; Choi, J.K.; Oh, Y.; Won, J.S. Potential uses of TerraSAR-X for mapping herbaceous halophytes over salt marsh and tidal flats. *Estuar. Coast. Shelf Sci.* **2012**, *115*, 366–376. [CrossRef]

85. Ullmann, T.; Schmitt, A.; Roth, A.; Duffe, J.; Dech, S.; Hubberten, H.W.; Baumhauer, R. Land cover characterization and classification of arctic tundra environments by means of polarized synthetic aperture X-and C-Band Radar (PolSAR) and Landsat 8 multispectral imagery—Richards Island, Canada. *Remote Sens.* **2014**, *6*, 8565–8593. [CrossRef]

86. Rangoonwala, A.; Enwright, N.M.; Ramsey III, E.; Spruce, J.P. Radar and optical mapping of surge persistence and marsh dieback along the New Jersey Mid-Atlantic coast after Hurricane Sandy. *Int. J. Remote Sens.* **2016**, *37*, 1692–1713. [CrossRef]

87. Zhou, C.; Gong, H.; Chen, B.; Li, J.; Gao, M.; Zhu, F.; Chen, W.; Liang, Y. InSAR time-series analysis of land subsidence under different land use types in the Eastern Beijing Plain, China. *Remote Sens.* **2017**, *9*, 380. [CrossRef]

remote sensing

MDPI

Article

Mitigation of Tropospheric Delay in SAR and InSAR Using NWP Data: Its Validation and Application Examples

Xiaoying Cong [1,2], Ulrich Balss [3], Fernando Rodriguez Gonzalez [3] and Michael Eineder [3,*]

[1] Previously with Technische Universität München, Arcisstr. 21, D-80333 Munich, Germany;
    xiaoying.cong@gmail.com
[2] Currently with ADC Automotive Distance Control Systems GmbH, Peter-Dornier-Str. 10,
    D-88131 Lindau, Germany
[3] DLR, Remote Sensing Technology Institute, Muenchener Str. 20, D-82234 Wessling, Germany;
    ulrich.balss@dlr.de (U.B.); Fernando.RodriguezGonzalez@dlr.de (F.R.G.)
*   Correspondence: Michael.Eineder@dlr.de; Tel.: +49-8153-28-1396

Received: 17 July 2018; Accepted: 11 September 2018; Published: 21 September 2018

Abstract: The neutral atmospheric delay has a great impact on synthetic aperture radar (SAR) absolute ranging and on differential interferometry. In this paper, we demonstrate its effective mitigation by means of the direction integration method using two products from the European Centre for Medium-Range Weather Forecast: ERA-Interim and operational data. Firstly, we shortly review the modeling of the neutral atmospheric delay for the direct integration method, focusing on the different refractivity models and constant coefficients available. Secondly, a thorough validation of the method is performed using two approaches. In the first approach, numerical weather prediction (NWP) derived zenith path delay (ZPD) is validated against ZPD from permanent GNSS (global navigation satellite system) stations on a global scale, demonstrating a mean accuracy of 14.5 mm for ERA-Interim. Local analysis shows a 1 mm improvement using operational data. In the second approach, NWP derived slant path delay (SPD) is validated against SAR SPD measured on corner reflectors in more than 300 TerraSAR-X High Resolution SpotLight acquisitions, demonstrating an accuracy in the centimeter range for both ERA-Interim and operational data. Finally, the application of this accurate delay estimate for the mitigation of the impact of the neutral atmosphere on SAR absolute ranging and on differential interferometry, both for individual interferograms and multi-temporal processing, is demonstrated.

Keywords: SAR; SAR interferometry; atmospheric propagation delay; persistent scatterer interferometry; numerical weather prediction; stratified atmospheric delay; zenith path delay; slant path delay

---

1. Introduction

The atmosphere affects the propagation of radar signals and causes delays in the range direction. There are two main effects: the dispersive delay caused by the ionosphere and the neutral delay caused mainly by the lower atmosphere. The first effect depends on the radar frequency and has an impact of several centimeters in X-band. Fortunately, this effect can be determined and compensated efficiently using total electron content (TEC) maps based on multi-frequency measurements from regional/global networks of global navigation satellite system (GNSS) [1,2] or using the split-spectrum method in interferograms [3,4]. The neutral atmospheric delay has an average value in the meter range and exhibits variations over space and time in the decimeter-level. It is thus mandatory to mitigate it in order to derive accurate absolute and/or relative synthetic aperture radar (SAR) range measurements.

Several methods have been developed for the mitigation of the neutral atmospheric delay exploiting external measurements. In [5,6], the delay is approximated using surface meteorological measurements, achieving an accuracy of several decimeters. The main limitation of such ground-based techniques as well as those based on space-borne water vapor measurements is the sparse spatial and temporal distribution of data [7,8]. Therefore, several methods based on regional or global numerical weather prediction (NWP) products have been proposed [9,10]. In [11–13], meso-scale weather models are applied with input from NWP data in order to simulate the weather situation at the exact acquisition time and with a higher spatial resolution. Their accuracy depends on the quality of the input NWP data and varies from centimeter to decimeter range.

The integration method has been introduced in [10,14,15]. It calculates the neutral atmospheric delay using global reanalysis data provided by the European Centre for Medium-Range Weather Forecasts (ECMWF). Its absolute accuracy has been validated against GPS zenith path delay (ZPD) measurements. In [14,15] the application of global NWP products to mitigate neutral atmospheric delay effects in InSAR measurement has been demonstrated. In both studies, the neutral atmospheric delay was firstly integrated in the zenith direction and then projected in the slant direction. In contrast, the direct integration method introduced in [10] integrates the refractivity along the actual propagation path of the radar echoes.

In this paper we focus on two aspects: firstly, a validation of the integrated delay using global NWP products and, secondly, an analysis of the impact of the neutral atmospheric delay on absolute range measurements and differential interferometry and its mitigation. The paper is organized in five sections. In Section 2, the methodology is introduced, discussing the different models and their coefficients. The validation experiments are presented in Section 3. In Section 4 the applications are demonstrated: firstly for absolute ranging measurements and secondly for interferometric measurements. Finally, a short conclusion is provided in Section 5.

2. Methodology

2.1. Atmospheric Path Delay in the Neutral Atmosphere

In the lower atmosphere (<80 km), gravitation and air pressure regulate the vertical atmospheric motion in a hydrostatic equilibrium. Under this condition, the troposphere tends to be stratified and air density is a function of temperature and pressure, which are themselves highly dependent on height. Their relationship can be expressed as Equation (1) [16]:

$$\frac{\partial P}{\partial z} = -\frac{Pg}{R_{ideal}T} = -\frac{Pg}{R_d T_v}, \tag{1}$$

where P is pressure in Pascal (Pa); z is geometric height in meter (m); T is temperature in Kelvin (K); g is the gravitational acceleration; R_{ideal} is a constant for 1 kg of an ideal gas, e.g., $R_d = 287.0 \text{ J K}^{-1}\text{kg}^{-1}$ is the gas constant for 1 kg of dry air and $R_w = 461.51 \text{ J K}^{-1}\text{kg}^{-1}$ the gas constant for 1 kg of water vapor; T_v is the virtual temperature, which is used to derive the temperature lapse rate of moist air [16].

In this section, the air refractivity equation for the neutral atmosphere as well as its alternative expressions associated with different coefficient constants are presented. Based on the statistical analysis in Section 2.1.3, an approximated expression of air refractivity has been selected for delay integration.

2.1.1. Air Refractivity and Neutral Atmospheric Propagation Delay

The refractive index n varies along the radar echo propagation path. The refractivity $(n-1)$ describes the variation with respect to vacuum due to the atmosphere. For convenience, scaled-up refractivity $N = (n - 1) \times 10^6$ is introduced. Neglecting non-ideal gas effects and the effect of the ionosphere, air refractivity N can be written as [17–19]:

$$N = k_1 \underbrace{\frac{P_d}{T}}_{\text{dry air}} + k_2 \underbrace{\frac{e}{T}}_{\text{wet air}} + k_3 \frac{e}{T^2} + \underbrace{1.45 \cdot W_{cl}}_{\text{liquid water}}, \tag{2}$$

where P_d is the partial pressure of dry air in hPa; e is the partial pressure of water vapor in hPa; T is the absolute temperature in K; W_{cl} is the cloud water content in g/m^3. The constants k_1, k_2 and k_3 are derived from laboratory measurements and under different approximations, which are discussed in Section 2.1.3. Note that this equation does not account for the refractivity of the ionosphere, since it is not part of the neutral atmospheric delay.

Ignoring the effect of the clouds, two expressions are available in the literature: under assumption of an ideal gas Equation (3) and of non-ideal gases Equation (4):

$$N_{atmo,ideal} = k_1 \frac{P}{T} + (k_2 - k_1) \frac{e}{T} + k_3 \frac{e}{T^2}, \tag{3}$$

$$N_{atmo,non-ideal} = k_1 \frac{P}{T_v} + (k_2 - \varepsilon k_1) \frac{e}{Z_w T} + k_3 \frac{e}{Z_w T^2}, \tag{4}$$

where the constant ε is defined as $R_d / R_w = 0.622$ and Z_w is the compressibility of water vapor.

2.1.2. Integration of the Neutral Atmospheric Delay and Its Alternative Expressions

The total neutral atmospheric delay can be defined as the difference between the radio path length and the straight-line distance [20], which is divided into two parts: the signal delay along the propagation path and the geometric delay due to the bending effect. In this paper, the ray-path bending effect is neglected due to the rather steep incidence angle (typically 20–45°) of SAR acquisitions. Therefore, the neutral atmospheric delay is approximated as the integral of air refractivity along the straight-line path from Earth's surface to the upper limit of the available NWP data. The integration in zenith direction L_{atmo} and in slant direction L_s can be expressed:

$$L_{atmo} = 10^{-6} \int_{z_S}^{z_{atmo}} (N(z)) dz, \quad L_s = 10^{-6} \int_{\vec{r}_S}^{\vec{r}_{atmo}} (N(r)) dr. \tag{5}$$

where z_S is the height of Earth's surface; z_{atmo} represents the upper limit of the integration path; $\vec{r}_S$ is the starting point of integration; $\vec{r}_{atmo}$ indicates the end point of the integration path.

Different assumptions have been performed for different applications. Thus the integration has been formulated according to different expressions. Taking the integration in zenith direction as an example, three groups of expressions have been defined under the assumptions of: (1) ideal gases; (2) non-ideal gases; and (3) non-ideal gases based on an approximated equation. For each of them the zenith delay can be written as a combination of hydrostatic $L_{h,*}$ and wet delay $L_{w,*}$:

(1) $L_{atmo,ideal} = L_{h,ideal} + L_{w,ideal}$;
(2) $L_{atmo,non-ideal} = L_{h,non-ideal} + L_{w,non-ideal}$;
(3) $L_{atmo,approx} = L_{h,approx} + L_{w,approx}$.

For an ideal gas, the hydrostatic delay $L_{h,ideal}$ can be approximated as a function of surface pressure P_S and the wet delay $L_{w,ideal}$ is given by the last two terms of Equation (3), thus

$$L_{atmo,ideal} = 10^{-6} \frac{k_1 R_d P_S}{g_m} + 10^{-6} \int_{z_S}^{z_{atmo}} \left((k_2 - k_1) \frac{e(z)}{T(z)} + k_3 \frac{e(z)}{T(z)^2} \right) dz, \tag{6}$$

where P_S is given by the last layer (first level) of the NWP product and g_m is the mean gravity acceleration at the mass center of the atmospheric column. In [21] an approximate form for g_m is provided as a function of latitude ϕ and geometric height z:

$$g_m \approx g_0(1 - 0.00266cos(2\phi) - 0.28 \cdot 10^{-6}z), \tag{7}$$

where the constant g_0 is equal to 9.7840 m/s². For non-ideal gases, according to Equation (4) the integration can be formulated be as:

$$L_{atmo,non-ideal} = 10^{-6} \int_{z_S}^{z_{atmo}} \left(k_1 \frac{P(z)}{T_v(z)} + k_2' \frac{e(z)}{Z_w T(z)} + k_3 \frac{e(z)}{Z_w T(z)^2} \right) dz, \tag{8}$$

where $k_2' = (k_2 - \varepsilon k_1)$.

Finally an approximate non-ideal gas hydrostatic equation can be derived as

$$L_{atmo,approx} = 10^{-6} \int_{z_S}^{z_{atmo}} \left(k_1 \frac{P(z)}{T(z)} + k_2' \frac{e(z)}{T(z)} + k_3 \frac{e(z)}{T(z)^2} \right) dz. \tag{9}$$

2.1.3. Coefficient Constants

The coefficient constants k_1, k_2 and k_3 are empirical values which can be retrieved from laboratory experiments, such as in [18,22]. For geodetic applications, different sets of coefficient constants have been proposed in [21,23,24]. Recently, [19,25] suggest more accurate coefficient constants for precise geodetic applications and GPS radio occultation. The coefficient constants proposed in the literature for the different integral equations are summarized in Table 1. In order to compare them, the ZPD at the Wettzell GNSS station has been integrated for each of the available NWP products from August 2013, generating thus one delay every six hours for each of them. The ZPD integrated according to Equation (9) and the coefficient constants from [18] are used as reference. The mean value and the standard deviation of the residual ZPDs after compensating the reference ZPD are summarized in Table 1.

Table 1. Summary of the coefficient constants k_1, k_2 and k_3 from [18,19,21–25]. They are divided into three groups: (1) ideal gases; (2) non-ideal gases; (3) non-ideal gases based on an approximated equation. Differences relative to the reference ZPD ($L_{atmo,approx}$ with rounded coefficient constants [18]) are calculated for NWP products during August 2013 at the Wettzell GNSS station. The mean value and the standard deviation of the differences are presented.

Authors (Year)	k_1 (K/hPa)	k_2 (K/hPa)	$k_3 \times 10^{-5}$ (K²/ hPa)	ZPD Diff. (mm) Mean	Std
Ideal Gas					
Smith and Weintraubt [18]	77.607 ± 0.01	71.600 ± 8.5	3.747 ± 0.031	-0.2	0.1
Saastamoinen [21]	77.624	64.70 ± 0.08	3.719	-0.9	0.4
Rüeger [25] (best available)	77.695	71.970	3.75406	2.4	<0.1
Rüeger [25] (best average)	77.6890	71.2952	3.75463	2.3	<0.1
Healy [19]	77.643	71.2952	3.75463	1.1	<0.1
Non-ideal Gases					
Thayer [22]	77.60 ± 0.014	64.80 ± 0.08	3.776 ± 0.004	-2.2	0.6
Davis et al. [23]	77.604 ± 0.008	65.27 ± 10	3.776 ± 0.03	-2.1	0.6
Healy [19]	77.643	71.2952	3.75463	-1.7	0.8
Non-ideal Gases—Approximated Equation					
Bevis et al. [24]	77.60 ± 0.05	70.40 ± 2.2	3.739 ± 0.012	-3.6	1.0

Regarding the statistical analysis of ZPD differences presented in Table 1, the mean value varies from −3.6 to 2.4 mm and the standard deviation from <0.1 mm to 1.0 mm. In general, the reference ZPD is close to the ZPDs under the assumption of an ideal gas, since their standard deviations are all <0.5 mm. The minimum offset of −0.2 mm is observed for the ideal gas using the same constants [18] as the reference ZPD. The maximum standard deviation (1.0 mm) and mean value (−3.6 mm) are yielded by the approximated formula for non-ideal gases, been the difference to the reference ZPD thus caused by the different constant values. The next closest constants to [18] are provided by [19] under the assumption of an ideal gas, where the standard deviation is less than 0.1 mm and the offset is about 1.1 mm. In summary, for an ideal gas, the discrepancies in terms of mean value are mainly caused by the difference in k_1; for non-ideal gases, the differences are primarily induced by different formulae of the hydrostatic delay $L_{h,*}$.

The approximated integral of Equation (9) and the rounded coefficient constants [18], which were used as reference, have been selected for our application. Therefore, after replacing these constants, the approximated air refractivity is given by

$$N_{atmo} = 0.776 \cdot \frac{P}{T} + 0.2333 \cdot \frac{e}{T} + 3.75 \times 10^3 \frac{e}{T^2}. \tag{10}$$

2.2. Direct Integration Method Using Global NWP Products

The direct integration method has been introduced in [10,26]. It has three processing steps: preprocessing and preparation, horizontal interpolation and direct integration. According to the input scene parameters, such as acquisition time, scene location and satellite orbit information, the corresponding NWP data sets are downloaded from the ECMWF product server. In the first step, the NWP products are preprocessed to a regular latitude-longitude grid and each model level is related to a geometrical height [27]. A proper horizontal interpolation is required due to the coarse horizontal resolution of NWP data. Finally, the integration starts from a given location (for instance, a GPS receiver or a SAR image pixel) and ends at the top of NWP data. The meteorological parameters at each integration step are interpolated or extrapolated along the integration path in zenith or in slant range direction in order to evaluate and integrate refractivity.

2.3. Terminology: Tropospheric Delay

The neutral atmospheric delay is integrated until the top level of the NWP data, which depending on the ECMWF product has a geometric altitude of approximately 65 km (0.2 hPa for ERA-Interim) (https://www.ecmwf.int/en/forecasts/documentation-and-support/60-model-levels) or 79 km (0.02 hPa for operational data) (https://www.ecmwf.int/en/forecasts/documentation-and-support/137-model-levels). Thus the integration is performed across the troposphere, the stratosphere and the mesosphere. Commonly, in the SAR community this neutral atmospheric delay is called *tropospheric delay*, since the troposphere has the highest contribution to the neutral atmospheric delay. In the following we will follow this convention, namely we will use the term tropospheric delay instead of neutral atmospheric delay.

3. Validation of Integrated Tropospheric Delay

3.1. ECMWF Products

In the last decades, major improvements in NWP accuracy have been achieved thanks mainly to increased computational power and increased observation coverage. Nowadays, the NWP models provide a variety of atmospheric parameters on a global scale with high horizontal and vertical resolution. It provides a unique possibility to derive the necessary 3-dimensional atmospheric parameters for delay integration at a given analysis time, namely P, T and e.

Two kinds of NWP products generated by the ECMWF Integrated Forecast System (IFS) have been selected for integration: reanalysis data sets and operational data (OP). Reanalysis data are produced with a fixed IFS version and at a fixed resolution. ERA-Interim is the current ECMWF global atmospheric reanalysis, providing data sets from 1979 up to a 3 months delay behind real time [28]. It will be fully replaced by ERA5 reanalysis by the end of 2018 [29], the next generation of reanalysis data set. ERA5 reanalysis products are available hourly compared to the 6-hour interval of ERA-Interim and OP. Reanalysis data sets can be accessed via the ECMWF public data service. The operational forecast based on the up-to-date IFS version enables a near real-time analysis with a wider range of atmospheric parameters only for registered users. The basic product specifications, such as horizontal and vertical resolution and IFS release cycle are presented in Table 2. The selection of NWP products depends on access authorization, data availability and the application.

In this study, both reanalysis and operational data are used. The validation is focused on ERA-Interim and OP. ERA-Interim data are downloaded from the ECMWF public data sets (http://apps.ecmwf.int/datasets/data/interim-full-daily/levtype=ml/). OP are acquired from archive catalogue.

Table 2. Summary of ECMWF reanalysis data sets (ERA-Interim and ERA5) and OP specifications: horizontal resolution in latitude/longitude; vertical resolution model level (ML); IFS release cycle and date [28]. Only releases with major changes, either in spatial resolution or ML, are reported.

ECMWF Products	Horizontal Resolution (deg) [1]	Vertical Resolution [1]	IFS Release [2]	Release Date
	0.100	137-Level	*Cycle 41r2*	8 March 2016
	0.125	137-Level	*Cycle 38r2*	26 June 2013
Operational	0.125	91-Level	*Cycle 38r1*	11 March 2008
	0.225	91-Level	*Cycle 30r1*	1 February 2006
	0.350	60-Level	*Cycle 23r3*	21 November 2000
ERA-Interim	0.750	60-Level	*Cycle 31r2*	1 January 2006
ERA5	0.280	137-Level	*Cycle 41r2*	1 January 2018

[1] https://www.ecmwf.int/en/forecasts/documentation-and-support; [2] https://www.ecmwf.int/en/forecasts/documentation-and-support/changes-ecmwf-model.

3.2. Validation Approaches

The ECMWF IFS performs each data assimilation with millions of observations from an extensive range of inputs, which are not error-free. Therefore, it is important to determine the accuracy and the reliability of NWP products for the calculation of the integrated tropospheric delay. According to the air refractivity equation (Equation (10)), the accuracy of the delay depends on the accuracies of P, T and e. An overview of the general assessment of ERA-Interim products in comparison with GPS ZPD is provided in [28]. Compared to the accuracy of P and T, the accuracy of water vapor (related to e) is limited due to the modeling complexity and to the accuracy of the input data [30–33].

A validation approach is proposed in order to evaluate the accuracy of ECMWF integrated tropospheric delay using GPS and corner reflector (CR) observations as reference data sets. A sketch of the validation concept is depicted in Figure 1. The reference tropospheric delays are ZPD derived from GPS measurements (ZPD_{GPS}) and SPD estimated from CR observations (SPD_{CR}) using the *Imaging Geodesy* method [34]. The tropospheric delays are integrated both in zenith (ZPD_{ECMWF}) and in slant range (SPD_{ECMWF}) direction using ECMWF products in order to compare them to the reference data sets. In Section 3.3, ZPD_{ECMWF} from both ERA-Interim and OP are compared to ZPD_{GPS} from the International GNSS Service (IGS) [35]. In Section 3.4, a comparison of SPD_{ECMWF} from both ERA-Interim and OP with SPD_{CR} at two GNSS stations is presented.

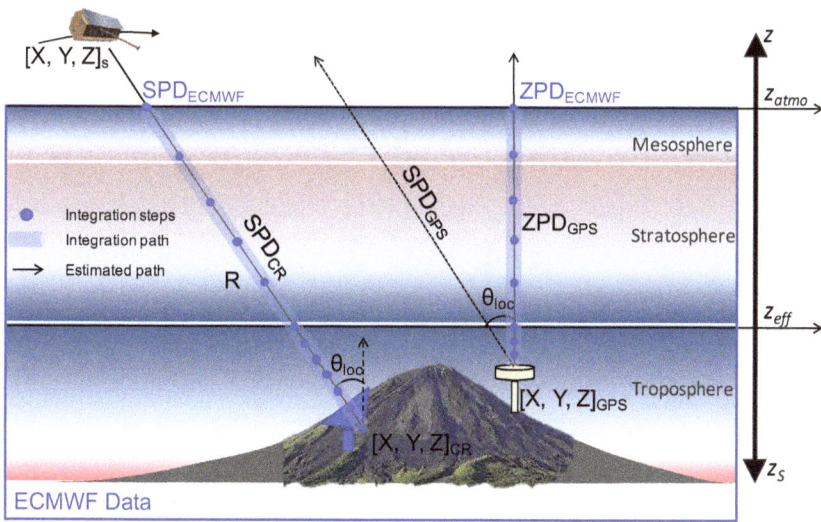

Figure 1. Sketch of validation methods. Types of delay measurements: (1) ZPD from GPS measurements (ZPD$_{GPS}$) estimated at the GPS coordinates $[X, Y, Z]_{GPS}$ and SPD projected with the local incidence angle θ_{loc} (SPD$_{GPS}$); (2) SPD from CR measurements (SPD$_{CR}$) at the CR coordinates $[X, Y, Z]_{CR}$; (3) integrated ECMWF delays: ZPD$_{ECMWF}$ at $[X, Y, Z]_{GPS}$ and SPD$_{ECMWF}$ at $[X, Y, Z]_{CR}$. The integration steps are visualized as points over the integration path, with a smaller step under the effective height z_{eff} and a bigger step over z_{eff} until z_{atmo}.

3.3. Validation of Integrated ZPD with GPS ZPD

For precise positioning purposes the tropospheric delay must be estimated in the GPS analysis [21,23]. GPS ZPDs are not assimilated at the ECMWF [36]. Hence, ECMWF ZPDs are independent from GPS ZPDs. GPS ZPD has been widely used for comparison against the integrated total delay and/or the integrated delay of water vapor (ZWD) derived from radiosondes, microwave radiometers and NWP products [37–39]. The accuracy of GPS ZPD measurements has been proven to be of about 5 mm [38].

3.3.1. Time-Series Comparison Using ERA-Interim and OP Data Sets at the Wettzell Station

A time series of ECMWF ZPD has been generated using 1460 ERA-Interim and OP data fields at the Wettzell station in 2015. For validation purposes, the corresponding GPS ZPD time series has been gathered from IGS. Unfortunately, GPS ZPDs are not always available during this year. There is a 17 days data-gap. The single ECMWF ZPD for each analysis time (0, 6, 12 and 18 h UTC) is integrated from the station coordinates to the top of the NWP data. Furthermore, empirical ZPDs are calculated based on the GPT2w model [40].

The comparison is presented in Figure 2. GPS ZPD varies about 29 cm along the year, from 2.101 m in winter to 2.391 m in summer. Both integrated ZPDs using ERA-Interim and OP exhibit a good agreement with GPS ZPD. The differences between integrated ZPDs and GPS ZPD are depicted in Figure 2b. The standard deviation of differences is about 10.9 mm for ERA-Interim and 9.0 mm for OP. The offsets are both under 5 mm, 4.9 mm for ERA-Interim and 1.6 mm for OP. The empirical model can capture the seasonal trend of the delay, but the standard deviation of the residual difference is 32.4 mm, approximately three times higher than that of integrated delays.

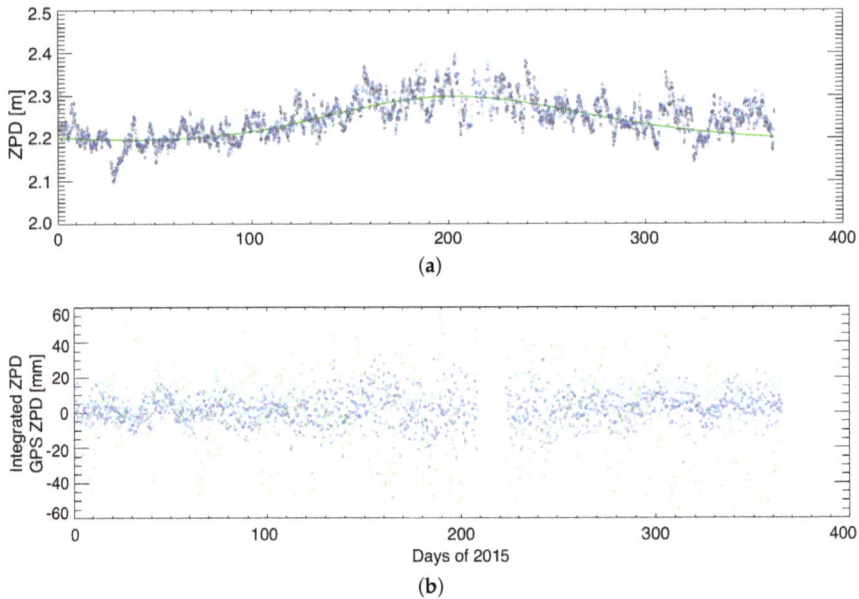

Figure 2. Time-series analysis of integrated and approximated ZPD during 2015 at each analysis time (0, 6, 12 and 18 h UTC) compared to GPS ZPD at the Wettzell station. (**a**) Absolute ZPD: integrated using ERA-Interim (cyan) and OP (blue), empirical from GPT2w model (green) and GPS (black). (**b**) Difference between integrated and empirical ZPDs and GPS ZPD with same color coding.

3.3.2. Global Validation of ERA-Interim Products

In order to test the quality of ECMWF products globally, the validation of ERA-Interim products is extended to a global scale using 268 permanent stations from the IGS network as reference. The time-series validation approach presented in Section 3.3.1 is carried out for the GPS stations of the IGS network over the time period from 2010 to 2015. The average value of the ZPD over time is presented in Figure 3a. The maximum delay is 2.672 m and the minimum is 1.515 m. The average over the 268 stations is 2.345 m. The magnitude of the temporal variation of the delay can be represented by the standard deviation of the ZPD over time for a given station, which varies from 15.1 mm to 98.1 mm and has an average value over the stations of 50.7 mm.

The difference between integrated ZPD and GPS ZPD is calculated in order to evaluate the accuracy of the ERA-Interim integrated delay. The standard deviation of the differences is depicted in Figure 3b. The average standard deviation over the 268 GPS stations is 14.5 mm, with a minimum of 5.1 mm and a maximum of 38.0 mm. The accuracy of the integrated delay is directly related to the water vapor content distribution. In higher latitudes the water vapor content is lower than in lower latitudes due to the lower temperatures at higher latitudes [41]. In order to visualize this dependency of the integrated delay accuracy on latitude, a scatter plot with the absolute latitude as X-axis and the standard deviation of the difference as Y-axis is shown in Figure 4 (right) color coded according to the station height. The accuracy of the integrated delay in higher latitudes (>55°) is better than 1.5 cm. In lower latitudes (<30°), the accuracy is mostly between 1.5 and 3.0 cm due to the higher water vapor content. No obvious dependency on height can be observed. An analogous scatter plot for the mean integrated ZPD is shown in Figure 4 (left), where a clear dependency on absolute latitude as well as on height can be observed.

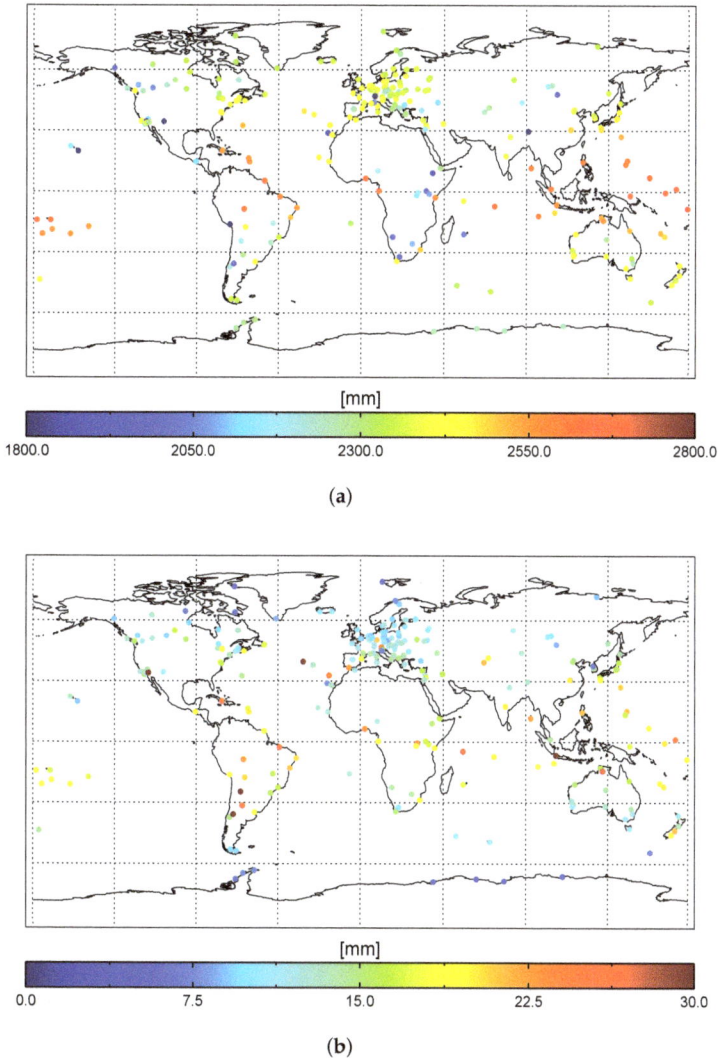

Figure 3. Global validation of ERA-Interim products from 2010 to 2015 based on 268 GPS stations of the IGS network: (**a**) mean value of the integrated ZPD; (**b**) standard deviation of the difference between the integrated ZPD and GPS ZPD.

3.4. Validation of Integrated SPD with SPD Estimated from CR Measurements

As introduced in [34], the mismatch between the measured absolute range of a target in a SAR image and its predicted position (calculated using the orbit information and its known absolute position) is caused not only by the tropospheric delay and measurement noise, but also by the ionospheric delay as well as delays caused by geodetic effects, such as solid Earth tides (SET), ocean tidal loading (OTL), atmospheric pressure loading (APL) and continental drift (CD). Atmospheric (tropospheric and ionospheric) delays and geodetic effects must be considered in order to achieve centimeter-level ranging accuracy [1,26,34]. The tropospheric delay in SAR range measurements (SAR

SPD) can be estimated if all the other effects have been accurately corrected for on the range mismatch. This SAR SPD can be used for validation of the ECMWF SPD. Furthermore, if there is a permanent GPS station nearby then the slant tropospheric delay GPS SPD, here calculated from the GPS ZPD using the secant of the SAR incidence angle as mapping function, can be used as an additional reference for validation. In this section, both SAR SPD and GPS SPD are used as reference for validation of the ECMWF integrated SPD.

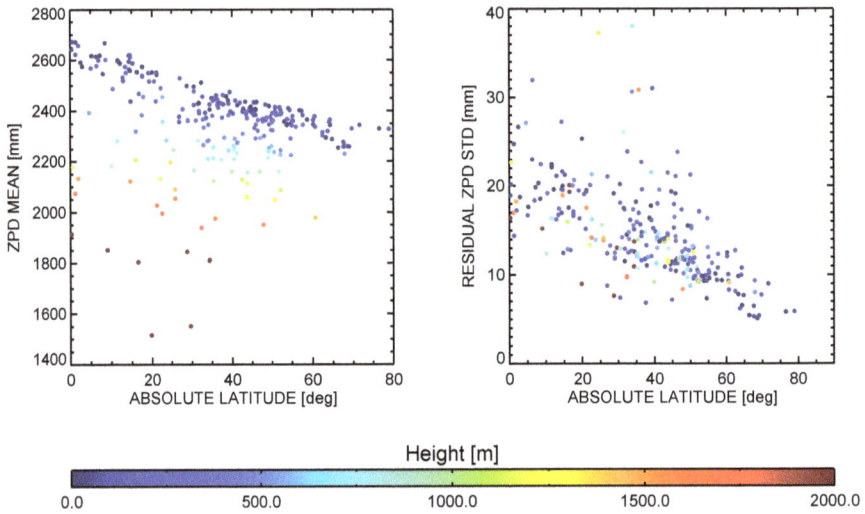

Figure 4. Dependency on absolute latitude of mean ERA-Interim ZPD and standard deviation of ZPD residuals with respect to GPS ZPD evaluated at IGS stations. (**Left**) Mean integrated ZPD as a function of absolute latitude. (**Right**) Standard deviation of the difference between integrated and GPS ZPD as a function of absolute latitude. The color coding is used to indicate the height of the IGS stations.

For this experiment, three trihedral CRs with 1.5 m edge length were installed at two test sites close to permanent GNSS stations: two at the Wettzell geodetic observatory (WTZR) in Germany, with WTZR 01 oriented for ascending (ASC) orbits and WTZR 02 for descending (DSC) orbits, and one at the Metsähovi observatory in Finland (METS) for DSC orbits [42–44]. High resolution SAR images from the two German satellites TerraSAR-X (TSX-1) and TanDEM-X (TDX-1) have been acquired regularly: 176 scenes from July 2012 to October 2015 at WTZR and 155 scenes from December 2013 to February 2016 at METS. The acquisitions were acquired around 17:00 UTC at WTZR 01 and around 05:00 UTC at WTZR 02. At METS, SAR scenes were acquired around 05:00 UTC. ERA-Interim and OP data sets have been collected around the acquisition time, prepared according to the CR coordinate and integrated directly from the phase center of the CR to the upper limit of the ECMWF data in the direction towards the satellite position. In Figure 5, GPS SPD and ECMWF SPD show a clear agreement with SAR SPD with a correlation coefficient of 99.8% in WTZR. Measurements are color coded according to the incidence angle class, i.e. the acquisition geometry. The impact of tropospheric delay increases in absolute value with increasing incidence angle.

Visually, GPS SPD has a better agreement with SAR SPD than the integrated delay based on ECMWF data sets (see Figure 5). We summarize the statistics of the comparison in Table 3 for each CR. Thanks to the higher temporal resolution (and to the fact that a GPS station is locally available), the best accuracy for the three CRs is achieved using GPS SPD as tropospheric delay corrections, with an accuracy that varies from 9.7 to 13.6 mm. The correction accuracy based on both ECMWF data sets are similar, having OP data a slightly better performance (around 1 mm) than ERA-Interim. The maximum mean offsets (up to −27.6 mm) are observed in WTZR 01, which are mainly caused by the local ground subsidence that occurred after the installation of the CR [44].

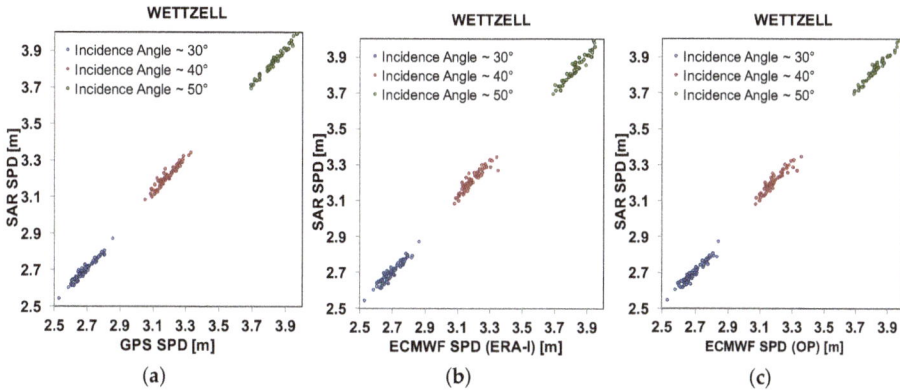

Figure 5. Correlation analysis of absolute localization residuals of both CRs in WTZR: (**a**) scatter plot of the mapped range delay based on GPS ZPD (GPS SPD) and the SAR range residuals (SAR SPD). (**b**) scatter plot of the range delays integrated using ERA-Interim data (ECMWF SPD) and SAR SPD. (**c**) scatter plot of the range delays integrated using OP data (ECMWF SPD) and SAR SPD. Each dot represents a SAR acquisition, which is color coded according to the different incidence angle classes: ∼30° in blue; ∼40° in red and ∼50° in green.

Table 3. Validation results using SAR SPDs as reference.

CR ID	Input Data	Mean (mm)	STD (mm)
	GPS ZPD	−20.5	9.7
WTZR 01 (ASC)	ERA-Interim	−27.6	16.7
	Operational	−25.9	15.7
	GPS ZPD	−8.6	13.6
WTZR 02 (DSC)	ERA-Interim	−8.2	17.6
	Operational	−9.6	16.2
	GPS ZPD	−0.6	11.8
METS (DSC)	ERA-Interim	−2.4	15.2
	Operational	−3.2	14.7

4. Applications

The absolute accuracy of the tropospheric delay calculated with the direct integration method using ECMWF data has been validated against both GPS and CR measurements, demonstrating an accuracy in the centimeter range (Section 3). Note that the accuracy estimates also contain errors in the reference data used (GPS ZPD [38] and SAR SPD [1,44]).

In this section, we present several applications using the direct integration method to mitigate the tropospheric impact on SAR absolute range measurements (Section 4.1) and on SAR interferometry (Section 4.2).

4.1. Application to Absolute Ranging Measurements

In the framework of the DLR@Uni Munich Aerospace Project "Geodetic Earth Observation", five CRs have been installed in three test sites [43,44]: two CRs at the geodetic observatory in Wettzell (WTZR) in Germany, two at GARS O'Higgins (OHIS) in the Antarctic Peninsula and one at Metsähovi

(METS) in Finland. The two CRs in WTZR and the CR in METS have been used in Section 3.4 for ECMWF SPD validation purposes. From 2007 to 2014 over four hundreds of High Resolution SpotLight (HRSL) images have been acquired by both the TSX-1 and the TDX-1 satellites from ASC and DSC orbits. The acquisitions started firstly in July 2011 in WTZR, then in March 2013 in OHIS and finally in October 2013 in METS. Due to the near-polar orbit, most acquisitions are gathered in OHIS, where 10 different viewing geometries are available.

As discussed in Section 3.4, discrepancies between expected and measured absolute ranges are caused by geodetic and atmospheric effects [1,26,34]. Assuming that geodetic effects and ionospheric delay have been already compensated as described in [43,44], the remaining tropospheric delay can be compensated using either the IGS SPD (mapped ZPD) or the integrated ECMWF SPD using ERA-Interim data sets. A statistical analysis of the ranging residuals after these corrections has been performed. Only the data sets acquired before January 2015 have been evaluated. The results are reported in Table 4.

The mean values of the residuals are similar when using either IGS SPD or ECMWF SPD. However, the standard deviation varies from 11.2 mm at WTZR (ASC) using IGS SPD to 20.4 mm at OHIS (ASC) using ECMWF SPD. At WTZR, the difference of the standard deviation between IGS SPD and ECMWF SPD is more than 5 mm, namely 7.6 mm for ASC and 6.7 mm for DSC. At METS, the difference decreases to 3.9 mm. Moreover, ECMWF SPD has a slightly better performance in OHIS ASC. These accuracy differences are mainly influenced by the different temperatures at each test site, which can be expressed as a function of latitude. For instance, test sites located at higher latitudes receive less sunlight than those at lower latitudes. Therefore, since the average temperature is lower at higher latitudes (e.g., METS and OHIS) the atmosphere can hold less water vapor, which leads to smaller water vapor variability and thus can be better modeled by NWP model (see Section 3.3.2) [16].

Table 4. Statistical analysis of absolute range residuals using IGS SPD (mapped ZPD) and integrated ECMWF SPD at test sites: Wettzell (WTZR) in Germany, GARS O'Higgins (OHIS) at the Antarctic Peninsula and Metsähovi (METS) in Finland.

Test Site Code	Crossing Orbit Direction	IGS SPD		ECMWF SPD	
		Mean (mm)	STD (mm)	Mean (mm)	STD [mm]
WTZR	ASC	−17.1	11.2	−21.5	18.8
	DSC	−8.8	13.6	−9.3	20.3
OHIS	ASC	−7.1	20.3	−5.3	20.4
	DSC	4.9	17.2	3.4	16.3
METS	DSC	−3.2	11.4	−4.5	15.3

4.2. Application to Interferometric Measurements

For SAR Interferometry (InSAR), the relative tropospheric delay between two acquisitions is of interest. According to its height dependency, the tropospheric delay can be divided into two parts: a height-dependent delay dominated by vertical stratification, the so-called stratified tropospheric delay, and a height-independent delay induced by turbulent mixing [6,17], which is described as a spatially correlated and temporally decorrelated effect. Let us illustrate these two delay components by means of interferograms of the Hierro Island in Spain. Firstly, in Figure 6c a height-dependent tropospheric delay can be clearly observed, which is solely due to different meteorological conditions. Secondly, the tropospheric delay effect induced by the turbulent atmosphere can be seen in Figure 6d, where in contrast to Figure 6c no correlation with height can be observed.

Two application test cases have been selected for interferometric applications. Firstly, the impact of tropospheric delay mitigation on persistent scatterer interferometry (PSI) estimation is demonstrated by using stacks of TerraSAR-X HRSL images of the Stromboli Volcano (Section 4.2.1). Secondly, an application to a stripe of Sentinel-1 interferograms in a mountainous area with a coverage of around 250×750 km^2 is presented (Section 4.2.2).

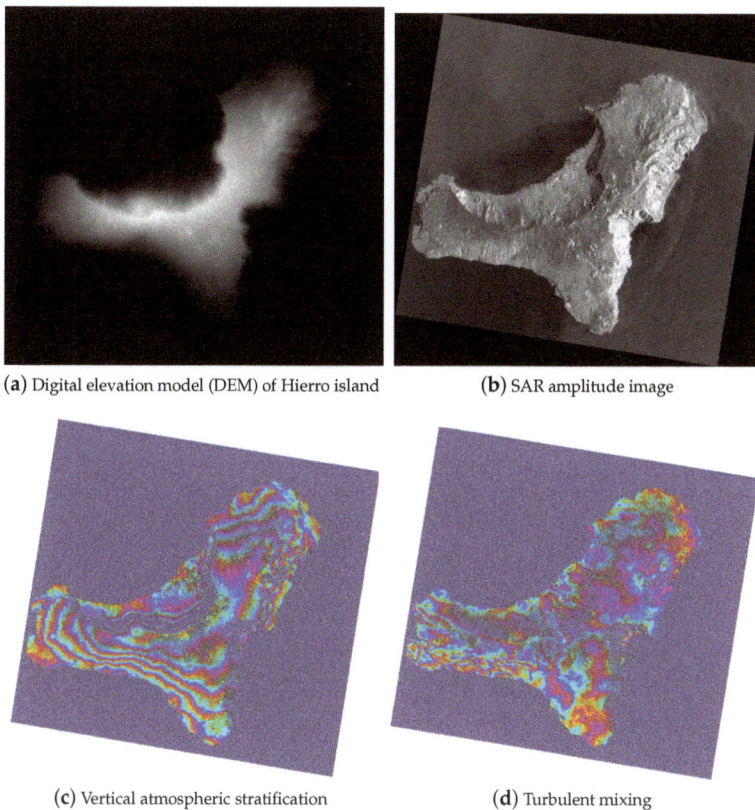

(**a**) Digital elevation model (DEM) of Hierro island

(**b**) SAR amplitude image

(**c**) Vertical atmospheric stratification

(**d**) Turbulent mixing

Figure 6. Tropospheric propagation delay effects on SAR interferometry due to vertical stratification and turbulent mixing: an example in the Hierro Island (Spain) using TerraSAR-X StripMap images. (**a**) Digital elevation model (DEM) of Hierro Island, with elevation varying from sea level (black) to about 1542 m (white). (**b**) SAR intensity image acquired on 10 October 2011. (**c**) Differential interferogram with master image acquired on 10 October 2011 and slave image on 21 October 2011 with an effective baseline of about 58 m. (**d**) Differential interferogram with master image acquired on 4 December 2011 and slave image on 15 December 2011 with an effective baseline of about 80 m. In (**c**) the interferometric phase is dominated by vertical stratification, whereas in (**d**) by turbulent mixing. Note that both are 11-day repeat pass interferograms.

4.2.1. PSI Processing—Test Site: Stromboli Volcano, Italy

The Stromboli Volcano is one of the most active volcanoes. The Sciara del Fuoco (SdF) depression, the nested horseshoe-shaped scar opening located on the north-east flank, exhibits the most dynamic changes in Stromboli, which are caused by lava flows, rock falls, landslides, lava accumulation, condensation, etc. Current volcanic activities are concentrated within the summit crater zone, which is located in the upper part of the SdF. This summit crater zone has remained at its current position. Nevertheless, the number and size of vents located inside it are varying over time due to the changing magma level within the conduit. The two most recent major effusive eruptions, which occurred in 2002–2003 and in 2007, induced large morphological changes on the SdF.

In the period from January to October 2008, 68 TerraSAR-X HRSL SAR images have been acquired from ascending and descending orbits with different incidence angles, which constitutes about 1 acquisition every 3.6 days. The resulting four stacks have been processed using the PSI technique. For both ascending and descending geometries, on average around 3×10^5 persistent scatterers (PSs) were detected per stack when selecting PSs with a signal-to-clutter ratio (SCR) larger than 3.0 [45]. The final estimation results are presented in Figure 7, having selected as valid PSs those with an overall model test (OMT) smaller than 1.0 [46]. A clear correlation between deformation estimates and height values is observed in the deformation maps as well as in the scatter plots depicted in Figure 7 [26].

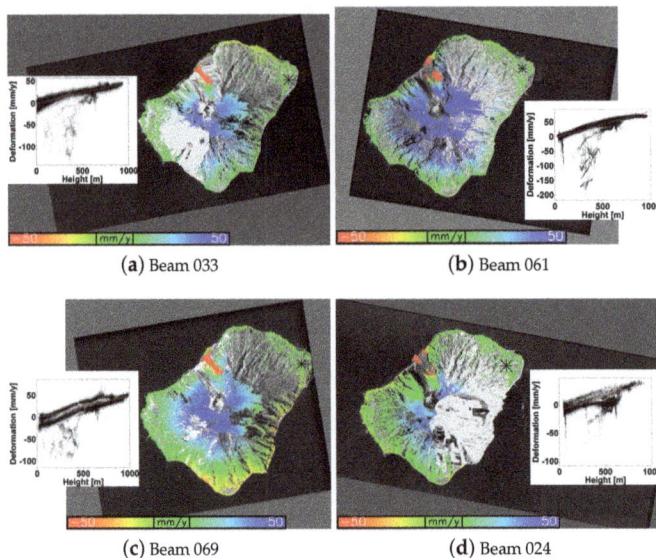

(**a**) Beam 033 (**b**) Beam 061

(**c**) Beam 069 (**d**) Beam 024

Figure 7. PS deformation maps without tropospheric correction of four TerraSAR-X HRSL stacks over the Stromboli Island with the geocoded average amplitude image as background. These stacks cover the time period January to October 2008 and contain (**a**) 18, (**b**) 17, (**c**) 16 and (**d**) 17 acquisitions. A scatter plot of deformation versus height is also illustrated for each of them. A linear fit is presented as a red line on each scatter plot. A clear correlation between deformation estimates and height is observed.

At Stromboli the deformation and residual topography signals are overlaid with a strong stratified tropospheric delay in the differential interferometric phases. This is the cause for the height-correlated deformation signals that can be observed in Figure 7. Therefore it is necessary firstly to compensate the tropospheric delay and then to apply time series processing (PSI estimation) in order to reliably estimate deformation and residual topography.

In order to mitigate the height-correlated stratified tropospheric delay, the differential tropospheric delay is firstly calculated for the PSs for each scene of the reference network using ERA-Interim data, then it is interpolated for all the other PSs and finally it is corrected on the differential interferometric phases. After this correction PSI processing is performed again. The resulting PS deformation maps after tropospheric delay mitigation are presented in Figure 8. In comparison to Figure 7, the height-dependent components have been effectively removed. Similar deformation patterns in SdF can be observed both in ascending and descending orbits. Some mismatches near craters are caused by horizontal deformation components, which are differently projected in the ascending and the descending geometries.

In order to assess the impact of the stratified tropospheric delay in the PSI estimates, the difference between the original estimates and the estimates after tropospheric delay compensation at identical PSs has been calculated. Note that an offset must be corrected for both difference of deformation estimates $\Delta Defo_i = Defo_{ori,i} - Defo_{atmo,i}$ and difference of residual topography $\Delta Topo_i = Topo_{ori,i} - Topo_{atmo,i}$ due to different reference points. Their maximum and minimum values as well as their correlation with height are summarized in Table 5. On the one hand, all deformation differences have a positive correlation (>85%) with height due to the temporal distribution (from January to October). The maximum relative deformation difference observed is 74.6 mm/y in beam 061. On the other hand, differences of residual topography have both positive and negative correlation with height due to the different baseline distributions, exhibiting an absolute correlation higher than 90% except for beam 033. The maximum relative residual topography difference observed is 61.7 m in beam 069.

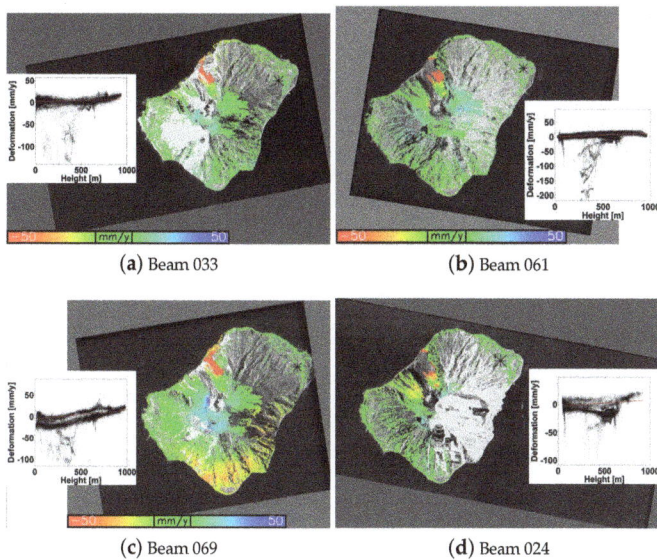

(**a**) Beam 033

(**b**) Beam 061

(**c**) Beam 069

(**d**) Beam 024

Figure 8. PS deformation maps after mitigation of the differential tropospheric delay for the stacks introduced in Figure 7. A comparison with Figure 7 clearly shows that the compensation of ECMWF based integrated tropospheric differential delay corrections has effectively mitigated the impact of tropospheric stratification.

Table 5. Summary of statistical analysis of the differences of two PSI processing results without and with tropospheric delay correction: maximum and minimum value of deformation differences and residual topography differences as well as their correlation with height. An offset compensation has been performed on both differences in order to account for the different reference points.

Beam	$\Delta Defo$ (mm/year)			$\Delta Topo$ (m)		
Nr.	Max	Min	Corr. (-)	Max	Min	Corr. (-)
033	49.2	−6.6	92.1%	5.9	−32.0	−14.5%
061	64.9	−9.7	98.2%	37.6	2.7	−92.4%
069	33.2	−5.0	87.8%	56.7	−5.0	97.4%
024	25.0	−4.5	96.3%	11.6	−2.8	94.6%

4.2.2. Wide Area Interferometry—Sentinel-1 Interferogram

In this section, the OP-based integrated tropospheric delay as well as ionospheric delay (based on CODE TEC maps) and SET correction have been applied on four contiguous Sentinel-1 interferograms in order to demonstrate the application of integrated tropospheric delay in a large-scale. The Sentinel-1 images have been acquired by the Sentinel-1A satellite in the Interferometric Wide Swath Mode (IW) on 14 and 26 July 2016 at around 05:30 UTC. The resulting swath has a size of approximately 250 km by 750 km and spans from 8.5°E to 13.0°E in longitude and 42.5°N to 48.8°N in latitude. The effective baseline is around 21 m. In the acquired region the elevation varies from sea level up to almost 4000 m.

The interferograms have been processed using the Integrated Wide Area Processor (IWAP) [47,48] without and with compensation of atmospheric and geodetic delays. The results are shown in Figure 9. It can be clearly observed that the interferometric phase of the interferogram without corrections (see left column of Figure 9) is dominated by the differential stratified tropospheric delay. The tropospheric delay of the scenes varies from 1.79 to 3.47 m for the master acquisition and from 1.86 to 3.65 m for the slave acquisition. These big variations are due to the very big height variation within the scene as well as to the high incidence angle in far range. The maximum differential delay between the two acquisitions is of about 28 cm and the minimum of about 6 cm. After compensation of the tropospheric differential phases, the height-dependent phase has been significantly mitigated (see right column of Figure 9). Other atmospheric effects with lower magnitude are thus visible, such as the turbulent atmosphere, which is not contained in the ECMWF data but could be potentially modeled using meso-scale weather assimilation techniques [49,50].

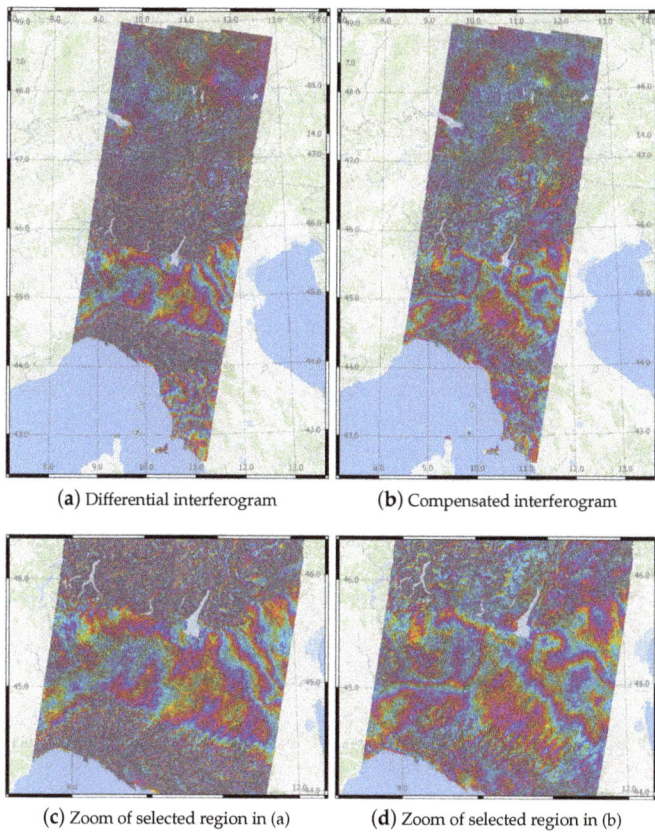

(**a**) Differential interferogram   (**b**) Compensated interferogram

(**c**) Zoom of selected region in (a)   (**d**) Zoom of selected region in (b)

Figure 9. Eight Sentinel-1 images acquired in Interferometric Wide Swath Mode (IW) on 14 and 26 July 2016 at around 05:30 UTC. (**a**) Differential interferogram. (**b**) Atmospheric delays (tropospheric and ionospheric) and SET compensated interferogram. Zooms of the selected region around Po Valley provided in (**c,d**). Contains modified Copernicus Sentinel data 2016. Background: Data © OpenStreetMap contributors, Rendering © DLR/EOC.

5. Conclusions

The propagation of radar echoes across the troposphere (in fact, troposphere, stratosphere and mesosphere) causes a signal delay due to atmospheric refractivity. This delay has a mean value in the meter range and variations over time in the decimeter range. As a consequence, it has a great impact on both SAR absolute ranging measurements and differential interferometry. In this paper, we have shortly reviewed the modeling of the tropospheric delay for the direct integration method based on NWP products, focusing on the different refractivity models available and presenting our model choice. Its accuracy is evaluated using two NWP products provided by ECMWF: ERA-Interim and OP data. These products are provided at four analysis times per day (at 0, 6, 12 and 18 h UTC respectively) with a spatial resolution from approximately 80 km to less than 10 km. The accuracy of the integrated tropospheric delay using both products has been assessed using GPS ZPD of IGS stations and SAR SPD derived from TerraSAR-X HRSL CR measurements as reference.

In the zenith direction, the accuracy of the integrated tropospheric delay using ERA-Interim and OP is evaluated firstly against GPS ZPD at the Wettzell GNSS station. The standard deviation of the differences between the integrated ECMWF ZPD and GPS ZPD is about 10.9 mm for ERA-Interim and

9.0 mm for OP. Secondly, the validation of ERA-Interim products has been extended to a global scale using 268 permanent GNSS stations from the IGS network. The standard deviation of the difference between ECMWF ZPD and GPS ZPD varies from 5.1 to 38.0 mm. The accuracy of the integrated tropospheric delay at higher latitudes ($>55°$) is better than 15.0 mm, whereas that at lower latitudes ($<30°$) is mostly between 15.0 and 30.0 mm. This variation of the accuracy is mainly correlated with the total water vapor content and depends on several factors such as location, station height and analysis time.

In the range direction, the accuracy of the integrated delay from both ECMWF data sets has been assessed using SAR SPD measurements as reference. More than 300 TerraSAR-X HRSL images of three CRs installed at the Wettzell geodetic observatory and at the Metsähovi observatory have been acquired. The standard deviation of the differences between ECMWF SPD and SAR SPD is approximately 15 mm for both ERA-Interim and OP, whereas their mean values vary from -2.4 to -27.6 mm.

Finally, the application of the direct integration method to tropospheric delay mitigation for SAR absolute range measurements and differential interferometry has been demonstrated. The impact on absolute range measurements has been evaluated using 5 CRs located at three far-distributed test sites. A ranging accuracy of 15.3 to 20.4 mm has been achieved. The accuracy using locally available GNSS station is better (from 11.2 to 20.3 mm). Nevertheless, using ECMWF data allows the exploitation of SAR absolute ranging with a centimeter-level accuracy without the requirement of a locally available GNSS station.

The impact of tropospheric delay mitigation on differential interferometry has been demonstrated for two test cases: TerraSAR-X PSI processing in the Stromboli Island and wide area Sentinel-1 differential interferometry. The differential interferograms without tropospheric delay correction exhibit a dominant phase component correlated with height which is caused by tropospheric stratification. This leads to height-dependent errors in the PSI estimates. The proposed direct integration method has been successfully applied in order to correct this stratified tropospheric phase component on the interferometric phase. As a result, the impact of the tropospheric delay on PSI estimation in the Stromboli test site has been effectively mitigated. The difference between PSI estimates without and with tropospheric delay mitigation is up to 74.6 mm/year of relative deformation and up to 61.7 m of relative residual topography. Due to the reduction of the height-dependent deformation component caused by the tropospheric delay it is now possible to observe the ground deformation patterns around the crater zones as well as to clearly delimitate the area with active deformation in the SdF. The mitigation of tropospheric stratification has been further demonstrated for wide area Sentinel-1 differential interferometry.

Author Contributions: X.C. conceived, designed and developed the techniques and experiments reported in this paper under supervision from M.E. X.C. carried out the experiments with inputs from the co-authors: U.B. provided the corner reflector SAR slant path delay measurements; F.R.G. performed the Sentinel-1 InSAR processing. X.C. wrote this paper.

Acknowledgments: The work was partially funded by the German Federal Ministry of Education and Research BMBF through its project "Exupéry: Managing Volcanic Unrest—The Volcano Fast Response System" and by the German Helmholtz Association through its DLR@Uni Alliance project "Munich Aerospace—High-resolution Geodetic Earth Observation: Correction Methods and Validation".

Conflicts of Interest: The authors declare no conflict of interest.

Abbreviations

The following abbreviations are used in this manuscript:

APL	Atmospheric pressure loading
ASC	Ascending
CD	Continental drift
CR	Corner reflector
DEM	Digital elevation model
DLR	German Aerospace Center
DSC	Descending
ECMWF	European Centre for Medium-Range Weather Forecasts
GNSS	Global navigation satellite system
GPS	Global positioning system
HRSL	High Resolution SpotLight
ID	Ionospheric delay
IFS	Integrated forecast system
IGS	International GNSS Service
InSAR	SAR interferometry
METS	CR test site in Metaähovi (Finland)
ML	Model level
NWP	Numerical weather prediction
OHIS	CR test site in GARS O'Higgins at the Antarctic Peninsula
OMT	Overall model test
OP	Operational data
OTL	Ocean tidal loading.
PSI	Persistent Scatterer (PS) interferometry
SAR	Synthetic aperture radar
SCR	Signal-to-Clutter Ratio
SET	Solid Earth tide
SPD	Slant range path delay
SWD	Slant range wet delay
TEC	Total electron content.
TDX-1	German TanDEM-X satellite
TSX-1	German TerraSAR-X satellite
WTZR	Wettzell GNSS station in EPN (Germany)
ZHD	Zenith hydrostatic delay
ZPD	Zenith path delay
ZWD	Zenith wet delay

References

1. Balss, U.; Cong, X.Y.; Brcic, R.; Rexer, M.; Minet, C.; Breit, H.; Fritz, T. High precision measurement on the absolute localization accuracy of TerraSAR-X. In Proceedings of the 2012 IEEE International Geoscience and Remote Sensing Symposium, Munich, Germany, 22–27 July 2012; pp. 1625–1628.
2. Gisinger, C. Atmospheric Corrections for TerraSAR-X Derived from GNSS Observations. Master's Thesis, Technische Universität München, Munich, Germany, 2012; pp. 1–117.
3. Bamler, R.; Eineder, M. Split-band interferometry versus absolute ranging with wideband SAR systems. In Proceedings of the 2004 IEEE International Geoscience and Remote Sensing Symposium, Anchorage, AK, USA, 20–24 September 2004; pp. 980–984.
4. Brcic, R.; Parizzi, A.; Eineder, M.; Bamler, R.; Meyer, F. Ionospheric effects in SAR interferometry: An analysis and comparison of methods for their estimation. In Proceedings of the 2011 IEEE International Geoscience and Remote Sensing Symposium, Vancouver, BC, Canada, 24–29 July 2011; pp. 1497–1500.
5. Askne, J.; Nordius, H. Estimation of tropospheric delay for microwaves from surface weather data. *Radio Sci.* **1987**, 22, 379–386. [CrossRef]

6. Delacourt, C.; Briole, P.; Achache, J. Tropospheric corrections of SAR interferograms with strong topography application to Etna. *Geophys. Res. Lett.* **1998**, *25*, 2849–2852. [CrossRef]

7. Li, Z.; Muller, J.-P.; Cross, P.; Fielding, E.J. Interferometric synthetic aperture radar (InSAR) atmospheric correction: GPS, Moderate Resolution Imaging Spectroradiometer (MODIS), and InSAR integration. *J. Geophys. Res. Solid Earth* **2005**, *110*, 2156–2202. [CrossRef]

8. Cimini, D.; Pierdicca, N.; Pichelli, E.; Ferretti, R.; Mattioli, V.; Bonafoni, S.; Montopoli, M.; Perissin, D. On the accuracy of integrated water vapor observations and the potential for mitigating electromagnetic path delay error in InSAR. *Atmos. Meas. Tech. Dis.* **2012**, *5*, 1015–1030. [CrossRef]

9. Jehle, M.; Perler, D.; Small, D.; Schubert, A.; Meier, E. Estimation of atmospheric path delays in TerraSAR-X data using models vs. measurements. *Sensors* **2008**, *8*, 8479–8491. [CrossRef] [PubMed]

10. Cong, X.; Balss, U.; Eineder, M.; Fritz, T. Imaging Geodesy-Centimeter-level anging accuracy with TerraSAR-X: An update. *IEEE Geosci. Remote Sens. Lett.* **2012**, *9*, 948–952. [CrossRef]

11. Nico, G.; Tomé, R.; Catalão, J.; Miranda, P.M.A. On the use of the WRF Model to mitigate tropospheric phase delay effects in SAR interferograms. *IEEE Geosci. Remote Sens. Lett.* **2011**, *49*, 948–952. [CrossRef]

12. Catalão, J.; Nico, G.; Hanssen, R.F.; Catita, C. Merging GPS and atmospherically corrected InSAR data to map 3-D terrain displacement velocity. *IEEE Geosci. Remote Sens. Lett.* **2011**, *49*, 2354–2360. [CrossRef]

13. Mateus, P.; Nico, G.; Tomé, R.; Catalão, J.; Miranda, P.M. Experimental study on the atmospheric delay based on GPS, SAR interferometry, and numerical weather model data. *IEEE Geosci. Remote Sens.* **2013**, *51*, 6–11. [CrossRef]

14. Doin, M.-P.; Lasserre, C.; Peltzer, G.; Cavalié, O.; Doubre, C. Corrections of stratified tropospheric delays in SAR interferometry: Validation with global atmospheric models. *J. Appl. Geophys.* **2009**, *69*, 35–50. [CrossRef]

15. Jolivet, R.; Agram, P.S.; Lin, N.Y.; Simons, M.; Doin, M.-P.; Peltzer, G.; Li, Z. Improving InSAR geodesy using global atmospheric models. *J. Geophys. Res. Solid Earth* **2014**, *119*, 2324–2341. [CrossRef]

16. Wallace, J.M.; Hobbs, P.V. *Atmospheric Science: An Introductory Survey*; Elsevier Inc.: New York, NY, USA, 2006; 505p, ISBN 9780127329512.

17. Hanssen, R.F. *Radar Interferometry: Data Interpretation and Error Analysis*; Springer: Berlin, Germany, 2001; ISBN 9780306476334.

18. Smith, E.K.; Weintraubt, S. The constants in the equation for atmospheric refractive index at radio frequencies. *Proc. IRE* **1953**, *41*, 1035–1037. [CrossRef]

19. Healy, S.B. Refractivity coefficients used in the assimilation of GPS radio occultation measurements. *J. Geophys. Res.* **2011**, *116*, 1–10. [CrossRef]

20. Nafisi, V.; Urquhart, L.; Santos, M.C.; Nievinski, F.G.; Bohm, J.; Wijaya, D.D.; Zus, F. Comparison of ray-tracing packages for troposphere delays. *IEEE Geosci. Remote Sens.* **2012**, *50*, 469–481. [CrossRef]

21. Saastamoinen, J. Atmospheric correction for the troposphere and stratosphere in radio ranging satellites. *Geophys. Monogr. Ser.* **1972**, *15*, 247–251.

22. Thayer, G.D. An improved equation for the radio refractive index of air. *Radio Sci.* **1974**, *9*, 803–807. [CrossRef]

23. Davis, J.L.; Herring, T.A.; Shapiro, I.I.; Rogers, A.E.E.; Elgered, G. Geodesy by radio interferometry: Effects of atmospheric modeling errors on estimates of baseline length. *Radio Sci.* **1985**, *20*, 1593–1607. [CrossRef]

24. Bevis, M.; Businger, S.; Chiswell, S.; Herring, T.A.; Anthes, R.A.; Rocken, C.; Ware, R.H. An improved equation for the radio refractive index of air. *J. Appl. Meteorol.* **1994**, *33*, 379–386. [CrossRef]

25. Rüeger, J.M. Refractive index formulae for radio waves. In Integration of Techniques and Corrections to Achieve Accurate Engineering. In Proceedings of the XXII FIG International Congress ACSM/SPRS Annual Conference, Washington, DC, USA, 19–26 April 2002; pp. 19–26.

26. Cong, X. SAR Interferometry for Volcano Monitoring: 3D-PSI Analysis and Mitigation of Atmospheric Refractivity. Doctoral dissertation, Technische Universität München, Munich, Germany, 2014.

27. Povic, J. ETA Model in Weather Forecast. Royal Institute of Technology. 2006. Available online: www.nada. kth.se/utbildning/grukth/exjobb/rapportlistor/2006/rapporter06/popovic_jelena_06054.pdf (accessed on 23 July 2018).

28. Dee, D.P.; Uppala, S.M.; Simmons, A.J.; Berrisford, P.; Poli, P.; Kobayashi, S.; Andrae, U.; Balmaseda, M.A.; Balsamo, G.; Bauer, D.P.; et al. The ERA-Interim reanalysis: Configuration and performance of the data assimilation system. *Q. J. R. Meteorol. Soc.* **2011**, *137*, 553–597. [CrossRef]

29. Hersbach, H., Dee. D. ERA5 reanalysis is in production. *ECMWF Newslett.* **2016**, *147*, 7.

30. Bock, O.; Keil, C.; Richard, E.; Flamant, C.; Bouin, M.-N. Validation of precipitable water from ECMWF model analyses with GPS and radiosonde data during the MAP SOP. *Q. J. R. Meteorol. Soc.* **2005**, *131*, 3013–3036. [CrossRef]

31. Flentje, H.; Dörnbrack, A.; Fix, A.; Ehret, G.; Hólm, E. Evaluation of ECMWF water vapour fields by airborne differential absorption lidar measurements: A case study between Brazil and Europe. *Atmos. Chem. Phys.* **2007**, *7*, 5033–5042. [CrossRef]

32. Schäfler, A.; Dörnbrack, A.; Kiemle, C.; Rahm, S.; Wirth, M. Tropospheric water vapor transport as determined from airborne Lidar measurements. *J. Atmos. Ocean. Technol.* **2010**, *27*, 2017–2030. [CrossRef]

33. Schäfler, A.; Dörnbrack, A.; Wernli, H.; Kiemle, C.; Pfahl, S. Airborne lidar observations in the inflow region of a warm conveyor belt. *Q. J. R. Meteorol. Soc.* **2011**, *137*, 1257–1272. [CrossRef]

34. Eineder, M.; Minet, C.; Steigenberger, P.; Cong, X.; Fritz, T. Imaging geodesy-Toward centimeter-level ranging accuracy with TerraSAR-X. *IEEE Geosci. Remote Sens.* **2011**, *49*, 661–671. [CrossRef]

35. Dow, J.M.; Neilan, R.; Rizos, C. The international GNSS service in a changing landscape of global navigation satellite systems. *J. Geod.* **2009**, *83*, 191–198. [CrossRef]

36. Bevis, M.; Businger, S.; Herring, T.A.; Rocken, C.; Anthes, R.A.; Ware, R.H. GPS meteorology: Remote sensing of atmospheric water vapor using the global positioning system. *J. Geophys. Res.* **1992**, *97*, 787–801. [CrossRef]

37. Tregoning, P.; Boers, R.; O'Brien, D. Accuracy of absolute precipitable water vapor estimates from GPS observations. *J. Geophys. Res.* **1998**, *103*, 28701–28710. [CrossRef]

38. Niell, A.E. Improved atmospheric mapping functions for VLBI and GPS. *Earth Planets Sci. Lett.* **2000**, *52*, 699–702. [CrossRef]

39. Kacmarík, M.; Douša, J.; Dick, G.; Zus, F.; Brenot, H.; Möller, G.; Pottiaux, E.; Kapłon, J.; Hordyniec, P.; Václavovic, P.; et al. Inter-technique validation of tropospheric slant total delays. *Atmos. Meas. Tech.* **2017**, *10*, 2183–2208. [CrossRef]

40. Böhm, J.; Möller, G.; Schindelegger, M.; Pain, G.; Weber, R. Development of an improved empirical model for slant delays in the troposphere (GPT2w). *GPS Solut.* **2005**, *19*, 433–441. [CrossRef]

41. DGordon, N.D.; Jonko, A.K.; Forster, P.M.; Shell, K.M. An observationally based constraint on the water-vapor feedback. *J. Geophs. Res. Atmos.* **2013**, *118*, 12–435.

42. Balss, U.; Breit, H.; Fritz, T.; Steinbrecher, U.; Gisinger, C.; Eineder, M. Analysis of internal timings and clock rates of TerraSAR-X. In Proceedings of the 2014 IEEE Geoscience and Remote Sensing Symposium, Quebec City, QC, Canada, 13–18 July 2014; pp. 2671–2674.

43. Balss, U.; Gisinger, C.; Cong, X.Y.; Brcic, R.; Hackel, S.; Eineder, M. Precise Measurements on the Absolute Localization Accuracy of TerraSAR-X on the Base of Far-Distributed Test Sites. In Proceedings of the 10th European Conference on Synthetic Aperture Radar (EUSAR 2014), Berlin, Germany, 3–5 June 2014; pp. 993–996.

44. Balss, U.; Gisinger, C.; Eineder, M. Measurements on the Absolute 2-D and 3-D Localization Accuracy of TerraSAR-X. *Remote Sens.* **2018**, *10*, 1–21. [CrossRef]

45. Gernhardt, S.M. High Precision 3D Localization and Motion Analysis of Persistent Scatterers Using Meter-Resolution Radar Satellite Data. Doctoral dissertation, Technische Universität München, Munich, Germany, 2011.

46. Kampes, B.M. *Radar Interferometry: Persistent Scatterer Technique*; Springer: Berlin, Germany, 2006.

47. Rodriguze Gonzalez, F.; Adam, N.; Parizzi, A.; Brcic, R. The Integrated Wide Area Processor (IWAP): A processor for wide area persistent scatterer interferometry. *ESA Liv. Planet Symp.* **2013**, *722*, 353.

48. Eineder, M.; Balss, U.; Suchandt, S.; Gisinger, C.; Cong, X.; Runge, H. A definition of next-generation SAR products for geodetic applications. In Proceedings of the 2015 IEEE International Geoscience and Remote Sensing Symposium (IGARSS), Milan, Italy, 26–31 July 2015; pp. 1638–1641.

49. Pichelli, E.; Ferretti, R.; Cimini, D.; Panegrossi, G.; Perissin, D.; Pierdicca, N.; Rommen, B. InSAR water vapor data assimilation into mesoscale model MM5: Technique and pilot study. *IEEE J. Sel. Top. Appl. Earth Obs. Remote Sens.* **2015**, *8*, 3859–3875. [CrossRef]

50. Mateus, P.; Miranda, P.M.A.; Nico, G.; Catalão, J.; Pinto, P.; Tomé, R. Assimilating InSAR maps of water vapor to improve heavy rainfall forecasts: A case study with two successive storms. *J. Geophys. Res.* **2018**, *123*, 1638–1641. [CrossRef]

remote sensing

MDPI

Article

Where We Live—A Summary of the Achievements and Planned Evolution of the Global Urban Footprint

Thomas Esch [1,*], Felix Bachofer [1], Wieke Heldens [1], Andreas Hirner [1], Mattia Marconcini [1], Daniela Palacios-Lopez [1], Achim Roth [1], Soner Üreyen [1], Julian Zeidler [1], Stefan Dech [1] and Noel Gorelick [2]

[1] German Aerospace Center (DLR), Earth Observation Center (EOC), German Remote Sensing Data Center (DFD), Oberpfaffenhofen, D-82234 Weßling, Germany; felix.bachofer@dlr.de (F.B.); wieke.heldens@dlr.de (W.H.); andreas.hirner@dlr.de (A.H.); mattia.marconcini@dlr.de (M.M.); daniela.palacioslopez@dlr.de (D.P.-L.); achim.roth@dlr.de (A.R.); soner.uereyen@dlr.de (S.Ü.); julian.zeidler@dlr.de (J.Z.); stefan.dech@dlr.de (S.D.)
[2] Google Switzerland GmbH, 8002 Zurich, Switzerland; gorelick@google.com
* Correspondence: thomas.esch@dlr.de; Tel.: +49-(0)-8153-283-721

Received: 30 April 2018; Accepted: 6 June 2018; Published: 7 June 2018

Abstract: The TerraSAR-X (TSX) mission provides a distinguished collection of high resolution satellite images that shows great promise for a global monitoring of human settlements. Hence, the German Aerospace Center (DLR) has developed the Urban Footprint Processor (UFP) that represents an operational framework for the mapping of built-up areas based on a mass processing and analysis of TSX imagery. The UFP includes functionalities for data management, feature extraction, unsupervised classification, mosaicking, and post-editing. Based on >180.000 TSX StripMap scenes, the UFP was used in 2016 to derive a global map of human presence on Earth in a so far unique spatial resolution of 12 m per grid cell: the Global Urban Footprint (GUF). This work provides a comprehensive summary of the major achievements related to the Global Urban Footprint initiative, with dedicated sections focusing on aspects such as UFP methodology, basic product characteristics (specification, accuracy, global figures on urbanization derived from GUF), the user community, and the already initiated future roadmap of follow-on activities and products. The active community of >250 institutions already working with the GUF data documents the relevance and suitability of the GUF initiative and the underlying high-resolution SAR imagery with respect to the provision of key information on the human presence on earth and the global human settlements properties and patterns, respectively.

Keywords: global; urban footprint; processing; validation; community survey; sustainability

1. Introduction

Settlements and urban areas characterize the cores of human activity. Population growth and the related urbanization as well as climate change represent the most relevant developments for the human presence on the planet that challenge our ecologic, societal and economic systems at a global scale. The global population is prospected to increase to 9.8 billion in 2050, and the persistent urbanization process will lead to a constantly growing share of the urban population. In 2007, the urban population exceeded 50% for the first time in history and this proportion will most likely rise to more than 66% in 2050 according to the 2014 revision of the World Urbanization Prospects [1–3]. However, the phenomenon of population growth and urbanization is not distributed evenly on our planet since 90% of the total population growth until 2050 will take place in Asia and Africa.

The massive and dynamic growth of population and urban agglomerations plays a key role for environmental monitoring and sustainability frameworks. Therefore, a precise mapping of the current and future human settlements pattern and dynamics—in urban as well as in rural areas and from local to global scale—is essential. Detailed and reliable information on global human settlements can directly contribute to the monitoring and decision making regarding the 2030 Agenda for Sustainable Development and provides relevant data for the Sustainable Development Goals (SDG), specifically SDG 11 [4]. The Urban Sustainability Framework (USF) of the World Bank points out the relevance of data on urban development for sustainable planning processes [5]. Besides those well-known frameworks, many other assessment tools exist on regional and national level to measure sustainability in urban development [6–8].

Earth observation (EO) is capable of providing information on human settlements and urban agglomerations on a global scale and allows monitoring their development. Potere et al. [9,10] provide comprehensive overview global information products on human settlements, which often are derived of medium resolution multispectral imagery like the MODIS 500 and the GlobCover 2009 land cover maps [11,12]. Potere et al. [9] report enormous variations in the total global urban extent of the different products and identify three main reasons: (i) the varying production dates of the maps, (ii) different spatial resolutions of the utilized data, and (iii) the diverse class descriptions of urban land use. In addition, the accuracy in detecting and classifying small and scattered settlements is low for all compared products. Miyazaki et al. [13] demonstrated the integrated analysis of ASTER multispectral images and existing GIS data to process a global HR settlement mask. Ban et al. [10] utilized a robust processing chain for a method based on spatial indices and Grey Level Co-occurrence Matrix (GLCM) textures developed by Gamba and Lisni [14] on ENVISAT ASAR C-band data. Wieland and Pittore [15] applied an object-based analysis on Support Vector Machine (SVM)-based pattern recognition on Landsat-8 images over large areas. The Global Human Settlement Layer (GHSL) was proposed by the European Joint Research Center (JRC). The built-up areas are semi-automatically extracted from several Landsat for three timesteps [16]. More recently, they tested also the application of Sentinel-1 C-band radar data for the GHSL [17]. Liu et al. [18] derived multitemporal urban settlement products for the period of 1990 to 2010 by utilizing the Landsat archive in the Google Earth Engine (GEE) and deriving the Normalized Urban Areas Composite Index (NUACI). The World Urban Database and Access Portal Tools (WUDAPT) initiative focuses on a global urban database for Local Climate Zones (LCZ) classifications of urban areas [19,20].

The TerraSAR-X (TSX) and Tandem-X (TDX) mission and image archive provides another source of high-resolution satellite images with global coverage. Two years prior to the launch of TSX X-band microwave satellite system, Roth et al. [21] already raised the question "TerraSAR-X—How can high-resolution SAR data support the observation of urban areas?" in 2005. They assumed that TSX will enable the detection and mapping of buildings and transport infrastructure, as well as the assessment of disaster damage and detection of ground motions applying differential interferometry and persistent scatterer techniques in urban areas. Succeeding research of various groups worldwide proofed these assumptions correct [22–26]. In preparation for the TSX mission, Esch [27] validated the potential of high-resolution X-band data for the automated mapping of settlement areas. In the following years, the methodology was improved and operationalized [28–30], finally boosted by the launch of the TDX mission [31]. The methodology was then applied at a global scale and resulted in the Global Urban Footprint (GUF) dataset with a spatial resolution of 12 m [32,33]. Considering this evolution and the more than 10 years of experience in the SAR-based mapping of built-up areas, this paper aims at providing a summary of the major achievements and intended future developments of the GUF initiative and product, respectively. Hence, a first section details the data base and methodological concept of the UFP processing and analysis framework (Section 2.1). Next, the basic product characteristics of the generated GUF dataset and the first-time results of a global validation campaign (Section 2.2) are presented, followed by first statistics on the global human settlements distribution derived from the GUF layer (Section 2.3). Section 3 then focuses on the user perspective by analyzing the 300 requests for GUF data that have been submitted since the first release of the product in November 2016 (Section 3). Finally, the conclusions are drawn, and an outlook on the GUF follow-on activities and products is given. Considering this content, the presented contribution is expected to provide an overview of the state-of-the-art global settlement mapping and the potentials of global, continental and regional analyses of the urban system and development.

2. Global Urban Footprint—Pushing the Limits of Mapping Human Settlements from Space

Strengthened by the promising results of diverse studies on the use of SAR data for detecting and delineating human settlements, the German Aerospace Center (DLR, Cologne, Germany) initiated the Global Urban Footprint campaign [31]. This initiative aimed at the generation of a so far unprecedented worldwide map of human settlements that would for the first time also include a significant proportion of the small-scale rural settlements in addition to the comparably large structures of cities and urban clusters already covered by other data sets available at that time.

2.1. Data Base and Processing Framework

In the context of the GUF campaign, the constellation of the identically constructed TerraSAR-X and TanDEM-X satellites was utilized to collect a global coverage of SAR imagery within a comparably short period of time. In detail, a total of 182,249 single look complex (SSC) images was acquired in StripMap mode with 3 m ground resolution between 2011 and 2012 (93% of the global coverage), completed by some final additions collected in 2013–2014. The volume of this input data set adds up to 308 TB. Considering all auxiliary data used and intermediate products generated during GUF processing, the UFP framework had to handle >20 million files with a total volume of >400 TB.

2.1.1. Urban Footprint Processor

The production of the GUF layer is based on a fully automatic, generic and autonomous processing environment orchestrating an extensive suite of processing and analysis modules: the Urban Footprint Processor (UFP). Basically, the UFP consists of five main technical modules covering functionalities for data management, feature extraction, unsupervised classification, mosaicking, and automatic post-editing. The systems design and implementation of the UFP was first described by Esch et al. [33]. Additional modifications are geared towards the elimination of false positives during post-editing and are detailed in Esch et al. [32].

2.1.2. Data Management

The UFP is implemented at DLR's German Remote Sensing Data Center (DFD, Oberpfaffenhofen, Germany) and deployed on two basic processing platforms: a Sun cluster on the one hand and a Calvalus cluster on the other. The Sun Fire X4640 machine with eight CPUs is used for feature extraction and unsupervised classification. Processing on the Sun cluster is orchestrated by DLR's Processing System Management (PSM) described in further detail by Böttcher et al. [34]. Digital Elevation Model (DEM) data required in the GUF analysis process are obtained through the W42 Raster Data Repository, which provides a best-of-DEM for any given area from different sources such as SRTM or ASTER [35]. The Calvalus cluster [36] based on Apache Hadoop consists of 50 compute nodes and is employed to for the mosaicking and post editing modules of the UFP.

2.1.3. Feature Extraction

Characteristic properties of SAR data in built environments arise from the high local image heterogeneity that originates from intense backscatter plus shadow effects around vertical structures. This texture directly relates to the presence of buildings or any structure with a distinct vertical component. To define this local image heterogeneity or texture the UFP calculates the so-called speckle divergence feature, which is defined as the ratio between the local standard deviation and local mean of the backscatter computed in a given local neighborhood. A detailed description of the feature extraction algorithm is provided in Esch et al. [33].

2.1.4. Unsupervised Classification

The classification procedure couples an analysis of the original backscatter amplitude data and the derived local texture image. For that purpose, an unsupervised classification method based on advanced Support Vector Data Description (SVDD) one-class classification was implemented as described in detail in Esch et al. [33]. For each single scene, the approach identifies the optimal settings for the classification by using training samples that are automatically identified based on thresholds derived from image statistics of the amplitude and texture data, respectively.

The SVDD technique aims at (i) determining the hypersphere with minimum radius enclosing all the training samples available for the built-up class and (ii) finally associating all the unknown samples falling inside the boundary with it. This approach allows increased generalization and obtains a more consistent and reliable GUF map [33]. The outcome of the classification procedure is a binary raster layer indicating the class built-up area and the category non-built-up for any other region. The resulting GUF masks reflect the detailed building distribution and not the impervious surface that is usually resulting from built-up area extractions based on (multi-)spectral satellite imagery.

In order to compensate effects of over- or underestimation caused by specific land cover types, environmental conditions or acquisition constellations, six additional GUF raw versions with systematically altered classification settings were generated. Therefore, the speckle divergence threshold initially defined by the automated estimation procedure is systematically increased, respectively decreased three times, by 200 DN with each step, resulting in three increasingly stricter (levels 3, 2, 1) as well as three more relaxed (levels 5, 6, 7) classification versions. The initial classification result based on the automatically defined threshold is represented by level 4. If the underlying speckle divergence threshold for the level 4 was 2000, the corresponding values are 2200 for level 3, 2400 for level 2, 2600 for level 1, and 1800 for level 5, 1600 for level 6 and 1400 for level 7, respectively. Hence, the different GUF versions can be considered as varying confidence levels—from level 3 to level 1 the completeness will decrease while reliability increases (increasing error of omission, decreasing error of commission) and vice versa from level 5 to level 7 [32].

2.1.5. Mosaicking

In order to provide more manageable working units for the post-editing procedure, all seven individual GUF raw masks or confidence levels, respectively, were merged from their original image geometry to tiles of $5° \times 5°$ geographical latitude and longitude. During mosaicking the overlapping areas of several individual scenes were aggregated by means of a majority vote for the binary classification assignment of each single pixel (built-up, non-built-up) and for each individual GUF level. If a certain area of the mosaic is, for instance, covered by four different scenes, there are four classifications available for each confidence level in that overlap region. Separately for each individual confidence level, the pixels of the overlap areas are therefore finally assigned the dominating classification outcome (e.g., if the pixel is three times labelled as "built-up" and only in one scene as "non-built", the pixel is finally assigned as "built-up" in corresponding GUF level of the mosaic). As a result, each tile comprises seven GUF bands in the geometric resolution of 0.4″ (12 m), or approaching the poles, in correspondingly lower longitudinal resolutions—e.g., 0.6″ between 50°N and 60°N [32].

2.1.6. Automated Post-Editing

The automated post-editing stage of the GUF production is split-up into two phases. Image segmentation based on Chang et al. [37] transfers all clusters of connected pixels classified as built-up in each of the seven GUF raw raster layers (confidence levels) into individual image objects and a corresponding set of descriptive attributes. After that, a rule-based approach implemented in Python selects the appropriate local GUF confidence level and finally removes all GUF segments from the resulting collection that most likely represent false alarms.

Nine global reference layers were used for the optimal GUF level definition as well as for the formulation of exclusion or inclusion criteria in the context of false alarms identification [32]. Two of them, TimeScan-ASAR (DLR-TSA) and TimeScan-Landsat (DLR-TSL), are GeoTIFF in float formatting [38], and the others (OSM-Settlements, OSM-Roads, GL30-Settlements, DLR-ReliefMap, DLR-RoadCluster, CIL, and NLCD) are binary masks derived from defined thresholds or specific classes of selected source data sets, namely Open Street Map, GL30, SRTM/ASTER, Copernicus Imperviousness Layer, and US National Land Cover Dataset. The reference layers are merged by summing up the number of positive reference counts: a value of 1 representing a built-up area is assigned if at least two out of seven binary masks are positive; otherwise, it is discarded and set to 0.

Next, for each single GUF object at each confidence level, it is calculated to what part of the reference layer intersects with the GUF object. The optimal GUF version (level 1–7) is determined for an entire 5° × 5° tile by analyzing the overlap calculated above. Empirical tests on the basis of local ground truth data available for various globally distributed test regions indicated that overestimation usually starts as soon as two-thirds of the GUF object area corresponds with the reference data. The optimal confidence level is chosen when the majority of all GUF objects of a tile satisfy the 66%-rule.

The final step of the post-editing phase includes a procedure to identify and eliminate false alarms that might still be present in the previously selected best-fitting GUF layer by applying exclusion masks, thresholds regarding NDVI, and mean temporal backscatter. GUF features with high NDVI (e.g., highly textured forests), low temporal backscatter (e.g., rice fields), and an overlap exceeding a certain percentage with exclusion features (e.g., water areas) are removed.

It should be stressed that no correction procedure ever introduces features that do not originate from the original SAR data and classification procedure.

2.2. GUF Product Specification and Validation

The GUF data produced by the UFP processor is provided as 8-bit, LZW-compressed GeoTIFF with a value of 255 indicating built-up area, a value of 0 representing all non-built-up areas, and no data assigned by value 128. Near the equator, the geometric resolution of the GUF product is 0.4″ (which corresponds to 12 m per pixel), whereas the resolution decreases toward the poles (0.6″ between 50°–60°N/S, 0.8″ from 60°–70°N/S, and 1.2″ > 80°N/S). The projection is Geographic coordinates (lat/lon). Due to the TerraSAR-X/TanDEM-X data policy of the underlying Public Private Partnership between DLR and Airbus DS, the full resolution GUF layer is freely available for scientific use, whereas any nonscientific/non-commercial application can use a GUF version with a resolution of 2.8″ (84 m near the equator). For commercial GUF use cases, Airbus DS has to be contacted. The generalized GUF version in 2.8″ is directly derived from the 0.4″ version by assigning a value of 255 (= built-up) to all pixels whose coverage contains a proportion of >25% GUF area as defined by the original 0.4″ data.

The very high accuracy and reliability of the GUF layer have already been discussed in [32] for a set of 12 different test locations. Nevertheless, this mostly included largely urbanized areas, hence not fully characterizing the many existing settlement patterns scattered around the globe. To overcome this limitation, a new broader and more comprehensive validation set has been produced in collaboration with Google for a collection of 50 globally distributed test sites (tiles of 1° × 1° lat/lon each) including a total of ~900,000 reference points. This activity is part of a currently on-going campaign for assessing the quality of the current state-of-the-art global human settlements maps, in particular, the novel World Settlement Footprint (WSF) 2015. The WSF2015 is a 10 m spatial resolution global map outlining the 2015 settlement extent that has been newly generated at DLR by jointly exploiting multitemporal optical Landsat-8 and radar Sentinel-1 imagery [39]. To include a representative population of settlement patterns in the global accuracy assessment, the 50 tiles have been selected in a way that there are tiles assigned to all continents and that they cover the full range of potential "settlement densities" (defined by the ratio between the number of settlements within each tile—i.e., disjoint clusters of pixel assigned as built-up, with the WSF2015 being used as a basis—and their total area). For all selected tiles, 2000 locations were randomly extracted and used as the center of a 3 × 3 block sampling units whose nine cells show a size of 10 × 10 m each. Next, each cell has been labelled by crowd-sourcing via photointerpretation of 2015 very high resolution (VHR) imagery available from Google Earth. In particular, it has been marked as settlement only if it intersects a building defined—according to United Nations [40]—as any structure having a roof supported by columns or walls and intended for the shelter, housing, or enclosure of any individual, animal, process, equipment, goods, or materials of any kind. Overall, due to the lack in some few cases of reference cloud-free VHR data, it was possible to label 892,926 cells, of which 137,910 (15.44%) have been marked as settlement, and the remaining 755,016 (84.56%) as non-settlement.

Table 1 reports the kappa coefficient, as well as percentage overall (OA%), user's (UA%) and producer's (PA%) accuracies exhibited against the above-mentioned validation set by: the GUF 0.4″ and GUF 2.8″, and a selection of comparable, frequently referenced layers such as GHSL, GL30, and the MODIS500. Here, given the different spatial resolution of the datasets, each cell of the block sample units has been considered as settlement if the intersection with the given settlement layer at hand (GUF 0.4″, GUF 2.8″, GHSL, GL30, or MODIS500, respectively) showed any built-up assignment. It is worth noting that, despite changes might have occurred over time, we reasonably expect that these do not sensibly affect the figures exhibited by the GUF and the GHSL (derived from imagery acquired in 2012–2013 and 2014, respectively). Instead, results shown by GL30 and, especially, MODIS500 (generated from 2010 and 2001–2002 imagery, respectively) might be more significantly affected. A more comprehensive documentation and discussion of the global (cross-)validation campaign and its findings—also including additional layers such as CIESIN's HBase [41] and the multi-temporal urban land products presented by Liu et al. [18]—is currently being conducted by the authors and a dedicated publication is planned for 2018.

Table 1. Accuracy assessment results derived based on ~900,000 reference points labelled by crowd-sourcing via photointerpretation of 2015 VHR imagery from Google Earth.

	Kappa	OA%	PA% Settlement	PA% Non-Settlement	UA% Settlement	UA% Non-Settlement
GUF 0.4″	0.637	90.23	72.19	93.35	67.08	94.85
GUF 2.8″	0.600	88.96	71.27	92.19	62.49	94.61
GHSL	0.472	87.98	45.31	95.77	66.18	90.55
GL30	0.459	87.85	43.60	95.93	66.20	90.30
MODIS500	0.246	84.84	22.71	96.19	52.11	87.20

As one can immediately notice, the difference between the GUF 0.4″ and GUF 2.8″ is very limited for all six considered accuracy measures (i.e., on average lower than 2%), hence confirming that the spatially reduced version shows a level of detail and quality almost identical to the full-resolution product [32]. All layers exhibit relatively high OA% (with the GUF 0.4″ being the only one exceeding 90); nevertheless, such measure is not particularly meaningful in our case where the vast majority of reference cells belongs to the non-settlement class (i.e., an "empty" map where no pixel is marked as settlement would result in 84.56 OA%). Instead, when analyzing the Kappa coefficient (which is derived also by accounting for both omission and commission errors with respect to simple agreement calculation) it is evident that the GUF outperforms the other three layers (with an increase for the GUF 0.4″ higher than 0.16 with respect to both GHSL and GL30 and about 0.4 with respect to MODIS500). This occurs especially in the light of the better capability in detecting true built-up areas, corresponding to a PA% for the settlement class greater than 70 (i.e., about +25, +27, and +50 with respect to GHSL, GL30, and MODIS500, respectively, which exhibited severe underestimation issues).

2.3. The State of Global Urbanization—First Figures Derived from the GUF Data

A global analysis of the GUF layer indicates a total global built-up area of 834,260 km² which represents approximately 0.64% of the Earth's land surface. At continental scale Europe shows the highest proportion of built-up area per land surface (1.52%), whereas in Australia only 0.12% of the land mass is covered by settlements. In Asia 0.71%, North America 0.46%, South America 0.31%, and Africa 0.25% of the land surface falls upon built-up area. Figure 1a depicts the percentage of GUF pixels for a raster of 5° × 5° degrees (lat/lon) tiles. Therefore, the highest share of GUF pixels (11.44%) is detected for the tile covering the larger urban region of the city of Beijing, China. In general, the analysis shows that the highest densities of built-up area per tile occur in China, Central-Europe, the north-east USA and India. The pattern over Europe reveals continuous high percentages of >1% for almost all tiles except for those covering Iceland and Scandinavia, with the highest value observed for the tile covering the Ruhr region in Germany and parts of Netherland and Belgium. Figure 1b illustrates the percentage of settlement area (defined by GUF coverage) for the administrative units of countries. The use of these national boundaries as an alternative to the 5 × 5° lat/lon tiles leads to a spatial homogenization of the information (at least within each country) so that the distribution of the urban agglomerations and the low density rural areas are finally blurred. However, this representation is more appropriate for (future) studies that aim at combining the information on the human settlement area with other data sources such as socio-economic or population-related statistics which are normally collected and related to administrative units. The scatterplot provided in Figure 2 highlights that—as expected—the city-states show the highest share of settlement area in relation to the country area, whereas large-area states feature a comparably low proportion of settlement area compared to the total area of the country.

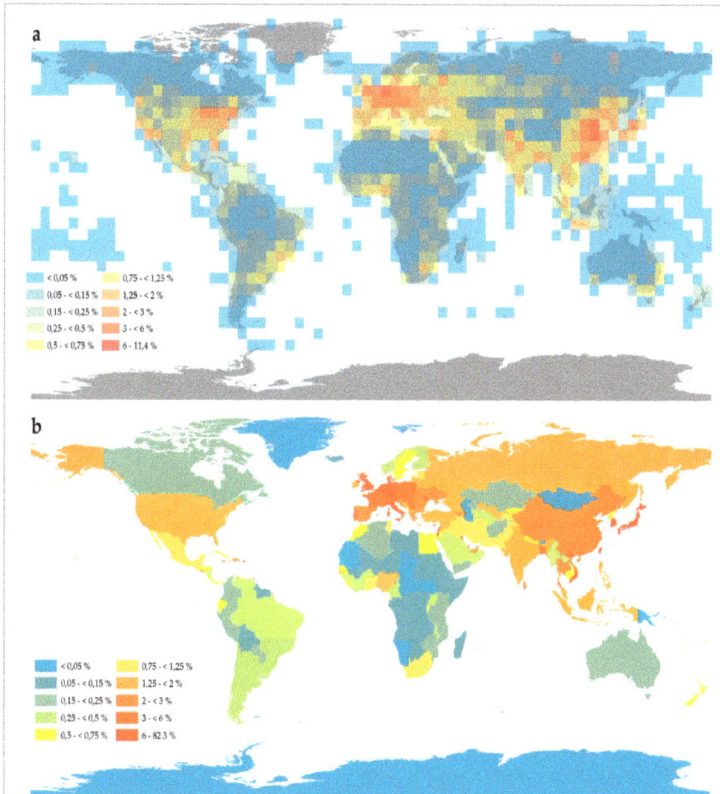

Figure 1. Percentage of built-up area derived from global Global Urban Footprint (GUF) layer (**a**) per 5° × 5° lat/lon tile and (**b**) per country.

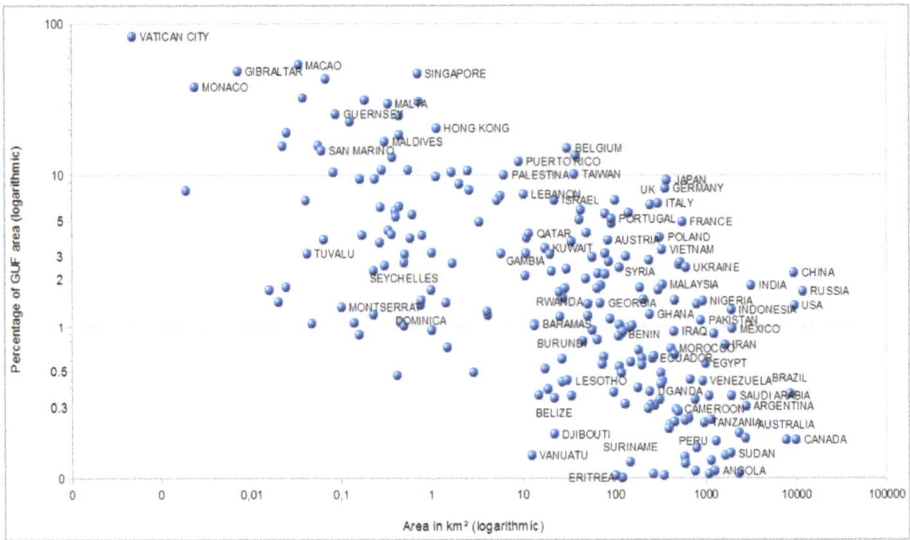

Figure 2. Scatterplot based on the GUF data showing the relation between the total administrative area and the percentage of settlement area for selected countries.

3. The User Perspective—Precise Data for Evidence-Based Planning and Decision Making

Since the release of the Global Urban Footprint (GUF) dataset in November 2016, more than 300 institutions have requested the corresponding thematic data (record date: 31 December 2017). Due to the data policy defined in the PPP between DLR and Airbus DS, a pre-requisite to get access to the GUF layer is the submission of a filled License Agreement and Order Form that also includes a short questionnaire. Based on an analysis of the questionnaires received in the first year of GUF delivery, the team was able to develop a comprehensive picture of the requirements and application scenarios—and therewith the societal benefit—of the new global dataset. The results of this survey are presented in this section.

In general, 260 out of the 300 data requests were positively evaluated and the users could be provided with the GUF data. Only four requests had to be rejected because the use scenario described a commercial application or challenged the commercial license exclusively granted to Airbus DS. Another 40 requests could not be successfully processed yet due to a missing or incomplete order form, an unspecified area of interest, etc. From all requests, 182 aimed at the 0.4″ version (scientific use), and 103 wanted access to the 2.8″ dataset. 33 users requested the GUF in both resolutions.

Figure 3a indicates that 72% of the GUF users are affiliated with a university or a research institute, 21% with non-profit NGOs, public organizations and international or European institutions, and only 4% with private companies. From these users, 62% of the requests had a clear scientific objective, and 17% a non-profit background. Eight percent of the users indicated both options. For the remaining 10%, no use type was specified. The provenience of the institutions requesting the GUF is shown in Figure 4a, with most users being located in the US (23%), followed by Germany (14%), UK (7%), Italy (6%), and China (6%).

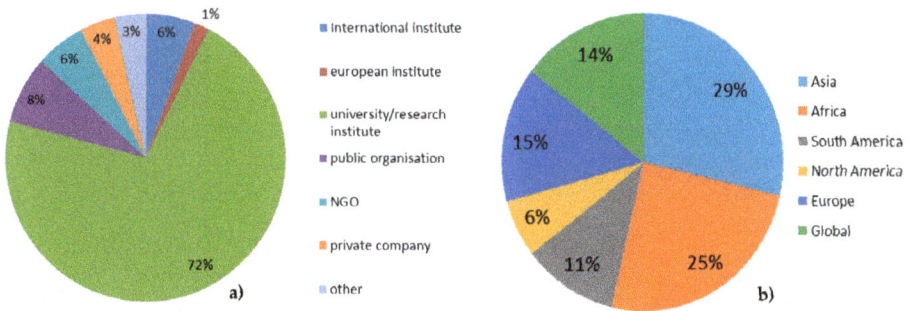

Figure 3. Categorized affiliation of GUF users (**a**) and location of the area of interest by continent (**b**).

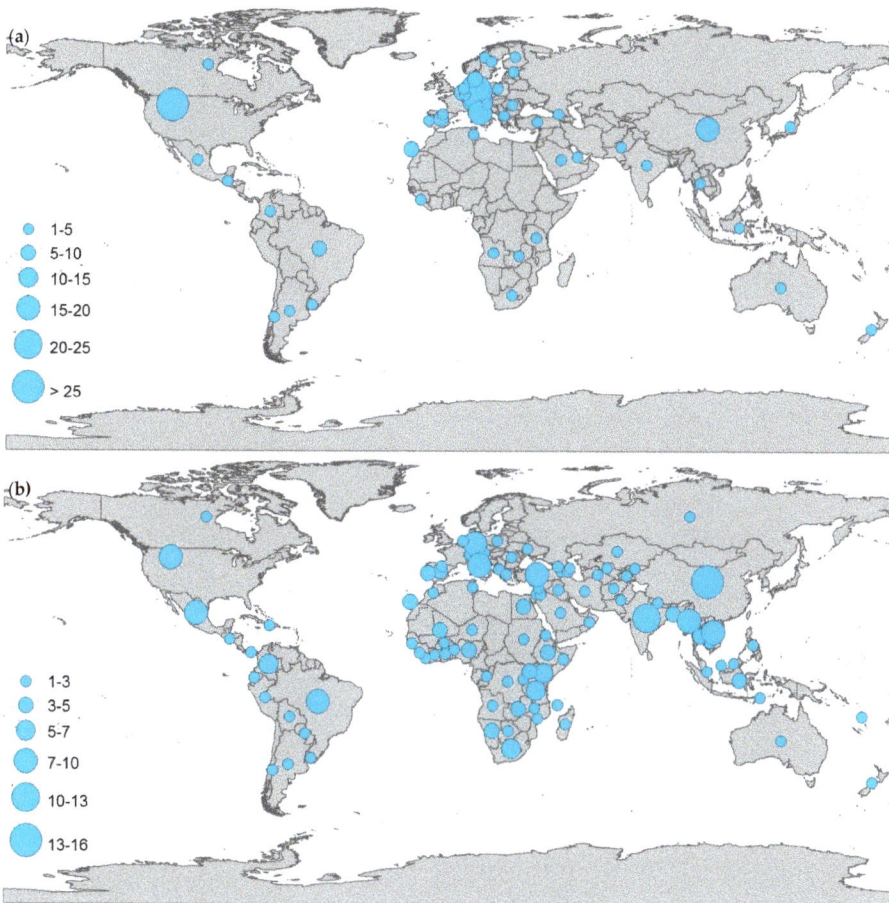

Figure 4. Visualization of the amount and origin of GUF users (**a**) per country and the distribution of the related areas of interest (**b**).

Summarized per continent, Figure 3b provides statistics related to the location of the 399 area(s) of interest (AOI) that have been defined by the users. The statistics reveal that the GUF layer is most often requested for regions in Asia (29%) and Africa (25%). This is not surprising since these two continents show the highest growth rates of urban expansion, often coupled with a lack of accurate and/or up-to-date data on the actual settlements location and distribution, in particular in rural areas. Most users requested the GUF for a country or even a smaller regional extent AOIs (291 times). In addition, 47 requests related to entire continents with Africa, South America, and Europe as top three (13, 12, and 11 times, respectively). A global GUF coverage was asked for in 58 cases. Figure 4 visualizes the amount and origin of GUF requests (a) and areas of interest (b) per country.

The questionnaire in the GUF License Agreement and Order Form also contained a question about the intended overall use of the data, with multiple answers being possible for a selection of pre-defined sectors (which could in case be supplement by individual user-defined fields). Generally, the average number of different use scenarios aimed at by a user comes up to 2.13. Therefore, the main intended use is that of a primary spatial/statistical analysis based on the GUF (148 cases), followed by using the GUF data as an input for further follow-on processing such as modelling (125 cases) or simply as spatial reference or background to generate new maps or figures (80 cases). More details on the specific fields of application behind these general use cases are provided by Figure 5. Also using the user input for a selected list of options provided in the questionnaire, an average of 4.26 different applications were indicated by each user. Most popular are "land use and land use change" and "urban growth and urban sprawl". This is followed by other applications such as "ecosystems and environmental protection", "population estimation", or "urban/regional planning". Additionally, many users selected application fields such as "disease modelling and health care", "climate modelling", "Biodiversity", or "infrastructure planning".

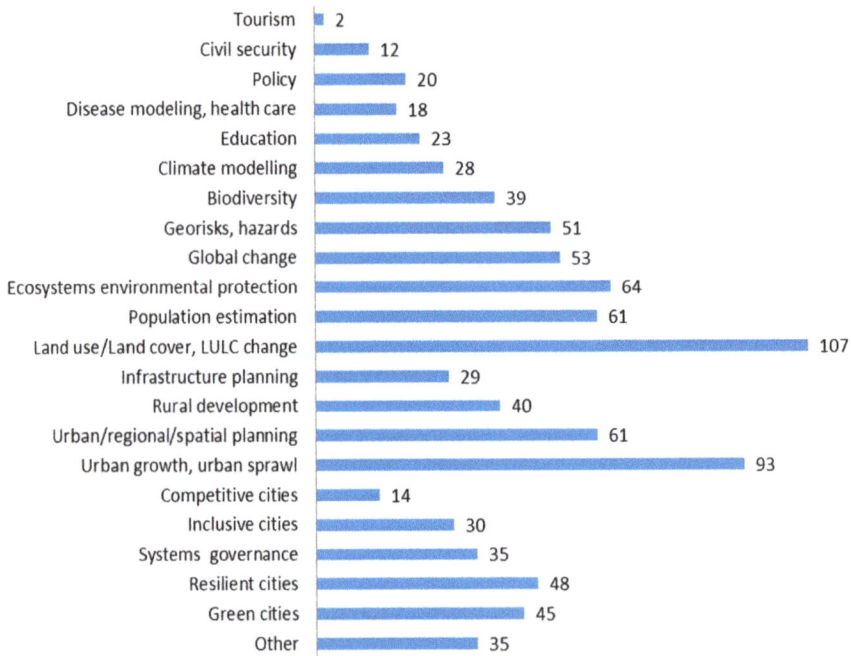

Figure 5. Fields of GUF application as indicated by the user community.

In addition to the statistics that could be derived from the License Agreement and Order Form, the GUF team finally conducted a first general analysis of scientific publications that refer to the GUF data and the underlying techniques. This study shows that, so far, a total of 470 publications cite the primary

DLR publications related to GUF data or methodology, respectively. Soon the GUF team plans to conduct a more dedicated survey of these citations, in particular with respect to the key characteristics of the underlying studies (e.g., thematic or technical sector, background of organization, etc.).

4. Evolution of the Product Portfolio and Future Updating Capability

Based on the lessons learnt from the GUF development and operations and considering the comprehensive feedback of the GUF user community, DLR has already started a systematic enhancement and extension of the current product and service portfolio. On the one hand, this includes an enhancement of the semantic and thematic scope of the layer. Here, first experimental developments include extensions of the GUF in form of the GUF-NetS, GUF-DenS, and GUF 3D.

The GUF-NetS aims at providing dedicated parameters and information tailored for an effective characterization of settlements properties and patterns such as their sizes and shapes, but also their spatial arrangement and relative properties with respect to the network of neighboring settlements. Such information is, for instance, needed for the differentiation of urban and rural areas, infrastructure planning, or disease modeling. The GUF-NetS is derived from the binary GUF layer (see Figure 6), and its parameters include attributes related to the geometry of the settlement patches (e.g., area, perimeter, shape index) as well as metrics describing the settlement pattern—meaning the spatial arrangement the built-up patches in their local or regional neighborhood (e.g., sparsely spread small settlements versus contiguous arrangement of large built-up patches). The major objective of this approach is to enable the users to analyze the settlement properties and patterns in high spatial detail at continental or even global scale. The GUF-NetS processing starts with a segmentation of the binary GUF mask that transforms all adjacent GUF pixels assigned as "built-up"—and thus forming a settlement patch—into an individual object. Then various geometric properties are derived for each single settlement object which are then added as attributes to the related polygon. These geometric parameters include area, perimeter, eccentricity, equivalent diameter, solidity and shape index. Next, a virtual spatial network is created with each node representing the centroid of an extracted GUF object and the edges connecting all neighboring objects that lie within a defined distance from each other. For each edge of the network a specific set of weights can be calculated—e.g., centroid distance, minimum distance, number of crossed edges or the local significance as a function of the distance between two connected nodes and the area of the two corresponding settlement objects. Finally, a collection of indexes such as betweenness, closeness, or eigenvector centrality is computed that describes the relevance of the individual settlements within the spatial network of their surrounding settlement objects. A detailed description of the GUF + NetS methodology is given by Esch et al. [42].

The GUF-DenS is an enhanced GUF version that specifies the built-up density or—as an inverse, the urban greenness—in form of the percent impervious surface within the area assigned as settlements by the conventional GUF. This information is particularly interesting for urban climate or hydrological modelling or approaches of population disaggregation. Technically, the GUF-DenS results from a combination of the GUF mask and imperviousness/greenness information modelled from temporal characteristics of the Normalized Difference Vegetation Index (NDVI) provided by a TimeScan dataset derived from Landsat imagery [32]. Assuming a strong inverse relation between vegetated and sealed surfaces, the intensity of vegetation cover defined by the NDVI can be used as a proxy for the percent impervious surface [43]. Figure 7 shows a subset of the GUF-DenS layer representing the impervious surface area for Johannesburg (South Africa). The layer is a continuous raster with values between 0 and 100, where red tone indicates high and green tone low density of impervious surface area. The global GUF-DenS product can be inspected at the Urban Thematic Exploitation Platform (https://urban-tep.eo.esa.int).

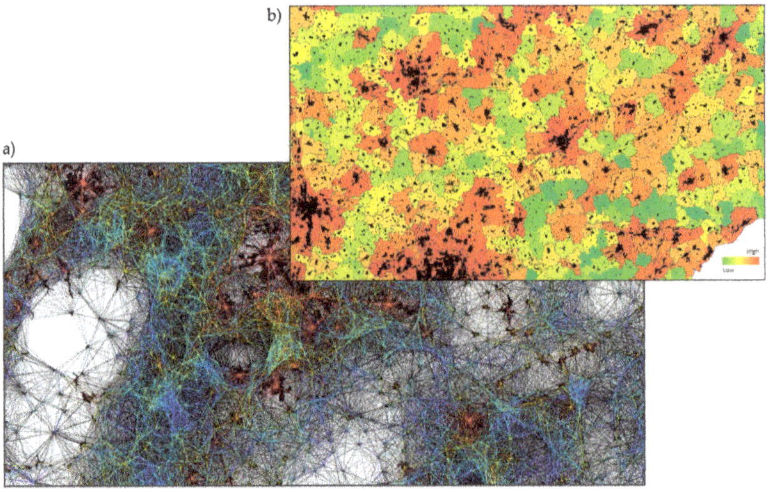

Figure 6. Example of GUF + NetS product (**a**) providing additional information for an analysis of settlements properties and patterns—e.g., local betweeness derived for administrative municipal units as shown in (**b**).

Figure 7. GUF-DenS layer covering Johannesburg in South Africa and subset comparing the layer with high resolution imagery.

The GUF-3D layer will define the average building height within the built-up area indicated by the GUF mask based on an analysis of DEM data. Compared to the GUF or GUF DenS, this new layer can provide additional valuable information such as it is particularly required for improved population or urban climate modelling. The methodological approach along with some first results obtained on the basis of TanDEM-X DEM information are detailed in Marconcini et al. [44]. In particular, the basic idea is to first identify ground pixels (i.e., whose elevation corresponds to that of the terrain) by iteratively analyzing the relative change in elevation within local neighborhoods of growing size. Next, these are used as seeds for Natural Neighbor (NN) interpolation [45] for retrieving a Digital Terrain Model (DTM) of the study area. Specifically, the NN algorithm has been chosen since it does not require any input parameter and due to its proven effectiveness in the presence of irregularly distributed data (making it then particularly suitable for large-scale application). The final building height is retrieved by subtracting the computed DTM from the original DEM within the areas labelled as positive in the GUF layer. However, to compensate for noise in the TanDEM-X DEM over built-up areas, a median filter is first applied and then the average height is derived over a grid of 120 × 120 m. As an example, in Figure 8 we report the results obtained for an area enclosing the city of Dongying (China) located at the delta of the Yellow River. In particular, Figure 8a shows a quicklook of the corresponding TanDEM-X DEM, whereas Figure 8b,c depict the building height estimated with the implemented methodology in 2D and 3D, respectively.

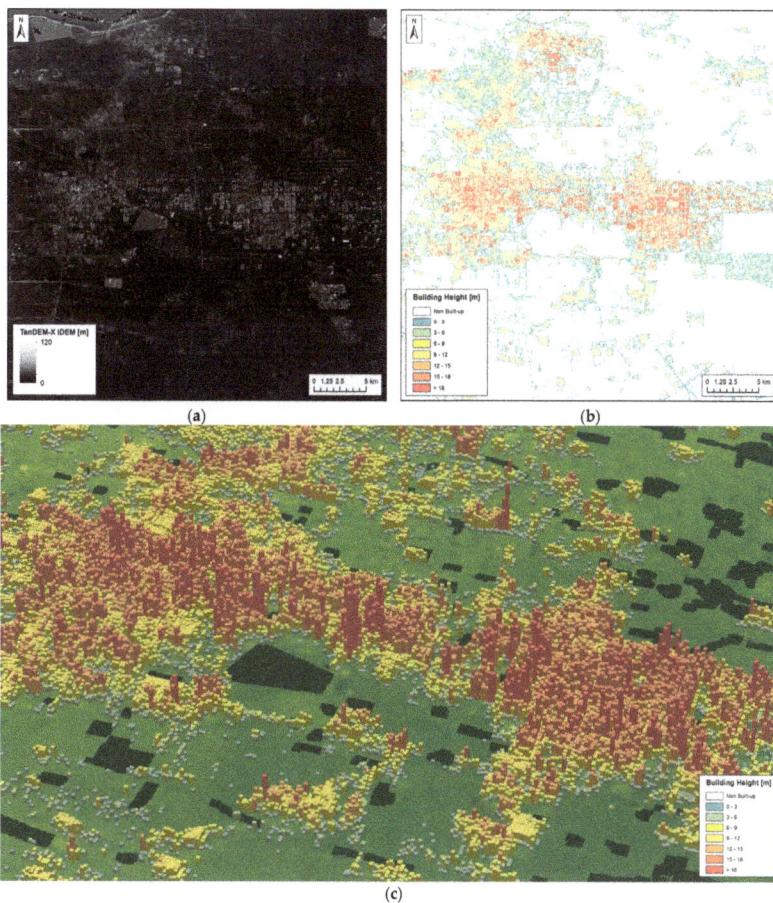

Figure 8. Dongying (China)—quicklook of the TanDEM-X DEM (**a**) and corresponding building height estimated with the implemented methodology in 2D (**b**) and 3D (**c**), respectively.

A key requirement for effective and serviceable global urban monitoring using EO is the capability of future updates (and related mapping campaigns) in regular intervals of 1–3 years. In order to meet this requirement, DLR is currently adapting the methodologies for the GUF, GUF-NetS, and GUF-DenS generation to the use of the fully open and free satellite data provided by the European Sentinel-1 (SAR) and Sentinel-2 (multispectral) as well as the US Landsat (multispectral) missions. This activity will lead to the provision of a new suite of global layers under the label "World Settlement Footprint (WSF)", starting in 2018 with a release of the WSF 2015 (equivalent of binary GUF, based on a joint analysis of multi-temporal Sentinel-1 and Landsat-8 data for the year 2015) and followed by WSF-2015-NetS and WSF-2015-DenS versions. In addition, a WSF-Evolution product will be produced that provides detailed information about the spatiotemporal development from 1985–2015 for each human settlement identified in the WSF-2015. The corresponding analysis is based on a processing of multitemporal mass data collections of the Landsat archive using the Google Earth Engine [46].

5. Conclusions and Outlook

With its spatial resolution of 12 m, the Global Urban Footprint dataset currently represents the most detailed and consistent global inventory of human settlements. At the same time, the outcome of the worldwide quality assessment confirms a high accuracy of the GUF map as it had already been indicated at regional scales by previous studies [32,47,48]. With a Kappa coefficient of 0.6373, the GUF 0.4" shows values about 0.16 higher with respect to both the state-of-the-art GHSL and GL30 layer derived from optical data, which exhibited severe underestimation issues (see Section 2.2). Here, a direct comparison revealed that the layers based on multispectral data most frequently suffer from the spectral similarity between bare soils or rock (e.g., mountain ranges, deserts, beaches, agricultural areas) on the one hand and urban impervious surface types on the other hand.

In contrary, built-up areas show a comparably distinct and globally consistent appearance in radar images. Due to the SAR imaging principle, the backscattering—and the image characteristics—are rather determined by structural and geometric features of the observed surfaces and objects than by chemical or physiologic properties as in case of optical/multispectral data. These geometric features are also quite distinct in the case of human settlements with their unique side-by-side accumulation of man-made vertical structures (inducing bright corner reflections and dark shadow regions) and low-signal areas due to specular reflection at smooth impervious surfaces. Accordingly, false positive alarms in the GUF classification mainly arise from confusions with structurally similar surface types such as rugged terrain in deserts or mountainous regions. However, Esch et al. [33] showed that these could effectively be eliminated during the post-editing phase by the inclusion of correction layers that are partially derived from multispectral satellite data that provided complementary features for an improved discrimination between real built-up areas and other critical non-built-up regions (see Section 2.1.6).

In addition, the GUF campaign benefitted from the SAR-specific advantage of a comparably fast collection of the required global database of satellite imagery: actually, 93% of the input SAR data had been acquired within a period of just two years—a significant improvement compared to the use of optical data. Here, the actual acquisition dates of the single scenes in fact frequently differ by several years while still showing local data gaps due to cloud coverage. Nevertheless, with the TimeScan approach, Esch et al. [38] have recently introduced a methodology based on multitemporal data collections that helps to compensate for this limiting factor in multispectral data. Indeed, a global TimeScan layer derived from >450.000 Landsat images acquired within a two-year period was already successfully used as correction layer for the false alarms identification and removal in the context of the GUF post-processing (see Section 2.1.6). Moreover, the new TimeScan technology forms the basis for a currently ongoing GUF-update for the year 2015 (WSF product), which has been described in more detail in Section 4.

Due to the characteristics and qualities described afore, the GUF product could already attract a large community of users within the first year after its official release. Thereby the systematic community survey documents that the activated users come from diverse sectors and scientific disciplines with their use scenarios covering a wide span of different applications. In combination with the direct feedback from the users, it can therefore be reasoned that the GUF initiative—and therewith also the underlying TerraSAR-X mission—has successfully provided a new dimension of valuable data, facts, and figures on the global phenomenon of urbanization. This new empirical evidence is supposed to substantially help address future key societal and environmental challenges such as rapid urban expansion, population growth, poverty reduction, loss of biodiversity, increasing carbon emissions, and ongoing climate change. Currently, every interested user can request the global GUF data via email (guf@dlr.de) free of charge at full spatial resolution of 0.4″ (12 m) for any scientific application and the generalized version at 2.8″ (84 m) for any non-profit use. For commercial use cases, Airbus DS has to be contacted (http://www.intelligence-airbusds.com). More information about the access procedure is available at DLR's GUF website (http://www.dlr.de/guf). In addition, the GUF layer is provided via a Web Mapping Service (WMS) from DLR's Earth Observation Center (https://geoservice.dlr.de/web/maps/eoc:guf:4326) and the Urban Thematic Exploitation Platform (https://urban-tep.eo.esa.int/) which is funded by the European Space Agency (ESA, Frascati, Italy). To further increase the societal benefit of the GUF, DLR actively shares the GUF layer and derived metrics with several international activities and networks—amongst others including the "Task GI-17—Global Urban Observation and Information" [49] and the "Human Planet Initiative" [50] of the Group on Earth Observations (GEO) or the POPGRID data collective [51].

However, in order to further increase the efficiency in reaching and supporting the science and policy communities developing the concrete strategies for the design, implementation, and management of sustainable urban environments, it is actually necessary to provide them with actionable information derived from a variety of sources in one single place. For this purpose, the ESA-funded "Urban Thematic Exploitation Platform (U-TEP)" has recently been set up [52]. U-TEP represents a web-based enabling instrument (https://urban-tep.eo.esa.int/) in form of an open, collaborative virtual environment that combines high-performance access to multi-source data repositories (e.g., EO, statistics, surveying data, volunteered geographic information, social media data) with efficient processing, analysis, and visualization functionalities and mechanisms for the effective development and sharing of methods and knowledge. For the future, it is planned that the entire suite of WSF products will be available at the U-TEP platform along with dedicated analytics tools to address major urban challenges such as urban growth/transformation monitoring, livable cities, resilience and disaster risk reduction, poverty eradication, health issues and diseases, world heritage conservation, biodiversity/ecosystem services, energy and supplies, traffic, and climate change adaptation and mitigation.

Author Contributions: Conceptualization, T.E.; Data curation, W.H. and A.H.; Formal analysis, F.B., D.P.L., S.Ü. and J.Z.; Funding acquisition, T.E., A.R. and S.D.; Methodology, T.E., M.M., A.R. and J.Z.; Resources, A.R., S.D. and N.G.; Software, A.H. and J.Z.; Supervision, T.E., M.M. and S.D.; Validation, D.P.L., M.M., S.Ü. and N.G.; Writing—original draft, T.E., F.B., M.M. and S.Ü.; T.E. and F.B. revised and edited the manuscript.

Funding: The European Space Agency (ESA) for funding the project Urban Thematic Exploitation Platform TEP Urban (ESRIN/Contract No. 4000113707/15/I-NB). We also thank the World Bank and the Swiss State Secretariat for Economic Affairs (SECO) for their financial support in the context of the GUF quality assessment and enhancement activities (Contract 7174285).

Acknowledgments: The authors would like to thank the TerraSAR-X Science Team for providing the global SAR data used to derive the GUF layer. We also thank the Google Maps Team for the in-kind crowd-sourced labelling of the validation dataset used for assessing the quality of the different layers analyzed in this study.

Conflicts of Interest: The authors declare no conflict of interest.

References

1. IPCC. Human settlements, infrastructure and spatial planning. In *Climate Change 2014: Mitigation of Climate Change*; Intergovernmental Panel on Climate Change: Geneva, Switzerland, 2014; p. 127.

2.   United Nations, Department of Economic and Social Affairs, Population Division. *World Urbanization Prospects: The 2014 Revision*; United Nations, Department of Economic and Social Affairs, Population Division: New York, NY, USA, 2015; p. 517.

3.   United Nations, Department of Economic and Social Affairs, Population Division. *World Population Prospects: The 2017 Revision, Key Findings and Advance Tables*; United Nations: New York, NY, USA, 2017; p. 52.

4.   United Nations. *Transforming Our World: The 2030 Agenda for Sustainable Development*; General Assembly United Nations: New York, NY, USA, 2015.

5.   GPSC. *Urban Sustainability Framework; Global Platform for Sustainable Cities (GPSC)*; World Bank: Washington DC, USA, 2018; p. 143.

6.   Abu Bakar, A.H.; Cheen, K.S. A framework for assessing the sustainable urban development. *Procedia Soc. Behav. Sci.* **2013**, *85*, 484–492. [CrossRef]

7.   Grafakos, S.; Gianoli, A.; Tsatsou, A. Towards the development of an integrated sustainability and resilience benefits assessment framework of urban green growth interventions. *Sustainability* **2016**, *8*, 461. [CrossRef]

8.   Lützkendorf, T.; Balouktsi, M. Assessing a sustainable urban development: Typology of indicators and sources of information. *Procedia Environ. Sci.* **2017**, *38*, 546–553. [CrossRef]

9.   Potere, D.; Schneider, A.; Angel, S.; Civco, D. Mapping urban areas on a global scale: Which of the eight maps now available is more accurate? *Int. J. Remote Sens.* **2009**, *30*, 6531–6558. [CrossRef]

10.   Ban, Y.; Jacob, A.; Gamba, P. Spaceborne sar data for global urban mapping at 30 m resolution using a robust urban extractor. *ISPRS J. Photogramm. Remote Sens.* **2015**, *103*, 28–37. [CrossRef]

11.   Schneider, A.; Friedl, M.A.; Potere, D. Mapping global urban areas using modis 500-m data: New methods and datasets based on 'urban ecoregions'. *Remote Sens. Environ.* **2010**, *114*, 1733–1746. [CrossRef]

12.   Arino, O.; Ramos Perez, J.J.; Kalogirou, V.; Bontemps, S.; Defourny, P.; Van Bogaert, E. Global land cover map for 2009 (globcover 2009). *PANGAEA* **2012**. [CrossRef]

13.   Miyazaki, H.; Shao, X.; Iwao, K.; Shibasaki, R. An automated method for global urban area mapping by integrating aster satellite images and gis data. *IEEE J. Sel. Top. Appl. Earth Obs. Remote Sens.* **2013**, *6*, 1004–1019. [CrossRef]

14.   Gamba, P.; Lisini, G. A robust approach to global urban area extent extraction using ASAR Wide Swath Mode data. In Proceedings of the 2012 Tyrrhenian Workshop on Advances in Radar and Remote Sensing (TyWRRS), Naples, Italy, 12–14 September 2012; pp. 1–5.

15.   Wieland, M.; Pittore, M. Large-area settlement pattern recognition from landsat-8 data. *ISPRS J. Photogramm. Remote Sens.* **2016**, *119*, 294–308. [CrossRef]

16.   Pesaresi, M.; Ehrlich, D.; Ferri, S.; Florczyk, A.; Freire, S.; Halkia, M.; Julea, A.; Kemper, T.; Soille, P.; Syrris, V. *Operating Procedure for the Production of the Global Human Settlement Layer from Landsat Data of the Epochs 1975, 1990, 2000, and 2014*; Joint Research Centre—European Commission: Luxembourg, 2016.

17.   Corbane, C.; Pesaresi, M.; Politis, P.; Syrris, V.; Florczyk, A.J.; Soille, P.; Maffenini, L.; Burger, A.; Vasilev, V.; Rodriguez, D.; et al. Big earth data analytics on sentinel-1 and landsat imagery in support to global human settlements mapping. *Big Earth Data* **2017**, *1*, 118–144. [CrossRef]

18.   Liu, X.; Hu, G.; Chen, Y.; Li, X.; Xu, X.; Li, S.; Pei, F.; Wang, S. High-resolution multi-temporal mapping of global urban land using landsat images based on the google earth engine platform. *Remote Sens. Environ.* **2018**, *209*, 227–239. [CrossRef]

19.   WUDAPT. World Urban Database and Assess Poatal Tools—Wudapt. Available online: http://www.wudapt. org/ (accessed on 22 May 2018).

20.   Bechtel, B.; Alexander, P.; Böhner, J.; Ching, J.; Conrad, O.; Feddema, J.; Mills, G.; See, L.; Stewart, I. Mapping local climate zones for a worldwide database of the form and function of cities. *ISPRS Int. J. Geo-Inf.* **2015**, *4*, 199–219. [CrossRef]

21.   Roth, A.; Hoffmann, J.; Esch, T. Terrasar-x: How can high resolution Sar data support the observation of urban areas? In Proceedings of the ISPRS Conference 3rd International Symposium Remote Sensing and Data Fusion Over Urban Areas (URBAN 2005), Tempe, AZ, USA, 14–16 March 2005.

22.   Kuenzer, C.; Guo, H.; Schlegel, I.; Tuan, V.; Li, X.; Dech, S. Varying scale and capability of envisat asar-wsm, terrasar-x scansar and terrasar-x stripmap data to assess urban flood situations: A case study of the mekong delta in can tho province. *Remote Sens.* **2013**, *5*, 5122–5142. [CrossRef]

23.   Bai, L.; Jiang, L.; Wang, H.; Sun, Q. Spatiotemporal characterization of land subsidence and uplift (2009–2010) over wuhan in central china revealed by terrasar-x insar analysis. *Remote Sens.* **2016**, *8*, 350. [CrossRef]

24. Gui, R.; Xu, X.; Dong, H.; Song, C.; Pu, F. Individual building extraction from terrasar-x images based on ontological semantic analysis. *Remote Sens.* **2016**, *8*, 708. [CrossRef]

25. Qin, X.; Yang, T.; Yang, M.; Zhang, L.; Liao, M. Health diagnosis of major transportation infrastructures in shanghai metropolis using high-resolution persistent scatterer interferometry. *Sensors* **2017**, *17*, 2770. [CrossRef] [PubMed]

26. Weissgerber, F.; Colin-Koeniguer, E.; Nicolas, J.-M.; Trouvé, N. 3d monitoring of buildings using terrasar-x insar, dinsar and polsar capacities. *Remote Sens.* **2017**, *9*, 1010. [CrossRef]

27. Esch, T. Automatisierte Analyse von Siedlungsflächen auf der Basis Höchstauflösender Radardaten (Automated Analysis of Urban Areas Based on High Resolution Sar Images). Ph.D. Dissertation, Bayerische Julius-Maximilians Universität Würzburg, Würzburg, Germany, 2006.

28. Esch, T.; Schenk, A.; Ullmann, T.; Thiel, M.; Roth, A.; Dech, S. Characterization of land cover types in terrasar-x images by combined analysis of speckle statistics and intensity information. *IEEE Trans. Geosci. Remote Sens.* **2011**, *49*, 1911–1925. [CrossRef]

29. Esch, T.; Schenk, A.; Thiel, M. Monitoring of urban environments with terrasar-x data. In Proceedings of the IGARSS 2007, Barcelona, Spain, 23–27 July 2007.

30. Esch, T.; Thiel, M.; Schenk, A.; Roth, A.; Muller, A.; Dech, S. Delineation of urban footprints from terrasar-x data by analyzing speckle characteristics and intensity information. *IEEE Trans. Geosci. Remote Sens.* **2010**, *48*, 905–916. [CrossRef]

31. Esch, T.; Taubenböck, H.; Roth, A.; Heldens, W.; Felbier, A.; Thiel, M.; Schmidt, M.; Müller, A.; Dech, S. Tandem-x mission—new perspectives for the inventory and monitoring of global settlement patterns. *J. Appl. Remote Sens.* **2012**, *6*. [CrossRef]

32. Esch, T.; Heldens, W.; Hirner, A.; Keil, M.; Marconcini, M.; Roth, A.; Zeidler, J.; Dech, S.; Strano, E. Breaking new ground in mapping human settlements from space—The global urban footprint. *ISPRS J. Photogramm. Remote Sens.* **2017**, *134*, 30–42. [CrossRef]

33. Esch, T.; Marconcini, M.; Felbier, A.; Roth, A.; Heldens, W.; Huber, M.; Schwinger, M.; Taubenbock, H.; Muller, A.; Dech, S. Urban footprint processor—Fully automated processing chain generating settlement masks from global data of the tandem-x mission. *IEEE Geosci. Remote Sens. Lett.* **2013**, *10*, 1617–1621. [CrossRef]

34. Böttcher, M.; Reißig, R.; Mikusch, E.; Reck, C. Processing management tools for earth observation products at DLR-DFD. In *Data Systems in Aero-Space (DASIA)*; Agency, E.S., Ed.; European Space Agency: Nice, France, 2001; p. 5.

35. Habermeyer, M.; Marschalk, U.; Roth, A. W42—A scalable spatial database system for holding digital elevation models. In Proceedings of the 2009 17th International Conference on Geoinformatics, Fairfax, VA, USA, 12–14 August 2009; pp. 1–6.

36. Fomferra, N.; Böttcher, M.; Zühlke, M.; Brockmann, C.; Kwiatkowska, E. Calvalus: Full-mission eo cal/val, processing and exploitation services. In Proceedings of the 2012 IEEE International Geoscience and Remote Sensing Symposium, Munich, Germany, 22–27 July 2012; pp. 5278–5281.

37. Chang, F.; Chen, C.-J.; Lu, C.-J. A linear-time component-labeling algorithm using contour tracing technique. *Comput. Vis. Image Underst.* **2004**, *93*, 206–220. [CrossRef]

38. Esch, T.; Üreyen, S.; Zeidler, J.; Metz–Marconcini, A.; Hirner, A.; Asamer, H.; Tum, M.; Böttcher, M.; Kuchar, S.; Svaton, V.; et al. Exploiting big earth data from space—First experiences with the timescan processing chain. *Big Earth Data* **2018**, *2*, 36–55. [CrossRef]

39. Marconcini, M.; Üreyen, S.; Esch, T.; Metz-Marconcini, A.; Zeidler, J.; Palacios-Lopez, D.; Strano, E.; Gorelick, N. Exploiting satellite big data for outlining settlements globally—The world settlement footprint. 2015. Scientific Data in preparation.

40. United Nations. *Principles and Recommendations for Population and Housing Censuses, Revision 1*; United Nations: New York, NY, USA, 1998; p. 335.

41. Wang, P.; Huang, C.; Brown de Colstoun, E.C.; Tilton, J.C.; Tan, B. *Global Human Built-up and Settlement Extent (Hbase) Dataset from Landsat*; NASA Socioeconomic Data and Applications Center (SEDAC): Palisades, NY, USA, 2017.

42. Esch, T.; Marconcini, M.; Marmanis, D.; Zeidler, J.; Elsayed, S.; Metz, A.; Müller, A.; Dech, S. Dimensioning urbanization—An advanced procedure for characterizing human settlement properties and patterns using spatial network analysis. *Appl. Geogr.* **2014**, *55*, 212–228. [CrossRef]

43. Esch, T.; Himmler, V.; Schorcht, G.; Thiel, M.; Wehrmann, T.; Bachofer, F.; Conrad, C.; Schmidt, M.; Dech, S. Large-area assessment of impervious surface based on integrated analysis of single-date landsat-7 images and geospatial vector data. *Remote Sens. Environ.* **2009**, *113*, 1678–1690. [CrossRef]

44. Marconcini, M.; Marmanis, D.; Esch, T.; Felbier, A. A novel method for building height estmation using tandem-x data. In Proceedings of the 2014 IEEE Geoscience and Remote Sensing Symposium, Quebec City, QC, Canada, 13–18 July 2014; pp. 4804–4807.

45. Sibson, R. A brief description of natural neighbor interpolation. In *Interpolating Multivariate Data*; John Wiley & Sons: New York, NY, USA, 1981; pp. 21–36.

46. Gorelick, N.; Hancher, M.; Dixon, M.; Ilyushchenko, S.; Thau, D.; Moore, R. Google earth engine: Planetary-scale geospatial analysis for everyone. *Remote Sens. Environ.* **2017**, *202*, 18–27. [CrossRef]

47. Klotz, M.; Kemper, T.; Geiß, C.; Esch, T.; Taubenböck, H. How good is the map? A multi-scale cross-comparison framework for global settlement layers: Evidence from central europe. *Remote Sens. Environ.* **2016**, *178*, 191–212. [CrossRef]

48. Mück, M.; Klotz, M.; Taubenböck, H. Validation of the dlr global urban footprint in rural areas: A case study for Burkina Faso. In Proceedings of the 2017 Joint Urban Remote Sensing Event (JURSE), Dubai, UAE, 6–8 March 2017; p. 4.

49. Group on Earth Observations. Global Urban Observation and Information. Available online: https://www.earthobservations.org/activity.php?id=125 (accessed on 30 April 2018).

50. Group on Earth Observations. Geo Human Planet Initiative: Spatial Modeling of Impact, Exposure and Access to Resources. Available online: https://www.earthobservations.org/activity.php?id=119 (accessed on 30 April 2018).

51. POPGRID. Popgrid—Data Collaborative Enhanced Population, Settlement and Infrastructure Data. Available online: https://sites.google.com/ciesin.columbia.edu/popgrid (accessed on 30 April 2018).

52. Esch, T.; Uereyen, S.; Asamer, H.; Hirner, A.; Marconcini, M.; Metz, A.; Zeidler, J.; Boettcher, M.; Permana, H.; Brito, F.; et al. Earth observation-supported service platform for the development and provision of thematic information on the built environment—The tep-urban project. In Proceedings of the 2017 Joint Urban Remote Sensing Event (JURSE), Dubai, UAE, 6–8 March 2017; pp. 1–4.

remote sensing

MDPI

Article

Combining TerraSAR-X and Landsat Images for Emergency Response in Urban Environments

Shiran Havivi [1,*], Ilan Schvartzman [2], Shimrit Maman [3], Stanley R. Rotman [2] and Dan G. Blumberg [1]

[1] Earth and Planetary Image Facility, Department of Geography and Environmental Development, Ben-Gurion University of the Negev, P.O. Box 653, Beer-Sheva 84105, Israel; blumberg@bgu.ac.il

[2] Department of Electrical and Computer Engineering, Ben-Gurion University of the Negev, P.O. Box 653, Beer-Sheva 84105, Israel; ilan.schvartzman@gmail.com (I.S.); srotman@bgu.ac.il (S.R.R.)

[3] Homeland Security Research Institute, Ben-Gurion University of the Negev, P.O. Box 653, Beer-Sheva 84105, Israel; tiroshs@gmail.com

* Correspondence: havivi@post.bgu.ac.il; Tel.: +972-52-358-9213

Received: 29 March 2018; Accepted: 17 May 2018; Published: 21 May 2018

Abstract: Rapid damage mapping following a disaster event, especially in an urban environment, is critical to ensure that the emergency response in the affected area is rapid and efficient. This work presents a new method for mapping damage assessment in urban environments. Based on combining SAR and optical data, the method is applicable as support during initial emergency planning and rescue operations. The study focuses on the urban areas affected by the Tohoku earthquake and subsequent tsunami event in Japan that occurred on 11 March 2011. High-resolution TerraSAR-X (TSX) images of before and after the event, and a Landsat 5 image before the event were acquired. The affected areas were analyzed with the SAR data using only one interferometric SAR (InSAR) coherence map. To increase the damage mapping accuracy, the normalized difference vegetation index (NDVI) was applied. The generated map, with a grid size of 50 m, provides a quantitative assessment of the nature and distribution of the damage. The damage mapping shows detailed information about the affected area, with high overall accuracy (89%), and high Kappa coefficient (82%) and, as expected, it shows total destruction along the coastline compared to the inland region.

Keywords: InSAR coherence; NDVI; damage assessment; density map; tsunami; earthquake; GIS

1. Introduction

Natural and man-made disasters often cause severe economic and physical damage, the effects of which are potentially far more devastating in urban areas [1]. In the wake of a disaster, the effectiveness of the emergency response depends critically on the extent to which the damage assessment is both rapid and precise. To that end, remote sensing imagery, which can be used to assess the change over time of a given geographical area, and can rapidly map the damage with high accuracy and spatial coverage and at low cost [2]. Both optical and radar remote sensing data are increasingly being used for damage assessment [1,3–5].

The conventional technologies used in passive optical satellites operate mostly in the visible and infrared spectrum and are mainly used to map damage by visual interpretation [6–8] and to assess the vegetation status via the normalized difference vegetation index (NDVI). Conversely, synthetic aperture radar (SAR) is an active microwave sensor with day/night and all-weather operational capabilities that is highly sensitive to surface roughness, geometrical structures (e.g., built-up areas) and canopy volume in vegetated areas [9].

Changes in urban environments over time can be detected by using multi-temporal SAR intensity images. More detailed information can be obtained via interferometric SAR (InSAR) data, which are based on the differences in the phase information of the returning signals of two temporal images [10]. To detect and assess the damage caused by natural disasters, most published studies use either InSAR coherence [11–15] or the SAR intensity correlation [3,16,17], or both [18–20]. The accurate mapping that is essential to the planning and execution of an emergency response to a natural disaster can be obtained by using very high resolution (VHR) SAR data, such as that generated by the TerraSAR-X satellite. Its extremely high spatial resolution enables TerraSAR-X to discriminate between buildings and other elements more precisely than is possible using lower resolution technology [5,21].

Assessments of the level of similarity between two SAR images can be provided by InSAR coherence change detection by using image amplitude and phase data. Coherence change detection is capable of detecting very subtle scene changes, expressed as loss of coherence, which may remain undetected using intensity data alone [22]. Loss of coherence measurements rely on the typically high coherence values exhibited by urban environments over time. Thus the scene changes (i.e., damage to the urban environment) caused by a natural disaster reflect loss of coherence that can be interpreted quantitatively in terms of the severity and extent of the damage [2]. To date, coherence results have mostly been analyzed by using the normalized difference (ND) of the coherence change [14,18,19,23]. For such an analysis, three SAR images are required (two from before and one from after the event) to generate pre- and co-event coherence maps. Among the principal drawbacks of this approach are its long processing and analysis times and its high costs. Furthermore, the accuracy of the results of the analysis can be low due to temporal decorrelation, which can occur when the dates of the SAR images are separated by long time intervals. Lastly, two pre-event images are not always available for change detection analyses.

In this study, we propose a new mapping approach for damage assessment based on the combination of SAR and multispectral data. The novelty of the proposed technique is set firstly in its ability to be implemented rapidly and accurately for urban damage assessment by a relatively small number of both optical and radar images. Secondly, the use of spatial analysis to produce a damage assessment map based on the coherence and NDVI imagery results was not yet tested. We demonstrate the effectiveness of this combined approach by applying it to the Tohoku earthquake and the subsequent tsunami event that occurred in Japan in 2011.

2. Research Area

On 11 March 2011, the magnitude 9 Tohoku earthquake occurred. With its epicenter located off Japan's eastern coast (38.322°N/142.369°E; Figure 1), the earthquake also triggered a destructive tsunami with a wave height that exceeded 10 m. Penetrating as far as 5 km inland when it struck the Japanese coast [24], the tsunami wave caused loss of human life and destruction of infrastructure. The extensive damage, assessed and reported by [25], included nearly 360,000 totally or partially destroyed houses. Although the seismic motion along the coast and inland accounted for some of the damage to housing, the tsunami was much more destructive, as it washed away many of the houses along the coast, and, in some places, it completely wiped out entire areas [25].

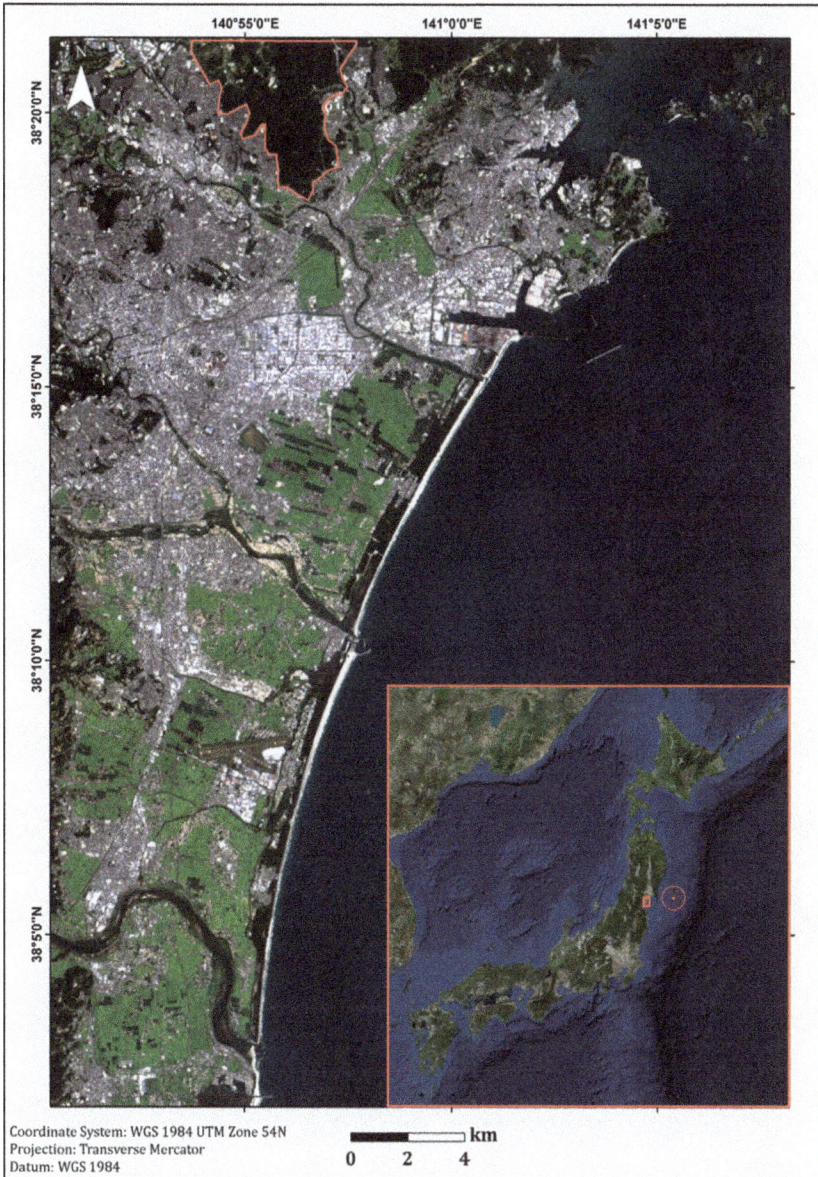

Figure 1. Pre-event Landsat 5 image (in visible bands 1—blue, 2—green, and 3—red) of the research area from 24 August 2010. Inset: overview of the area most affected by the Tohoku earthquake and the earthquake's epicenter located at 38.322°N/142.369°E.

3. Materials and Methods

The methodology followed to create a damage assessment map is illustrated in Figure 2.

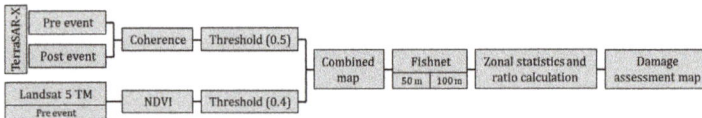

Figure 2. Research outline.

3.1. Remote Sensing Data and Pre-Processing

Two types of images, SAR and optical, were used to map the area affected by the earthquake and tsunami. SAR images comprised three high-resolution X-band (3.1 cm) pre- and post-event TerraSAR-X (TSX) images that were single-look complex (SLC) in a descending orbit direction. Data were acquired in StripMap mode with single HH polarization, an incidence angle of 37°, and pixel spacing of 2 m on the vertical axis and 0.9 m on the horizontal axis. Vegetation cover was mapped with a multispectral image taken by Landsat5 Thematic Mapper (TM) before the event. With moderate spatial resolution of 30 m/pixel (apart from band 6, with spatial resolution of 120 m/pixel), this multispectral image comprises seven spectral bands in the following ranges: VIS (bands 1–3), NIR (bands 4–5), TIR (band 6), and MIR (band 7), as specified in Table 1.

Table 1. Satellite images used in this study.

Satellite	Sensor Type	Acquisition Date		Resolution	Spectral Properties
TerraSAR-X	SAR	Pre-event	21 September 2008	2 m	SLC
		Pre-event	20 October 2010		X-band
		Post-event	12 March 2011		HH polarization
Landsat5 TM	Multispectral	Pre-event	24 August 2010	30 m	7 bands

The SAR images were co-registered, according to the 20 October 2010 image, which was selected as the reference image ('master') for both periods (before and after the event). The error RMS was up to 0.1 pixel, and an eight-point truncated sinc kernel was used for the interpolation [26,27]. Lastly, the Landsat5 image was georeferenced and resampled (to the TSX spatial resolution) according to the reference image.

3.2. InSAR Coherence

Based on both amplitude and phase, coherence change detection determines the stability of an area and shows where changes have occurred, depicting even subtle changes with levels of accuracy ranging from several millimeters to centimeters [9]. It, thus, quantifies the resemblance of a subset of the image between two points in time. The local coherence γ (for each pixel) was calculated according to the formula:

$$\gamma = \frac{\left|\sum_{k=1}^{N} f_k g^*_k\right|}{\sqrt{\sum_{k=1}^{N} |f_k|^2 \sum_{k=1}^{N} |g_k|^2}}, \quad 0 \leq \gamma \leq 1 \tag{1}$$

where f_k and g_k are the complex SAR images, and $f_k g^*_k$ is the expectation value operator. N represents the size of the estimation window. Due to the high resolution of the TSX images, a reliable accuracy of the coherence could be achieved by calculating with a relatively small window size of 3×5 pixels in the azimuth and range directions, respectively. This window size contains enough pixels to ensure a minimum loss of information by providing a high spatial resolution and high contrast coherence image.

The coherence index γ is an absolute value that represents a measure of the similarity between the two SAR images. The value of γ ranges between 0 and 1, wherein pixels with a good correlation are represented by high values, and pixels with no match are represented by low values [9,22]. Natural disaster-driven surface changes, such as damage to buildings and the presence of vegetation, lead to a loss of coherence.

3.3. Normalized Difference Vegetation Index (NDVI)

To overcome the loss of coherence caused by changes in vegetation cover, a vegetation mask was applied by using the NDVI to identify (and remove) vegetated areas from the coherence map. One of the main advantages of the NDVI is its ability to distinguish vegetated from non-vegetated areas in the image. To do so, it exploits the contrast between the absorption of the red light (Band 3, 0.63–0.69 μm) by chlorophyll and the strong reflection of near-infrared radiation (Band 4, 0.76–0.90 μm), which ensures that the difference between the red and infrared reflectance values is prominent and, therefore, that it yields a high ratio. The NDVI is defined as:

$$\text{NDVI} = \frac{\text{NIR} - \text{red}}{\text{NIR} + \text{red}} \qquad (2)$$

and generates values ranging from −1 to 1 (healthier/denser vegetation has higher values).

3.4. Combining SAR and Multi-Spectral Data

3.4.1. Setting Thresholds

To better focus on the man-made structures that were damaged by the earthquake and tsunami, a threshold was determined for the co-event coherence map (Figure 3) and the NDVI (Figure 4). A threshold value of ≥0.4 was chosen for the NDVI, and, therefore, only the non-vegetated areas remained. Lower NDVI thresholds overestimated the vegetated area by displaying some of the built areas as vegetation. A threshold value of 0.5 for the co-event coherence map was determined according to [27], which showed it to be the lowest coherence value in an urban area for the same time interval (144 days) depicted between two images. By dividing the results into two classes, a binary map was generated in which ≤0.5 was used for pixels that experienced change and >0.5 was used for those that experienced slight or no change. The coherence binary map was then overlain and masked by the threshold NDVI layer, after which both sets of results (thresholds of 0.4 and 0.5) were combined into one damage assessment map (Figure 5).

3.4.2. Damage Assessment Map

The generated map was classified into three classes: damaged, undamaged, and vegetated (Figures 5 and 6). This map, however, provides only a general view of the affected area. To obtain more detailed information about the damage, two grids with cell sizes of 50 m and 100 m were created using Create Fishnet (ArcGIS 10.5, ESRI, Redlands, CA, USA). The damaged area in each cell was calculated by using Zonal statistics (ArcGIS 10.5, ESRI, Redlands, CA, USA). The ratio between the damaged area and the grid cell area was then calculated for each cell in the two grids and expressed as a percentage. This process generated damage assessment map in which the damaged area is presented per unit area (50 m grid) (Figure 7).

3.5. Accuracy Assessment

Validation was derived from optical images from 14 March 2011 using Google Earth (high spatial resolution imagery). Overall accuracy, user's (UA) and producer's (PA) accuracies and Kappa coefficient were used to define the accuracy. Independent samples were randomly tested versus the imagery results (over 400 random distributed points). This was applied for both the entire scene and specific smaller sections were tested, such as inland and along the coastline (Figure 5).

4. Results and Discussion

To detect and identify the level of change in the study area, the results were presented according to the protocol outlined in Figure 2. The interferometric coherence maps of both periods, i.e., pre-event pair (left image) and co-event pair (right image), are presented in Figure 3. Light tones (high coherence values) indicate areas with slight or no change, while dark tones (low coherence values) mark areas that have undergone change. A simple visual examination shows that the most noticeable differences in both maps are evident between the coastline and inland areas. Along the coastline, changes are easily observed before and after the event, but the co-event map shows that most of the coastline area has changed. In contrast to the coastline, the inland areas show high stability, with no changes or slight changes in both maps.

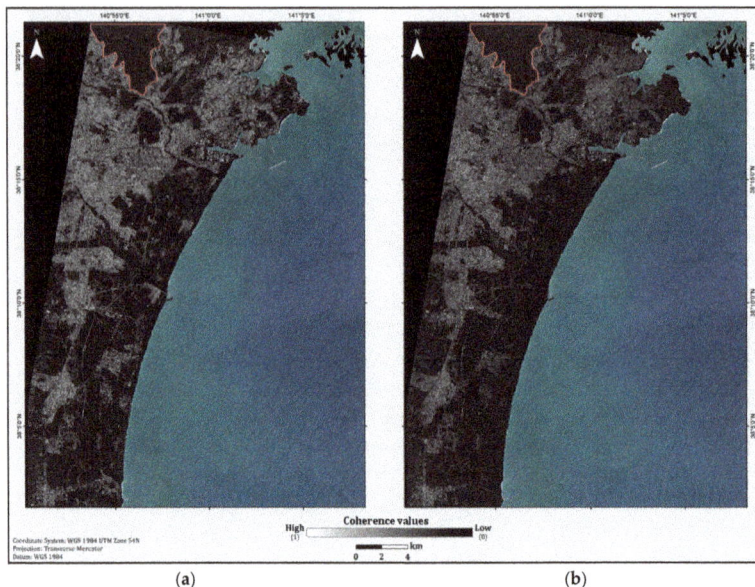

(a)   (b)

Figure 3. InSAR coherence maps overlying the Landsat 5 image: (**a**) pre-event coherence map; and (**b**) co-event coherence map. Light tones (high coherence values) represent no change. Dark tones represent changes (low coherence values). Red polygon represents low coherence values due to vegetation cover, see text for further explanation.

In general, a clear distinction exists in the level of change evident along the coastline, which experienced the most significant damage and its coherence values are correspondingly low, compared to that in the inland area. In contrast to the coastline, inland areas suffered much less damage overall, which is reflected in their generally high coherence values, but certain inland areas exhibited unusually low coherence. For example, the region located to the northwest of the study area (Figure 3, area outlined in red) displays low coherence values both pre- and co-event. Situated too far inland, this area was not affected by the tsunami.

In regions that have experienced major natural disasters, data collected during the period prior to the event contributes valuable information toward understanding the pre-event nature of the land cover. The inland regions of interest that display low coherence are suspected to have exceptionally high vegetation cover, the presence of which markedly influences the results, causing an apparent loss of coherence [27–29]. Observations of the X-band are the most strongly affected [30], as that wavelength penetrates the upper part of the vegetation cover, such as the canopy and branches [30], and causes volumetric scattering. Furthermore, even slight, wind-generated movement of vegetation causes a loss of coherence, as shown previously by [27].

To determine with certainty that the low coherence values were the result of changes due to the tsunami event and not due to changes in vegetation cover caused either by winds or phenology, a vegetation mask was applied using the NDVI values extracted from the Landsat image. The obtained pre-event NDVI map is presented in Figure 4, where light tones signify vegetated areas, and dark tones signify non-vegetated areas, such as urban areas or man-made structures.

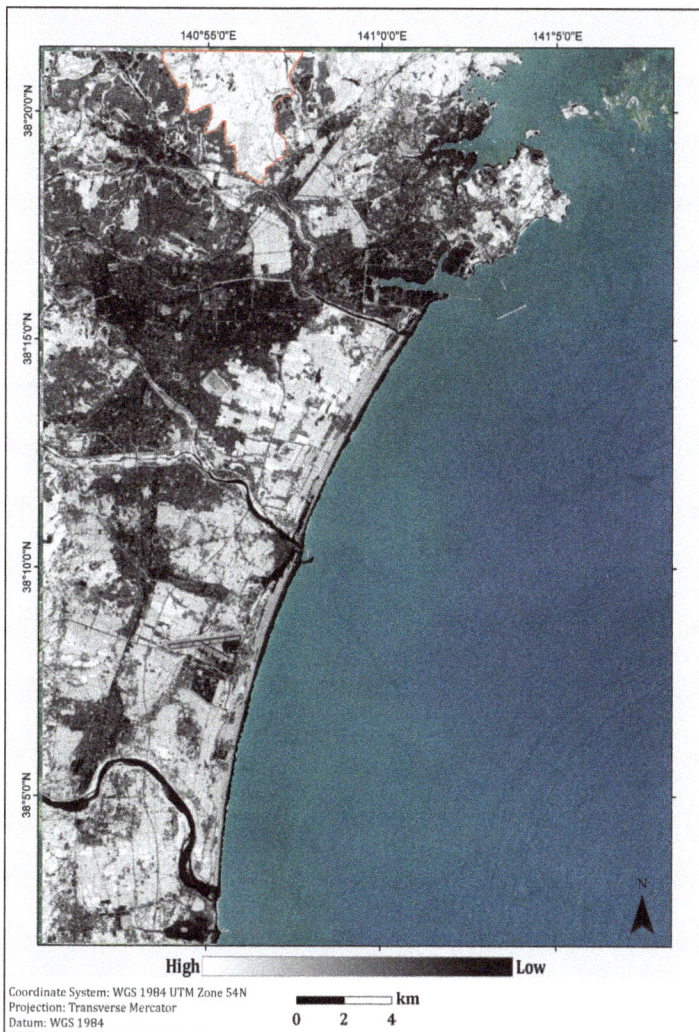

Figure 4. NDVI map. Light shades represent vegetated areas, observed mostly along the coastline. Dark shades represent urban areas or man-made structures, and they mostly appear in the inland areas. The red polygon represents a suspected area that was found to be a vegetated area.

Once the results for both the coherence and NDVI were obtained, thresholds were applied accordingly, and the two layers were then used to mask the co-event coherence results. A schematic of the procedure and the resulting combined map obtained after masking for the two thresholds are shown in Figure 5. By eliminating the extracted vegetation data, the combined map enables damaged buildings to be distinguished from undamaged structures in the wake of the earthquake and tsunami events. The combined map (Figure 5) could, therefore, be divided into three major classes: (1) vegetation; (2) damaged buildings; and (3) undamaged buildings.

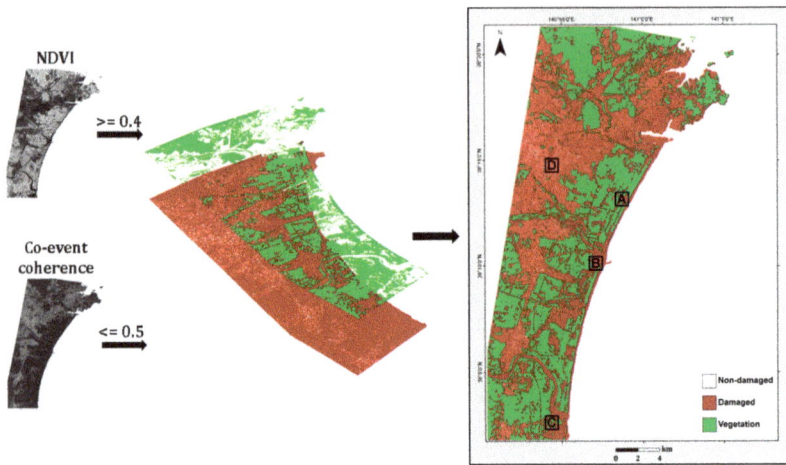

Figure 5. NDVI and co-event coherence maps generated with a threshold of 0.4 and 0.5, respectively. The combined map displays three classes: vegetated areas (green), damaged (dark red), and undamaged (white) areas. Insets A, B, C, and D are enlarged in Figure 6.

The extent of the observed damage depicted in the combined map was further validated by optical images from 14 March 2011 using Google Earth (Figure 6). Note that on that date (the same day as the post-event SAR acquisition), most of the study area was covered by clouds, a condition that supports the use of SAR imagery rather than optical images for damage assessment. Validation of the combined map was tested on sections of the coastline (Figure 6), where the severe damage that occurred due especially to the tsunami wave is clearly evident. This coastal area was characterized by extensive vegetation cover, mostly agricultural fields, which are clearly visible in the pre-event images (Figure 6). After the event, most of these regions were flooded. Despite the severe damage that occurred in these regions, because they were recognized as vegetated land cover in the NDVI map, they were extracted from the final analysis. The results for the built area, the most damaged of which are depicted in red, show that only a few structures remained unchanged (white dots), i.e., were slightly damaged or not damaged at all. In contrast to the coastline, the inland areas exhibit progressively less damage and greater numbers of stable structures, which appear white and represent a very small part of the area (Figure 6D).

Based on the combined map and after the removal of vegetation cover, a damage assessment map was generated. This map (Figure 7) addresses only the built area and is presented with a cell grid size of 50 m. A density map of the damage indicates the zones that were most strongly affected by the 2011 Tohoku earthquake and tsunami event. A damage density of 100% indicates total destruction. Dark colors represent a high probability of damage, light colors represent a lower probability of damage, and green represents vegetated areas empty of structures, and, as such, these areas were removed from the analysis and are displayed in this map for the sake of clarity. As was previously mentioned, the most severe damage was observed along the coastline, and, accordingly, the colors of the grid cells in that area indicate high damage density (dark colors). As the tsunami wave washed away most of the structures along the coastline, very few structures there were left standing [25]. The lower density of damage observed in more inland areas indicates that the buildings there suffered much less damage. According to [25], the relatively low level of damage observed inland was caused mainly by the earlier seismic events that preceded the Tohoku earthquake. Considering both the inland and coastal regions, damage density values in the range of 1% to 40%, indicating undamaged or slightly damaged structures, are scarcely observed. This outcome is due primarily to the high sensitivity of the coherence method to changes, especially to subtle changes on the scale of millimeters to centimeters [9].

Results accuracies (producer's and user's accuracy and Kappa coefficient) are summarized in Table 2. High overall accuracy (89%) and a high Kappa coefficient (82%) were achieved for the entire research area, as for the four enlarged areas in Figure 9 (insets A, B, C, and D). User's and producer's accuracies of the damage class were consistently high, ranging from 76% to 100%. In specific regions, such as Figure 9B,C, the UA and PA displays values of zero for the no/slight damage class. This is a result of the severe damage observed in the entire scene (Figure 6B,C). Despite the random distribution of the points, most of them were located in damaged areas.

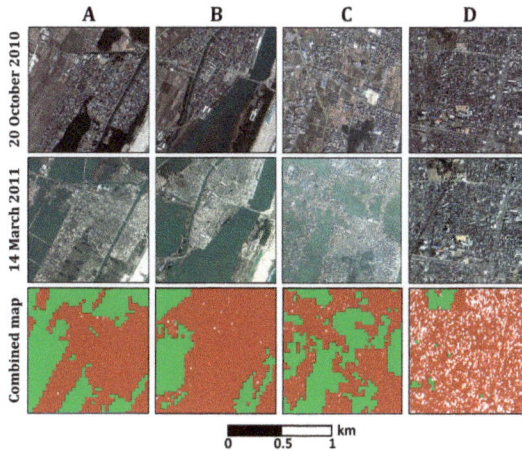

Figure 6. Insets (**A–D**) showing selected areas from the entire scene (Figure 5). Pre-event optical image; post-event optical image; combined map: vegetation is represented in green, undamaged areas are represented in white, and damaged areas are represented in red.

Table 2. Accuracy assessment (in percentage) for the damage assessment map: overall accuracy, Kappa coefficient, producer's accuracy (PA), and user's accuracy (UA).

		Entire Scene		A		B		C		D	
		PA	UA	PA	UA	PA	UA	PA	UA	PA	UA
	Vegetation	89	97	90	100	92	85	96	88	50	100
	Damage	96	76	100	90	95	97	88	88	100	80
50 m	No/slight damage	87	92	50	100	0	0	0	0	95	95
	Overall accuracy	89		94		94		88		92	
	Kappa coefficient	82		88		84		77		79	
100 m	Overall accuracy	82									
	Kappa coefficient	68									

As urban environments are characterized by very tall, densely-situated buildings with correspondingly high population concentrations, it is of the utmost importance that information about the level of damage be obtained rapidly. This necessitates extracting from the data vital information about the severity of the damage and the most likely locations in built-up areas where rescue missions will be needed in as short a time as possible. However, the choice of grid resolution should be dictated by the desired process run time, as the finer the resolution, e.g., 10 m or lower, the longer the calculation time. To reduce the calculation time in this study, we tested two grid sizes, 100 m and 50 m (Figures 8 and 9). These grid sizes were chosen after examining the average building sizes of several built regions and the dominant terrain in the study area, most of which is relatively flat and lacking any significant changes in topography. For the 50 m grid size, each cell of the grid contained a small number of buildings (approximately 1–8 buildings) together with their surroundings, such as the streets (Figure 8). The 50 m grid size density map provides detailed information about the damaged area, showing the variance of damage severity and its distribution in the field. In contrast, the 100 m grid lacks this level of detail about the damage and, instead, shows the area as having undergone more homogenous change, with little variance in damage severity. This cell size was tested because there are buildings in the research area that exceed 100 m (Figure 8). The results of the test of grid size show that because it differentiates between the levels of damage and spatially represents the area more accurately, the 50 m grid size is optimal. Furthermore, the overall accuracy and Kappa coefficient were found to be lower for the 100 m grid, further supporting our decision to use a 50 m grid (Table 2). For more inland, mountainous regions, however, a smaller cell size should be used, to better account for the influence of changes in topography.

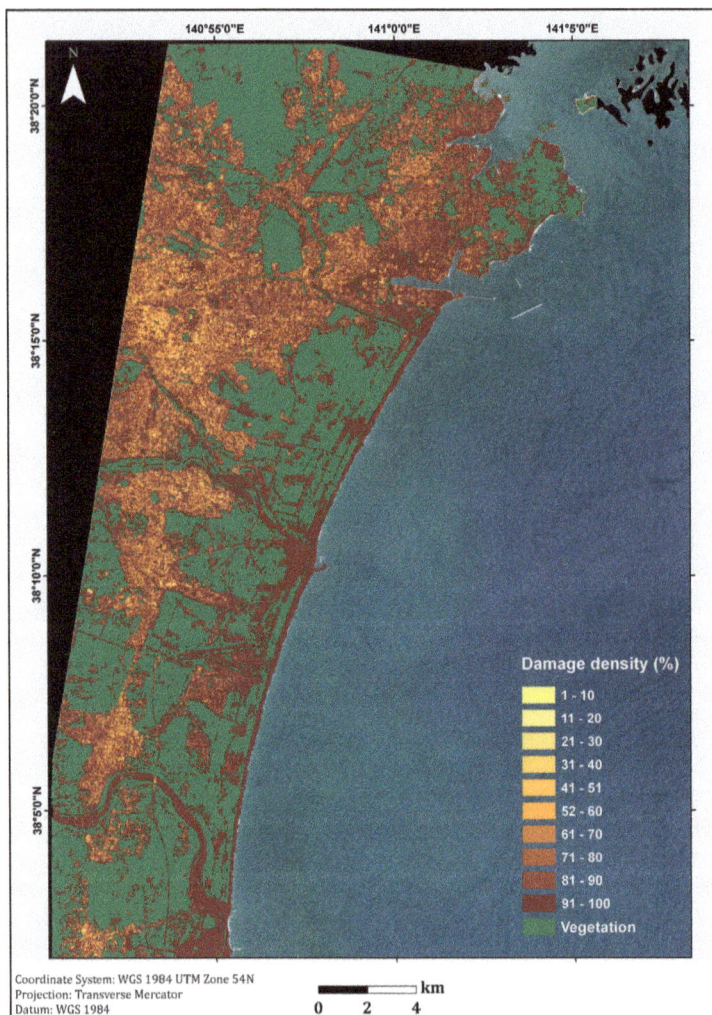

Figure 7. Damage density map per unit area of 50 m. Light colors represent lower probability of damage, dark colors represent higher probability of damage. Green represents vegetation.

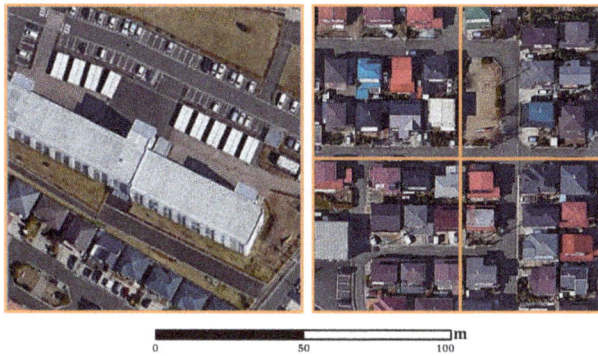

Figure 8. Grid size of 100 m vs. 50 m, overlying an optical image.

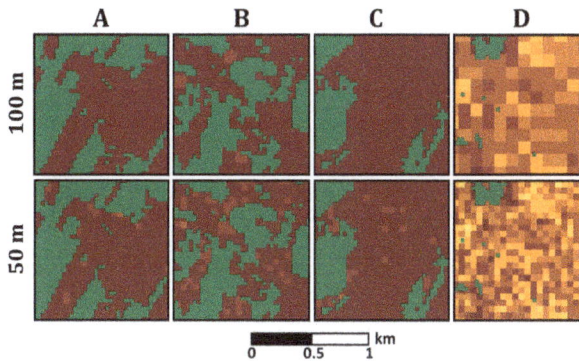

Figure 9. Damage density maps with grid size of 100 m vs. 50 m. Insets (**A–D**) showing selected areas from the entire scene, display the differences of the damage distribution between the two grids.

In this study, the damage level was divided into 10 classes to ensure that it was depicted in sufficient detail to obtain valuable information. Previous studies classified the building damage into several basic grades (usually three or four classes) at the block level [3,11,23]. Since the proposed approach in this study quantitatively evaluated the extent of change to the landscape, it enables the damage level to be divided into a greater number of classes to obtain finer detail. Thus, the detected damage levels range from no-damage to total destruction, with an additional eight levels of damage between these two extremes (Figure 7). The high level of detail of this approach facilitates the coordination of the rescue teams' capabilities in the field. Moreover, due to the capabilities of TerraSAR-X, especially its high spatial resolution, its cell grid sizes can be easily expanded or reduced without using any ancillary data (such as city blocks).

5. Conclusions

This study presented a novel method for mapping the urban damage in the wake of a natural disaster by combining SAR and optical data. To illustrate the methodology, we used a pair of TerraSAR-X images and a Landsat-5 image to analyze the urban damage that occurred from the 2011 Tohoku earthquake and tsunami in Japan. Due to the high spatial resolution of TerraSAR-X, damage assessment mapping could be applied by using only one coherence map comprising pre- and post-event images. Integration of the co-event coherence map with the NDVI improved the accuracy of the damage assessment by eliminating areas that were falsely identified as damaged and focusing only on the urban environment. The 50 m grid size damage assessment map was better than either the 100 m grid size or the smaller size grid in terms of the level of accurate detail it provided (high overall accuracy and Kappa coefficient, 89% and 82% respectively) about the locations and states of damaged buildings and structures, valuable information that could be exploited to aid rescue missions.

The remote sensing method tested in this research is particularly useful for the immediate response after a natural or man-made disaster event, as it can be implemented easily to rapidly identify and assess the damage to buildings and other man-made structures in urban environments. Moreover, it requires fewer resources, such as time and cost, and it provides real-time information that is that can be easily interpreted to gain a clear understanding of the damage.

Author Contributions: All authors contributed equally to the development of this research.

Acknowledgments: The authors wish to thank the Ministry of Science, Technology, and Space of Israel and the Italian Space Agency (ASI) for their partial support of this study. In addition, the authors thank German Aerospace Centre (DLR) for TerraSAR-X data, which was provided free of charge in the framework of the funded research proposal ID: GEO0053.

Conflicts of Interest: The authors declare no conflict of interest.

References

1. Oštir, K.; Veljanovski, T.; Podobnikar, T.; Stančič, Z. Application of satellite remote sensing in natural hazard management: The Mount Mangart landslide case study. *Int. J. Remote Sens.* **2003**, *24*, 3983–4002. [CrossRef]
2. Plank, S. Rapid damage assessment by means of multi-temporal SAR—A comprehensive review and outlook to Sentinel-1. *Remote Sens.* **2014**, *6*, 4870–4906. [CrossRef]
3. Uprety, P.; Yamazaki, F. Use of high-resolution SAR intensity images for damage detection from the 2010 Haiti earthquake. In Proceedings of the 2012 IEEE International Geoscience and Remote Sensing Symposium (IGARSS), Munich, Germany, 22–27 July 2012; pp. 6829–6832.
4. Joyce, K.E.; Belliss, S.E.; Samsonov, S.V.; McNeill, S.J.; Glassey, P.J. A review of the status of satellite remote sensing and image processing techniques for mapping natural hazards and disasters. *Prog. Phys. Geogr.* **2009**, *33*, 183–207. [CrossRef]
5. Dong, L.; Shan, J. A comprehensive review of earthquake-induced building damage detection with remote sensing techniques. *ISPRS J. Photogramm. Remote Sens.* **2013**, *84*, 85–99. [CrossRef]
6. Yamazaki, F.; Kouchi, K.; Kohiyama, M.; Muraoka, N.; Matsuoka, M. Earthquake damage detection using high-resolution satellite images. In Proceedings of the 2004 IEEE International Geoscience and Remote Sensing Symposium, IGARSS'04, Anchorage, AK, USA, 20–24 September 2004; pp. 2280–2283.
7. Saito, K.; Spence, R.J.; Going, C.; Markus, M. Using high-resolution satellite images for post-earthquake building damage assessment: A study following the 26 January 2001 Gujarat earthquake. *Earthq. Spectra* **2004**, *20*, 145–169. [CrossRef]
8. Saito, K.; Spence, R.; de C Foley, T.A. Visual damage assessment using high-resolution satellite images following the 2003 Bam, Iran, earthquake. *Earthq. Spectra* **2005**, *21*, 309–318. [CrossRef]
9. Ferretti, A.; Monti-Guarnieri, A.; Prati, C.; Rocca, F.; Massonet, D. *InSAR Principles-Guidelines for SAR Interferometry Processing and Interpretation*; ESA Publications: Noordwijk, The Netherlands, 2007.
10. Chini, M. Earthquake damage mapping techniques using SAR and optical remote sensing satellite data. In *Advances in Geoscience and Remote Sensing*; InTech: London, UK, 2009.
11. Hoffmann, J. Mapping damage during the Bam (Iran) earthquake using interferometric coherence. *Int. J. Remote Sens.* **2007**, *28*, 1199–1216. [CrossRef]
12. Fielding, E.J.; Talebian, M.; Rosen, P.A.; Nazari, H.; Jackson, J.A.; Ghorashi, M.; Walker, R. Surface ruptures and building damage of the 2003 Bam, Iran, earthquake mapped by satellite synthetic aperture radar interferometric correlation. *J. Geophy. Res. Solid Earth* **2005**, *110*. [CrossRef]
13. Yamazaki, F. Applications of remote sensing and GIS for damage assessment. In Proceedings of the 8th International Conference on Structural Safety and Reliability, Newport Beach, CA, USA, 17–22 June 2001; pp. 1–12.

14. Watanabe, M.; Thapa, R.B.; Ohsumi, T.; Fujiwara, H.; Yonezawa, C.; Tomii, N.; Suzuki, S. Detection of damaged urban areas using interferometric SAR coherence change with PALSAR-2. *Earth Planets Space* **2016**, *68*, 131. [CrossRef]

15. Milisavljevic, N.; Closson, D.; Holecz, F.; Collivignarelli, F.; Pasquali, P. An approach for detecting changes related to natural disasters using Synthetic Aperture Radar data. *Int. Arch. Photogramm. Remote Sens. Spat. Inform. Sci.* **2015**, *40*, 819–826. [CrossRef]

16. Romaniello, V.; Piscini, A.; Bignami, C.; Anniballe, R.; Stramondo, S. Earthquake damage mapping by using remotely sensed data: The Haiti case study. *J. Appl. Remote Sens.* **2017**, *11*, 016042. [CrossRef]

17. Matsuoka, M.; Yamazaki, F. Use of satellite SAR intensity imagery for detecting building areas damaged due to earthquakes. *Earthq. Spectra* **2004**, *20*, 975–994. [CrossRef]

18. Yonezawa, C.; Takeuchi, S. Decorrelation of SAR data by urban damages caused by the 1995 Hyogoken-nanbu earthquake. *Int. J. Remote Sens.* **2001**, *22*, 1585–1600. [CrossRef]

19. Arciniegas, G.A.; Bijker, W.; Kerle, N.; Tolpekin, V.A. Coherence-and amplitude-based analysis of seismogenic damage in Bam, Iran, using Envisat ASAR data. *IEEE Trans. Geosci. Remote Sens.* **2007**, *45*, 1571–1581. [CrossRef]

20. Liao, M.; Jiang, L.; Lin, H.; Huang, B.; Gong, J. Urban change detection based on coherence and intensity characteristics of SAR imagery. *Photogramm. Eng. Remote Sens.* **2008**, *74*, 999–1006. [CrossRef]

21. Dell'Acqua, F.; Gamba, P. Remote sensing and earthquake damage assessment: Experiences, limits, and perspectives. *Proc. IEEE* **2012**, *100*, 2876–2890. [CrossRef]

22. Preiss, M.; Stacy, N.J. Coherent change detection: Theoretical description and experimental results. *J. Am. Dent. Assoc.* **2006**, *38*, 365–372.

23. Tamkuan, N.; Nagai, M. Fusion of multi-temporal interferometric coherence and optical image data for the 2016 kumamoto earthquake damage assessment. *ISPRS Int. J. Geo-Inf.* **2017**, *6*, 188. [CrossRef]

24. Mori, N.; Takahashi, T.; Yasuda, T.; Yanagisawa, H. Survey of 2011 Tohoku earthquake tsunami inundation and run-up. *Geophys. Res. Lett.* **2011**, *38*. [CrossRef]

25. Kazama, M.; Noda, T. Damage statistics (Summary of the 2011 off the Pacific Coast of Tohoku Earthquake damage). *Soils Found.* **2012**, *52*, 780–792. [CrossRef]

26. Hanssen, F.R. *RADAR Interferometry: Data Interpretation and Error Analysis*, 1st ed.; Springer: Dordrecht, The Netherlands, 2002.

27. Havivi, S.; Amir, D.; Schvartzman, I.; August, Y.; Maman, S.; Rotman, S.R.; Blumberg, D.G. Mapping dune dynamics by InSAR coherence. *Earth Surf. Process Landf.* **2017**, *43*, 1229–1240. [CrossRef]

28. Zebker, H.A.; Villasenor, J. Decorrelation in interferometric radar echoes. *IEEE Trans. Geosci. Remote Sens.* **1992**, *30*, 950–959. [CrossRef]

29. Rosen, P.A.; Hensley, S.; Joughin, I.R.; Li, F.K.; Madsen, S.N.; Rodriguez, E.; Goldstein, R.M. Synthetic aperture radar interferometry. *Proc. IEEE* **2000**, *88*, 333–382. [CrossRef]

30. Hong, S.; Wdowinski, S.; Kim, S. Evaluation of TerraSAR-X observations for wetland InSAR application. *IEEE Trans. Geosci. Remote Sens.* **2010**, *48*, 864–873. [CrossRef]

![remote sensing logo] *remote sensing*

MDPI

Article

TerraSAR-X Time Series Fill a Gap in Spaceborne Snowmelt Monitoring of Small Arctic Catchments—A Case Study on Qikiqtaruk (Herschel Island), Canada

Samuel Stettner [1,*], **Hugues Lantuit** [1,2], **Birgit Heim** [1], **Jayson Eppler** [3], **Achim Roth** [4], **Annett Bartsch** [5] **and Bernhard Rabus** [3]

[1] Alfred Wegener Institute, Helmholtz Centre for Polar and Marine Research, Telegrafenberg A45, 14473 Potsdam, Germany; hugues.lantuit@awi.de (H.L.); birgit.heim@awi.de (B.H.)
[2] Institute of Earth and Environmental Science, University of Potsdam, Karl-Liebknecht-Str., 24-25, 14476 Potsdam-Golm, Germany
[3] Synthetic Aperture Radar Laboratory, Simon Fraser University, 8888 University Dr. Burnaby, BC V5A 1S6, Canada; jayson_eppler@sfu.ca (J.E.); bernhard_t_rabus@sfu.ca (B.R.)
[4] Department Land Surfaces, German Aerospace Center Oberpfaffenhofen, 82234 Weßling, Germany; achim.roth@dlr.de
[5] b.geos, Industriestrasse 1, 2100 Korneuburg, Austria; annett.bartsch@bgeos.com
* Correspondence: samuel.stettner@awi.de; Tel.: +49-331-288-20114

Received: 4 June 2018; Accepted: 20 July 2018; Published: 21 July 2018

Abstract: The timing of snowmelt is an important turning point in the seasonal cycle of small Arctic catchments. The TerraSAR-X (TSX) satellite mission is a synthetic aperture radar system (SAR) with high potential to measure the high spatiotemporal variability of snow cover extent (SCE) and fractional snow cover (FSC) on the small catchment scale. We investigate the performance of multi-polarized and multi-pass TSX X-Band SAR data in monitoring SCE and FSC in small Arctic tundra catchments of Qikiqtaruk (Herschel Island) off the Yukon Coast in the Western Canadian Arctic. We applied a threshold based segmentation on ratio images between TSX images with wet snow and a dry snow reference, and tested the performance of two different thresholds. We quantitatively compared TSX- and Landsat 8-derived SCE maps using confusion matrices and analyzed the spatiotemporal dynamics of snowmelt from 2015 to 2017 using TSX, Landsat 8 and in situ time lapse data. Our data showed that the quality of SCE maps from TSX X-Band data is strongly influenced by polarization and to a lesser degree by incidence angle. VH polarized TSX data performed best in deriving SCE when compared to Landsat 8. TSX derived SCE maps from VH polarization detected late lying snow patches that were not detected by Landsat 8. Results of a local assessment of TSX FSC against the in situ data showed that TSX FSC accurately captured the temporal dynamics of different snow melt regimes that were related to topographic characteristics of the studied catchments. Both in situ and TSX FSC showed a longer snowmelt period in a catchment with higher contributions of steep valleys and a shorter snowmelt period in a catchment with higher contributions of upland terrain. Landsat 8 had fundamental data gaps during the snowmelt period in all 3 years due to cloud cover. The results also revealed that by choosing a positive threshold of 1 dB, detection of ice layers due to diurnal temperature variations resulted in a more accurate estimation of snow cover than a negative threshold that detects wet snow alone. We find that TSX X-Band data in VH polarization performs at a comparable quality to Landsat 8 in deriving SCE maps when a positive threshold is used. We conclude that TSX data polarization can be used to accurately monitor snowmelt events at high temporal and spatial resolution, overcoming limitations of Landsat 8, which due to cloud related data gaps generally only indicated the onset and end of snowmelt.

Keywords: Snow Cover Extent (SCE); TerraSAR-X; Landsat; wet snow; small Arctic catchments; satellite time series

---

1. Introduction

The evolution of snowmelt is a crucial component in the seasonal cycle of Arctic ecosystems; affecting temporal and spatial patterns of hydrology, vegetation, and biogeochemical processes. Snow also influences the ground thermal regime by insulating permafrost-affected soils from cold temperatures in winter and from warm temperatures in spring [1–3]. Deeper and prolonged winter snow cover can increase permafrost temperatures and over time can lead to increased active layer thickness, soil nutrient availability, and shifts in vegetation composition [4–7]. Late lying snow patches affect the soil moisture content and thermal properties of the active layer late in the season and create unique vegetation communities beneath and in their vicinity [8]. Both prolonged winter snow and late lying snow patch dynamics directly affect heterotrophic soil respiration and consequently carbon cycling [6,9–13].

The time between the onset and end of snowmelt initiates the hydrological year, drives vegetation phenology, and marks an increase in soil biogeochemical activity [14–16]. In small Arctic catchments, snowmelt is often the most important hydrological driver and generates the majority of annual discharge [17]. The timing of snowmelt, as opposed to temperature, also drives the onset of vegetation phenology and influences subsequent phenological phases and overall fitness of individual plants [18,19]. On a regional scale, spring snow cover in May and June in the Northern Hemisphere has decreased drastically in the last 30 years following trends of increasing air temperatures and reductions in sea ice extent and duration [20]. On a local scale, changes in winter precipitation in the Arctic are expected to be highly variable in space and time [21]. The spatial variability of snow cover extent (SCE) and the temporal variability of snowmelt, expressed on a catchment scale through changes in fractional snow cover (FSC), is inherently high due to low vegetation and snow redistribution by strong and prevailing wind patterns [22,23]. Adding to the uncertainty in changes to SCE and snowmelt is the observed expansion of tall shrubs across the Arctic, which will greatly impact the distribution and depth of snow [24,25]. Consequently, the monitoring of snowmelt at high spatial (30 m) and temporal scales (daily) is important to better understand the impacts of changing SCE on the abiotic and biotic functioning of small catchments in a rapidly changing Arctic. In situ knowledge about snow properties such as snow depth, snow density and the snow water equivalent at the time of image acquisition, are important parameters when undertaking snow related microwave satellite studies [26]. However, the snowmelt period is a logistically challenging time in Arctic regions to conduct in situ work because of unstable sea and river ice conditions as well as very wet ground surfaces due to snowmelt and permafrost thaw.

Currently, in Arctic regions there is no operational product available that captures snow cover in simultaneously high temporal and spatial resolution. Snow cover products from remote sensing data sources are predominantly derived from optical and microwave sensors. To detect snow, optical sensors rely on the high proportion of reflected radiation in the visible spectrum in contrast to the very low reflection in the near infrared part of the electromagnetic spectrum. Common optical sensors for snow cover retrieval include the Moderate Resolution Imaging Spectroradiometer (MODIS), the Advanced Very High Resolution Radiometer (AVHRR), and the Landsat series [27]. MODIS has a spatial resolution of up to 250 m and theoretically delivers a daily snow product. AVHRR also delivers daily snow cover information with a 1 km spatial resolution. Landsat 8 acquires imagery at a spatial resolution of 30 m and with a revisit time of 16 days [28]. The theoretical temporal resolution of MODIS and AVHRR would be sufficient to track the temporal dynamic of snowmelt, but their spatial resolution is not sufficient to capture snowmelt at the small Arctic catchment scale. Landsat 8 offers a spatial resolution sufficient for analysis at the small catchment scale and in Arctic regions converging satellite orbit paths increase the temporal resolution, which would allow monitoring of the highly dynamic phenomenon of snowmelt [29]. However, the retrieval of snow cover from optical sensors in Arctic regions is often challenging due to complete and fragmented cloud cover which introduces gaps in the time series or errors in snow detection [30]. This limits optical time series spatially and temporally, consequently prohibiting fine-scale mapping of the rapid and spatially variable phenomenon of snowmelt. Microwave satellite systems operate largely unaffected by atmospheric distortions from water vapor and clouds and also are independent of solar illumination; as a result they can acquire data at all times. Passive microwave systems can detect snow cover and snow properties at high temporal resolution [31], but the acquired imagery is at a coarse km scale spatial resolution, which is unsuitable for small catchment analyses. Recent active microwave satellite missions operating in Synthetic Aperture Radar (SAR) modes can obtain imagery in sufficient spatial and temporal resolution for small catchment based analysis.

The German active microwave TerraSAR-X (TSX) satellite mission has high potential to address the temporal limitations of operational optical and spatial limitations of other microwave missions by reliably providing imagery in high spatial and reasonable temporal resolution through combining different orbits and viewing geometries. The TSX satellites operate at a polar sun-synchronous orbit with an orbital revisit time of 11 days during which they revisit the same orbital location two times with the same viewing geometry. Satellite missions with polar orbits in general have a higher density of orbits and resulting coverage in Arctic regions because the flight paths converge from the equator to the poles. A fixed antenna beam is used to acquire imagery in the StripMap imaging mode of TSX at a spatial ground range resolution of 1.7 to 3.5 m with incidence angles between 20° and 45°. In the StripMap mode the swath has the dimensions of 30 km and up to 1500 km in cross track and along-track, respectively [32]. TSX is a SAR system that emits pulses in the microwave length of the electromagnetic spectrum, which propagate through the atmosphere. The SAR antenna receives the pulse echoes after being scattered by objects on the Earth's surface, transmitting the physical structure and dielectric properties of the surface. The amplitude of these echoes determines the backscatter intensity and can be used to describe and classify different surfaces. However, a prominent feature of SAR imagery is the speckle effect which introduces a variation to the image texture in a granular pattern that does not arise from the imaged surface but from the SAR system itself. This signal is often regarded as unwanted noise and makes averaging necessary to retrieve the actual mean backscatter amplitude of the surface.

Air, ice and at times liquid water make up a snowpack. The main parameters differentiating the backscatter of dry snow and soil background is the mass of snow or snow water equivalent (SWE), the size of the snow grains, and the roughness and dielectric properties of the soil [33,34]. As air does not influence the transmitted microwave signal [35,36], the propagation and backscatter of microwaves from a snowpack depend on the dielectric constants of ice and water, which are very different and can be used to map snow volume and cover [37,38]. In general, wet snow is easily classified due to free water within the snowpack strongly attenuating the microwave signal [39,40]. The ability of X-band SAR systems like TSX or COSMO-SkyMed to map wet snow due to attenuation of the microwave signal by free water within the snowpack, has been widely reported [38,40–43]. The presence of wet snow in a snowpack is indicative of the onset of snowmelt and, therefore, offers an opportunity to monitor snowmelt dynamics in high spatial and temporal detail.

In addition to wet snow detection, re-freezing of snow layers can be detected using SAR data. Previous research has shown that the formation of ice crusts is highly visible in Ku-Band data [44–46]. A metamorphism of snow crystals due to compaction, sintering and temperature change, in turn influences the dielectric properties of the snowpack. During the snowmelt period, diurnal changes in temperature from below to above freezing can result in the formation of ice layers and can greatly influence backscatter signals [47]. The detection of both wet snow and ice crusts by SAR data expands the utility of this approach in monitoring snow melt dynamics in the Arctic which are strongly affected by diurnal temperature changes and, therefore, ice layer formation.

Our research goal is to resolve the spatiotemporal patterns of the highly dynamic seasonal snowmelt in small Arctic catchments by combining time-series of multi-orbit and multi-polarization TSX data with optical remote sensing data and in situ observations. We investigate snowmelt dynamics of 3 years in small Arctic catchments at the long-term Canadian Arctic terrestrial observatory Qikiqtaruk (Herschel Island) that is also part of the World Meteorological Organization's Polar Space Task Group initiated TSX long-term monitoring dedicated to permafrost applications [48]. We address the following research questions: (1) How does TSX perform in mapping SCE in small Arctic tundra catchments compared to Landsat 8 optical satellite data; (2) How does TSX resolve temporal dynamics of snowmelt compared to in situ observations, and (3) What are the spatiotemporal dynamics of fractional snow cover from 2015 to 2017 in the small catchments of Qikiqtaruk? In order to answer these questions, we performed a quantitative comparison of snow cover extent (SCE) from Landsat 8 with SCE products derived from three different TSX polarizations at different incidence angles. In a second step, we validate catchment based FSC derived from TSX and Landsat 8 with in situ observations and relate the results to catchment characteristics.

2. Study Area

Qikiqtaruk (Herschel Island; 69°34′N; 138°55′W) is located off the northwestern Yukon coast in the Western Canadian Arctic, approximately 2 km from the mainland (Figure 1). The climate of Qikiqtaruk is polar continental with mean annual air temperatures between −9.9 °C and −11 °C and mean annual precipitation between 161 and 254 mm year^{-1} [49]. The dominant wind direction is northwest and storms are frequently observed in late August and September [50]. The island is characterized by rolling hills with a maximum elevation of 183 m above sea level and a polygonal tundra that is dissected by a variety of different valley types [51,52]. There is only one larger water body in the center of the Island, referred to by the local community as Water lake.

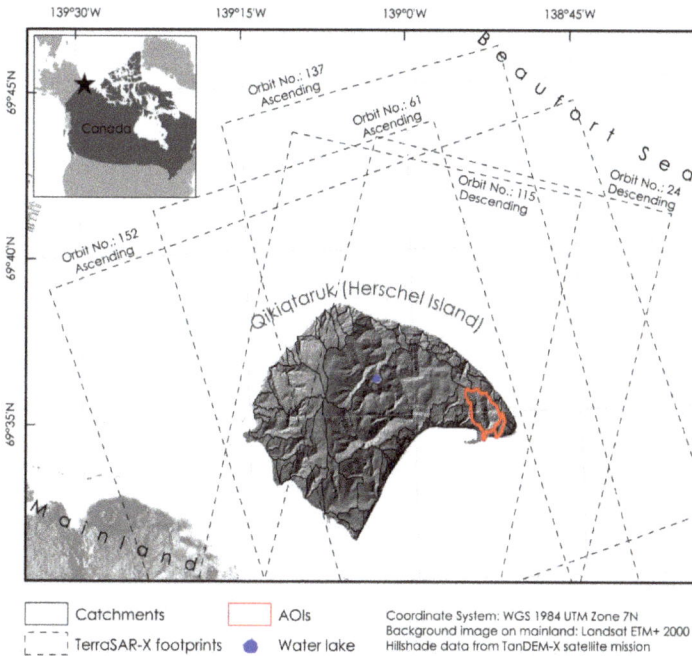

Figure 1. Location of Qikiqtaruk in the southwestern Beaufort Sea Region and footprints of TerraSAR-X imagery.

The vegetation on Herschel Island belongs to the lowland tundra and can be assigned to subzone E of the circumarctic vegetation map (CAVM) [53] or the Low Arctic, respectively [54–56]. The lowland tundra vegetation type makes up the majority of the Arctic tundra. The vegetation is dominated by graminoids and dwarf shrubs, with a relatively species-rich forb flora and a well-developed moss layer [55–57]. The sediments are unconsolidated and mostly fine-grained glacigenic material with marine origin [58,59]. The thickness of the active layer generally ranges between 40 cm and 60 cm in summer, depending on topography [60,61]. Permafrost on Qikiqtaruk is continuous and can be extremely ice-rich with mean ice volumes ranging between 30 and 60 vol%, and up to values >90 vol%, when underlain by massive ground ice beds [62–64]. While mean annual air temperatures have been stable in the Western Canadian Arctic between 1926 and 1970, a total increase of 2.7 °C was observed here between 1970 and 2005 [60]. The effect of this increase of temperature is also exhibited on Qikiqtaruk Island where a deepening of the active layer by 15 to 25 cm has been documented between 1985 and 2005 [60].

Seasonal snow cover typically starts to develop in September and lasts until June. Most snow falls in autumn before sea ice forms and when the ocean still provides a source of liquid water [60]. During winter, strong winds from northwest or northeastern directions affect the snow distribution. Because the tundra vegetation on elevated areas is sparse and trapping capacities are low, the strong winds blow much of the upland surfaces clear of snow and consequently large snow drifts develop in topographic depressions such as valleys and gullies [49]. These snowdrifts can last through the summer, in particular when protected from melting by an insulating layer of plant detritus. This layer can develop during winter when the strong winds transport and deposit plant detritus together with snow.

Ramage et al. [65] identified 40 hydrologic catchments on Qikiqtaruk Island that drain into the surrounding ocean. Our study focuses on the catchments around the Ice Creek catchment in the southeastern part of the island (Figure 1). The Ice Creek catchment consists of two sub-catchments each approximately 1.5 km^2 in size and draining from north to south into a fluvial plain. The maximum height of the catchment is 95 m above sea level in the north. Both catchments are cut by smaller valleys and gullies and characterized by rapid gully erosion as well as mass movements ranging from solifluction to rapid active layer detachments [66]. The occurrence of ice-rich permafrost and permafrost disturbances on the island as well as its location in the lowland tundra type and the long-term ongoing research make Qikiqtaruk a representative study site with respect to the pan-arctic scale.

3. Data & Methods

In order to create maps of snow cover extent (SCE) and fractional snow cover (FSC), we used dense time-series of TSX as well as Landsat 8 imagery as input and in situ time-lapse camera as well as meteorological data for the assessment of TSX derived SCE and FSC. The workflow that we followed is presented in Figure 2.

3.1. SAR Satellite Data

We used dual cross-(VH/VV) and co-polarized (HH/VV) multi-orbit TSX time-series with acquisition dates between April and July from 2015 to 2017. TSX uses a right looking active phased array antenna that operates at an X-Band center frequency of 9.65 Ghz with an orbital revisit time of 11 days. Using multi-orbit data from ascending and descending orbit headings allowed us to increase the revisit time to observe snowmelt in high temporal resolution. The incidence angles of the orbits at the image scene center varied between 25° and 39° with varying pixel spacing between 1.9 to 2.8 m in range and 6.6 m in azimuth. Though shadowing and layover can affect the quality of TSX data, the relief of Qikiqtaruk is low, meaning the influence of layover and shadow is generally minimal inland. We masked out the coastal areas with the steep cliffs since they are not representing delineated catchments and could introduce errors of shadow and layover. Examples of TSX backscatter and Landsat 8 images are shown in Figure 3.

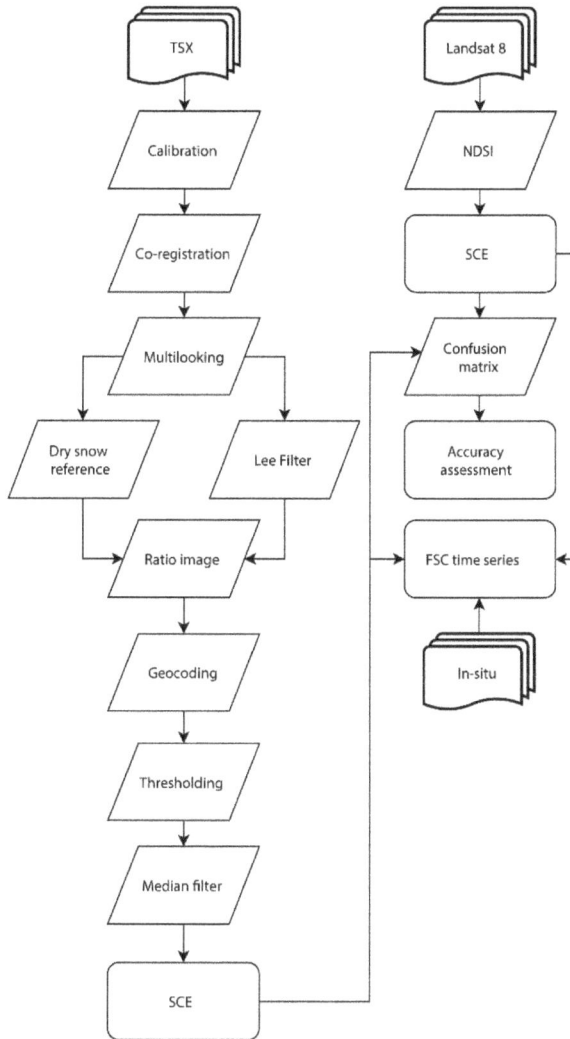

Figure 2. Data processing scheme for the optical Landsat 8 and TSX data. Layered objects represent input data, diamonds processing steps and rounded rectangles results. FSC = Fractional Snow Cover; NDSI = Normalized Difference Snow Index; SCE = Snow Cover Extent.

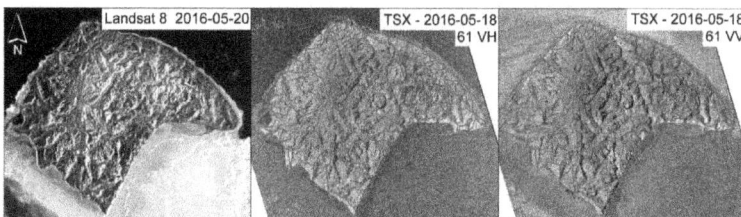

Figure 3. Examples of Landsat 8 (**left**), TerraSAR-X in VH (**center**), and TSX in VV (**right**) polarization.

3.2. Optical Satellite Data

We used all available cloud-free Landsat 8 imagery from 2015 to 2017 between the months of April and July. In total, we used 20 Landsat 8 Operational Land Imager (OLI) acquisitions downloaded in processing Level-1 TP. Level-1 TP is the Standard Terrain Correction of the Landsat products, including radiometrically calibrated images that were orthorectified using ground control points and digital elevation models. We downloaded all acquisitions using the bulk download application from the USGS earth explorer service. The Landsat 8 OLI visible, near-infrared and shortwave infrared bands have a 30 m spatial resolution. We created quasi-true RGB composite imagery from every acquisition as a visual reference and additionally used the green and short wave infrared bands for SCE generation (see section "Snow Cover Extent generation from Landsat 8").

3.3. In situ Time-Lapse Camera Data

Time-lapse cameras were set up in the framework of the Helmholtz Young Investigators Group "COPER" and the "ShrubTundra" research projects in the spring of 2016 and 2017 as part of ongoing hydrological and phenological monitoring on the island. Five cameras acquired images at an hourly resolution in a northward facing aspect with a landscape field of view. One camera was set up for monitoring a hydrological flume at the outlet of the western Ice Creek catchment and acquired imagery at a three hourly resolution from the 24th of April to the 20th of July 2016 (Camera ID: TL2). Four phenological cameras (PC) were set up for the monitoring of phenology of vegetation communities at locations around the Ice Creek catchment (Camera IDs: PC2, PC3, PC5 and PC6). While camera TL2 represents snow dynamics of deeply incised valley locations, cameras PC2, PC3, PC5 and PC6 represent snow dynamics of upland tundra topography and vegetation of Qikiqtaruk. The time-lapse camera data were used for a ground based quality assessment of a TSX derived FSC time series. Photos that correspond to the TSX acquisition dates and time of day were used in the analysis when available, otherwise the closest date and time were chosen. The greatest difference between photos and TSX acquisition was a single day. The cameras were subject to disturbance by wildlife, active layer thaw and weather and, therefore, were not always operational, causing data gaps up to several days. Because of the remoteness of the study site, detailed in situ data on snow properties were not available to assess the impact of snow genesis on the backscatter signal. The time lapse cameras that we use for estimation of snow cover extent are a cost-effective tool in this environment. Visual estimations of percent cover are a common approach in ecological studies [67]. The results present very valuable insights into the melt dynamics in the camera footprint that could not be obtained otherwise and can be related to the satellite based measurements.

In addition, meteorological records from an automated weather station run by Environment Canada on Simpson Point on the southeastern spit of the island (World Meteorological Organization ID: 715010) were used to assess the FSC time series in 2015, 2016, and 2017. The data is available at a daily temporal resolution, recording mean, minimum and maximum daily air temperature.

3.4. Snow Cover Extent from TerraSAR-X

To generate SCE maps from TSX we adapted the workflow suggested by Nagler and Rott [40] and calculated ratio images of backscatter intensity calibrated to radar brightness in sigma nought ($\sigma 0$) as well as speckle-filtered, see details below, between an averaged dry snow reference image from TSX winter scenes (See Table S1) and melting snow scenes from spring/summer with the same orbital configuration. The reduced backscatter signal of melting snow, as well as the increased backscatter of frozen ice layers, is the basis for a threshold segmentation that efficiently differentiates between wet snow or ice layers and dry snow or no snow. While this method was originally developed for C-Band of ERS-1 data and was further improved and adapted for ERS-2, RADARSAT-1, ENVISAT ASAR and Sentinel-1 data [40,68,69], it was also applied for X-Band data from TSX and COSMOSkyMed data [38,70,71].

The backscatter intensity of images acquired during the spring show notable differences between acquisitions caused by drifting sea ice and snowmelt that reduce the quality of co-registration when using cross-correlation. For a given acquisition orbit, all images therefore were co-registered to sub-resolution accuracy with respect to a pre-selected master image using the highly accurate orbital information from TSX and ellipsoidal heights. In order to avoid early season effects of snowmelt with resulting high soil moisture variations and beginning vegetation dynamics, we chose mid-summer acquisitions (July) as the master images when environmental conditions had stabilized.

By applying 3×9 multilooking to the intensity images, we address the speckle effect inherent to SAR imagery obtaining a roughly square pixel size of around 20 m (Table 1). We further reduced the effect of speckle by applying a Lee filter with a window size 5×5 pixel [72].

Table 1. Information on TSX orbits used in this study. RA = Range, AZ = Azimuth. The incidence angle θ refers to the image scene center. Winter scenes are defined by TSX acquisition dates with expected snowmelt before the 1st of May; spring/summer scenes are defined as all TSX scenes with dates after the 1st of May.

	Acquisition Time						No. of Scenes	
Orbit No.	UTC	Local Time	Orbit Heading	Incidence Angle θ	Pixel Spacing RA/AZ	Polarization	Winter	Summer
24	16:08	9:08	Descending	31°	2.3/6.6	HH/VV	7	22
61	2:26	19:26	Ascending	32°	2.2/6.6	VV/VH	5	25
115	15:59	8:59	Descending	39°	1.9/6.6	VV/VH	4	16
137	2:35	19:35	Ascending	39°	1.9/6.6	HH/VV	2	20
152	2:18	19:18	Ascending	25°	2.8/6.6	HH/VV	8	29

We tested the Lee [72] and Frost [73] filters to address the speckle effect within the intensity images for their performance for SCE generation. Figure 4 shows the optical reference data from Landsat 8 and the results from TSX with a two-day-later acquisition date and differing filter techniques. The differences between the Frost and the Lee Filter were insignificant; however, the processing time for the moving window calculation of the Frost filter was much longer. We therefore decided to use the Lee Filter with a 5×5 window for all SCE generation, since it satisfactorily removes speckle related effects from the SCE maps.

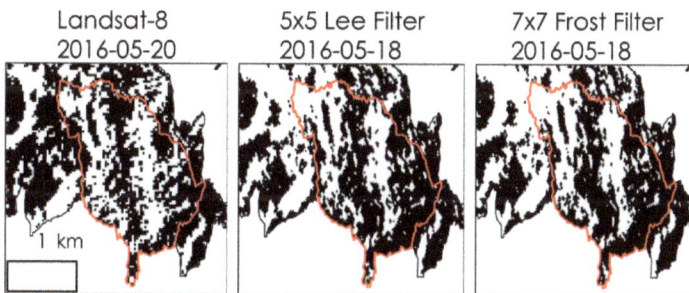

Figure 4. Comparison of filter methods for SCE generation. White pixels represent snow. The red line represents the Ice Creek catchment limit on the southeastern part of Qikiqtaruk.

We computed the σ0-ratios between the melting snow images and the dry snow reference separately for each polarization channel before transforming the ratio images to logarithmic scale (dB). We then geocoded the ratio images using the intermediate DEM product from the TanDEM-X mission with a spatial resolution of 12.5 m and a vertical accuracy of 2 m [74]. The DEM was created by DLR using several TerraSAR-X scenes from winter 2010 and 2011 and the Height Error Map provided with the DEM product states a mean error of 0.72 ± 0.2 m within the area of the hydrological catchments. We tested the application of two thresholds on TSX ratio images. We first used a threshold of −2 dB for VH and HH and −2.3 dB for VV data as reported in Schellenberger [39], with values below the threshold showing the presence of wet snow. We also applied a threshold of 1 dB on all TSX ratio images, with values below 1 representing the presence of snow with ice layers. We filtered the thresholding results using a median filter with a window size of 5 × 5 pixels.

3.5. Snow Cover Extent from Landsat 8

We applied the spectral band ratio of the Normalized Difference Snow Index (NDSI) to generate SCE maps from Landsat. The NDSI is a commonly applied ratio index for snow detection from optical sensor data [75], using the contrast in the visible green versus the shortwave infrared band reflectance. We applied the NDSI with OLI bands 3 and 6, and the threshold technique with NDSI > 0.4 to binary-classify snow presence and absence with greater or less than 50% snow cover at pixel-level, respectively [76]. In optical remote sensing pixels with NDSI values greater than 0.4, they have been found to be more than 50% snow covered [77].

3.6. Accuracy Assessment of TerraSAR-X Snow Cover Extent

Confusion or error matrices are commonly used to assess the quality of classified spatial data from different data sources, at the same time [78]. In order to estimate the quality of TSX derived SCE, we created confusion matrices between a TSX SCE and the corresponding Landsat 8 SCE. We chose corresponding Landsat 8 SCE with acquisition dates not further apart than ±3 days from the TSX SCE. We used 5000 accuracy assessment points for every pair, which were proportionally distributed between the classes of "snow" and "no snow" in a stratified random approach. We averaged the final results from 100 iterations of this accuracy assessment. The results include the users, producers and overall accuracies. These values give an indication of the reliability of the TSX maps, with users accuracy indicating that TSX is detecting snow where Landsat 8 does not, and producers accuracy indicating that TSX is not detecting snow when Landsat 8 is, overall accuracy is a weighted measure that indicates the general occurrence of errors.

3.7. Fractional Snow Cover Time Series Analysis

We calculated the fractional wet snow cover (hereafter referred to as FSC) from the generated SCE maps in percent (%) for (1) all small catchments on the island and (2) for the entire island by merging all individual catchments into a single area. We resampled and aligned the TSX SCE to the 30 m spatial resolution of the Landsat 8 SCE. The footprints of the TSX orbits 152, 137 and 24 do not cover the entire western part of the island. We calculated the fraction of snow pixels within a catchment for every SCE with regard to the total number of pixels within the catchments. We used the WGS UTM Zone 7N coordinate system for all map outputs. To validate the time series of TSX FSC, we analyzed in situ data from the time lapse cameras in and around the Ice Creek catchment (Table S2). For each time-lapse image, the snow cover within the field of view was estimated interactively by an independent visual interpretation of the photo by three different individuals. We then averaged the three estimations for every image acquisition. Additionally, for the cameras observing phenology in 2017 we averaged the estimations of the four available cameras per date in order to obtain a valid representation of the upland tundra type of the island.

4. Results

4.1. Evaluation of Backscatter Time Series

The variation of backscatter in winter and spring at a site within the Ice Creek catchment is shown in Figure 5. Variations of backscatter before May in the expected pre-melt phase are small. In all years and polarizations, a drop of backscatter intensity is recorded in early to mid-May during expected peak snowmelt. This is followed by a strong and rapid increase of backscatter that goes above the backscatter signature of the stable pre-melt phase.

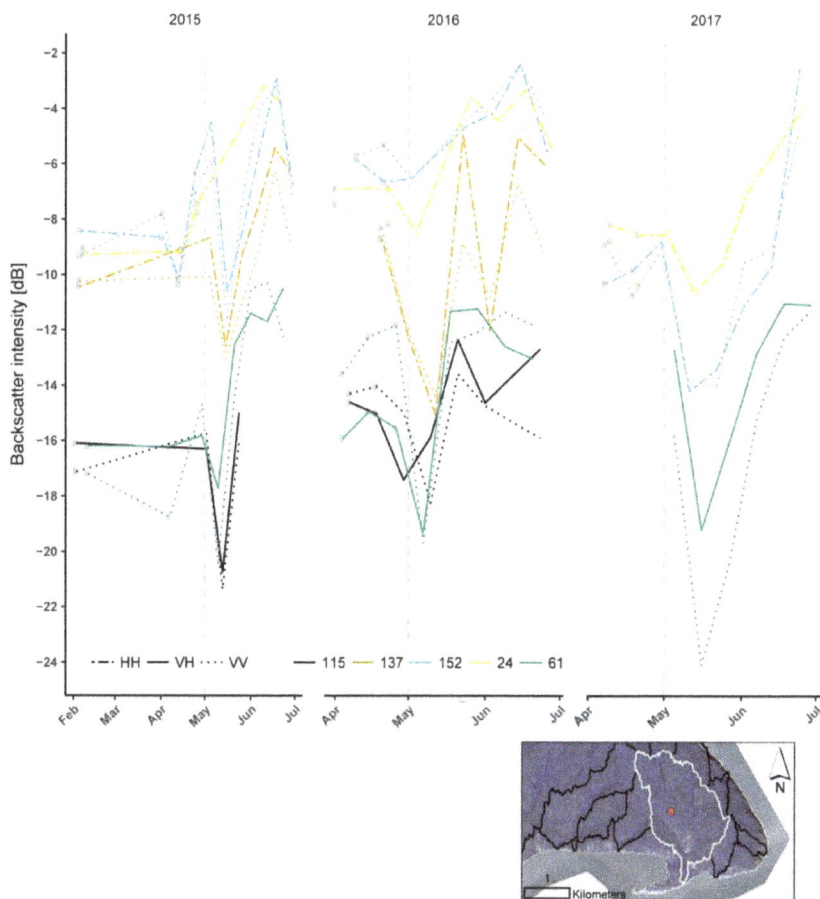

Figure 5. Backscatter time series of TSX orbits in the Western Ice Creek catchment. The red dot indicates the footprint of the extracted backscatter.

4.2. Evaluation of TSX Snow Cover Extent

The visual assessment of the TSX derived SCE from two different thresholds and from three polarization channels demonstrates the influence of polarization in X-Band on the application of monitoring SCE in an Arctic environment (Figures 6–9 and Figures S1–S6). Quantitative comparison of seven VH-, eight VV- and 15 HH based TSX derived SCE maps with paired optical SCE maps showed that the VH polarization channel performed best and most consistently in detecting snow cover during the snowmelt period (Table 2). We achieved the highest overall accuracies when using VH polarized images. The incidence angle and time difference between paired TSX and optical acquisitions did not show a clear effect on the overall accuracy. The overall accuracy of the SCE maps from VH data generally showed a higher agreement later in the snowmelt season except for the TSX acquisition from the 10th of May 2016. The other polarizations did not show a clear influence of time on the overall accuracy of SCE (Table 2).

Table 2. Confusion matrix results from the comparison of TerraSAR-X and Landsat 8-derived SCE during snow melt for the three polarizations and for the two applied thresholds (TH) −2 and 1 dB. U = Users accuracy, P = Producers accuracy, O = Overall accuracy.

					U		P		O	
TSX Date	Polarization	Orbit	Incidence Angle	Landsat Date	TH-2	TH1	TH-2	TH1	TH-2	TH1
15 May 2015	HH	137	39	16 May 2015	0.78	0.78	0.61	0.68	0.68	0.68
16 May 2015	HH	152	25	16 May 2015	0.78	0.73	0.79	0.86	0.75	0.72
26 May 2015	HH	137	39	25 May 2015	0.95	0.95	0.89	0.91	0.86	0.87
28 June 2016	HH	24	31	28 June 2016	1.00	1.00	0.78	0.78	0.78	0.78
24 June 2017	HH	152	25	24 June 2017	0.98	0.99	0.87	0.88	0.85	0.86
26 June 2017	HH	24	31	24 June 2017	0.96	0.96	0.36	0.37	0.36	0.37
7 July 2017	HH	24	31	8 July 2017	1.00	1.00	0.59	0.59	0.59	0.59
24 May 2015	VH	115	39	25 May 2015	0.98	0.98	0.89	0.89	0.88	0.88
23 June 2015	VH	61	32	26 June 2015	1.00	1.00	0.98	0.98	0.98	0.98
18 May 2016	VH	61	32	20 May 2016	0.86	0.86	0.67	0.68	0.74	0.74
21 May 2016	VH	115	39	20 May 2016	0.71	0.72	0.97	0.97	0.75	0.76
12 July 2016	VH	61	32	14 July 2016	1.00	1.00	0.98	0.98	0.98	0.98
15 July 2016	VH	115	39	14 July 2016	1.00	1.00	1.00	1.00	1.00	1.00
15 May 2015	VV	137	39	16 May 2015	0.80	0.77	0.53	0.53	0.65	0.63
16 May 2015	VV	152	25	16 May 2015	0.78	0.72	0.77	0.85	0.74	0.71
24 May 2015	VV	115	39	25 May 2015	0.96	0.96	0.46	0.47	0.48	0.49
26 May 2015	VV	137	39	25 May 2015	0.95	0.95	0.89	0.88	0.86	0.84
23 June 2015	VV	61	32	26 June 2015	1.00	1.00	0.83	0.84	0.83	0.83
18 May 2016	VV	61	32	20 May 2016	0.80	0.81	0.53	0.55	0.64	0.65
21 May 2016	VV	115	39	20 May 2016	0.77	0.78	0.82	0.82	0.75	0.75
28 June 2016	VV	24	31	28 June 2016	1.00	1.00	0.58	0.58	0.58	0.58
12 July 2016	VV	61	32	14 July 2016	1.00	1.00	0.67	0.67	0.67	0.67
15 July 2016	VV	115	39	14 July 2016	1.00	1.00	0.91	0.92	0.91	0.92
24 June 2017	VV	152	25	24 June 2017	0.98	0.98	0.78	0.83	0.77	0.82
26 June 2017	VV	24	31	24 June 2017	0.95	0.96	0.27	0.28	0.27	0.28
7 July 2017	VV	24	31	8 July 2017	1.00	1.00	0.46	0.47	0.46	0.47

Results from confusion matrices before snow melt are shown in Table 3. The correspondence between optical and TSX derived SCE is very low for the users accuracy and high for the producers accuracy (Table 3).

Table 3. Confusion matrix results from the comparison of TerraSAR-X and Landsat 8-derived SCE before peak snow melt for the three polarizations. U = Users accuracy, P = Producers accuracy, O = Overall accuracy.

					U		P		O	
TSX Date	Polarization	Orbit	Incidence Angle	Landsat Date	TH-2	TH0	TH-2	TH0	TH-2	TH0
12 May 2016	HH	137	39	11 May 2016	0.05	0.05	0.82	0.89	0.91	0.78
10 May 2016	VH	115	39	11 May 2016	0.01	0.01	0.87	0.88	0.23	0.23
10 May 2016	VV	115	39	11 May 2016	0.01	0.02	0.91	0.89	0.33	0.32
12 May 2016	VV	137	39	11 May 2016	0.05	0.06	0.82	0.81	0.90	0.88

The results of SCE maps from optical and TSX data are presented in the following, with Figures 6 and 7 showing SCE maps derived from thresholds −2 dB and Figures 8 and 9 showing the results from the threshold 1 dB. The SCE maps from orbit 61 and 115 generated from the literature derived thresholds of −2 and −2.3 dB for VH and VV polarizations, respectively, showing a strong under estimation of TSX SCE compared to Landsat 8 SCE (Figures 6 and 7). Figure 6 is the comparison of VH and VV from orbit 61 from May 18th to 20th and demonstrates a clear underestimation of SCE. Wet snow cover was detected in low-lying areas in the delineated catchments in late May (Figure 8, second row).

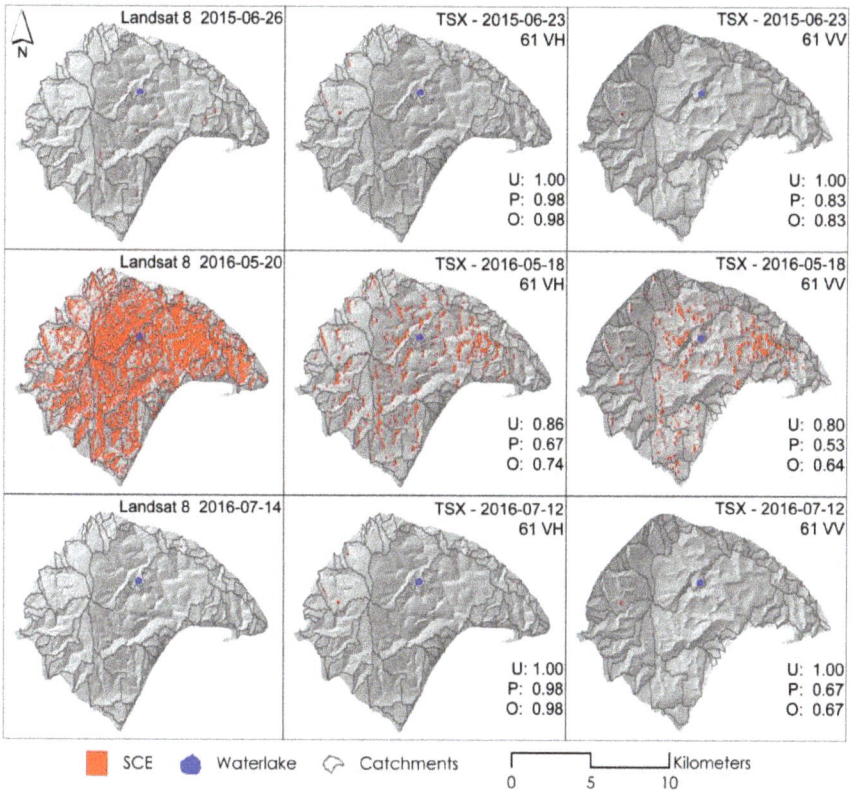

Figure 6. Comparison of Landsat 8 SCE (**left** panels) and corresponding TSX SCE derived using a threshold of −2 dB on the VH (**middle** panels) and −2.3 dB on the VV (**right** panels) polarized channels of orbit 61 from three dates in 2015 (first row) and 2016 (third and fourth row). Also shown are the results of the accuracy assessment, U = Users accuracy, P = Producers accuracy, O = Overall accuracy.

In Figure 7, the orbit 115 also shows a strong underestimation of SCE very similar to orbit 61. Snow is detected only in low-lying valley areas in late May. The pre-melt snow cover images in from 11 May 2016 demonstrate the invisibility of the dry snow cover in TSX for both thresholds.

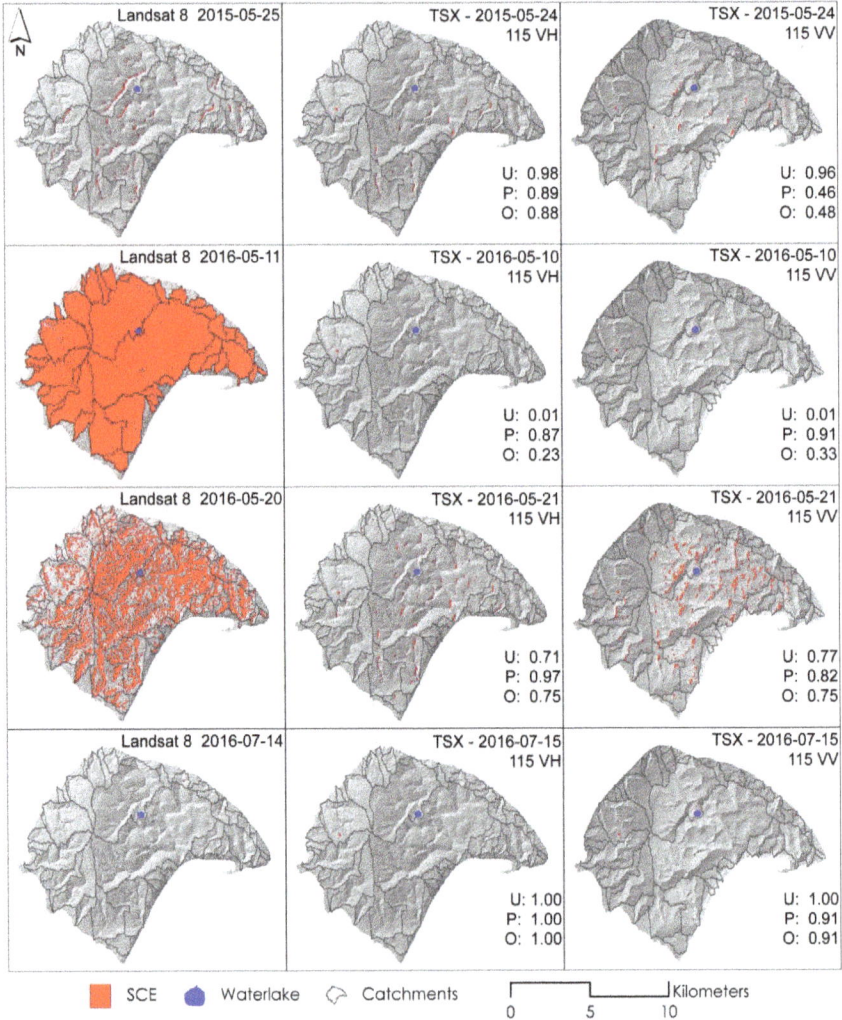

Figure 7. Comparison of Landsat 8 SCE (**left** panels) and corresponding TSX SCE derived using a threshold of −2 dB on the VH (**middle** panels) and −2.3 dB on the VV (**right** panels) polarized channels of orbit 115 from four dates in 2015 (first row) and 2016 (second, fourth and fifth row). Also shown are the results of the accuracy assessment, U = Users accuracy, P = Producers accuracy, O = Overall accuracy.

The comparison of TSX derived VH polarized SCE maps using the 1 dB threshold from orbits 61 and 115 and optical derived SCE maps highlight the spatial patterns of agreement between the two data sources (Figures 8 and 9). Figure 8 represents a comparison of VH and VV polarized TSX SCE from orbit 61 and the corresponding Landsat 8 SCE at three different stages of snowmelt in 2015 and 2016. The Landsat 8 SCE map from the 26th of June 2015 shows only a few remaining and small snow patches in low-lying valleys. On the 23rd of June 2015, the TSX in VH showed elongated single patches of snow distributed sparsely over the island while the VV channel detected larger connected areas of snow cover except in the central part of the island. During peak snowmelt, the TSX derived SCE from the 18th of May 2016 showed denser snow cover than the Landsat 8 SCE from the 20th of May

2016. In particular, TSX detected denser snow cover on the higher and flat terrain in the center of the island and northwest of the Ice Creek catchment (Figure 8, second row). The later stage snowmelt maps also show good agreement between TSX and optically derived products (Figure 8, third row). The general pattern of overestimation of VV is evident in all VV polarized SCE maps in Figure 8 and also in Figure 9.

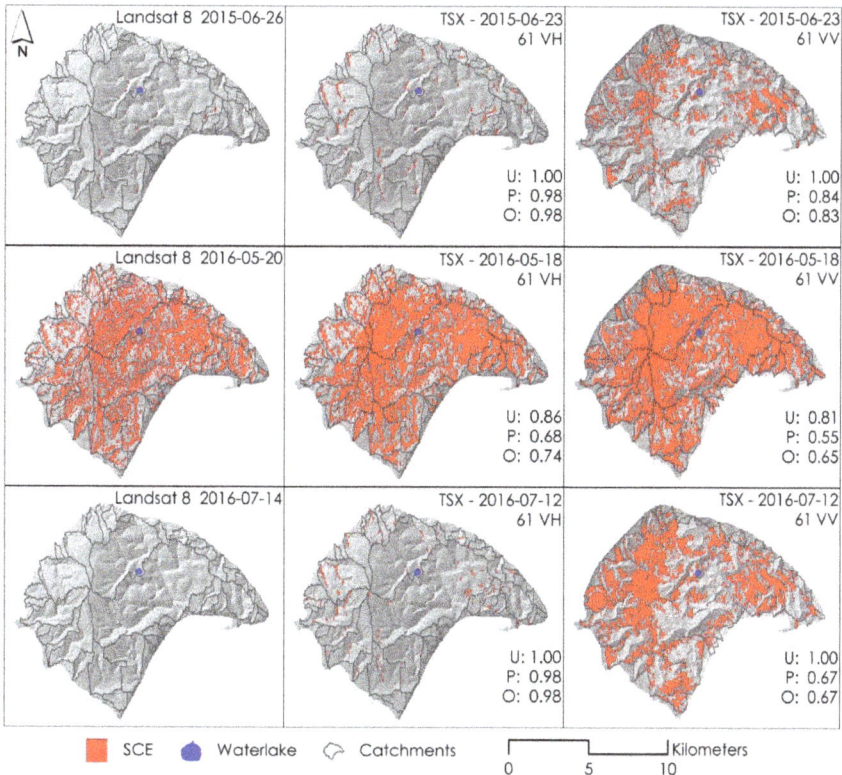

Figure 8. Comparison of Landsat 8 SCE (left panels) and corresponding TSX SCE derived using a threshold of 1 dB on the VH (middle panels) and VV (right panels) polarized channels of orbit 61 from three dates in 2015 (first row) and 2016 (third and fourth row). Also shown are the results of the accuracy assessment, U = Users accuracy, P = Producers accuracy, O = Overall accuracy.

In Figure 9, the 25 May 2015 Landsat 8 SCE map shows several elongated snow banks remaining on the island. The corresponding TSX result from a day earlier shows the same pattern in VH but additionally detects what appears to be wet snow patches with a more scattered distribution on the western part of the island. The VV channel again shows strong overestimation of SCE almost in all areas of the island. The early melt, as shown in the second row of Figure 9, is not captured well by TSX in both polarizations, while the overall accuracy is considerably higher in the late snowmelt season of 2016.

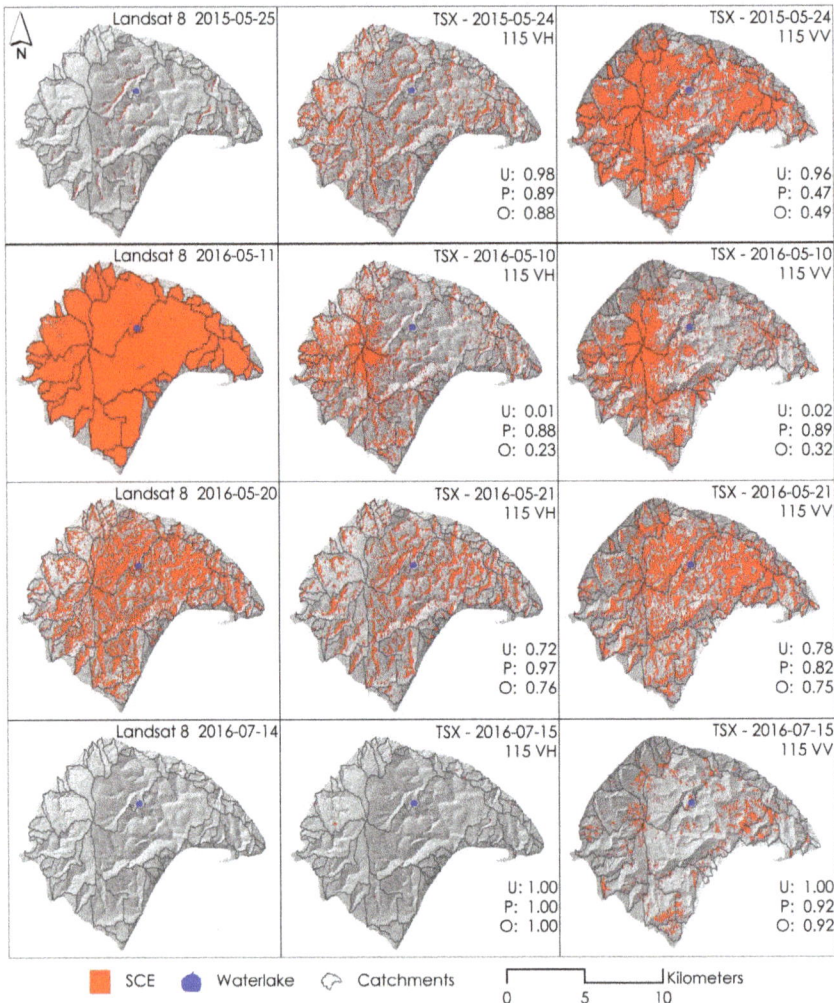

Figure 9. Comparison of Landsat 8 SCE (**left** panels) and corresponding TSX SCE derived using a threshold of 1 dB on the VH (**middle** panels) and VV (**right** panels) polarized channels of orbit 115 from four dates in 2015 (first row) and 2016 (second, fourth and fifth row). Also shown are the results of the accuracy assessment, U = Users accuracy, P = Producers accuracy, O = Overall accuracy.

4.3. Time Series of Fractional Snow Cover in All Catchments

Building on the results of the accuracy assessment, all orbits with VH polarization were chosen to examine the snowmelt dynamics in 2015, 2016, and 2017. Figure 10 shows a time series of FSC products calculated for the unified catchment area from Landsat 8 and VH polarized TSX derived SCE using the threshold of 1 dB as well as the corresponding minimum, mean and maximum daily air temperatures. In all three years, the TSX derived FSC showed good agreement with the optical Landsat 8 FSC and both data sources show a typical snowmelt pattern with increasing air temperatures (Figure 7). The agreement between TSX and Landsat 8 FSC data was lowest early in the snowmelt period (April to early May). Due to the independence of TSX to atmospheric conditions and cloud cover, the temporal resolution of TSX was higher in all years particularly in the phase of rapid snow cover decline from mid-May to June.

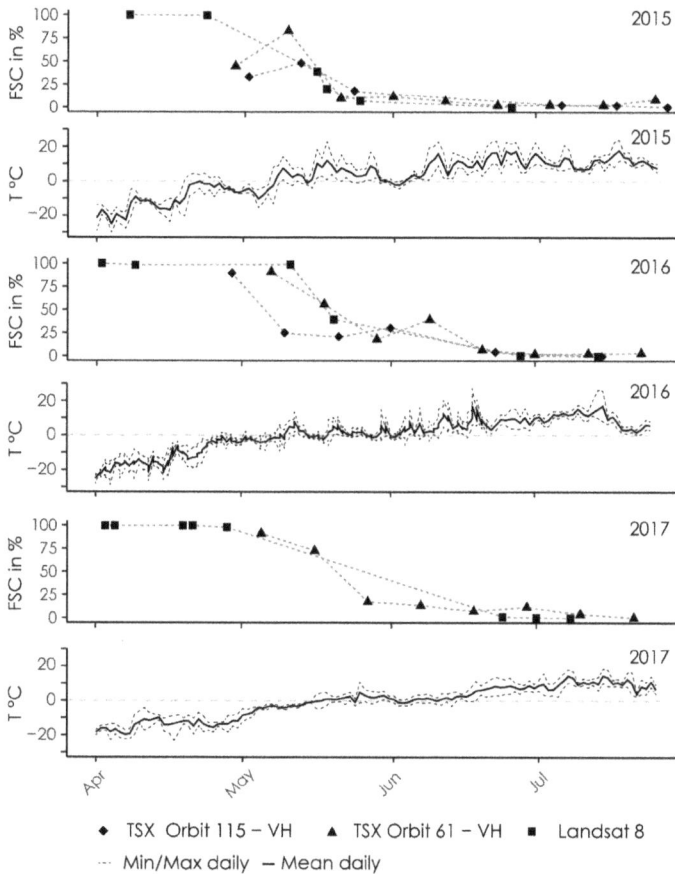

Figure 10. Time series of fractional snow cover (FSC) extent and air temperature of Qikiqtaruk from Landsat 8 and VH polarized TSX data from the orbits 115 and 61 for the years 2015, 2016, and 2017.

In 2015, the TSX FSC was below 50% in the beginning of May and increased to 75% in mid-May when there was no Landsat 8 acquisition. Maximum air temperatures were above 0 °C early in mid to late-April before falling below around −5 °C for about 2 weeks, which corresponds to low snow cover observed by TSX in early May. Both TSX datasets show an increase in FSC followed by a sharp drop to around 10% FSC at the beginning of June followed by a slow gradual decrease of FSC to 0% in July, which is visible in all datasets. In 2016, both TSX orbits showed nearly 100% FSC within the catchment areas in the beginning of May. TSX derived FSC calculated from orbit 61 then showed a strong gradual decrease to around 25% at the end of May while FSC calculated from orbit 115 showed a sharp drop to around 25% in early May. The FSC derived from Landsat 8 imagery shows a sharp drop from 100% FSC in early to about 30% in mid-May. In the late stage of snowmelt, all datasets show FSC of approximately 5%. The air temperature in 2016 was stable around 0 °C for almost the entire month of May, but shows two peaks in early and mid-May that correspond to sharp drops in snow cover in Landsat 8 and TSX orbit 61.

Air temperatures in 2017 showed a gradual increase to 0 °C by mid-May and a corresponding gradual decrease in TSX derived FSC. There was approximately a 2-month gap in successful Landsat 8 acquisitions in 2017.

4.4. Time Series of Fractional Snow Cover in Three Small Catchments

Figure 11 shows the time series of fractional snow cover in the Ice Creek catchment in 2016 as captured from different data sources. The six available acquisitions from Landsat 8 show the beginning snowmelt after the 9th of May with a sharp drop to about 50% within a week and a more gradual retreat of the snow cover to about 5% by the 27th of June. The SCE from orbit 61 recorded wet snow on the 9th of May at about 85% and showed a reduction of FSC at a relatively constant rate to about 20% in 3 weeks by the end of June. An increase of FSC to 30% is recorded with the next acquisition in early June, followed by a reduction to about 10% around the 20th of June followed by a slow decrease to 5% or less by the end of June. The SCE from orbit 115 shows a rapid decrease in FSC from around 80% to about 20% in a little more than a week between the 29th of April and the 9th of May. This coincides with the presented results in Figure 10 for the entire island and below 0 °C minimum air temperatures and the morning acquisition of this orbit. A slight increase to around 25% FSC is followed by a gradual decrease to less than 5% by the end of June. The in situ imagery, collected in the lower end of the ice creek valley within steeper valley topography, show snowmelt beginning around the 9th of May and a decrease of FSC to about 60% around the 17th of May in less than 2 weeks. After that, FSC decreases at a faster rate to about 25% in only a few days before a more gradual melt reduces FSC to 0% on the 1st of July.

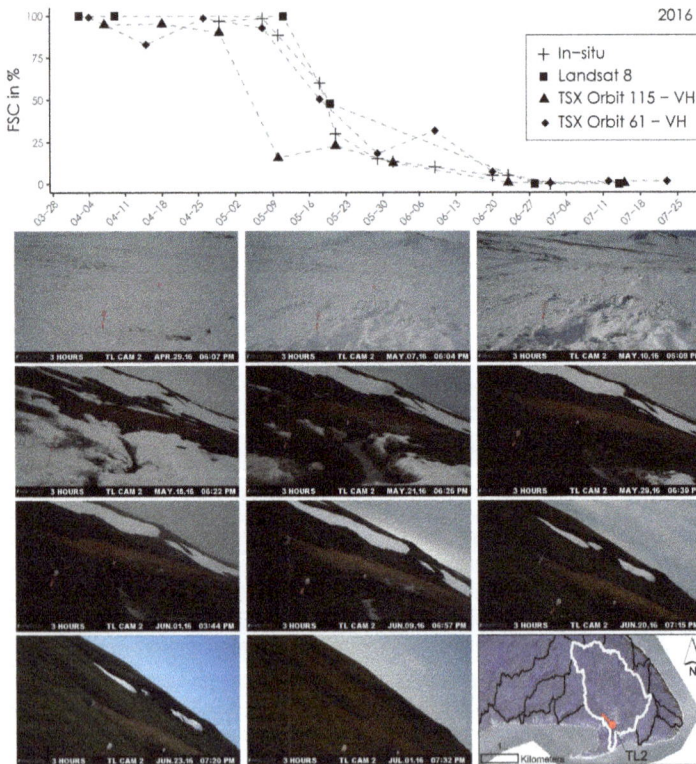

Figure 11. Top graph: Fractional snow cover from time lapse (in situ), Landsat 8 and TerraSAR-X (TSX) imagery in 2016 in the Ice Creek catchment (white outline in the map inlet on lower right). Dates on the x-axis show month-day. Time-lapse imagery is from Camera TL2 and is located in the lower Ice Creek. It's location (red dot) and viewing direction (red line) is indicated in the inset map on the lower right. Please note that the camera was unstable and moved between images because of ground thaw. Please note that the acquisition time of orbit 115 is in the morning, potentially affected by refreezing snow layers in early May.

Figure 12 shows the FSC time series for the year 2017 of a small catchment southeast of Ice Creek as observed from in situ Landsat 8 and TSX data. The Landsat 8-derived FSC is mainly from before or after the main snowmelt period. Therefore, the decrease in FSC appears gradual between the 1st of May and the middle of June from 100% to below 25. The TSX derived SCE from orbit 61 detects wet snow cover in early June with an FSC of approximately 60% and decreases until the end of May gradually to about 20%. Until the end of June, the FSC further decreases at a lower speed to about 5% and to 0% until the end of July. The in situ data collected from time lapse cameras that represent only upland areas of the watersheds show a rapid decrease in FSC from 80% to 20% in the first 2 weeks of May. At the end of May, FSC is below 5% and by early June no snow is detected.

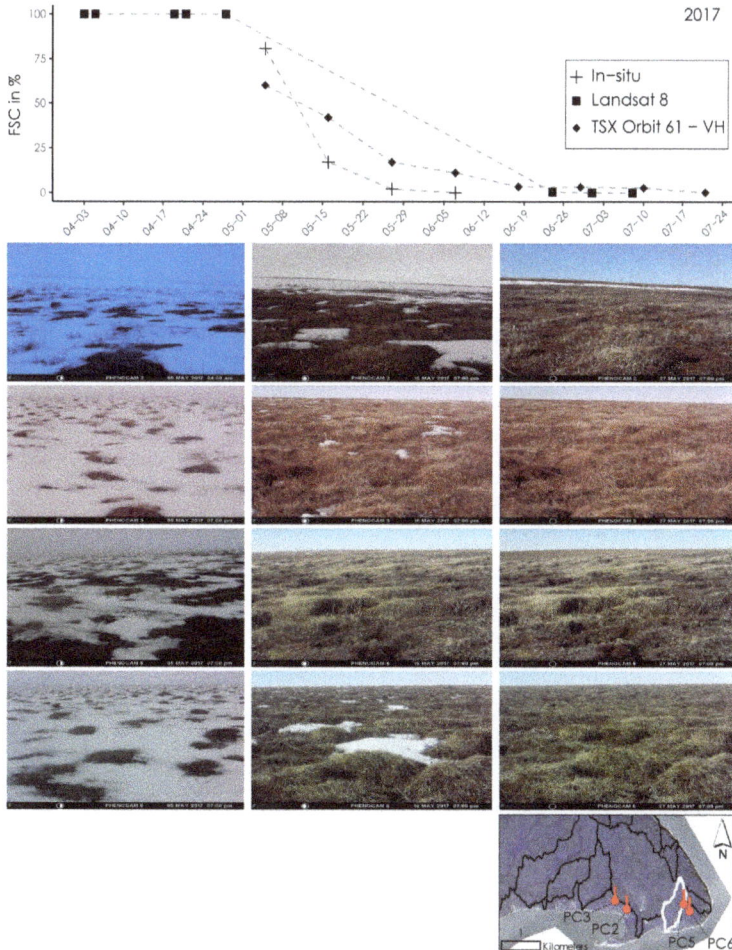

Figure 12. Top graph: Fractional snow cover from time lapse (in situ), Landsat 8 (L8) and TerraSAR-X (TSX) imagery in 2017 in a selected small Arctic catchment (white shape in the map inlet on lower right). Dates on the x-axis show month-day. Time-lapse imagery is from the cameras PC2 (first row) and PC3 (second row), PC5 (third row) and PC6 (fourth row), all representing flat upland tundra locations with low vegetation and tussocks. Dates of the images are the 5th of May (first column), 15/16th May (second column) and 27th of May (third column). Camera locations (red dots) and viewing directions (red lines) are shown in the inset map on the lower right.

5. Discussion

5.1. Spatiotemporal Monitoring of Snowmelt Dynamics Using TSX

The agreement between TSX and Landsat 8-derived SCE products supports X-Band SAR data as a complementary gap filling data source for detailed spatiotemporal monitoring of snowmelt dynamics at both the landscape and Arctic catchment scale. Overall, the cross polarized VH channel performed best in detecting wet snow and ice layers and also detected late-lying snow patches in protected areas. The better performance in the late season is likely due to ice crystals caused by compaction in the snow pack at this stage of melt. This indicates that the contribution of snow backscatter signals prevails within a pixel late in the season in SAR systems while the optical snow detection capabilities decrease with lower snow albedo and greater vegetation contribution within a pixel later in the season. This is an advantage of a high spatial resolution SAR system over a high resolution optical system for mapping late lying snow patches that are often located in steep topographic areas and create unique abiotic and biotic conditions [79].

Our results suggest that the extraction of wet snow alone is not sufficient to monitor Arctic snowmelt in Qikiqtaruk. The previously documented threshold of −2 dB for VH polarized imagery showed a strong underestimation of SCE when compared to the Landsat SCE [39]. We therefore suggest that the extraction of frozen ice layers is also required to accurately estimate snow cover [80]. A threshold of 1 dB, which includes both wet snow (negative ratio) and ice layers (positive ratio), seems to be a more appropriate threshold in Arctic ecosystems. The threshold of 1 dB may also result in the inclusion of noise in the SCE maps and a slight overestimation; however, the accuracy is notably better than mapping wet snow alone as the qualitative comparison with the Landsat SCE shows. The TSX product is delivering a wet and frozen snow product. This provides the potential to not only detect snow but also through the use of multiple thresholds, differentiate between the types of snow present.

The TSX data provides a significantly higher temporal resolution than that available with Landsat 8 data alone, revealing the rapid advancement of snowmelt shortly following onset. This provides a better temporal picture of snowmelt dynamics than what can be derived from optical data, which due to cloud cover and subsequent data gaps, generally indicate only the onset and end of snowmelt. A more complete picture of snow cover and snowmelt timing provides an opportunity to better understand the impacts on hydrology, vegetation, active layer and permafrost thermal regimes.

At the catchment scale, two different snowmelt dynamics were observed simultaneously with in situ and TSX data. In a representative area of the Ice Creek catchment with greater topographic variation including steeper slopes and late-lying snow patches, TSX data showed high correspondence to fractional snow cover estimates from time lapse cameras (Figure 11). Both datasets in these steeper topography catchments showed rapid snowmelt followed by a slower snowmelt when only the snow patches remained. When selecting a catchment type with less steep topography characteristic of upland tundra, the TSX FSC time series again showed correspondence to time-lapse imagery with a rapid advancement of snowmelt (Figure 12). Snow patches were also present in the upland tundra and, therefore, prolonged the FSC signal.

On the 9th of June 2016, the time lapse imagery within the lower Ice Creek catchment showed dense fog for 9 hours with no traces of snow on the ground after the fog lifted. The TSX SCE from that date (Figure 13) suggests that new snow cover developed in the upper catchment of the ice creek at higher elevation above the time lapse camera location. This highlights the ability of TSX to potentially capture short-lived snowfall events. Short-term freezing and snowfall events in the spring time can negatively impact early vegetation development and greatly impact hydrological discharge of Arctic catchments and are therefore ecologically important [16,81].

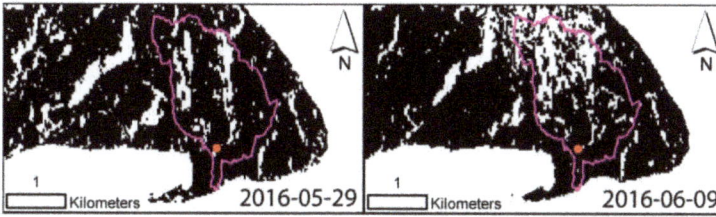

Figure 13. SCE from TerraSAR-X VH from the 29th of April 2016 (**left**) and from the 9th of May 2016 (**right**) for the Ice Creek catchment and surroundings. The red dot shows the position of the time lapse camera.

5.2. Technical Considerations for Using TSX for Wet Snow Detection

While our method improves the temporal resolution of snowmelt patterns, it faces potential limitations in low Arctic tundra environments. In the later phase of snowmelt, the HH and VV TSX derived FSC were highly variable compared to the Landsat 8-derived FSC. The observed variability indicates that both channels react to other surface features. Nagler et al. [69] also reported lower accuracies of VV compared to the VH channel in C-Band, an effect likely connected to low local incidence angles. In our study, the strong overestimations in VV are also likely influenced by local topography as the areas of false snow detection are predominantly on the flat tundra uplands on the western part of the island. These flat upland areas are likely the first to melt out, highlighting the sensitivity of VV to non-snow surface properties. Limitations in wet snow detection with the VV-polarized X-Band channel has been shown previously with false detection of wet snow in areas of water and (water saturated) bare ground [82]. Additionally, while all polarizations will react to attenuation of the microwave signal in wet snow, the cross polarized VH probably also reacts to a shift from volume scatter in dry snow to surface scatter on wet snow. This might increase the capability of VH to distinguish the active melt with liquid water concentrated at the snow surface from liquid water co-existing with the snow pack in vegetation or bare soils.

In addition to bare ground and water, previous research has also demonstrated the sensitivity of TSX (VV/HH polarized) data to the presence of Arctic shrubs and vegetation communities in summer [83], as well as in winter under a dry snow cover [84]. Under both conditions, backscatter is expected to increase with higher shrub density because the fraction of volume scattering increases with taller vegetation. In the case of shrubs that protrude through the snow, backscatter could increase through higher volume scatter and decrease the drop of backscatter between dry snow and wet snow images. As a consequence, the algorithm would detect no snow even though snow is still present between and underneath the shrubs.

Previous research has shown that refreezing of the surface layer can increase the backscatter signal because of the change in dielectric properties of the frozen layer [43,85]. This was also confirmed by our analysis of backscatter dynamics. Particularly, the X-band can be sensitive to refreezing snow layers because of its short wavelength and minimal penetration through the frozen layer [43,69]. In this context, the observed deviations between the 2015 as well as 2016 data from the 115 and 61 orbits in late April and early May most likely result from the difference in acquisition timing. Orbit 115 acquisitions are taking in the morning local time and 61 in the afternoon (see Table 1). Diurnal variations and resulting freeze thaw cycles are typical for the snowmelt period in the Arctic [12], with typical duration of periods with diurnal freeze thaw cycling in this region being up to 2 weeks [86]. The freezing snow layers can drastically lower the detection of wet SCE when using the wet snow detection threshold of -2 dB (see Figure 7: row 2) and, therefore, the derived FSC product if the acquisition time is during the minimum temperatures of the day (Figure 11). The cross polarized channels in C-Band show less angular dependence than the co-polarized channels and, therefore, perform more consistently in deriving snow cover [69].

Although there are some limitations for accurate mapping of snow cover with TSX, we were able to show that our 30 m aggregated TSX SCE product added valuable information to the characterization of 3 years of snowmelt data in Arctic catchments. In combination with optical data from Landsat 8, a complete picture of snowmelt can be drawn, including short term snow dynamics. The potential of full spatial resolution TSX in combination with an optimized speckle filter for SCE generation would open up more applications for fine scale monitoring of the impact of snow melt dynamics on ecosystem functioning in heterogeneous Arctic tundra environments.

6. Conclusions

The results of this study highlight the potential of TerraSAR-X X-Band to improve and complement existing optical based snow cover products by increasing the temporal resolution of snow cover measurements. We identified the VH channel as the best performing polarization channel. When we used common thresholds of −2 and −2.3 dB on TSX images, SCE was strongly underestimated when compared to Landsat 8 SCE maps, while a threshold of 1 dB produced very comparable results to Landsat 8 SCE. The VH polarization used with a threshold of 1 dB also showed an advantage by detecting reduced backscatter due to wet snow as well as increased backscatter due to ice layers. The positive threshold and detection of ice layers resulted in the detection of late lying snow patches that Landsat 8 did not capture due to the lower reflectance of old snow. Differences in the incidence angle did not seem to have a strong effect on the accuracy of the SCE, though local topography and resulting incidence angle likely led to false snow detection in the co-polarized channels. The TSX data provides a significantly higher temporal resolution than that available with Landsat 8 data alone. This provides a much more complete temporal picture of snowmelt dynamics than the optical data which, due to data gaps as a result of cloud cover, generally indicates only the onset and end of snowmelt. Using both in situ time lapse camera data and TSX imagery, we could show that depending on catchment topography, different temporal patterns of snowmelt exist. A studied catchment with a higher tundra upland contribution showed faster snowmelt than a catchment with a higher contribution of incised valleys. Overall, we conclude that a multi-source approach using conventional optical data in combination with high spatiotemporal resolution SAR in X-Band and in situ time lapse camera data is very well suited to study rapid snowmelt in small Arctic catchments.

Supplementary Materials: The following are available online at http://www.mdpi.com/2072-4292/10/7/1155/s1, Figure S1: Figure S1 Comparison of Landsat 8 SCE (left panels) and corresponding TSX SCE derived using a threshold of −2 and −2.3 dB on the HH (center panels) and VV (left panels) polarized channels of orbit 24, respectively with results of the accuracy assessment., Figure S2: Figure S2 Comparison of Landsat 8 SCE (left panels) and corresponding TSX SCE derived using a threshold of −2 and −2.3 dB on the HH (center panels) and VV (left panels) polarized channels of orbit 137 with results of the accuracy assessment., Figure S3: Figure S3 Comparison of Landsat 8 SCE (left panels) and corresponding TSX SCE derived using a threshold of −2 and −2.3 dB on the HH (center panels) and VV (left panels) polarized channels of orbit 152 with results of the accuracy assessment., Figure S4, Comparison of Landsat 8 SCE (left panels) and corresponding TSX SCE derived using a threshold of 1 dB on the HH (center panels) and VV (left panels) polarized channels of orbit 24, respectively with results of the accuracy assessment. Figure S5: Figure S5 Comparison of Landsat 8 SCE (left panels) and corresponding TSX SCE derived using a threshold of 1 dB on the HH (center panels) and VV (left panels) polarized channels of orbit 137 with results of the accuracy assessment., Figure S6: Figure S3 Comparison of Landsat 8 SCE (left panels) and corresponding TSX SCE derived using a threshold of 1 dB on the HH (center panels) and VV (left panels) polarized channels of orbit 152 with results of the accuracy assessment., Table S1: List of TSX scenes per relative orbit that were used for the averaged dry snow, Table S2: Fractional snow cover from independent visual estimations of in situ time lapse imagery.

Author Contributions: Conceptualization, S.S., H.L. and B.H.; Data curation, S.S. and A.R.; Formal analysis, S.S.; Investigation, S.S., J.E., A.B. and B.R.; Methodology, S.S., J.E., A.B. and B.R.; Resources, B.H.; Validation, S.S., H.L. and B.H.; Visualization, S.S.; Writing—original draft, S.S.; Writing—review & editing, H.L., B.H., J.E., A.R., A.B. and B.R.

Funding: The authors want to thank the German Helmholtz Alliance Earth System Dynamics (EDA) for funding of this project and access to the TSX datasets. Samuel Stettner and Hugues Lantuit were additionally supported through HGF COPER, Annett Bartsch through the European Space Agency project DUE GlobPermafrost (Contract Number 4000116196/15/I-NB) and Birgit Heim through the Helmholtz program for Regional Climate Change REKLIM.

Acknowledgments: The authors thank Isla Myers-Smith and Team Shrub setting up and sharing the data of the phenology time-lapse cameras, funding for this research was provided by NERC through the ShrubTundra (NE/M016323/1) standard grant. We thank George Tanski (AWI Potsdam), Jan Kahl (AWI Potsdam), Samuel McLeod (Herschel Ranger) and Edward McLeod (Herschel Ranger) for setting up the time lapse camera in spring 2016 possible. The authors thank Mike Kubanski from the SARlab in Vancouver. The authors want to thank the Rangers of Herschel Island and the Aurora Research Institute for making the field work in this remote area possible. We thank Nicholas Coops and the IRSS lab at University of British Columbia for providing an excellent working environment during the processing of the optical satellite imagery for this work. We thank the Inuvialuit People for the opportunity to conduct research on their traditional lands.

Conflicts of Interest: The authors declare no conflict of interest.

References

1. Ling, F.; Zhang, T. Impact of the timing and duration of seasonal snow cover on the active layer and permafrost in the Alaskan Arctic. *Permafr. Periglac. Process.* **2003**, *14*, 141–150. [CrossRef]
2. Zhang, T.; Stamnes, K. Impact of climatic factors on the active layer and permafrost at Barrow, Alaska. *Permafr. Periglac. Process.* **1998**, *9*, 229–246. [CrossRef]
3. Zhang, T.; Osterkamp, T.E.; Stamnes, K. Influence of the depth hoar layer of the seasonal snow cover on the ground thermal regime. *Water Resour. Res.* **1996**, *32*, 2075–2086. [CrossRef]
4. Johansson, M.; Callaghan, T.V.; Bosiö, J.; Åkerman, H.J.; Jackowicz-Korczynski, M.; Christensen, T.R. Rapid responses of permafrost and vegetation to experimentally increased snow cover in sub-arctic Sweden. *Environ. Res. Lett.* **2013**, *8*, 035025. [CrossRef]
5. Semenchuk, P.R.; Elberling, B.; Amtorp, C.; Winkler, J.; Rumpf, S.; Michelsen, A.; Cooper, E.J. Deeper snow alters soil nutrient availability and leaf nutrient status in high Arctic tundra. *Biogeochemistry* **2015**, *124*, 81–94. [CrossRef]
6. Schimel, J.P.; Bilbrough, C.; Welker, J.A.; Schimel, J.P.; Bilbrough, C.; Welker, J.M. Increased snow depth affects microbial activity and nitrogen mineralization in two Arctic tundra communities. *Soil Biol. Biochem.* **2004**, *36*, 217–227. [CrossRef]
7. Krab, E.J.; Roennefarth, J.; Becher, M.; Blume-Werry, G.; Keuper, F.; Klaminder, J.; Kreyling, J.; Makoto, K.; Milbau, A.; Dorrepaal, E. Winter warming effects on tundra shrub performance are species-specific and dependent on spring conditions. *J. Ecol.* **2018**, *106*, 599–612. [CrossRef]
8. Ballantyne, C.K. The Hydrologic Significance of Nivation Features in Permafrost Areas. *Geogr. Ann. Ser. A Phys. Geogr.* **1978**, *60*, 51–54. [CrossRef]
9. Schimel, J.P.; Kielland, K.; Chapin, F.S. Nutrient Availability and Uptake by Tundra Plants. In *Landscape Function and Disturbance in Arctic Tundra*; Springer: Berlin/Heidelberg, Germany, 1996; pp. 203–221.
10. Ostendorf, B.; Quinn, P.; Beven, K.; Tenhunen, J.D. Hydrological Controls on Ecosystem Gas Exchange in an Arctic Landscape. In *Landscape Function and Disturbance in Arctic Tundra*; Springer: Berlin/Heidelberg, Germany, 1996; pp. 369–386.
11. Hobbie, S.E.; Chapin, F.S. Winter regulation of tundra litter carbon and nitrogen dynamics. *Biogeochemistry* **1996**, *35*, 327–338. [CrossRef]
12. Bartsch, A.; Kidd, R.A.; Wagner, W.; Bartalis, Z. Temporal and spatial variability of the beginning and end of daily spring freeze/thaw cycles derived from scatterometer data. *Remote Sens. Environ.* **2007**, *106*, 360–374. [CrossRef]
13. Brooks, P.D.; Grogan, P.; Templer, P.H.; Groffman, P.; Öquist, M.G.; Schimel, J. Carbon and Nitrogen Cycling in Snow-Covered Environments. *Geogr. Compass* **2011**, *5*, 682–699. [CrossRef]
14. Woo, M. Hydrology of a small Canadian High Arctic basin during the snowmelt period. *Catena* **1976**, *3*, 155–168. [CrossRef]
15. Billings, W.D.; Mooney, H.A. The Ecology of Arctic Plants. *Biol. Rev.* **1968**, *43*, 481–529. [CrossRef]
16. Hinzman, L.D.; Kane, D.L.; Benson, C.S.; Everett, K.R. Energy Balance and Hydrological Processes in an Arctic Watershed. In *Landscape Function and Disturbance in Arctic Tundra*; Springer: Berlin/Heidelberg, Germany, 1996; pp. 131–154.
17. Pohl, S.; Marsh, P. Modelling the spatial-temporary variability of spring snowmelt in an arctic catchment. *Hydrol. Process.* **2006**, *20*, 1773–1792. [CrossRef]

18. Billings, W.D.; Bliss, L.C. An alpine snowbank environment and its effects on vegetation, plant development, and productivity. *Ecology* **1959**, *40*, 388–397. [CrossRef]

19. Bjorkman, A.D.; Elmendorf, S.C.; Beamish, A.L.; Vellend, M.; Henry, G.H.R. Contrasting effects of warming and increased snowfall on Arctic tundra plant phenology over the past two decades. *Glob. Chang. Biol.* **2015**, *21*, 4651–4661. [CrossRef] [PubMed]

20. Brown, R.D.; Robinson, D.A. Northern Hemisphere spring snow cover variability and change over 1922–2010 including an assessment of uncertainty. *Cryosphere* **2011**, *5*, 219–229. [CrossRef]

21. Weller, G.; Symon, C.; Arris, L.; Hill, B. Summary and synthesis of the ACIA. In *Arctic Climate Impact Assessment*; Campbridge University Press: New York, NY, USA, 2005; pp. 990–1020.

22. Clark, M.P.; Hendrikx, J.; Slater, A.G.; Kavetski, D.; Anderson, B.; Cullen, N.J.; Kerr, T.; Örn Hreinsson, E.; Woods, R.A. Representing spatial variability of snow water equivalent in hydrologic and land-surface models: A review. *Water Resour. Res.* **2011**, *47*. [CrossRef]

23. Liston, G.E.; Liston, G.E. Representing Subgrid Snow Cover Heterogeneities in Regional and Global Models. *J. Clim.* **2004**, *17*, 1381–1397. [CrossRef]

24. Sturm, M.; Holmgren, J.; McFadden, J.P.; Liston, G.E.; Chapin, F.S.; Racine, C.H.; Sturm, M.; Holmgren, J.; McFadden, J.P.; Liston, G.E.; et al. Snow–Shrub Interactions in Arctic Tundra: A Hypothesis with Climatic Implications. *J. Clim.* **2001**, *14*, 336–344. [CrossRef]

25. Myers-Smith, I.H.; Forbes, B.C.; Wilmking, M.; Hallinger, M.; Lantz, T.; Blok, D.; Tape, K.D.; Macias-Fauria, M.; Sass-Klaassen, U.; Lévesque, E.; et al. Shrub expansion in tundra ecosystems: Dynamics, impacts and research priorities. *Environ. Res. Lett.* **2011**, *6*, 045509. [CrossRef]

26. Strozzi, T.; Wegmuller, U.; Matzler, C. Mapping wet snowcovers with SAR interferometry. *Int. J. Remote Sens.* **1999**, *20*, 2395–2403. [CrossRef]

27. Dietz, A.J.; Kuenzer, C.; Gessner, U.; Dech, S.; Juergen, A.; Kuenzer, C.; Gessner, U.; Dech, S.; Dietz, A.J.; Kuenzer, C.; et al. Remote sensing of snow—A review of available methods. *Int. J. Remote Sens.* **2012**, *33*, 4094–4134. [CrossRef]

28. Irons, J.R.; Dwyer, J.L.; Barsi, J.A. The next Landsat satellite: The Landsat Data Continuity Mission. *Remote Sens. Environ.* **2012**, *122*, 11–21. [CrossRef]

29. Salomonson, V.V.; Appel, I. Estimating fractional snow cover from MODIS using the normalized difference snow index. *Remote Sens. Environ.* **2004**, *89*, 351–360. [CrossRef]

30. Stow, D.A.; Hope, A.; McGuire, D.; Verbyla, D.; Gamon, J.; Huemmrich, F.; Houston, S.; Racine, C.; Sturm, M.; Tape, K.; et al. Remote sensing of vegetation and land-cover change in Arctic Tundra Ecosystems. *Remote Sens. Environ.* **2004**, *89*, 281–308. [CrossRef]

31. Romanov, P.; Gutman, G.; Csiszar, I.; Romanov, P.; Gutman, G.; Csiszar, I. Automated Monitoring of Snow Cover over North America with Multispectral Satellite Data. *J. Appl. Meteorol.* **2000**, *39*, 1866–1880. [CrossRef]

32. Roth, A.; Eineder, M.; Schättler, B. TerraSAR-X: A new persepctive for applications requiring high resolution spaceborne SAR data. 2003. Available online: https://www.ipi.uni-hannover.de/fileadmin/institut/pdf/roth.pdf (accessed on 21 July 2018).

33. Rott, H.; Heidinger, M.; Nagler, T.; Cline, D.; Yueh, S. Retrieval of snow parameters from Ku-band and X-band radar backscatter measurements. In Proceedings of the IEEE International Geoscience and Remote Sensing Symposium, Cape Town, South Africa, 12–17 July 2009; pp. II-144–II-147.

34. Ulaby, F.T.; Stiles, W.H. The active and passive microwave response to snow parameters: 2. Water equivalent of dry snow. *J. Geophys. Res.* **1980**, *85*, 1045. [CrossRef]

35. Mätzler, C.; Wegmüller, U. Dielectric properties of freshwater ice at microwave frequencies. *J. Phys. D Appl. Phys.* **1987**, *20*, 1623–1630. [CrossRef]

36. Mätzler, C. Passive microwave signatures of landscapes in winter. *Meteorol. Atmos. Phys.* **1994**, *54*, 241–260. [CrossRef]

37. Leinss, S.; Parrella, G.; Hajnsek, I. Snow height determination by polarimetric phase differences in X-Band SAR Data. *IEEE J. Sel. Top. Appl. Earth Obs. Remote Sens.* **2014**, *7*, 3794–3810. [CrossRef]

38. Schellenberger, T.; Ventura, B.; Zebisch, M.; Notarnicola, C. Wet Snow Cover Mapping Algorithm Based on Multitemporal COSMO-SkyMed X-Band SAR Images. *IEEE J. Sel. Top. Appl. Earth Obs. Remote Sens.* **2012**, *5*, 1045–1053. [CrossRef]

39. Schellenberger, T.; Ventura, B.; Notarnicola, C.; Zebisch, M.; Nagler, T.; Rott, H. Exploitation of Cosmo-Skymed image time series for snow monitoring in alpine regions. In Proceedings of the 2011 IEEE International Geoscience and Remote Sensing Symposium, Vancouver, BC, Canada, 24–29 July 2011; pp. 3641–3644. [CrossRef]

40. Nagler, T.; Rott, H. Retrieval of wet snow by means of multitemporal SAR data. *IEEE Trans. Geosci. Remote Sens.* **2000**, *38*, 754–765. [CrossRef]

41. Rott, H.; Nagler, T. Snow and glacier investigations by ERS-1 SAR: First results. In Proceedings of the 1st ERS-1 Symposium: Space at the Service of our Environment, Cannes, France, 4–6 November 1992; pp. 577–582.

42. Nagler, T. Methods and Analysis of Synthetic Aperture Radar Data for ERS-1 and X-SAR for Snow and Glacier Applications. Ph.D. Thesis, University of Innsbruck, Innsbruck, Austria, 1996.

43. Floricioiu, D.; Rott, H. Seasonal and short-term variability of multifrequency, polarimetric radar backscatter of alpine terrain from SIR-C/X-SAR and AIRSAR data. *IEEE Trans. Geosci. Remote Sens.* **2001**, *39*, 2634–2648. [CrossRef]

44. Bartsch, A.; Kumpula, T.; Forbes, B.C.; Stammler, F. Detection of snow surface thawing and refreezing in the Eurasian Arctic with QuikSCAT: Implications for reindeer herding. *Ecol. Appl.* **2010**, *20*, 2346–2358. [CrossRef] [PubMed]

45. Kimball, J.S.; McDonald, K.C.; Keyser, A.R.; Frolking, S.; Running, S.W. Application of the NASA scatterometer (NSCAT) for determining the daily frozen and nonfrozen landscape of Alaska. *Remote Sens. Environ.* **2001**, *75*, 113–126. [CrossRef]

46. Wilson, R.R.; Bartsch, A.; Joly, K.; Reynolds, J.H.; Orlando, A.; Loya, W.M. Frequency, timing, extent, and size of winter thaw-refreeze events in Alaska 2001–2008 detected by remotely sensed microwave backscatter data. *Pol. Biol.* **2013**, *36*, 419–426. [CrossRef]

47. Bartsch, A. Monitoring of terrestrial hydrology at high latitudes with scatterometer data. In *Geoscience and Remote Sensing New Achievements*; Imperatore, P., Riccio, D., Eds.; InTech: Rijeka, Croatia, 2010; p. 64, ISBN 9789537619992.

48. Bartsch, A.; Allard, M.; Biskaborn, B.K.; Burba, G.; Christiansen, H.H.; Duguay, C.R.; Grosse, G.; Günther, F.; Heim, B.; Högström, E.; et al. Permafrost longterm monitoring sites (Arctic and Antarctic). In *Requirements for Monitoring of Permafrost in Polar Regions*; A Community White Paper Response to WMO Polar Space Task Group (PSTG), Version 4, 2014-10-09; Austrian Polar Res. Institute: Vienna, Austria, 2014.

49. Burn, C.R. *Herschel Island Qikiqtaryuk: A Natural and Cultural History*; University of Calgary Press: Calgary, Alberta, 2012; pp. 48–53.

50. Solomon, S.M. Spatial and temporal variability of shoreline change in the Beaufort-Mackenzie region, northwest territories, Canada. *Geo-Mar. Lett.* **2005**, *25*, 127–137. [CrossRef]

51. De Krom, V. Retrogressive Thaw Slumps and Active Layer Slides on Herschel Island, Yukon. Unpublished Master's Thesis, McGill University, Montréal, QC, Canada, 1990.

52. Rampton, V.N. Quaternary geology of the Yukon Coastal Plain. *Geol. Surv. Can. Bull.* **1982**, *49*. [CrossRef]

53. Walker, D.A.; Raynolds, M.K.; Daniëls, F.J.A.; Einarsson, E.; Elvebakk, A.; Gould, W.A.; Katenin, A.E.; Kholod, S.S.; Markon, C.J.; Melnikov, E.S. The circumpolar Arctic vegetation map. *J. Veg. Sci.* **2005**, *16*, 267–282. [CrossRef]

54. Bliss, L.C. Arctic ecosystems of North America. In *Polar and Alpine Tundra*; Elsevier: Amsterdam, Netherlands, 1997; pp. 551–683.

55. Smith, C.A.S.; Kennedy, C.E.; Hargrave, A.E.; McKenna, K.M. *Soil and Vegetation of Herschel Island, Yukon territory*; Land Resource Research Centre, Agriculture Canada: Ottawa, ON, Canada, 1989.

56. Myers-smith, I.H.; Hik, D.S.; Kennedy, C.; Cooley, D.; Johnstone, J.F.; Kenney, A.J.; Krebs, C.J. Expansion of Canopy-Forming Willows Over the Twentieth Century on Herschel Island, Yukon Territory, Canada. *AMBIO A J. Hum. Environ.* **2011**, *40*, 610–623. [CrossRef]

57. Kennedy, C.E.; Smith, C.A.S.; Cooley, D.A. Observations of change in the cover of polargrass, Arctagrostis latifolia, and arctic lupine, Lupinus arcticus. *Upl. Tundra Herschel Isl. Yukon Territ. Can. Field-Nat* **2001**, *115*, 323–328.

58. Blasco, S.M.; Fortin, G.; Hill, P.R.; O'Connor, M.J.; Brigham-Grette, J. The late Neogene and Quaternary stratigraphy of the Canadian Beaufort continental shelf. In *The Arctic Ocean Region*; Geological Society of America: Boulder, CO, USA, 1990; pp. 491–502.

59. Fritz, M.; Wetterich, S.; Schirrmeister, L.; Meyer, H.; Lantuit, H.; Preusser, F.; Pollard, W.H. Eastern Beringia and beyond: Late Wisconsinan and Holocene landscape dynamics along the Yukon Coastal Plain, Canada. *Palaeogeogr. Palaeoclimatol. Palaeoecol.* **2012**, *319–320*, 28–45. [CrossRef]

60. Burn, C.R.; Zhang, Y. Permafrost and climate change at Herschel Island (Qikiqtaruq), Yukon Territory, Canada. *J. Geophys. Res.* **2009**, *114*, F02001. [CrossRef]

61. Kokelj, S.V.; Smith, C.A.S.; Burn, C.R. Physical and chemical characteristics of the active layer and permafrost, Herschel Island, western Arctic Coast, Canada. *Permafr. Periglac. Process.* **2002**, *13*, 171–185. [CrossRef]

62. Couture, N.J.; Pollard, W.H. A Model for Quantifying Ground-Ice Volume, Yukon Coast, Western Arctic Canada. *Permafr. Periglac. Process.* **2017**, *28*, 534–542. [CrossRef]

63. Fritz, M.; Opel, T.; Tanski, G.; Herzschuh, U.; Meyer, H.; Eulenburg, A.; Lantuit, H. Dissolved organic carbon (DOC) in Arctic ground ice. *Cryosphere Discuss.* **2015**, *9*, 77–114. [CrossRef]

64. Lantuit, H.; Pollard, W.H.; Couture, N.; Fritz, M.; Schirrmeister, L.; Meyer, H.; Hubberten, H. Modern and Late Holocene Retrogressive Thaw Slump Activity on the Yukon Coastal Plain and Herschel Island, Yukon Territory, Canada. *Permafr. Periglac. Process.* **2012**, *51*, 39–51. [CrossRef]

65. Ramage, J.L.; Fortier, D.; Hugelius, G.; Lantuit, H.; Morgenstern, A. Dissecting valleys: Snapshot of carbon and nitrogen distribution in Arctic valleys. *Catena* **2018**, Submitt.

66. Obu, J.; Lantuit, H.; Myers-Smith, I.; Heim, B.; Wolter, J.; Fritz, M. Effect of Terrain Characteristics on Soil Organic Carbon and Total Nitrogen Stocks in Soils of Herschel Island, Western Canadian Arctic. *Permafr. Periglac. Process.* **2017**, *28*, 92–107. [CrossRef]

67. Meese, R.J.; Tomich, P.A. Dots on the rocks: A comparison of percent cover estimation methods. *J. Exp. Mar. Bio. Ecol.* **1992**, *165*, 59–73. [CrossRef]

68. Nagler, T.; Rott, H. Snow classification algorithm for Envisat ASAR. In Proceedings of the 2004 Envisat & ERS Symposium, Salzburg, Austria, 6–10 September 2004; Volume 572.

69. Nagler, T.; Rott, H.; Ripper, E.; Bippus, G.; Hetzenecker, M. Advancements for Snowmelt Monitoring by Means of Sentinel-1 SAR. *Remote Sens.* **2016**, *8*, 348. [CrossRef]

70. Wendleder, A.; Heilig, A.; Schmitt, A.; Mayer, C. Monitoring of Wet Snow and Accumulations at High Alpine Glaciers Using Radar Technologies. *ISPRS Int. Arch. Photogramm. Remote Sens. Spat. Inf. Sci.* **2015**, *40*, 1063–1068. [CrossRef]

71. Venkataraman, G. Snow cover area monitoring using multitemporal TerraSAR-X data. In Proceedings of the 3rd TerraSAR-X Science Team Meeting, Oberpfaffenhofen, Germany, 25–26 November 2008.

72. Lee, J.-S. A simple speckle smoothing algorithm for synthetic aperture radar images. *IEEE Trans. Syst. Man. Cybern.* **1983**, 85–89. [CrossRef]

73. Frost, V.S.; Stiles, J.A.; Shanmugan, K.S.; Holtzman, J.C. A model for radar images and its application to adaptive digital filtering of multiplicative noise. *IEEE Trans. Pattern Anal. Mach. Intell.* **1982**, 157–166. [CrossRef]

74. Krieger, G.; Moreira, A.; Fiedler, H.; Hajnsek, I.; Werner, M.; Younis, M.; Zink, M. TanDEM-X: A satellite formation for high-resolution SAR interferometry. *IEEE Trans. Geosci. Remote Sens.* **2007**, *45*, 3317–3341. [CrossRef]

75. Hall, D.K.; Riggs, G.A.; Salomonson, V.V.; DiGirolamo, N.E.; Bayr, K.J. MODIS snow-cover products. *Remote Sens. Environ.* **2002**, *83*, 181–194. [CrossRef]

76. Crawford, C.J.; Manson, S.M.; Bauer, M.E.; Hall, D.K. Multitemporal snow cover mapping in mountainous terrain for Landsat climate data record development. *Remote Sens. Environ.* **2013**, *135*, 224–233. [CrossRef]

77. Hall, D.K.; Riggs, G.A.; Salomonson, V.V. Development of methods for mapping global snow cover using moderate resolution imaging spectroradiometer data. *Remote Sens. Environ.* **1995**, *54*, 127–140. [CrossRef]

78. Foody, G.M. Status of land cover classification accuracy assessment. *Remote Sens. Environ.* **2002**, *80*, 185–201. [CrossRef]

79. Green, K.; Pickering, C.M. The decline of snowpatches in the Snowy Mountains of Australia: Importance of climate warming, variable snow, and wind. *Arct. Antarct. Alp. Res.* **2009**, *41*, 212–218. [CrossRef]

80. Bartsch, A.; Kumpula, T.; Forbes, B.C.; Stammler, F. Detection of snow surface thawing and refreezing in the Eurasian Arctic with QuikSCAT: implications for reindeer herding. *Ecol. Appl.* **2010**, *20*, 2346–2358. [CrossRef] [PubMed]

81. Inouye, D.W. Effects of climate change on phenology, frost damage, and floral abundance of montane wildflowers. *Ecology* **2008**, *89*, 353–362. [CrossRef] [PubMed]

82. Mora, C.; Jiménez, J.J.; Pina, P.; Catalão, J.; Vieira, G. Evaluation of single-band snow-patch mapping using high-resolution microwave remote sensing: An application in the maritime Antarctic. *Cryosphere* **2017**, *11*, 139–155. [CrossRef]

83. Ullmann, T.; Schmitt, A.; Roth, A.; Duffe, J.; Dech, S.; Hubberten, H.-W.; Baumhauer, R. Land Cover Characterization and Classification of Arctic Tundra Environments by means of polarized Synthetic Aperture X- and C-Band Radar (PolSAR) and Landsat 8 Multispectral Imagery—Richards Island, Canada. *Remote Sens.* **2014**, *6*, 1–26. [CrossRef]

84. Duguay, Y.; Bernier, M.; Lévesque, E.; Tremblay, B. Potential of C and X Band SAR for Shrub Growth Monitoring in Sub-Arctic Environments. *Remote Sens.* **2015**, *7*, 9410–9430. [CrossRef]

85. Reber, B.; Mätzler, C.; Schanda, E. Microwave signatures of snow crusts modelling and measurements. *Int. J. Remote Sens.* **1987**, *8*, 1649–1665. [CrossRef]

86. Bartsch, A. Ten Years of SeaWinds on QuikSCAT for Snow Applications. *Remote Sens.* **2010**, *2*, 1142–1156. [CrossRef]

MDPI

St. Alban-Anlage 66

4052 Basel

Switzerland

Tel. +41 61 683 77 34

Fax +41 61 302 89 18

www.mdpi.com

Remote Sensing Editorial Office

E-mail: remotesensing@mdpi.com

www.mdpi.com/journal/remotesensing